Anatomy and Physiology
An Easy Learner

Ethel Sloane
Department of Biological Sciences
University of Wisconsin–Milwaukee

Jones and Bartlett Publishers
Boston   London

Editorial, Sales, and Customer Service Offices
Jones and Bartlett Publishers
One Exeter Plaza
Boston, MA 02116
1-800-832-0034
1-617-859-3900

Jones and Bartlett Publishers International
PO Box 1498
London W6 7RS
England

Copyright © 1994 by Jones and Bartlett Publishers, Inc.

All rights reserved. No part of the material protected by this copyright notice may be reproduced or utilized in any form, electronic or mechanical, including photocopying, recording, or by any information storage and retrieval system, without written permission from the copyright owner.

The following illustrations are used with permission from Anderson, P.D. *Basic Human Anatomy and Physiology*. Boston: Jones and Bartlett Publishers, 1984: 1.1, 1.3, 1.4, 3.1, 3.2, 5.6, 7.2–7.17, 9.11, 9.16, 9.17, 9.23, 9.24, 9.25, 9.27, 9.31, 9.34, 9.35, 10.5, 11.2, 11.3, 13.1, 14.1–14.4, 16.1, 16.2, 18.3–18.10.

The following illustrations are used with permission from Farish, D.J. *Human Biology*. Boston: Jones and Bartlett Publishers, Inc., 1993: 2.7, 2.9, 8.5, 9.1, 9.20, 9.33, 10.4.

The following illustration is used with permission from Anderson, P.D. *Human Anatomy and Physiology Coloring Workbook and Study Guide*. Boston: Jones and Bartlett Publishers, Inc., 1990: 7.1.

Library of Congress Cataloging-in-Publication Data
Sloane, Ethel
    Anatomy and physiology: an easy learner/Ethel Sloane.
        p. cm.
    Includes bibliographical references and index.
    ISBN 0-86720-832-5
    1. Human physiology—Outlines, syllabi, etc.   2. Human anatomy—
Outlines, syllabi, etc.   I. Title.
        [DNLM: 1. Anatomy—outlines. 2. Physiology—outlines.   QS 18
S634a 1994]
QP41.S63   1994
612—dc20
DNLM/DLC
for Library of Congress   93-44843
                                                                                                          CIP

Acquisitions Editor: Joseph E. Burns
Manufacturing Buyer: Dana Cerrito
Typesetting: LeGwin Associates
Design: Colophon
Editorial Service: Colophon
Cover Design: Hannus Design Associates
Cover Illustration: Vincent Perez
Printing and Binding: Braun-Brumfield, Inc.

Printed in the United States of America

10 9 8 7 6 5 4 3 2 1   99 98 97 96 95 94

Contents

Preface and Acknowledgments vii

1. **OVERVIEW OF HUMAN ANATOMY AND PHYSIOLOGY** 1

   Introduction 1
   Taxonomic Classification of Humans 2
   Structural Levels of Body Organization 3
   Characteristics of Living Matter 4
   Homeostasis 4
   Structural Plan of the Body 5

2. **THE CHEMICAL BASIS OF LIFE** 11

   Introduction to Basic Chemistry 11
   Chemical Reactions 16
   Biochemistry 17
   Energy, Enzymes, and Enzyme Activity 28

3. **THE CELL** 34

   Introduction 34
   Components of a Cell 35
   Movement of Materials across Cell Membranes 41
   Cell Metabolism 45
   Cell Division 46
   Protein Synthesis 52

4. **HUMAN GENETICS** 58

   Introduction 58
   Terminology of Genetics 58
   Mendelian Laws of Genetics 61
   Human Inheritance 61
   Chromosomal Abnormalities 64

5. TISSUES 69

Introduction 69
Epithelial Tissues 69
Connective Tissues 74
Muscle Tissue 79
Nervous Tissue 80

6. INTEGUMENTARY SYSTEM 84

Introduction 84
Skin 85
Skin Derivatives 86
Role of Skin in Thermoregulation 88

7. SKELETAL SYSTEM 92

General Introduction to the Skeletal System 92
Skeletal Anatomy: Axial Skeleton 97
Skeletal Anatomy: Appendicular Skeleton 105
Articulations 113

8. THE MUSCULAR SYSTEM 119

Part I: General Structure and Physiology 119
Introduction 119
Skeletal Muscle 120
Smooth Muscle 127
Cardiac Muscle 128

Part II: Major Skeletal Muscles and Their Actions 129
General Principles 129
Muscle Tables 131

9. THE NERVOUS SYSTEM 155

Part I: Organization, Cells, and Nerve Impulses 155
Introduction 155
Cells of the Nervous System 156
The Nerve Impulse 158
Neuronal Pools and Circuits 164
Reflexes 165

Part II: Central Nervous System and Peripheral Nervous System 167
Brain 167
Spinal Cord 175
Peripheral Nervous System 178

Part III: Sensory Receptors 185
Classification of Sensory Receptors 185

Eye and the Sense of Vision   186
Ear: Senses of Hearing and Equilibrium   191
Gustation: The Sense of Taste   196
Olfaction: The Sense of Smell   197

10. ENDOCRINE SYSTEM   202

Introduction   202
Hypophysis (Pituitary Gland)   206
Thyroid Gland   211
Parathyroid Glands   212
Adrénal Glands   213
The Endocrine Pancreas   215
Pineal Gland   217
Thymus Gland   217

11. CIRCULATORY SYSTEM   221

Introduction   221
Blood   222
The Heart   231
Cardiac Physiology   235
Hemodynamics: Blood Flow and Blood Pressure   240
Circulatory Pathways   243
Capillary Exchange and the Lymphatic System   247

12. NONSPECIFIC DEFENSES AND THE IMMUNE SYSTEM   255

Introduction   255
Nonspecific Defenses   255
The Immune System: Specific Defenses   258

13. RESPIRATORY SYSTEM   269

Introduction   269
Functional Anatomy of the Respiratory Tract   269
Mechanics of Breathing (Pulmonary Ventilation)   273
Gas Exchange   275
Transport of Gases by the Blood   276
Control of Respiration   279
Respiratory Problems   280

14. DIGESTIVE SYSTEM   283

Introduction   283
Oral Cavity, Pharynx, and Esophagus   285
Stomach   288
Small Intestine   290
Large Intestine   296

15. METABOLISM, NUTRITION, AND BODY TEMPERATURE REGULATION 301

Introduction 301
Metabolism of Absorbed Carbohydrates 302
Metabolism of Fats 308
Protein Metabolism 310
Absorptive (Feasting) and Postabsorptive (Fasting) States 311
Metabolic Rate and Nutrition 312
Temperature Regulation 313

16. URINARY SYSTEM 320

Introduction 320
Formation of Urine 324
Concept of Clearance 326
Urine Concentration and Dilution Mechanisms 326
Characteristics of Urine 329
The Ureters, Urinary Bladder, and Urethra 330
Disorders of the Urinary System 332

17. FLUIDS, ELECTROLYTES, AND ACID-BASE BALANCE 336

Introduction 336
Body Fluids 336
Electrolyte Balance 339
Acid-Base Balance 341

18. THE REPRODUCTIVE SYSTEMS, PREGNANCY, AND DEVELOPMENT 348

Introduction 348
Embryonic Development 348
The Male Reproductive System 350
The Female Reproductive System 356
Fertilization 365
Contraception 366
Pregnancy and Early Development 367

Index 377

Preface and Acknowledgments

Anatomy and physiology courses, in common with basic science courses taught in universities across the country, are staggering under the burden of massive amounts of information accumulated in the last decade. Textbooks have become thicker with new material, detailed explanations, multicolored illustrations, and an abundance of study guides and other pedagogical aids. This book is designed to help students (and instructors) wend their way through the vast and complex amount of information in the textbooks by presenting the core facts of anatomy and physiology in outline form.

The treatment of the material, although it is condensed, is comprehensive and up to date. The book includes, in abbreviated form, all of the concepts in approximately the same sequence as those in the major anatomy and physiology texts on the market. The outline format, however, eliminates lengthy descriptions. It provides the essentials in an organized framework that encourages easier learning and rapid recall. The back-to-basics approach of this text can serve students currently enrolled in anatomy and physiology as a stand-alone primary text or as a companion book to a more detailed volume. The book also can be used as a review for class examinations and as generic lecture notes that allow students to do more listening and thinking and less writing during their lectures. In addition, this text can assist students in the review of anatomy and physiology for various board examinations, the Medical College Admissions Test, or the Graduate Record Examination. Each chapter ends with questions that are designed for self-evaluation of knowledge derived from the information presented. The questions are in the objective style most often used in anatomy and physiology courses and are followed by extensive explanations, which cite the location of the answer in the outline, reinforce the learning of the material, and assist the student in identifying information needing further study.

Anatomy and physiology is a dynamic and exciting discipline. But, as students strive to master it, they can get so bogged down in the intimidating amount of material to be assimilated that they lose confidence in their ability to learn and understand it. My hope is that this concise text helps to provide an easier path to learning and understanding— one that encourages students to see the whole picture quickly, so that they can experience both the joy of mastery and the excitement of knowing how the human body works.

I acknowledge with gratitude the consistent encouragement and valuable suggestions I received from Joseph E. Burns, Vice President of Jones and Bartlett. I also express my appreciation for the talented efforts of all the members of the editorial and production staff.

I give my sincere thanks to the following reviewers. Their willingness to read carefully huge portions of the manuscript and to provide detailed comments and advice helped to improve the book and make it more accurate and useful.

Paul Badarocco, Yuba College
Norman E. Conger, Coconino County Community College
Fran Garb, University of Wisconsin–Stout
Larry E. Hibbert, Ricks College
Gloria Hillert, Triton College
Paul Holmgren, Northern Arizona University
Constance Martin, Hunter College
Roxine McQuitty, Milwaukee Area Technical College, West
Neal Smatresk, University of Texas at Arlington

Overview of Human Anatomy and Physiology 1

I. **INTRODUCTION.** The study of human anatomy (structure) and human physiology (function) is closely interrelated. Because structure provides the basis for function, each of the sciences contributes to the understanding of the human body.

  A. **Anatomy** is the study of the structure of the body. The word **anatomy** is from the Greek **ana** and **tome**, meaning to take or cut apart.

    1. **Subdivisions of anatomy**
      a. **Gross or macroscopic anatomy** is the study of the structures of the body that can be examined by observation and dissection without the use of a microscope.
        (1) **Regional anatomy** is the study of all the anatomical features in a particular area of the body.
        (2) **Systemic anatomy** is the study of the body organ systems one by one.
      b. **Histology (microscopic) anatomy** is the study of cells, tissues, and organs of the body when seen with the **light microscope**, which is also called the **bright-field microscope** or the **compound microscope**. The best magnification possible with the light microscope is 1,000 to 2,000 times.
        (1) The light microscope is named for the light energy that provides the source of illumination. It consists of a **condenser** to focus rays of light on the specimen, a **stage** on which the specimen is placed, an **ocular lens**, and an **objective lens**.
        (2) Other microscopes used in the anatomy laboratory are the **phase contrast, dark field, interference, polarizing, ultraviolet,** and **fluorescence** microscopes.
      c. **Ultrascopic anatomy** studies cell ultrastructure by use of the **electron microscope (EM)**.
        (1) In the transmission electron microscope (TEM), the light beam of the light microscope is replaced by a beam of electrons that passes through the specimen in order to produce an image on a screen. Magnification of over 1,000,000 is possible with some electron microscopes.
        (2) In a scanning electron microscope (SEM), the beam of electrons scans the surface of the specimen and provides a three-dimensional image.
      d. **Cytology** is the microscopic study of the structure of individual cells.
      e. **Embryology and fetology** is the study of growth and development from conception to birth.

              f. **Developmental anatomy** is the study of growth and differentiation of structures throughout the life span of an organism.
              g. **Pathology (pathological anatomy)** is the study of body structures and changes associated with disease or injury.
              h. **Radiographic anatomy (radiology)** is the study of body structures using x-rays or other imaging techniques.
  B. **Physiology** is the study of the functions of the living body.
      1. Like anatomy, physiology includes specialized areas that consider the function of particular organ systems; for example, **neurophysiology, cardiac physiology,** or **reproductive physiology**.
      2. The study of physiology is based on cellular and molecular function. It requires knowledge of the principles of chemistry and physics.
  C. Some **historical figures** in anatomy and physiology include the following:
      1. **Hippocrates** (460–375 B.C.), the founder of the most ancient school of medicine in Greece, is commonly known as "the father of medicine." He provided scientific foundation for the field of medical practice and his name is associated with the Hippocratic oath, which is the ethical guide of the medical profession.
      2. **Aristotle** (384–322 B.C.), the first comparative anatomist, understood the relationship of structure and function. He undertook a systematic classification of animals.
      3. **Galen** (131–201), after Hippocrates, is considered the most important figure in the history of medicine; he was the first experimental physiologist. His book, **Uses of the Parts of the Body of Man**, showed how organs are perfectly constructed for and adapted to their function.
      4. **Leonardo da Vinci** (1451–1519) was an artist, engineer, inventor, and scientist who left extraordinary drawings of muscle action and cardiovascular activity.
      5. **Andreas Vesalius** (1514–1564) was a teacher and dissector whose work **De Humani Corporis Fabrica Libri Septem** (Seven Books on the Structure of the Human Body) laid the foundation for modern anatomy and physiology.
      6. **William Harvey** (1578–1657), one of the most distinguished anatomists of all time, discovered the circulation of the blood, a milestone in the history of medicine.

II. TAXONOMIC CLASSIFICATION OF HUMANS

  A. **Taxonomy** is the branch of biology that names and classifies all living things. Taxonomy is a part of **systematics**, the science that studies the diversity of organisms.
      1. The process of taxonomy classifies organisms into a hierarchy of units that is based on the degrees of similarity among them. Each level of classification is a **taxon**.
      2. Some of the characters used in classifying organisms into the units (taxa) include their gross morphology, developmental stages, probable evolutionary relationships, and the comparison of their proteins and gene structure.
      3. Each organism is assigned a two-part Latin name: the **genus** (plural, genera) and **species**. Species are grouped into genera, genera into **families**, families into **orders**, orders into **classes**, classes into **phyla**, and phyla into **kingdoms**.
      4. The broadest taxonomic unit is the kingdom. The five kingdoms include the **Monera** (microscopic organisms lacking a nucleus), **Protista** (one-celled nucleated organisms), **Plantae** (multicellular plants), **Fungi**, and **Animalia**. Humans are in the kingdom Animalia.

B. **Classification of a human**
   1. **Kingdom: Animalia.** All organisms in the animal kingdom obtain their nutrition by consuming and digesting other organisms.
   2. **Phylum: Chordata.** Humans, like fish, amphibians, reptiles, birds, and other mammals, are animals that at some time during their life cycle have the following three distinct characteristics that group them into the phylum:
      a. A **notochord** is a tough, flexible supporting rod located along the back.
      b. A **tubular (hollow) nerve cord** is located dorsally, or above the notochord.
      c. **Paired gill slits** (clefts) are located on the sides of the pharynx (throat region) at least sometime during development.
   3. **Subphylum: Vertebrata.** Vertebrates include all animals having an internal bony skeleton and an articulated (jointed) backbone. The vertebrates are divided into eight classes; humans are included in the class Mammalia.
   4. **Class: Mammalia.** Mammals are vertebrates that are **warm-blooded**. Additional characteristics of mammals include the following:
      a. **Hair,** an insulating and protective covering, helps to maintain a constant body temperature.
      b. **Mammary glands,** which produce milk, serve to nourish their young.
      c. A muscular **diaphragm** functions in respiration.
      d. A **four-chambered heart** separates oxygenated from unoxygenated blood.
      e. An intrauterine **placenta** nourishes the young during prenatal development.
      f. **Three ear ossicles** (bones) and **differentiated teeth** adapted to chewing a variety of foods also are characteristic to mammals.
   5. **Order: Primata.** Primates include lemurs, monkeys, and great apes as well as humans. They have **grasping hands,** an **opposable thumb,** claws modified to finger and toe **nails,** and relatively large, **well-developed brains.**
   6. **Family: Hominidae.** Hominidae include both the extinct and the living races of humans.
   7. **Genus: Homo** and **species: sapiens. Homo sapiens** includes all the ethnic varieties of humans.
C. **Uniquely human characteristics** include the following:
   1. Humans are **bipedal,** which is the ability to stand upright and walk on two feet.
   2. Humans have a **large skull** with a huge cranial capacity. In absolute weight, humans' brain size is among the heaviest of all existing animal brains.
   3. Humans have an **expanded pelvis,** which enables them to bear large-headed children.
   4. Humans have a **prominent chin, weak jaws,** and **small teeth** relative to other mammals.

III. **STRUCTURAL LEVELS OF BODY ORGANIZATION.** The structural organization of the human body progresses from the lowest level of organization (atoms and molecules) through higher, more complex levels to make up the total body.

   A. **Chemical level. Atoms,** such as hydrogen, oxygen, carbon, nitrogen, and sodium, combine to form **molecules,** such as water and salt, and macromolecules, such as carbohydrates, proteins, and fats.
   B. **Cells.** The cell is the fundamental unit of living things. Cellular structures, such as the nucleus, ribosomes, mitochondria, and lysosomes, perform the life-sustaining functions of the cell.
   C. **Tissues.** A group of cells of similar structure that perform the same functions is called a **tissue.** There are four basic tissues: **epithelial tissue, connective tissue, muscle tissue,** and **nervous tissue.** (Tissues are described in detail in Chapter 5.)

4 Overview of Human Anatomy and Physiology

D. **Organs.** Two or more tissues combine to form an organ, such as the stomach, kidney, or eye. An organ functions as a specialized physiological center for the body's activities.

E. **Organ systems.** When several organs combine to perform related functions, they constitute an organ system. The organ systems of the body are the **integumentary, skeletal, muscular, nervous, endocrine, cardiovascular (circulatory), lymphatic, respiratory, digestive, urinary,** and **reproductive** systems.

IV. **CHARACTERISTICS OF LIVING MATTER** are the properties that distinguish living material from nonliving things and that enable living cells to carry on the activities necessary for their survival. They include the following:

A. **Irritability** or responsiveness is the ability to respond to internal or external stimuli in the environment.

B. **Conductivity** is the ability to conduct or transmit the irritability (stimulus) from one site to another. This property is highly developed in nerve and muscle cells.

C. **Movement** is the result of shortening or contracting of cells and is a property particularly well developed in muscle cells.

D. **Growth** is an increase in the size of individual cells or an increase in the number of cells.

E. **Reproduction** is the ability of living things to replicate themselves.

F. **Metabolism** is the sum total of all the chemical reactions that take place in living material. Such chemical processes include **catabolism**, or the breakdown of complex molecules to simpler substances, and **anabolism**, the synthesis of complex macromolecules from simpler building block substances. (See also Chapter 15.) Metabolism in the body depends on the following processes:

1. **Digestion** is the process whereby complex foods (carbohydrates, proteins, and fats) are broken down into simple molecules (simple sugars, amino acids, fatty acids, and glycerol) for **absorption.**

2. **Respiration** refers to the processes whereby oxygen and carbon dioxide are exchanged between the body cells and the external environment.

3. **Cellular respiration** includes the processes whereby oxygen nutrients are utilized by the body cells to produce energy and carbon dioxide.

4. **Circulation** of body fluids transports oxygen and nutrients to the cells and removes the products of metabolism from the cells.

5. **Excretion** is the elimination of waste products of metabolism from the body.

V. **HOMEOSTASIS**

A. The **concept of homeostasis** (steady state) refers to the maintenance, within physiological limits, of relatively constant chemical and physical conditions within the environment of the cells of the organism. Chemical requirements to maintain constancy include the appropriate amounts of water, nutrients, and oxygen; the physical requirements include the appropriate temperature and atmospheric pressure.

B. **Mechanisms of homeostasis** involve almost all organ systems of the body. Although internal conditions vary constantly, the body is protected against extremes by self-regulating control mechanisms such as **feedback systems**, which refer to the feeding of information from a system (output) back into the system (input) to cause a response.

1. **Components** of a feedback system
   a. The **setpoint** is the normal physiologic value of any body variable, such as normal temperature, concentration of substances in the extracellular fluid, or blood acidity or alkalinity.

b. The **sensor (receptor)** detects a deviation from any given normal variable.
c. The **control center** receives information from the various sensors, integrates and processes the information, and determines the compensatory response to get back to the setpoint.
d. The **effectors** implement the response, which is continued until the set point is again reached.
2. Examples of feedback systems
   a. **Negative feedback mechanisms** are those in which the information returned to the system (input) **decreases** the change (output) in order to get back to the appropriate setpoint. One example is the way in which the blood glucose is maintained at a relatively constant level of 90 to 110 mg/100 ml blood.
      (1) After a meal, a rise in the blood glucose level stimulates the release of insulin from special cells in the pancreas.
      (2) Insulin facilitates the passage of glucose into the body cells and thus lowers the blood glucose level.
      (3) The lowered glucose level affects the insulin-releasing cells (negative feedback) to decrease the release of insulin and the blood glucose is maintained at the appropriate level.
   b. **Positive feedback mechanisms** are those in which the information returned to the system **increases or prolongs**, rather than decreases, the deviation from the original physiological state.
      (1) One example of positive feedback occurs when a nerve membrane is stimulated.
         (a) The stimulus changes the permeability of the membrane to sodium ions, which then flow across the membrane.
         (b) The influx of sodium ions further increases the membrane's permeability to sodium ions and results in even more sodium ions flowing inward. The result is the initiation of a nerve impulse.
      (2) Positive feedback also occurs in the mechanism of blood clotting. The initiation of the blood clotting process causes the release of chemicals that accelerate the blood clotting process.

VI. **STRUCTURAL PLAN OF THE BODY.** The study of anatomy requires positional and directional terminology and reference points.

A. **Planes** (sections) of the body are imaginary flat surfaces that pass through it to provide reference points (Figure 1–1).
   1. A **sagittal plane** divides the body into a right and a left side.
      a. A **midsagittal plane** divides the body into equal right and left halves.
      b. A **parasagittal plane** divides the body into unequal left and right parts.
   2. A **frontal or coronal plane** is one at right angles to the sagittal plane. It divides the body or organs into front and back parts.
   3. A **transverse (horizontal, cross-sectional) plane** divides the body or organs into upper and lower portions.
B. The **anatomical position** of the body is used as a reference so that all body parts can be described in relation to it. In anatomical position, the body is erect with eyes forward, feet together, arms at the sides, palms up with the thumbs pointing away from the body, and the little fingers pointing toward the body (Figure 1–2).
   1. The **anterior** of the body (**ventral** in animals) is the front or belly side. For example, the nose is anterior to the rest of the face.
   2. The **posterior** is the back side (**dorsal** in animals). For example, the buttocks are posterior to the abdomen.
   3. **Superior** is toward the head or uppermost part; it is also referred to as **cephalic, craniad, or rostral.** For example, the head is superior to the neck.

Figure 1–1. Body planes and directions.

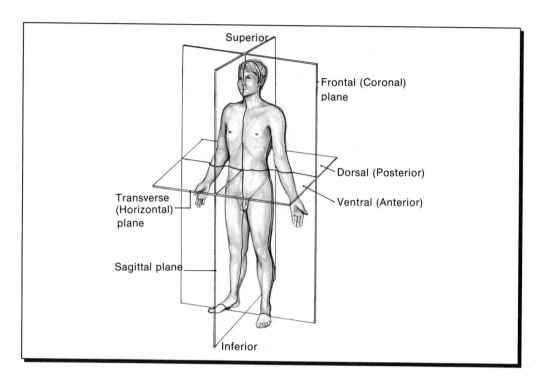

Figure 1–2. Direction terms in anatomical position.

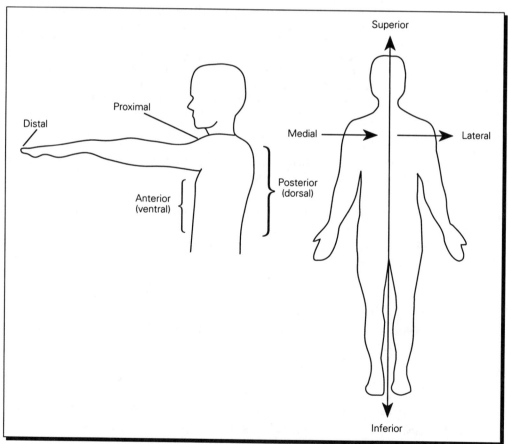

4. **Inferior** is away from the head and toward the lower parts of the body; it is also called **caudad**. For example, the chest is inferior to the neck.
5. **Medial** is any structure closest to the imaginary midline of the body. For example, the nose is medial to the eyes.
6. **Lateral** is to the side, away from the midline of the body. For example, the ears are lateral to the eyes.
   a. **Ipsilateral** means located on the same side.
   b. **Contralateral** means located on the opposite side.
7. **Proximal** refers to that part of a structure that is closer to the midline or, with reference to a limb, nearer to the origin or point of attachment closest to the trunk. For example, the elbow is proximal to the wrist.
8. **Distal** means farthest from the midline or away from the origin or the point of attachment to the trunk. For example, the foot is distal to the ankle.
9. **Superficial** means anything near the surface of the body. For example, skin is superficial to the muscles.
10. **Deep** means located internally, within the body. For example, the intestines lie deep to the abdominal muscles and skin.

C. **Body cavities** are the spaces within the **axial** part of the body that contain the internal organs or viscera. Two main cavities lie within the axial portion of the body: the **dorsal cavity** and the **ventral cavity** (Figure 1–3). The **appendicular** or limb portion of the body contains no cavities.
   1. The **dorsal body cavity** is located posteriorly (dorsally) and is subdivided into the cranial cavity and the spinal cavity.
      a. The **cranial cavity** is surrounded by bone and contains the brain.
      b. The **spinal (vertebral) cavity** is formed by the backbone and contains the spinal cord.
   2. The **ventral body cavity** is located anteriorly (ventrally) and is subdivided into the thoracic cavity and the abdominal cavity separated by the diaphragm.
      a. The **thoracic cavity** is the chest cavity. It contains the right and left pleural cavities (sacs) and the mediastinum.
         (1) The **pleural cavities** each contain a lung.
         (2) The **mediastinum** contains the heart, which is located within the **pericardial cavity**, the thymus gland, part of the esophagus, and many large blood vessels.
      b. The **abdominopelvic (peritoneal) cavity** contains the viscera of the abdomen and the pelvic areas.
      c. Additional small cavities in the head include the **oral cavity**, the **nasal cavity**, the **middle ear cavities**, and the **orbital cavities** for the eyes.

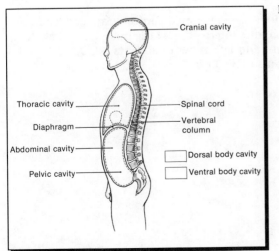

Figure 1–3. *Dorsal and ventral body cavities.*

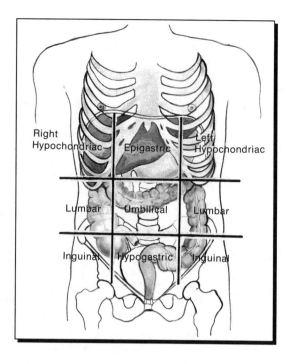

Figure 1–4. *Anterior view of the abdominopelvic cavity showing superficial organs.* Nine regions are delineated by four planes. The superior horizontal plane is just inferior to the ribs, the inferior horizontal plane is just superior to the hip bones, and the vertical planes are just medial to the nipples.

   3. **Serous membranes** line the thoracic and abdominopelvic cavities and also cover the organs within these cavities. A **parietal membrane lines** a cavity; a **visceral membrane** covers the organs.
      a. The **parietal pleura** lines the pleural cavities and the **visceral pleura** covers the lungs.
      b. The **parietal pericardium** lines the pericardial cavity and the **visceral pericardium** covers the heart.
      c. The **parietal peritoneum** lines the abdominopelvic cavity; the **visceral peritoneum** covers the abdominal organs and many of the pelvic organs.
   D. **Abdominal-pelvic regions.** Nine designations are used in anatomical studies to facilitate references to body structures and internal organs (Figure 1–4).
      1. The **umbilical region** is in the center of the abdomen.
      2. The **epigastric region** is immediately superior to the umbilical region.
      3. The **hypogastric region** is immediately inferior to the umbilical region.
      4. The **right and left hypochondriac regions** are lateral to the epigastric region.
      5. The **right and left lumbar regions** are lateral to the umbilical region.
      6. The **right and left inguinal (iliac) regions** are lateral to the hypogastric region.
   E. Four **abdominopelvic designations**, generally used clinically, are derived from imaginary horizontal and vertical lines intersecting at the umbilicus. These lines divide the abdomen into the **right and left upper quadrants** (RUQ, LUQ) and the **right and left lower quadrants** (RLQ, LLQ).

Study Questions

Directions: Each question below contains four suggested answers. Choose the **one best** answer to each question.

1. A pathologist who conducts an autopsy to determine the cause of death is most likely to have been trained primarily as a(an)
   - (A) physiologist
   - (B) anatomist
   - (C) biochemist
   - (D) chemist

2. Which of the following contributors to the science of anatomy and physiology is known as "the father of medicine?"
   - (A) Hippocrates
   - (B) Vesalius
   - (C) Aristotle
   - (D) Galen

3. Into which taxonomic group are apes, mice, and humans all classified?
   - (A) Primates
   - (B) Hominidae
   - (C) Rodents
   - (D) Vertebrates

4. All of the following are characteristics of the class Mammalia EXCEPT
   - (A) four-chambered heart
   - (B) bony spinal column
   - (C) warm-blooded
   - (D) feathers

5. The fundamental structural and functional component of a living organism is the
   - (A) atom
   - (B) cell
   - (C) organ
   - (D) organ system

6. The total of the chemical changes or reactions that occur within the body are referred to collectively as
   - (A) organization levels
   - (B) physiology
   - (C) metabolism
   - (D) homeostasis

7. All of the following statements are true concerning homeostasis EXCEPT
   - (A) Homeostasis implies that the internal environment is in a static, unchanging state.
   - (B) Feedback systems are involved in homeostatic mechanisms.
   - (C) Various organ systems participate in homeostasis in an interrelated fashion.
   - (D) Homeostasis is concerned with the control of the composition of the extracellular fluid that constantly bathes the outsides of the cells.

8. The brain is contained within the
   - (A) dorsal body cavity
   - (B) thoracic cavity
   - (C) spinal cavity
   - (D) cranial cavity

9. The serous membrane that covers the heart is the
   - (A) visceral peritoneum
   - (B) visceral pericardium
   - (C) parietal pleura
   - (D) parietal pericardium

10. A transverse line passing medially from the posterior border of the left lumbar region would cross closest to which of the following structures?
    - (A) left nipple
    - (B) right nipple
    - (C) navel
    - (D) heart

Questions 11–14: Match the word below with the phrase or word that best describes or is synonymous with it.
   - (A) proximal
   - (B) anterior
   - (C) sagittal
   - (D) horizontal

11. A plane that divides the body into right and left halves

12. Closest to the midline of the body or origin of the part

13. Ventral

14. A plane that divides the body into cranial and caudal parts

Answers and Explanations

1. **The answer is B.** (I A 1 g) Pathology is the branch of anatomy that studies body structures and changes associated with disease or injury.

2. **The answer is A.** (I C 1) Although all of the historical figures listed contributed to the principles of medical practice, Hippocrates, a Greek physician, founded the most ancient school of medicine and is regarded as "the father of medicine."

3. **The answer is D.** (II B 1, 2, 3) Both apes and humans are in the order Primata; humans are in the family Hominidae; and mice are in the order Rodentia. The subphylum Vertebrata includes all animals with an internal bony skeleton and a backbone.

4. **The answer is D** (II B 4 a–f) Feathers are characteristic of the class Aves (birds). Although birds, in common with mammals, are warm-blooded, have a four-chambered heart, and a backbone, only mammals have hair or fur as an insulating and protective covering.

5. **The answer is B.** (III B) The cell is the fundamental unit of a living organism. Both living and nonliving matter is composed of atoms and molecules. Cells form tissues, tissues form organs, and organs form organ systems.

6. **The answer is C.** (IV F) The sum total of all the chemical reactions that take place in living material is called metabolism. Physiology is the study of function. Homeostasis is the constancy of the internal environment surrounding the cells.

7. **The answer is A.** (V A, B, 1, 2) Although the components of the immediate environment of the cells must be kept constant within a physiological range, internal conditions constantly vary. The maintenance of a constant internal environment is achieved through the dynamic, active processes of homeostasis.

8. **The answer is D.** (VI C 1) The brain is located within the cranial cavity and the spinal cord is in the spinal cavity. Both the cranial and the spinal cavities are subdivisions of the dorsal cavity.

9. **The answer is B.** (VI C 3) Serous membranes that line a body cavity are parietal membranes; those that cover an organ are visceral membranes. The pericardial cavity is lined by the parietal pericardium and the visceral pericardium covers the heart.

10. **The answer is C.** (VI D 1–6) The right and left lumbar regions lie lateral to the umbilical region. A line drawn medially from the posterior border of the left lumbar region would pass closest to the navel.

11–14. **The answers are 11-C, 12–A, 11–B, 14–D.** (VI A B) A sagittal plane divides the body into right and left halves, while a horizontal or transverse plane divides it into upper (cranial) and lower (caudal) portions. Proximal means closest to the midline or point of attachment of a limb and anterior and ventral are synonymous.

The Chemical Basis of Life 2

I. **INTRODUCTION TO BASIC CHEMISTRY**

   A. **Chemistry** is the science concerned with the structure and composition of substances and with the changes, or chemical reactions, they undergo.

   B. **Matter** is anything that occupies space and has mass.

   C. **Energy** is the capacity to do work. Energy "moves" matter and can be defined only by the effect it has on matter.

   D. **Elements.** All matter is composed of elements, which are basic substances that cannot be broken down into simpler substances by chemical means.

   1. There are 92 naturally occurring elements and additional (approximately 16) synthetic elements.

   2. Each element is designated by its chemical symbol (called an atomic symbol): that is, one or two letters that represent its English or Latin name (e.g., H for hydrogen; O for oxygen; C for carbon; N for nitrogen; Cl for chlorine; Na for sodium).

   3. Of the naturally occurring elements, six—sulfur, phosphorus, oxygen, nitrogen, carbon, hydrogen—make up about 99% of living matter. The acronym for these elements is **SPONCH**.

   E. **Atoms.** An atom is the smallest unit of an element that retains its unique chemical properties.

   1. **Atomic structure.** Atoms are composed of three principle subatomic particles (i.e., **protons, electrons,** and **neutrons**), which differ in mass, electrical charge, and position in the atom. Protons and neutrons are located in the **atomic nucleus**; electrons move around the nucleus in an **electron field**.

      a. A **proton** carries one unit of a positive electrical charge.

      b. A **neutron** is an uncharged particle with about the same mass as a proton. Protons and neutrons make up most of the mass of an atom and are located in the **atomic nucleus**.

      c. An **electron** carries one unit of a negative electrical charge but a very small mass—only about 1/1,800 of the mass of a proton. Electrons are responsible for the binding between atoms.

      d. All atoms are considered electrically neutral because the number of protons equals the number of electrons.

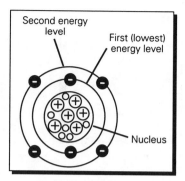

Figure 2–1. *The Bohr model of a carbon atom, developed by Niels Bohr in the early 1900s. Although it is not an accurate way to depict electron configurations, it is used because of its simplicity.*

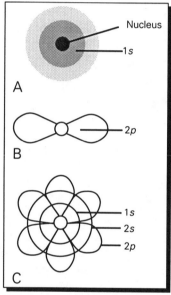

Figure 2–2. *Representation of atomic orbitals by density clouds. A, the first energy level is a single, spherical orbital that holds a maximum of two electrons that could be anywhere in the dotted area. It is designated 1s. B, One of the orbitals, designed 2p, in the second energy level. The two electrons travel in a pathway resembling a dumbbell. C, The second energy level has 4 orbitals, one spherical (2s) and three that are dumbbell shaped and extend out at right angles to each other.*

2. Atomic number and mass
   a. **Atomic number**
      (1) Each element has a fixed number of protons in its atomic nucleus; this number is the **atomic number.**
      (2) The atomic number is written as a subscript to the left of the chemical symbol. For example, ^{8}O indicates that each oxygen nucleus contains eight protons.
   b. **Atomic mass (mass number)** is the total number of protons and neutrons in the nucleus and is indicated by a superscript to the left of the chemical symbol. For example: $_{16}O$ indicates that the oxygen nucleus contains eight protons and eight neutrons.
   c. **Atomic weight** is the average atomic mass of the isotopes of an element.
   d. **Isotopes**
      (1) Isotopes are atomic forms of the same element in which the number of protons is constant (i.e., they have the same atomic number), but the number of neutrons varies (i.e., they have different atomic masses).
      (2) Elements usually occur in nature as a mixture of isotopes. Examples include: ^{12}C (carbon) and ^{14}C; ^{235}U (uranium), and ^{238}U.
   e. **Radioisotopes** are unstable isotopes with an excess of neutrons and generally are the heavier isotopes of many elements. They tend to decompose spontaneously into a more stable form by emitting energy or radiation in the form of subatomic particles.

3. **Electrons and orbitals**
   a. **Orbitals** are the regions of space where electrons occur in constant motion around the atomic nucleus. They are defined as the three-dimensional space where the electron is found 90% of the time.
      (1) These probability regions are called orbitals because of the way they were once shown in drawings—as flat circles that moved about the nucleus like planets move in their orbits around the sun (Figure 2–1). An orbital can contain zero, one, or two electrons.
      (2) Another way of depicting orbitals is by electron density clouds (Figure 2–2). At a given point in time, an electron pair may be anywhere within the cloud. The density of a shaded area is an indication of the probability that an electron is there at a given moment.
   b. **Shells**
      (1) The many orbitals in which electrons can travel are grouped into **electron shells** (electron energy levels), and the electrons present in a particular shell exist at that particular **energy level**. No more than two electrons can occupy the same orbital.
         (a) The **first shell**, or innermost (i.e., nearest the nucleus), has a single orbital and thus can accommodate a maximum of two electrons. The first shell is at the lowest energy level.
         (b) The **second shell** has four orbitals with two electrons in each, is at a higher energy level than the first shell, and can hold a maximum of eight electrons.
         (c) The **third and fourth shells** can contain more than eight electrons but are most stable and nonreactive when only eight are present.
      (2) The elements of biologic importance are the lighter elements (e.g., carbon, oxygen, hydrogen, nitrogen, and sodium) in which electrons occupy only the first three energy levels. They are said to conform to the **two-eight-eight** rule for numbers of electrons in the shells; that is, the maximum number of electrons that each of the first three shells can hold is two, eight, and eight, respectively.
      (3) The electrons farthest from the nucleus have the greatest potential energy (they successfully resist the attractive force of the nucleus) and are most likely to become involved with other atoms and molecules because of their distance from the nucleus.
         (a) The greater the distance of the energy level from the nucleus,

the greater the energy of the electrons in that level.
- (b) An electron that is provided with more energy can be moved to an orbital farther from the nucleus.
- (c) An electron that loses energy can move back to an orbital closer to the central nucleus.
- (d) A **quantum** is the exact amount of energy required to move an electron from one energy level to another.

c. **Valence.** Atoms fill their shells from the innermost shell outward. The chemical properties of an atom depend on the number of electrons found in its outermost shell (outermost energy level).
  (1) **Valence electrons** are located in the outermost shell (valence shell) of an atom.
  (2) If the **outermost shell is filled to capacity**, it is complete, or inactive. The atom is then **inert**; that is, it will not react with other atoms. Examples are helium (atomic number 2), neon (atomic number 10), and argon (atomic number 18) (Figure 2–3).
  (3) Atoms with **incomplete outermost shells** are chemically reactive and interact with other atoms so that each completes its valence shell.
     (a) In a chemical reaction, valence electrons may be shared by two atoms, or they may be completely transferred.
     (b) Such interactions result in the atoms being bound together in a **chemical bond**.

F. **Molecules.** A molecule is two or more atoms held together by forces of attraction called **chemical bonds**. Each element has its own characteristic bonding capacity, depending on the arrangement and number of its valence electrons in the outermost shell. There are three major types of bonds: **covalent** and **ionic** bonds (strong) and **hydrogen** bonds (weak).

1. **Covalent bonds** involve the sharing of a pair of electrons (valence electrons) between atoms.
   a. **Single covalent bonds**
     (1) Each atom of **hydrogen** has one electron in its shell. If a hydrogen atom gained an electron from another atom, it would have a complete shell but twice as much negative charge (two electrons) as positive charge (one proton). Instead, each hydrogen atom shares its electrons with the other atom so that each one has two electrons; thus, two hydrogen atoms combine to form a molecule of hydrogen gas (Figure 2–4).
     (2) In a water molecule, two hydrogen atoms are covalently bonded to an oxygen molecule. Oxygen has six valence electrons; it shares electrons with two hydrogen atoms to complete its outer shell with eight electrons.

$$H \cdot + H \cdot + :O: \longrightarrow H:O:H$$

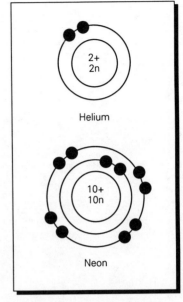

Figure 2–3. *The electron shell model of helium and neon. Electrons in each shell are depicted in pairs with the pairs separated from each other. Although it is known that electrons do not circle the nucleus in fixed, concentric paths, it is useful to depict them as though they do to convey other kinds of information.*

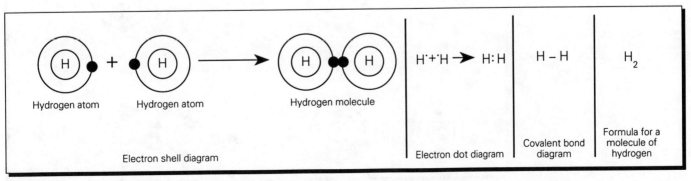

Figure 2–4.

14 The Chemical Basis of Life

(3) Carbon also forms covalent bonds. The carbon atom has four electrons in its outer energy shell. For example, one carbon and four hydrogen atoms can share electrons to form a molecule of methane (CH_4).

$$C + 4H \longrightarrow H-\underset{\underset{H}{|}}{\overset{\overset{H}{|}}{C}}-H$$

b. **Double bonds and triple bonds.** When one pair of electrons is shared between two atoms, the covalent bond is a single bond. When two pairs of electrons are shared, the bond is a double bond. Some atoms share three pairs of electrons to form triple bonds with one another (Figure 2-5).

c. **Polar and nonpolar covalent bonds.** In a covalent bond, the atoms of each element have a specific attraction for electrons, known as **electronegativity.** Electronegativity depends on the number of vacancies in the outer shell and the distance of the outer shell from the atomic nucleus.

(1) In **polar covalent bonds**, the shared electrons are pulled closer to the element with the greater electronegativity or affinity for electrons. Thus, the charge is asymmetrically distributed. An example is the water molecule in which the entire molecule is neutral, but because oxygen is much more electronegative than hydrogen, the shared electrons are drawn closer to the oxygen atom, creating a slight charge at each end of the molecule and polarity (Figure 2-6).

(2) In **nonpolar covalent bonds**, the two atoms are of the same element or do not differ much in electronegativity. The shared electrons are pulled with equal force between the atoms and the charge is distributed symmetrically. Examples are the hydrogen molecule or the methane molecule.

2. **Ionic bonds** represent extremes of polarity. They are formed when electrons are pulled completely from one atom (the more electronegative element) to another atom.

a. **Ions.** When an atom gains or loses electrons in its outer shell it becomes a **charged particle** called an **ion.**

(1) Cations. Atoms with one, two, or three valence electrons in their outer shell tend to **lose** the electrons to other atoms. These atoms, which are then known as cations, become positively charged because of the excess of protons in their nuclei. For example, potassium ion (K^+) and calcium ion (Ca^{++}).

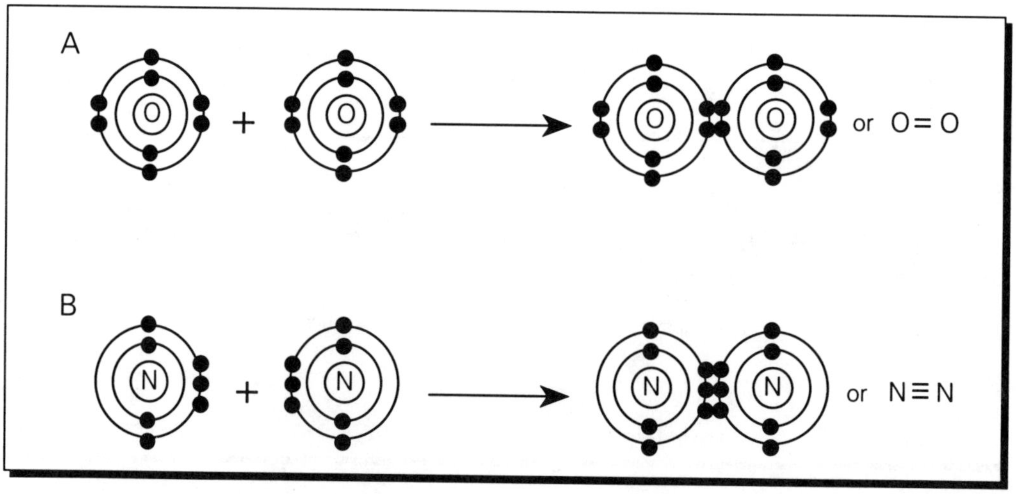

Figure 2-5. *A,* Double bonds in an oxygen molecule. *B,* Triple bonds in a nitrogen molecule.

(2) **Anions.** Atoms with five, six, or seven valence electrons in their outer shell tend to gain electrons and become negatively charged ions called **anions**. For example, chloride ion (Cl⁻) and sulfur ion (S⁼).
  b. **Ionic bonds between ionic compounds** are formed when anions and cations are bonded together by their attraction of opposite charges. For example, sodium chloride (NaCL), common table salt, is an example of an ionic compound.
  (1) Sodium (atomic number 11) has one electron in its outer shell (two in the first shell, eight in the second shell, and one in the third shell). Chlorine (atomic number 17) has seven electrons in its outer shell (two in the first shell, eight in the second shell, and seven in the third shell).
  (2) When sodium reacts with chlorine, it transfers its outermost one electron to chlorine.
   (a) The sodium ion now has 11 protons in its nucleus, 10 electrons orbiting the nucleus, and a net charge of +1.
   (b) The chlorine ion has 17 protons in its nucleus, 18 electrons circling the nucleus, and a net charge of -1.
  (3) The **two ions, Na⁺ and Cl⁻** are held together in an ionic bond as a result of their opposite charges as NaCl (table salt) (Figure 2–7).
  c. Ionic bonds can involve the transfer of more than one electron. For example, magnesium (Mg) has two electrons in its outer shell; therefore, each magnesium atom can supply valence electrons to two chlorine (Cl) atoms. Thus, the salt, magnesium chloride (MgCl$_2$) has two chloride ions for each magnesium ion.
  d. Ions also can be an entire covalent molecule with an electrical charge. For example, in ammonium chloride (NH$_4$Cl), the **cation is ammonium (NH$_4$⁺)**, and the **anion is chloride (Cl⁻)**.
  e. Ionic bonds are weaker than covalent bonds and tend to **dissociate** in water to form the separate component ions. The compound then is said to **dissolve** in water.
 3. **Hydrogen bonds** are weak bonds formed between the oppositely charged portions of neighboring polar molecules and thus are the result of the electrostatic attraction between them.
  a. Hydrogen bonds exist between polar covalent molecules that contain at least one hydrogen atom bonded to a highly electronegative atom such as oxygen or nitrogen.
  b. Water molecules form hydrogen bonds with other water molecules. Although the individual water molecules are electrically neutral as a whole, because of the phenomenon of polarity (see section I E 1 c), the hydrogens in the molecule have a net positive charge while the oxygen has a net negative charge.
 4. **Van der Waals interactions,** or forces, are attractive forces that are due to interaction of electron clouds between atoms and molecules that are very close together.

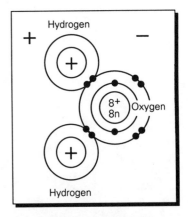

Figure 2–6. *Polar covalent bond in a molecule of water.* The two hydrogen atoms are nearer one end of the molecule giving that end a positive charge while the other end of the molecule has a negative charge. The unequal distribution of charge causes the entire molecule to have polarity, like a magnet.

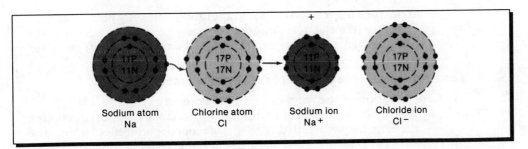

Figure 2–7. *Formation of an ionic bond.* Chlorine, which has a very strong attraction for an additional electron, reacts with sodium, which transfers its single valence electron to chlorine. The negative chlorine ion, Cl⁻, attracts the oppositely charged sodium ion, Na⁺.

5. **Hydrophobic (water-hating) interactions** are nonpolar attractions that occur between groups of nonpolar molecules.
   a. These molecules are insoluble in water and tend to clump together in the presence of water, thus minimizing their exposure to the water.
   b. A polar environment tends to exclude nonpolar compounds and push them out. An example is the reaction of oil droplets in water, which coalesce to form a layer.

II. **CHEMICAL REACTIONS** are interactions of atoms or molecules to form or break chemical bonds, which result in changes in the composition of matter. Chemical reactions do not create or destroy matter, but only rearrange it. Involved are reactants, which are the substances that combine, and products, which are the substances that are formed.

   A. **Types of chemical reactions**
      1. A **synthesis reaction** is the combining of two or more atoms, ions, or molecules to form a larger, more complex molecule. Synthesis reactions occur in the body and are collectively called **anabolic reactions**, or **anabolism**.
      2. A **decomposition reaction** is the reverse of a synthesis reaction; that is, larger molecules are broken down to form smaller molecules, ions, or atoms. Decomposition reactions that occur in the body are collectively called **catabolic reactions**, or **catabolism**.
      3. An **exchange reaction** is one in which both synthesis and decomposition are reciprocally involved. The reactants are decomposed to synthesize the products.
         a. **Reversibility of reactions.** Most chemical reactions are reversible. The **products** of the forward reaction can become the reactants of the reverse reaction. For example, hydrogen (H) and nitrogen (N) molecules combine to form ammonia (NH_3), but ammonia can decompose to form hydrogen and nitrogen. Reversibility is indicated by a double arrow in the equation:

            $$3H_2 + N_2 \rightleftharpoons 2\,NH_3$$

         b. **Chemical equilibrium** is reached when the rate of the forward and reverse reactions is the same, and the relative concentrations of reactants and products are stabilized. In the reaction of hydrogen and nitrogen to form ammonia, equilibrium occurs when ammonia decomposes as rapidly as it is formed.
      4. **Oxidation-reduction reactions (redox reactions)** involve the transfer of one or more electrons from one reactant to another.
         a. **Oxidation** is the chemical process in which an atom, ion, or molecule (reducing agent) loses electrons.
         b. **Reduction** is the chemical process in which an atom, ion, or molecule (oxidizing agent) gains electrons. (Oxygen is a powerful oxidizing agent in biologic systems; that is, it accepts electrons easily because it is so electronegative.)
         c. **Oxidation-reduction reactions** occur simultaneously because as one substance (the **electron donor or reducing agent**) loses electrons, the other substance (the **electron acceptor or oxidizing agent**) gains the electrons.
   B. **Factors that influence the rate of the chemical reaction**
      1. **Concentration of the reactants and the products**
         a. When there are high concentrations of reactants, chemical reactions occur more frequently; that is, there are more opportunities for the reactants to collide with one another and form products.
         b. Therefore, as the products accumulate, they too have greater opportunity to collide, and this results in greater frequency of the reverse reaction.

2. **Particle size.** At a given temperature and concentration, smaller particles move faster than larger particles and collide more frequently. Therefore, the smaller the reacting particle, the faster the rate of chemical reaction.
3. **Temperature.** Chemical reactions proceed more rapidly at higher temperatures and are slowed by low temperatures.
4. **Catalytic action.** Catalysts are substances that increase the rate of a reaction that would occur much more slowly in their absence. Catalysts do not participate in the reaction and remain unchanged after the reaction is over.
   a. **Enzymes** are biologic catalysts that act by temporarily uniting with the reacting molecules, which are called **substrates**.
   b. Enzymes reduce the amount of energy required to make and break chemical bonds (**activation energy**) by providing a site of action (reactive site) for the reactants.

III. BIOCHEMISTRY

A. **Composition and structure of living matter.** Most chemicals in the human body exist in the form of compounds, which are divided into two main groups: **organic** and **inorganic** compounds.
   1. **Inorganic compounds** lack carbon. For the most part, they are all chemicals in the body not classified as organic.
      a. **Water**, which comprises about 70% of total body weight, is the most important inorganic compound in the human body.
      b. Other inorganic compounds of biologic importance are small, simple compounds such as simple **acids**, **bases**, and **salts**.
      c. Carbon dioxide (CO_2) and compounds containing carbonate (CO_3^{2-}) are classified as inorganic, although they contain carbon.
   2. **Organic compounds** contain carbon atoms and comprise most of the thousands of chemical compounds found in living organisms. The chemistry of life is said to be carbon chemistry.

B. **Inorganic compounds**
   1. **Properties of water.** Water has the following **physical** and **chemical properties**:
      a. **Cohesiveness and adhesiveness.** Due to their hydrogen bonds, water molecules have a great tendency to stick together (cohesion) and also to stick to many other substances (adhesion). As a result of these two qualities, water has:
         (1) **High surface tension.** Because of the cohesiveness of its molecules, water molecules form a strong surface layer.
         (2) **Capillary action.** The tendency for water to rise in capillary (very thin tubes) results from its adhesive and cohesive properties.
      b. **High heat capacity.**
         (1) Water can absorb or release large amounts of heat with only a slight change in its own temperature.
         (2) Because of this property, sudden changes in body temperature from external or internal factors are prevented.
      c. **High heat of vaporization.**
         (1) When water evaporates (i.e., vaporizes to change from a liquid to a gas) it requires nearly twice as much heat as that needed to vaporize alcohol or many other liquids.
         (2) As perspiration evaporates from the skin, therefore, large amounts of heat are removed from the body and lost to the environment, providing an efficient cooling mechanism.
      d. **Water as a solvent**
         (1) Because of the polarity of the water molecule and its propensity to form hydrogen bonds, water is a solvent for many different kinds of of ionic molecules and polar nonionic molecules (**hydrophilic molecules**).

(a) Water molecules form oriented solvent (hydration) shells around negatively charged molecules and atoms by orienting their positive (hydrogen) ends toward the negative ion.
(b) The negative oxygen ends of the water molecules point outward and attract the positive ends of other water molecules.
(c) Around positively charged ions, the orientation of the water molecules is reversed. Depending on their orientation, the solvent shells attract either the positive or negative ends of more water molecule. (Figure 2–8).
(d) Thus, as the attraction between solute particles is weakened, a solution is formed.
(2) Water will not dissolve nonpolar molecules that are neutral; i.e., hydrophobic molecules. It pushes them out to form droplets, which coalesce to form a surface layer, as oil does on water.

2. **Measures of concentration**
   a. **Concentrations of solutions expressed as a percentage.** Two types of percent concentration are used in the biological sciences and in human physiology: **weight/volume percent** and **milligram percent**.
   (1) **Weight/volume percent concentration** is the ratio of the **weight of the solute** to the **volume of the solvent**. For example, a 10% salt solution contains 10 g (grams) of NaCl (sodium chloride) in 100 ml (milliliters) of solution.

   $$\% \text{ concentration} = \frac{\text{g of solute}}{100 \text{ ml of solution}} \times 100\%$$

   (2) **Milligram percent (mg%)** concentration is the number of **milligrams of solute per 100 ml of solution**. It is a measure of concentration used for very low solute concentrations. For example, the normal blood glucose level is 90 to 110 mg%.

   $$\text{mg}\% = \frac{\text{mg solute}}{100 \text{ ml of solution}} \times 100\%$$

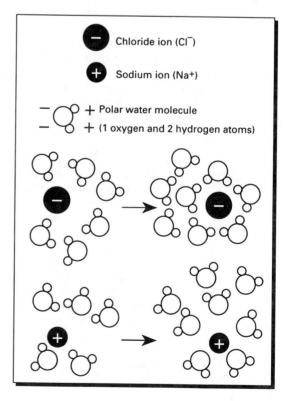

Figure 2–8. *Water is a solvent because of its polar structure.* The positive and negative ends of water orient to the negative and positive ions of chlorine and sodium to form solvent shells, which prevent the formation of sodium chloride.

b. **Concentrations expressed in terms of molarity or moles per liter.**
   (1) The **molarity** (M) of a solution is the number of moles (mol) of solute in 1 liter (L) of solution.
   (2) One **mole** of any substance is defined as the number of atoms in 12.000 grams of carbon (6.02×10^{23}), known as Avogadro's number. A mole, therefore, contains 6.02×10^{23} **particles**, whether they are atoms, molecules, ions, or electrons.
   (3) The weight of one mole depends on the nature of the particles.
      (a) The weight of one mole of **atoms** of an element is the mass in grams of the substance that is numerically equal to the atomic weight of that element. For example, the atomic weight of sodium is 23 g, so one mole of sodium will equal 23 g.
      (b) The weight of one mole of any **molecule** is equal to the **sum of the atomic weights** of all the atoms appearing in the chemical formula of the molecule. For example, the molecular weight of NaCl is 58.5 g, so one mole of NaCl equals 58.5 g.

$$23.0 + 35.5 = 58.5$$
$$\text{(atomic weight of Na)} \quad \text{(atomic weight of Cl)}$$

   (4) A 1 M sodium chloride solution is made by dissolving 58.5 g of sodium chloride in water to make 1 L of solution. Similarly, a 1 M glucose solution (molecular weight 180 g) is made by dissolving 180 g of glucose in water to make 1 L of solution.

3. **Acids, bases, and pH**
   a. **Ionization of water**
      (1) Water molecules have a slight tendency to ionize; that is, to dissociate into hydrogen ions (H^+) and hydroxide ions (OH^-).

$$H_2O \longrightarrow H^+ + OH^-$$

      (2) In pure water the concentration of hydrogen ions and hydroxide ions is equal at 10^{-7} M each; that is, there is one ten-millionth of a mole of hydrogen ions per liter of pure water and one ten-millionth of a mole of hydroxide ions.
   b. **Acids and bases**
      (1) A solution is acidic if it has a higher hydrogen ion concentration than hydroxide ion concentration. An acid is a substance that releases hydrogen ions (protons) when dissolved in water.
      (2) A solution is basic if its hydrogen ion concentration is lower than its hydroxide ion concentration. A base is a substance that releases hydroxide ions and accepts hydrogen ions, removing them from the solution, when dissolved in water.
   c. **pH scale** is used to describe the acidity or alkalinity (basicity) of a solution.
      (1) The pH scale, which runs from 0 to 14 (0 is the strongest acid, 14 is the strongest base, and 7 is neutral) expresses the range of hydrogen ion and hydroxide ion concentrations in a solution.
      (2) The pH of a solution is the **negative logarithm (base 10) of the hydrogen ion concentration** expressed in moles per liter. Thus,

$$pH = -\log[H^+]$$

      (a) A neutral solution (e.g., pure water) has a hydrogen ion concentration of 10^{-7} M, or

$$pH = -\log 10^{-7} = -(-7) = 7$$

      (b) Therefore, the **pH of a neutral solution is 7**, the midpoint of the scale.

Figure 2–9. *The pH scale illustrating the pH of various common solutions.*

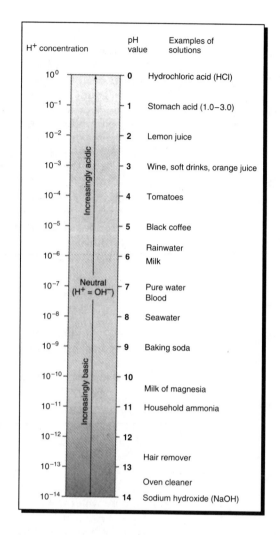

(3) Note that the pH declines as the hydrogen ion concentration increases. A solution with a pH of 2 has a concentration of hydrogen ions 10 times that of a solution with a pH of 3. Figure 2–9 illustrates the pH of some common aqueous solutions.

C. **Organic compounds**

1. **Components.** Carbon is the core component of organic compounds.
   a. The carbon atom with four electrons in its outermost shell can achieve stability by its sharing four electrons with other carbon atoms or with other elements to form up to four single covalent bonds, or by forming smaller numbers of double or triple bonds. In this way, carbon can form long chains or branched chains.
   b. Carbon most commonly bonds to itself or to hydrogen, oxygen, or nitrogen.
   c. **Functional groups** (i.e., groups of atoms that determine the characteristics, solubility, and reactivity of the organic molecule that contains the group) attach to the carbon backbone. Most organic compounds contain two or more functional groups. Functional groups found in biologic molecules include the following:
      (1) Hydrogen (–H) is found in almost all organic molecules. Bonds between carbon and hydrogen form nonpolar functional groups such as the **methyl group**, $-CH_3$.
      (2) The **hydroxyl (–OH) group** is found in carbohydrates, nucleic acids, some organic acids, alcohols, and steroids. Organic compounds containing hydroxyl groups are polar and soluble in water.

(3) The **carbonyl group (–CO or –COOH)** consists of a carbon atom joined to an oxygen atom by a double bond. Organic compounds known as sugars contain both carbonyl and hydroxyl groups.

(4) **Carboxyl group (–C–O–O–H or –COOH)** consists of an oxygen atom double-bonded to a carbon which is also bonded to a hydroxyl group. It is found in organic acids such as amino acids and fatty acids.

(5) The **amino (–NH₂) group** consists of a nitrogen atom bonded to two hydrogen atoms and to the carbon skeleton. It is found in organic compounds called amines, which are part of proteins and nucleic acids.

(6) The **phosphate (–PO₄) group** is found in nucleic acids, phospholipids, and organic phosphates. The organic phosphates store energy that can be transferred from one molecule to another by transfer of a phosphate group.

(7) The **sulfhydryl (–SH) group** consists of a sulfur atom bonded to a hydrogen atom. Compounds containing sulfhydryls are called thiols.

2. **Carbohydrates** (sugars and starches) are compounds composed of carbon, hydrogen, and oxygen. Characteristically, the molecules have twice as many hydrogen atoms as oxygen atoms, the same proportion as in water molecules. Carbohydrates are classified as **monosaccharides, disaccharides,** and **polysaccharides.**
   a. **Monosaccharides** (simple sugars) are the basic carbohydrate molecules. Some contain as few as three carbon atoms (**trioses**); others contain five (**pentoses**) or six (**hexoses**) carbon atoms.
      (1) All simple sugars contain a **carbonyl group.** If the double-bonded O is attached to the terminal carbon atom of a carbon chain (i.e., the carbon atom at the end), the sugar is an **aldehyde** sugar; if it is attached to a carbon atom at any other position than the end of the chain, it is called a **ketone** sugar (see Figure 2–10).
      (2) Glucose, fructose, and galactose (all hexoses) are simple sugars that have the same number and types of atoms ($C_6H_{12}O_6$) but differ in their structural three-dimensional arrangement and properties. Compounds related in this way are called **isomers.**
   b. **Disaccharides** (double sugars)
      (1) Disaccharides are formed by the linking together of two monosaccharides by **dehydration synthesis** (also known as the **condensation reaction**), which involves the removal of a molecule of water.
      (2) Disaccharides can be broken apart into their monosaccharide subunits by the addition of a molecule of water, a reaction known as hydrolysis.
      (3) Examples of disaccharides include the following:
         (a) **Sucrose,** or table sugar, contains glucose and fructose units.
         (b) **Lactose,** or milk sugar, contains glucose and galactose units.
         (c) **Maltose,** found in malt, is composed of two glucose molecules.
   c. **Polysaccharides** are polymers, long chains of molecules composed of similar units. They are made up of many monosaccharides joined together by dehydration linkages to form starch (in plants) or glycogen (in animals), important structural and energy storage compounds. Examples of polysaccharides include the following:
      (1) **Amylose** and **amylopectin** are digestible plant starches that form a portion of the human diet.
      (2) **Cellulose,** which is the most abundant polysaccharide in nature, is a structural component of cell walls. It is one component of indigestible "roughage," or fiber, in the human diet.
      (3) **Glycogen** is a highly branched glucose-storage polysaccharide found in the liver and skeletal muscle.

3. **Lipids** are a diverse group of molecules: all are insoluble in water but soluble in nonpolar solvents such as ether and chloroform. Lipids of biologic importance include the neutral fats, waxes, phospholipids, and steroids.

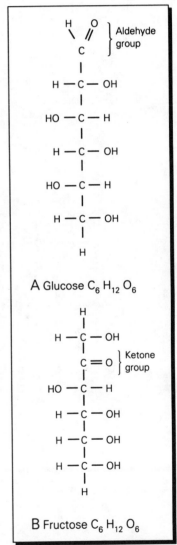

Figure 2–10. *Monosaccharide isomers depicted in chain form.* Both have the same chemical formula, $C_6H_{12}O_6$, but they are structurally different. *A,* Glucose is an aldehyde monosaccharide. *B,* Fructose is a ketone monosaccharide.

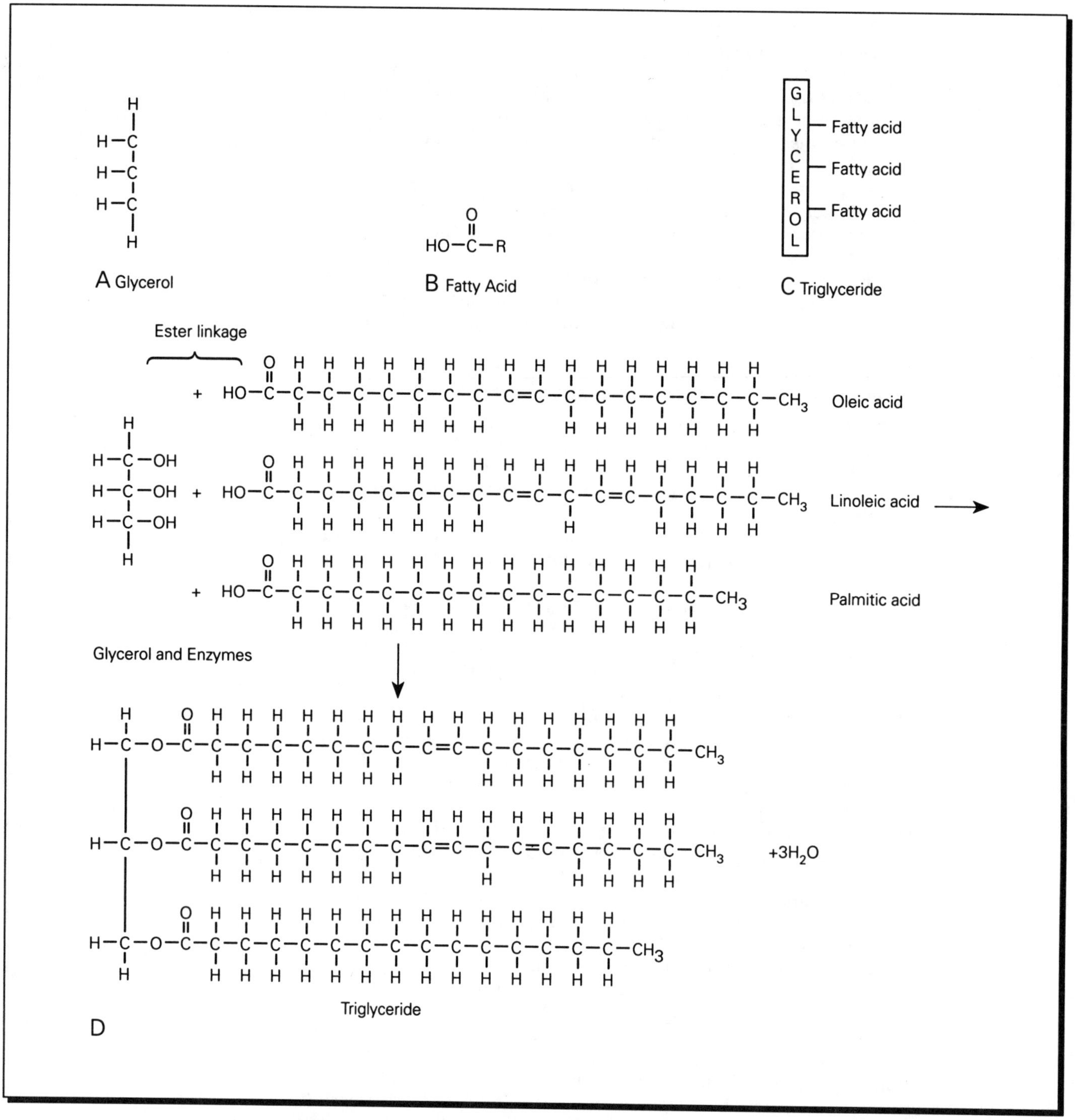

Figure 2-11. *Neutral fat. A,* Structure of glycerol. *B,* Structure of a fatty acid. All fatty acids contain the carboxyl (–COOH) group. The R represents the rest of the molecule, which varies with each type of fatty acid. *C,* A triglyceride. *D,* Synthesis of a mixed triglyceride. Specific dehydrating enzymes cause the linkage of three molecules of fatty acids to one molecule of glycerine. Simple triglycerides contain the same fatty acid; mixed triglycerides contain two or more different fatty acids. The triglyceride depicted is an unsaturated fat because two of its fatty acids contain double bonds between its carbon atoms.

a. Vegetable oils and animal fats are **neutral fats**, also called **triglycerides**. The neutral fats and waxes contain only carbon, hydrogen, and oxygen.
   (1) Neutral fats are compounds of fatty acids and glycerol. Three molecules of **fatty acids** (long chains of carbon and hydrogen atoms with a carboxyl group at one end) are covalently bonded to one molecule of **glycerol** (i.e., a three-carbon molecule with three hydroxyl side groups) through dehydration synthesis (Figure 2–11).
      (a) Fats tend to be solid at room temperature. The fatty acids in their molecules are long chains with the carbon atoms connected by single covalent bonds and with hydrogen atoms occupying all of the other available **bond positions** on the carbon atoms. The fat is called saturated; that is, it has as many hydrogen atoms as it can hold.
      (b) Oils tend to be liquid at room temperature. They exhibit varying degrees of **unsaturation** and **polyunsaturation** in that some of the carbon-to-carbon bonds are double covalent bonds and there are consequently fewer hydrogen atoms compared to the analogous saturated fats.
      (c) An oil can be converted to a fat by hydrogenating the oil; that is, by breaking the double bonds between carbon atoms replacing them with single covalent bonds and by adding hydrogen atoms to the remaining bond positions. Examples of hydrogenated fats are solid vegetable shortening, hydrogenated peanut butter, and margarine.
      (d) Most of the fatty acids that enter into the formation of edible fats and oils have long-chain carbon backbones. The most common are **stearic** and **oleic** acids, with 18 carbon atoms each, and **palmitic** acid, with 16 carbon atoms.
   (2) Waxes are similar to fats and oils except that the fatty acids in waxes are attached to a carbon chain alcohol that is not glycerol.
b. **Phospholipids** are a major constituent of cell membranes.
   (1) Structurally, phospholipids are similar to triglycerides with the exception that one of the three fatty acids is replaced by a phosphate group that has a short, polar, nitrogen-containing group at one end (Figure 2–12).
      (a) The nitrogen containing head of the molecule is polar, hydrophilic (water-attracting), and water soluble.
      (b) The other end of the molecule consists of two fatty acid tails, one saturated and one unsaturated, which are hydrophobic (water-repelling) and insoluble in water.
   (2) **Functionally**, the dual nature of phospholipids is significant in the structure of cell membranes.
      (a) The heads of the phospholipid molecules are in contact with the aqueous solutions at the surface of the cell membrane.

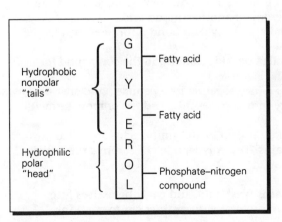

Figure 2–12. *Phospholipid.* The phosphate group forms a hydrophilic (water-attracting) polar head on the molecule; the two fatty acids form two hydrophobic (water-repelling) nonpolar tails. This gives phospholipids membrane-forming properties.

Figure 2–13. *A*, Model of phospholipid. *B*, Arrangement in cell membrane. The phospholipids are arranged in a double layer (bilayer) with their polar heads to the outside and their nonpolar tails toward the middle.

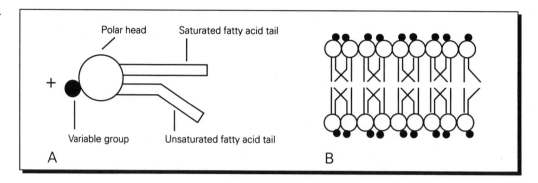

   (b) The tails point toward the center of the membrane, and hydrophobic interactions between the hydrocarbons help to hold the molecules of the membrane together, which then forms a boundary between the cell and the external environment (Figure 2–13).
  c. **Steroids** are large lipid molecules composed not of hydrocarbon chains but of four fused rings (the steroid nucleus) with various functional groups attached to them.
   (1) **Cholesterol**, which is a common component of the cells of animal membranes, is an important steroid; most other steroids are synthesized from cholesterol.
   (2) Examples of steroids in the body include male and female sex hormones (i.e., testosterone, estrogen, and progesterone), adrenal cortical hormones, and bile salts.
 4. **Proteins** are more complex chemically, but, like carbohydrates and lipids, they are composed of simple building-block compounds. All proteins contain carbon, oxygen, hydrogen, and nitrogen atoms and many proteins contain sulfur and phosphorus.
  a. **Amino acids** are the basic molecular units from which the long protein polymers are formed. There are 20 different amino acids found in proteins that are **fundamental to the structure and function of the human body.**
   (1) Each amino acid contains at least one acid **carboxyl (–COOH) group and** at least **one amino group** (–NH$_2$). Both groups are attached to the same carbon atom. Each amino acid also has a side chain, designated as an **R group.**
    (a) Amino acids differ in their R groups, which give them their unique characteristics and influence the properties of the proteins in which the amino acids are incorporated.
    (b) **Nonpolar** R groups result in the amino acids being relatively **insoluble in water.** R groups that are **polar** or electrically charged result in **water-soluble** amino acids.
   (2) Amino acids are linked together to form proteins by **condensation** (dehydration) reactions between the carboxyl group of one amino acid and the amino group of another amino acid.
    (a) Water is produced as the covalent bond is formed between the two amino acids.
    (b) The bonds are called **peptide bonds** and the compound formed is called a peptide (Figure 2–14).
    (c) Two amino acids joined together by a peptide bond are called **dipeptides,** three form a **tripeptide,** and ten or more form a **polypeptide.**
    (d) Long chains containing up to 100 amino acids are called **polypeptide chains.** The polypeptide chain forms the primary structure of a protein.
  b. **Protein structure**
   (1) The **polypeptide chains** twist, fold, and pack themselves into unique shapes to form proteins with distinct **conformations.**

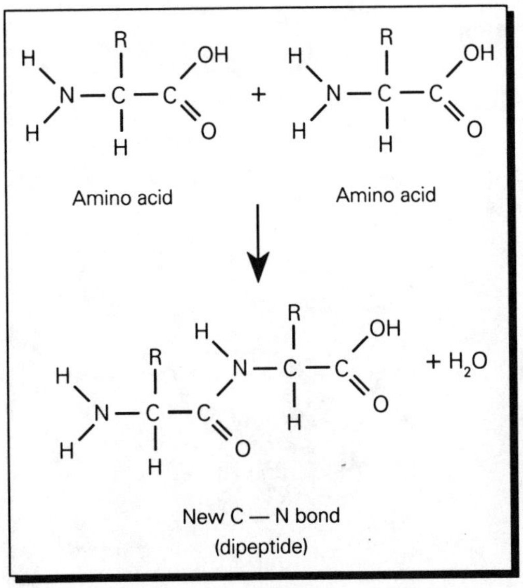

Figure 2-14. *The peptide bond joins two amino acids in the primary level of protein organization.* A peptide bond always occurs between the carboxyl and amino groups of adjacent amino acids.

   (a) **Structural or fibrous proteins** are arranged as long, linear macromolecules. Examples include collagen; myosin (muscle protein); fibrin; and the keratin of hair, nails, and skin.
   (b) **Globular proteins** are highly coiled and folded into a nearly spherical shape, similar to a tangled skein of yarn. Examples include enzymes, hormones, and blood proteins.
(2) There are **four levels of organization** to protein structure.
   (a) The **primary structure** is the polypeptide chains and the number and sequence of the amino acids in each chain.
   (b) The **secondary structure** is the coiling of the peptide chain into a spiral helix (like a Slinky toy) or another type of conformation.
      (i) The **alpha helix**, a uniform geometric coil with 3.6 amino acids occupying each turn of the helix, develops when hydrogen bonds form between amino acids in successive turns of the spiral. It is the basic shape of structural proteins in hair, skin, and nails.
      (ii) A pleated sheet structure is formed by hydrogen bonds holding adjacent chains together in a back-and-forth, or zig-zag, configuration. Such pleated sheets make up the core of globular proteins.
   (c) The **tertiary structure** is imposed on the regular secondary structure by less regular contortions, folds, and kinks of the polypeptide chains to form complex, three-dimensional shapes.
   (d) The **quaternary structure** is the complex arrangement assumed by two or more polypeptide chains, each with a primary, secondary, and tertiary structure, to form a huge, biologically active protein molecule (Figure 2-15).
      (i) **Hemoglobin** is an example of a globular protein with a quaternary structure. It consists of 574 amino acids arranged in four polypeptide chains.
      (ii) **Collagen** is an example of a fibrous protein with a quaternary structure. It has three polypeptide chains arranged in a triple helix, which is a supercoiled rope-like structure that gives collagen fibers great tensile strength.
   c. **Protein denaturation.** Proteins retain their conformational shape as long as their physical and chemical environments are maintained. If the environments are altered, proteins may unravel or denature; that is, they lose their secondary, tertiary, and quaternary structure and thereby lose their biologic activity.

Figure 2-15. *Four possible levels of protein structure.* The primary structure is the particular sequence of amino acids. The secondary structure may be an alpha helix or a beta pleated sheet. Twisting and bending of the coiled polypeptide chain gives the tertiary structure. Proteins composed of two or more polypeptide chains have a quaternary structure.

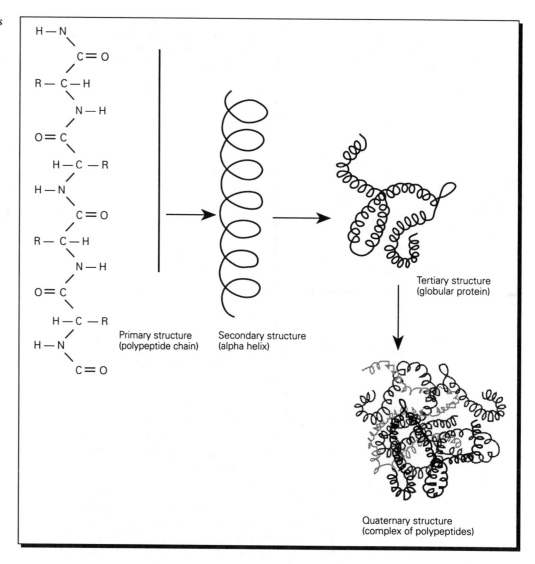

(1) **Protein conformation** is dependent on the hydrogen bonds, which are weak and very sensitive to changes in pH and temperature.
(2) Even brief exposure to high temperatures (i.e., those above 60° C) or exposure to strong acids or bases for long periods of time will cause denaturation through rupture of hydrogen bonds.
    (a) Some proteins can be restored to their original shape if they are denatured without becoming insoluble. For example, after only gentle heating, the protein can regain its original state if returned to normal temperature.
    (b) Extreme heat causes **irreversible denaturation**. The white of an egg (albumin) solidifies and becomes insoluble during heating.
        (i) Very high body temperatures can cause coagulation of cellular proteins.
        (ii) When body temperature rises above 106° to 108° F degeneration of cells, especially in the brain, begins to take place as a result of protein denaturation.
5. **Nucleic acids** are complex molecular structures composed of carbon, hydrogen, oxygen, nitrogen, and phosphorus. They are the molecules of inheritance and the regulators of protein function in the cells.
    a. The **two types of nucleic acids** are **deoxyribonucleic acid (DNA)** and **ribonucleic acid (RNA)**.

(1) DNA is found in the chromosomes of all living things and has the ability to replicate itself.
(2) RNA functions in the synthesis of proteins under the direction of DNA.

b. **Structure of nucleic acids.** DNA and RNA are composed of chains of subunits called **nucleotides**, linked together by dehydration synthesis.
  (1) Each nucleotide is composed of three parts: it contains a **nitrogenous base** joined to a **pentose** (five-carbon sugar), which then is bonded to a **phosphate group**.
    (a) Two kinds of nitrogenous bases are found in nucleotides (Figure 2–16):
      (i) **Pyrimidines** are single-ring molecules containing carbon, hydrogen, and nitrogen. The pyrimidines in nucleic acids are **cytosine (C)**, **thymine (T)**, and **uracil (U)**. Cytosine is found in DNA and RNA, thymine is found only in DNA, and uracil is found only in RNA.
      (ii) **Purines** are double-ring molecules. The purines are **adenine (A)** and **guanine (G)**, and they are found both in DNA and RNA.
    (b) The **pentose** connected to a nitrogenous base is **deoxyribose** in DNA and **ribose** in RNA.
  (2) In the DNA polymer, the nucleotides are linked sequentially, one on top of the other, into two opposing strands to form a **double helix** (Figure 2–17).
  (3) In the RNA polymer, the nucleotides are also one on top of the other, but RNA is single stranded; that is, the double helix does not form.

c. **Function of nucleic acids.** Some nucleotides exist in living cells as parts of other molecules. Examples include the following:
  (1) **Cyclic nucleotides**, which are intracellular messengers,
  (2) **Adenosine triphosphate**, a high-energy molecule that can store energy for later release, and
  (3) **Coenzymes**, which assist enzymes in their function as biologic catalysts (see section IV B).

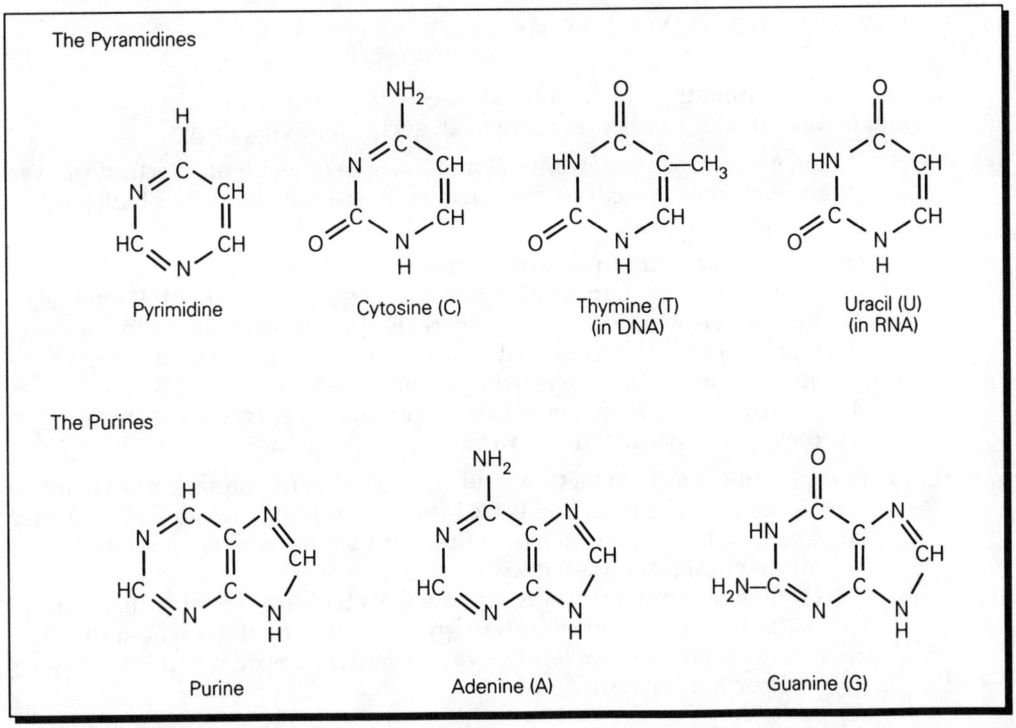

Figure 2–16. *Pyrimidines and purines.*

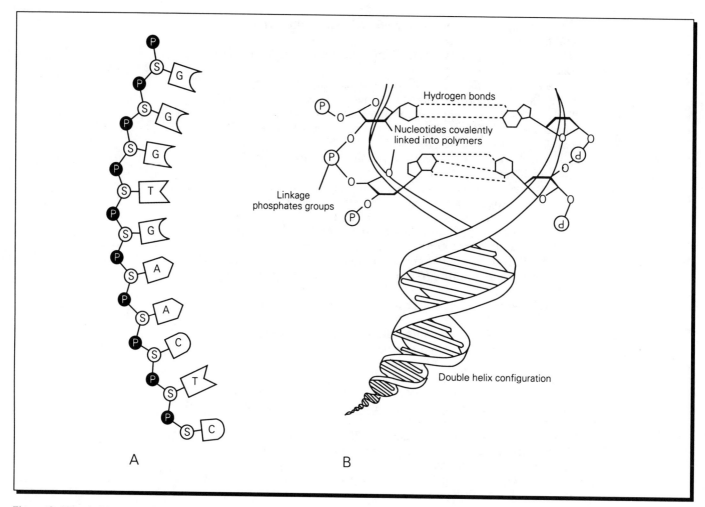

Figure 2-17. *A*, Diagram of a nucleic acid molecule. The nitrogenous bases, joined to sugar, which is bonded to phosphate, are arranged in various specific sequences. P, phosphate; S, sugar; G, guanine; A, adenine; T, thymine. *B*, In the DNA double helix, the nitrogenous bases face inward. The nucleotides of one polymer are attracted by weak hydrogen bonds to the nucleotides of the other to produce the helical formation.

IV. ENERGY, ENZYMES, AND ENZYME ACTIVITY

A. **Energy.** Light from the sun is captured by electrons, which are thereby excited into higher orbitals; this is the source of all energy for living things.

1. As these electrons change orbitals and move to more highly electronegative atoms in chemical reactions, the energy released can be used for biologic processes.
2. Energy is the capacity to do work. Energy has two forms.
   a. **Free energy**, the term used in biological contexts, is **potential energy** that can be obtained for doing work. Free energy is available in covalent bonds, in an electron excited into a higher orbital, and within the atomic nucleus to be released by a nuclear reactor.
   b. **Kinetic energy** is energy in movement; that is, energy associated with the actual process of doing work.
3. **Energy and chemical reactions.** When two reactants combine in a chemical reaction, old chemical bonds are broken, new bonds are formed, and another set of substances (which are known as the products) are produced.
   a. All chemical reactions fall within two categories.
      (1) In **exergonic reactions**, the reactants have more energy than the products and, therefore, **energy is released** by the reaction; that is, the products have less free energy in their covalent bonds than the reactants possessed.

$$A + B \longrightarrow C + D + \text{Energy}$$

(2) In **endergonic reactions**, the products contain more energy than the reactants. Endergonic reactions are "uphill" and require an input of energy into the low-energy reactants in order for the reaction to proceed.

$$A + B + Energy \longrightarrow C + D$$

b. **Coupled reactions.** Living organisms use the energy released by exergonic reactions (which are catabolic) in order to drive essential endergonic reactions such as protein synthesis, movement, and brain activity. Certain molecules within cells, such as adenosine triphosphate (ATP) serve as energy-carrier molecules that transfer energy from place to place.

4. **Activation energy.** All chemical reactions including exergonic reactions, require an initial input of energy, which is called activation energy. Such activation is necessary to bring the reacting molecules close together, break some of the preexisting covalent bonds, and allow forceful collisions of their electron clouds.
    a. The necessity for activation energy prevents spontaneous or uncontrolled reactions in the body. High-energy substances cannot break down easily and hence are more stable.
    b. The activation-energy barrier to reactions can be overcome by heating the reactions, but cellular proteins in the body will denature with increased heat.

B. **Enzymes are organic catalysts** and are globular proteins. They lower the activation energy barrier so that reactions can occur under the conditions normally found in living cells. Enzymes increase the rate of a reaction that would occur eventually, but much more slowly.
  1. **Mechanisms of enzyme action**
     a. An enzyme acts on a specific **substrate**.
     b. **Specificity** of enzymes. Each enzyme can distinguish its own substrate from other closely related compounds (including isomers) so that each type of enzyme catalyzes a particular reaction. The enzyme binds to its substrate or substrates and converts the substrate to the product or products of the reaction.

$$Substrate \xrightarrow{Enzyme} Product$$

  c. **Active site**
     (1) **Lock-and-key model of enzyme action.** Only a restricted area of the enzyme molecule actually binds to the substrate. This **receptor** region is the **active site**, typically a groove or pocket on the enzyme's surface compatible with the shape of the active site. The enzyme is the molecular lock into which only the molecular key of the substrate can fit (Figure 2–18).
     (2) **Induced fit model of enzyme action.** As the substrate binds to the enzyme at the active site, it may induce a slight structural change in the enzyme. This induced fit of enzyme and substrate may improve reactibility and help chemical bonds to break (Figure 2–19).

Figure 2–18. *Lock-and-key mechanism of enzyme action.* The substrate fits the active sites of the enzyme molecule. After the products separate from the enzyme, its active site is free to catalyze the production of more products.

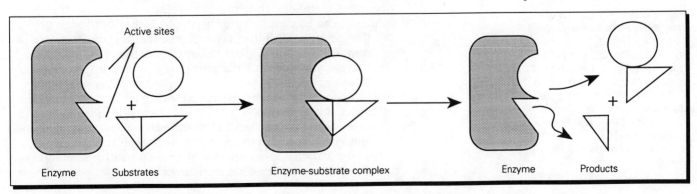

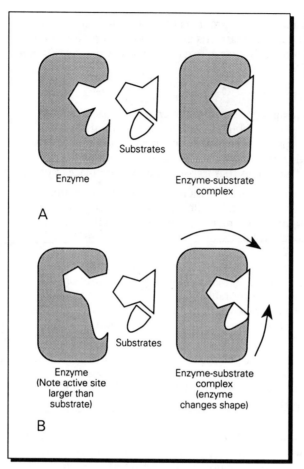

Figure 2–19. Comparison of *A*, lock-and-key model and *B*, induced-fit model of enzyme action. The induced-fit concept is a modification of the original lock-and-key model.

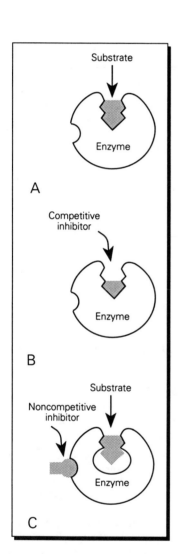

Figure 2–20. *Comparison of a competitive and a noncompetitive enzyme inhibitor. A,* Enzyme-substrate complex; *B,* a competitive inhibitor prevents the substrate from binding to the active site; *C,* a noncompetitive inhibitor does not prevent the substrate from binding but alters the shape of the enzyme so that the active site is no longer functional.

   d. **The enzyme-substrate complex** undergoes internal rearrangement, which forms the products. The enzyme releases the products, and its active site is then available to more substrate.
 2. **Factors affecting enzyme activity**
    a. **Temperature and pH.** Each enzyme in the human body has an optimal temperature (between 35°C and 40° C) and an optimal pH (which ranges from 6 to 8). Exceptions are certain digestive enzymes (such as pepsin), which function best at a pH as low as 1 to 2.
    b. **Cofactors and coenzymes.** Some enzymes require cofactors, which are inorganic helpers such as metal atoms (e.g., zinc, iron, and copper), or coenzymes, which are organic nonprotein molecules such as vitamins.
    c. **Enzyme inhibitors**
       (1) Certain chemicals can selectively inhibit the catalytic action of specific enzymes. Although some toxic chemicals (e.g., mercury and cyanide) may be deadly because of their inhibitory effect on enzymes, selective inhibition of enzymes is a **normal and essential process of metabolic control** within cells.
       (2) Inhibitors can be classified into two types (Figure 2–20):
          (a) **Competitive inhibitors** resemble the substrate molecule and compete with the substrate for entry into the active site of the enzyme, thus blocking the active site and reducing the enzyme's productivity.
          (b) **Noncompetitive inhibitors** block action without entering the active sites. The inhibitor binds to the enzyme in the same vicinity as the active site, causing the enzyme to change its shape and interfering with its receptivity to the substrate. Enzymes whose activities are controlled by noncompetitive inhibition are called **allosteric enzymes** (meaning change of shape).

Study Questions

Directions: Each question below contains four suggested answers. Choose the **one best** response to each question.

1. The number of electrons that surround the nucleus to result in an electrically neutral atom is determined by
    (A) the number of neutrons in the nucleus
    (B) the atomic mass of the atom
    (C) the number of protons in the nucleus
    (D) the number of electron shells in the atom

2. The almost spherical region of space around an atomic nucleus in which a given electron is found 90% of the time is its
    (A) orbital
    (B) parameter of probability
    (C) quantum
    (D) density habitat

3. How many electrons would be found in the third shell of an atom of phosphorus (atomic number 15)?
    (A) 0
    (B) 3
    (C) 5
    (D) 8

4. Which of the following elements would be the least chemically reactive?
    (A) an element whose atoms have seven electrons in the second shell
    (B) an element whose atoms are one electron short of having all of the capacity of the third shell filled with electrons
    (C) an element whose atoms have two electrons in the second shell
    (D) an element whose atoms have the outermost shell filled to capacity

5. Two atoms sharing one pair of electrons between them form a molecule held together by a(an)
    (A) covalent bond
    (B) hydrogen bond
    (C) ionic bond
    (D) electronegative bond

6. All of the following statements are true about water EXCEPT
    (A) Water can absorb or release heat with only a slight change in its own temperature.
    (B) Water is a solvent for polar molecules and nonpolar molecules.
    (C) Water requires a larger amount of heat to evaporate when compared with most liquids.
    (D) Water molecules are held together by hydrogen bonds.

7. Which of the following will result in a 1-molar solution of a substance?
    (A) one mole of the substance dissolved in 100 milliliters of solution
    (B) the largest amount of the substance that can be dissolved in one liter of solution
    (C) one gram molecular weight of the substance dissolved in one liter of solution
    (D) one milligram of the substance in 100 milliliters of solution

8. A solution that has a pH of 7
    (A) has a hydrogen ion concentration of 10^{-7} moles per liter
    (B) is strongly acidic
    (C) is weakly acidic
    (D) has more hydroxide ions than hydrogen ions per liter

9. The basic molecules that make up carbohydrates are
    (A) fatty acids and glycerol
    (B) monosaccharides
    (C) amino acids
    (D) nucleic acids

10. Saturated fats have hydrocarbon chains that contain
    (A) very little hydrogen
    (B) amino functional groups
    (C) no carbon-to-carbon double bonds
    (D) nitrogen

Questions 11 and 12 refer to the two molecules depicted below.

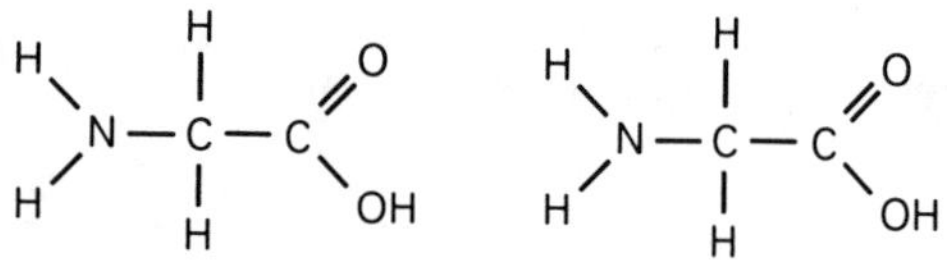

11. When a dehydration reaction takes place between the two molecules depicted, what kind of bond is formed between them?

    (A) ionic bond
    (B) peptide bond
    (C) hydrogen bond
    (D) acidic bond

12. If a long chain consisting of 90 of the molecules shown were bonded together in sequence, the resulting single molecule would be a

    (A) polypeptide
    (B) polysaccharide
    (C) carbohydrate
    (D) protein

13. An enzyme effective at a pH of 7.4 is experimentally exposed to a pH of 3 for 30 minutes. As a result, the enzyme is no longer able to speed up a chemical reaction. Decreasing the pH most likely caused all of the following EXCEPT

    (A) denaturation of the enzyme
    (B) alteration of the tertiary structure of the enzyme
    (C) rupture of the hydrogen bonds of the enzyme
    (D) rearrangement of the fatty acids of the enzyme

14. Both DNA and RNA

    (A) contain the same four nitrogenous bases
    (B) are single-stranded molecules
    (C) contain the same pentose sugar
    (D) contain phosphate groups

15. A certain antibiotic drug enters and binds with the active site of an enzyme whose normal activity is to catalyze the growth of harmful bacteria in the body. Therefore, the antibiotic prevents the action of the enzyme and could best be described as a

    (A) substrate
    (B) competitive inhibitor
    (C) noncompetitive inhibitor
    (D) coenzyme

Answers and Explanations

1. **The answer is C.** (I E 1 d) Atoms are electrically neutral because the number of positively charged protons in the nucleus equals the number of negatively charged electrons surrounding the nucleus. Neutrons are uncharged particles.

2. **The answer is A.** (I E 3 a) Orbitals are the three-dimensional regions of space where electrons occur in constant motion around the nucleus.

3. **The answer is C.** (I E 3 b 1,2) The first shell holds a maximum of two electrons and the second shell a maximum of eight electrons. Because phosphorus has an atomic number of 15 (15 protons), the third shell must contain five electrons.

4. **The answer is D.** (I E 3 c) If the outermost shell is filled to capacity, the atom is inert and will not react with other atoms.

5. **The answer is A.** (I F 1,2,3,4) Covalent bonds involve the sharing of a pair of valence electrons between atoms.

6. **The answer is B.** (III B 1 a,b,c,d) Water will not dissolve nonpolar hydrophobic molecules and pushes them out to form a surface layer.

7. **The answer is C.** (III B 2) A 1-molar solution of any substance is made by dissolving the molecular weight of the substance expressed in grams in enough water to bring the total volume of the solution up to 1 liter.

8. **The answer is B.** (III B 3) pH is the negative logarithm of the hydrogen ion concentration expressed in moles per liter; hence, a pH of 7 is a hydrogen ion concentration of 10^{-7} moles per liter.

9. **The answer is B.** (III C 2 a) Monosaccharides are the building blocks of carbohydrates. Fatty acids and glycerol are the building blocks of a neutral fat; amino acids are the simplest molecules of a protein, and nucleic acids the complex molecules of DNA and RNA.

10. **The answer is C.** (III C 3) The fatty acids in a saturated fat are long chains with the carbon atoms connected by single covalent bonds and hydrogen atoms occupying all of the other available bond positions. In unsaturated fats, some of the carbon-to-carbon bonds are double covalent bonds and thus there are fewer hydrogen atoms attached to the carbon atoms.

11. **The answer is B.** (III C 4 a (2)) A dehydration (condensation) reaction between the carboxyl group of one amino acid and the amino group of another amino acid forms a peptide bond.

12. **The answer is A.** (III C 4 a (2) (d)) Long chains containing up to 100 amino acids are called polypeptides.

13. **The answer is D.** (III C 4 b (2) (c)) Proteins do not contain fatty acids. Exposure to chemical agents, strong acids or bases, or excessive heat can cause denaturation of a protein, which can involve alteration of its conformation and disruption of its hydrogen bonds.

14. **The answer is D.** (III C 5 b) DNA contains thymine; RNA contains uracil. DNA is double-stranded; RNA is single stranded. DNA contains deoxyribose; RNA contains ribose.

15. **The answer is B.** (IV B 2 c) A competitive inhibitor competes with the enzyme's substrate for entry into the active site of the enzyme, thus blocking the active site and reducing the activity of the enzyme.

3 The Cell

I. **INTRODUCTION.** The cell is the smallest structural and functional living unit of the body. Most of the chemical reactions that sustain life occur inside cells. Cells and intercellular substances make up the tissues of the body.

   A. **Cell numbers.** There are trillions of cells in the human body. For example, the total number of only the red blood cells in an average-sized human is 25 trillion.

   B. **Cell shapes**

   1. The **fundamental shape** of isolated cells is round, such as in a blood cell, fat cell, or egg cell.
   2. The basic spheroidal shape is usually altered by **cell specialization** based on its function. For example, a **nerve cell** is **star-shaped** with long processes and a **smooth muscle cell** is **spindle shaped**.
   3. **Flattening** of cells occurs from contact with a surface. **Faceting** of cells occurs as a result of pressures from many surfaces.

   C. **Cell size**

   1. Human body cells are **microscopic** and range from 10 μm to 30 μm in diameter.
   2. The size of cells is restricted from growing too large because cells must maintain a surface area (the plasma membrane) sufficient to accommodate exchange of nutrients and wastes.
      a. Mathematically, as the size of a sphere increases, its volume grows faster than its surface area. Therefore, the larger a cell grows, the less membrane it has in relation to its cytoplasmic mass. Growth stops when the surface area is no longer sufficient to provide for the cell's internal activities.
      b. Large cells may extend their size limitation by modifications of the plasma membrane. For example, the available surface area may be increased through multiple invaginations or projections, which are called microvilli.

   D. **Cell functions** are similar in all cells.

   1. **Cells maintain a selective barrier** (plasma membrane) between their cytoplasm and the extracellular environment. All substances that enter or leave the cell must pass across the barrier. Derivatives of the plasma membrane, through a series of complex infoldings, subdivide the cell's interior and compartmentalize it for specific activities.

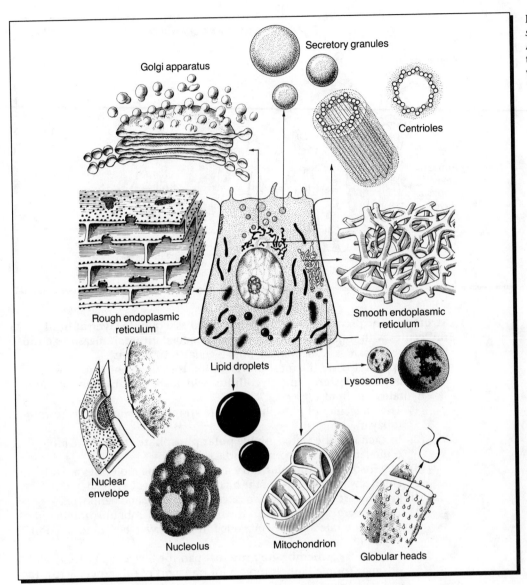

Figure 3–1. *A generalized cell is shown in the center of the diagram. Around it are enlargements of the various components as seen with the electron microscope.*

2. **Cells contain hereditary material** carrying encoded instructions for the synthesis of most of the cellular components. This hereditary material is duplicated prior to cell reproduction so that each new cell carries a full set of instructions.
3. **Cells carry out metabolic activities**, which are catalyzed chemical reactions that result in the synthesis and breakdown of organic molecules.

II. COMPONENTS OF A CELL

A. **Overview.** A generalized body cell has four principal divisions: the **plasma membrane** (plasmalemma, cell membrane); the **cytoplasm**, which is the protoplasm of the cell; a variety of cytoplasmic **organelles**, which are permanent structures that perform specific metabolic functions; and the **nucleus**, in which the genetic material is located (Figure 3–1).

B. The **plasma (cell) membrane** separates the interior of the cell from the extracellular environment. The **fluid mosaic model** is the currently accepted concept of the membrane (Figure 3–2).

Figure 3–2. *The fluid mosaic model of membrane structure.*

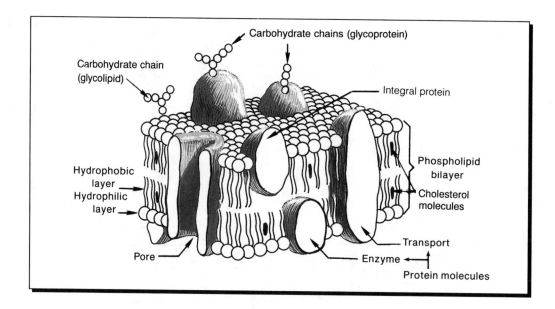

1. **Structure.** The plasma membrane is organized as a **double layer of lipid molecules** with some **globular proteins** embedded into it. It measures 6 nm to 10 nm in thickness. (A nanometer [nm] is equal to 10^{-9} meters).
   a. **Phospholipids** are the most prevalent lipids in the membrane. Other lipids are **cholesterol** and **glycolipids**, which are combinations of carbohydrates and lipids.
      (1) The molecules of phospholipids are arranged in two parallel rows (bilayer).
      (2) In each row the water-soluble **polar phosphate ends** of the molecules are directed to either surface.
      (3) The **insoluble nonpolar fatty acid tails** of the molecules are oriented toward the center of the bilayer. (See Figures 2–12 and 2–13.)
   b. The **proteins** are divided into two categories: integral and peripheral.
      (1) The **integral proteins** form the majority of membrane proteins. They penetrate and are embedded in the bilayer, bound to the nonpolar tail regions.
         (a) The **transmembrane proteins** span the bilayer completely and may form channels (pores) for transport of substances across the membrane.
         (b) Integral proteins also may lie partly submerged in one side or the other. They have several functions.
            i. Some integral proteins serve as cell surface enzymes.
            ii. Integral proteins bound to carbohydrates may form receptor sites for chemical messages from other cells, such as endocrine glands.
            iii. Some also function as markers, or antigens, which identify cell types.
      (2) The **peripheral proteins** are loosely bound to the membrane surface and can be easily removed from it. Their functions are not as well known as those of integral proteins. They may be involved in **structural support and changes in membrane shape** during cell division or cell movement.
   c. **Carbohydrates** may be linked to lipid or protein molecules. The resulting **glycolipids** or **glycoproteins** are believed to provide **surface recognition sites** for cell-to-cell interactions, such as keeping red blood cells apart or allowing the association of similar cells to form a tissue.
2. **Functions of the plasma membrane.** In addition to the receptor site and cell communication functions described above, the plasma membrane serves as a **selectively permeable** barrier that regulates the passage of materials into and out of the cell.

C. **Components of the cytoplasm**
  1. **Organelles.** Organelles are permanent components of the cytoplasm. Most organelles are enclosed by a membrane similar to the plasma membrane, which separates them from the surrounding cytoplasmic environment and allows the compartmentalizing of their metabolic activities.
      a. **Mitochondria** occur in almost all cells but are not present in red blood cells. Their number in a cell type is related to the energy consumption of the cell.
          (1) **Structure**
              (a) Mitochondria appear as rods or filaments that are in constant motion in a living cell.
              (b) Each mitochondrion consists of a smooth outer membrane and an inner membrane thrown up into folds called **cristae**. The cristae project like shelves into the mitochondrion and increase the surface area of the inner membrane.
              (c) The space between the cristae encloses the **matrix**, which contains proteins, DNA, RNA, and ribosomes.
          (2) **Function**
              (a) Mitochondria are often called the **powerhouses** of the cell because their most important function is the **production of energy** in the form of ATP.
              (b) The energy results from the **breakdown of nutrients** such as glucose, amino acids, and fatty acids.
              (c) The enzymes necessary for chemical release of energy are localized in the mitochondrial matrix and small particles on the cristae.
      b. **Ribosomes**
          (1) **Structure**
              (a) Ribosomes are small (25 nm in diameter), dark-staining granules composed of ribosomal RNA and almost 80 different proteins.
              (b) They occur as individual granules or in clusters called **polyribosomes.**
              (c) They may be free in the cytoplasm (free ribosomes) or attached to the membranes of the endoplasmic reticulum.
          (2) **Function**
              (a) Ribosomes are the site of **protein synthesis.**
              (b) Free ribosomes are involved in the synthesis of proteins for the cell's own use; for example, in the renewal of enzymes and membranes. Attached ribosomes are the site of synthesis of proteins that are secretory products to be released from the cell.
      c. **Endoplasmic reticulum (ER)**
          (1) **Structure**
              (a) The ER consists of a network of membrane-bounded flattened **cavities (cisternae)**, which are continuous with the plasma membrane and the nuclear membrane.
              (b) There are two types of ER: **rough (granular) ER**, which has membranes with attached ribosomes, and **smooth (agranular) ER**, which lacks ribosomes. In cells where both types are present, rough ER is continuous with smooth ER.
          (2) **Function**
              (a) The ER is the major site of **synthesis of cell products** and also functions in their **transport** and **storage.**
              (b) Rough ER is prominent in cells specialized for protein secretions such as digestive enzymes.
              (c) Smooth ER is abundant in the cells of a number of endocrine glands that synthesize hormones and in liver cells, where it is involved in the synthesis of lipids and cholesterol and the breakdown of glycogen.
              (d) In muscle cells, smooth ER is called sarcoplasmic reticulum and participates in the contraction process (see Chapter 8).

Figure 3–3. *Diagram of the function of the Golgi apparatus in processing and packaging proteins synthesized on the endoplasmic reticulum (ER), which are destined for secretion, incorporation into the plasma membrane, and lysosome enzymes.*

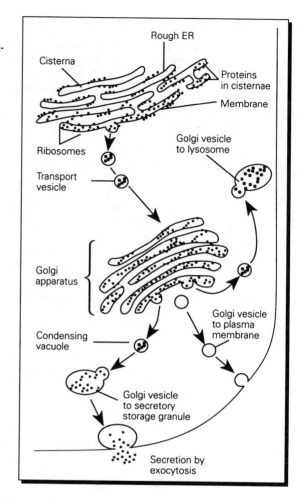

d. The **Golgi apparatus** is present in most cells but most developed and studied in glandular cells.
  (1) **Structure**
    (a) The Golgi apparatus consists of 3 to 7 membrane-bounded, **flattened sacs**, or cisternae, each of which is slightly curved. The sacs are stacked up like inverted saucers.
    (b) The convex surface of the stack faces the ER and nucleus; the concave surface faces the external surface of the cell.
    (c) There are usually numerous **transport vesicles** around the periphery of the stack and a few larger **condensing** vacuoles at one pole (Figure 3–3).
  (2) **Function**
    (a) The Golgi apparatus is the site of **accumulation, concentration, packaging**, and **chemical modification** of the secretory products synthesized on the rough ER.
      i. The transport vesicles pinch off from the ER and carry the secretions to the Golgi apparatus, where the secretions fuse with its cisternae.
      ii. The large condensing vacuoles concentrate the secretion and package them to become secretory granules.
      iii. Secretory (zymogen) granules, which are large, densely packed, membrane-bounded structures, unload their contents via exocytosis upon nervous or hormonal stimulation.
      iv. The Golgi apparatus also chemically modifies the molecules synthesized in the ER for incorporation into the plasma membrane. It adds fatty acid residues to certain proteins to convert them to lipoproteins, and it synthesizes and attaches carbohydrate side chains to proteins to form glycoproteins.

- (b) The Golgi apparatus processes proteins that function intracellularly, such as the lysosome enzymes.
- e. **Lysosomes** are found in all cells except red blood cells and completely keratinized skin cells at the surface of the body.
  - (1) **Structure**
    - (a) Lysosomes are small membrane-bounded vesicles filled with nearly 50 **hydrolytic enzymes**, which are capable of breaking down almost all types of macromolecules (proteins, lipids, carbohydrates, nucleic acids, etc.).
    - (b) **Primary lysosomes** contain enzymes exclusively; secondary lysosomes contain enzymes and the material being degraded.
  - (2) **Function**
    - (a) The major function of lysosomes is **intracellular digestion**. They play a role in both normal and pathologic processes.
    - (b) In phagocytic cells, a potentially harmful agent such as a bacterium, virus, or toxin is engulfed by the cell. It fuses with a primary lysosome to form a secondary lysosome and is then digested.
    - (c) Lysosomes also take part in **normal cellular growth and repair** by removing damaged or excess cellular components. The digested products are recycled by the cell to allow renewal and reconstruction of cell contents.
    - (d) Cell damage by a number of physical or chemical influences may cause destruction of the lysosome membrane and escape of the enzymes into the cytoplasm. The resulting **autolysis** (auto = self), or cell digestion, is why lysosomes are called **suicide bags** for cells.
    - (e) Some metabolic diseases, known as **storage diseases** (Tay-Sachs disease, Gaucher's disease, Fabry's disease), are caused by a congenital (existing at birth) absence of one of the lysosomal enzymes. As a result, an abnormal accumulation of substances interferes with the normal function of cells.
- f. **Peroxisomes (microbodies)**
  - (1) **Structure.** Peroxisomes are small, spherical, membrane-bounded organelles that contain destructive enzymes.
  - (2) **Function.** Peroxisomes function to protect the cell from the damaging effects of hydrogen peroxide. They may also function in lipid metabolism.
- g. The **nucleus** is the largest organelle. It is present in all cells of the body except mature red blood cells, which lost their nuclei as they developed. Generally, each cell has a single nucleus, but some giant cells, such as megakaryocytes of bone marrow, osteoclasts of bone, and skeletal muscle cells, may have several nuclei.
  - (1) **Structure**
    - (a) The **nuclear envelope** consists of a double membrane separated by the **perinuclear space**.
      - i. The inner membrane is smooth. The outer membrane often contains ribosomes and is continuous with the surrounding ER.
      - ii. The inner and outer membranes fuse at irregular intervals around the nucleus to form **nuclear pores**, which allow for exchange of materials between the nucleus and the cytoplasm.
    - (b) **Chromatin** appears as irregular clumps or granules of strongly basophilic, or blue-staining, material dispersed throughout the nucleus.
      - i. Chromatin is composed of **coiled strands of DNA** bound to basic proteins called **histones**, varying amounts of RNA, and other nonhistone proteins and enzyme systems.
      - ii. In a dividing cell, the chromatin is condensed and coiled into discrete units, the **chromosomes**. Human cells contain 23 pairs of chromosomes.

(c) The **nucleoplasm** is the matrix that surrounds the chromatin. It is composed of proteins, metabolites, and ions.
(d) The **nucleolus** is a spherical structure composed of RNA and protein. The size of the nucleolus and the number present varies in different cell types. It is missing in cells that do not synthesize protein, such as spermatozoa.

(2) Function
(a) The nucleus is **essential for all cellular activities**.
(b) The nucleus contains the **genetic material** of the cell (DNA), which encodes the information to control protein synthesis and cell reproduction, the two most essential functions of the cell.

2. **Microfilaments, microtubules, centrioles, and cilia and flagella.** In addition to the membrane-enclosed organelles, the cytoplasm contains a complex network of structural components.

   a. **Microfilaments**
      (1) Structure
         (a) Microfilaments are solid thread-like cylinders made of protein and found in a variety of sites within the cell.
         (b) They are often present in bundles called **fibrils** located just under the plasma membrane.
      (2) Function
         (a) Microfilaments are responsible for **contractility** of cells, which is a property of all cells but is especially well developed in muscle cells.
         (b) Contractility is responsible for cell **locomotion** and **movements associated with phagocytosis, pinocytosis, and cell division.**

   b. **Microtubules**
      (1) Structure
         (a) Microtubules are hollow tubes, 20 nm to 25 nm in length, present everywhere in the cytoplasm in all cells.
         (b) They are composed of protein **tubulin** molecules.
      (2) Function
         (a) Microtubules contribute to the **cytoskeleton**, or supporting elements, of the cell.
         (b) They also are involved in cell division, cell movements, and the transport of materials from one area of the cell to another.

   c. **Centrioles**
      (1) Structure
         (a) In a nondividing cell, two centrioles are located near the nucleus and Golgi apparatus in a specialized region called the **centrosome**.
         (b) The two members of the pair of centrioles, arranged perpendicular to each other, are referred to as a **diplosome**.
         (c) The wall of each centriole consists of nine sets of microtubules, each composed of three subunits known as **triplets**.
      (2) Function
         (a) Centrioles function in **cell division** and also as the site of the **formation** of **cilia** and **flagella**.
         (b) Centrioles are self-replicating and divide prior to cell division. Following replication, each original centriole and its duplicate migrate to opposite nuclear poles where they induce the formation of the spindle apparatus during cell division.
         (c) **Basal bodies** are forms of centrioles located just within the plasma membrane in cells that have cilia and flagella. The basal body directs the formation of the microtubules that make up the cilia and flagella.

   d. **Cilia and flagella**
      (1) Structure
         (a) Both cilia and flagella are motile processes that extend out from the cell surface.

- (b) They are composed of longitudinal microtubules, which are arranged as two single tubules surrounded by a ring of nine regularly spaced double tubules.
- (c) Cilia are short and numerous on the cell surface, projecting out like eyelashes. Groups of cilia oscillate backward and forward in an asynchronous wave, appearing like a field of wheat moving in the wind.
- (d) Flagella are longer and whiplike. As a rule, there is only one flagellum per cell. A flagellum has an undulating movement propagated throughout its length. The longest flagellum, at 55 nm, is the tail of a spermatozoan.
- (2) **Function**
  - (a) Both cilia and flagella function in **movement**.
  - (b) Cilia are able to move fluid or a layer of mucus over the surface of the cells on which they occur, while the flagellum of the sperm cell propels the cell.
- e. **Cytoplasmic inclusions** are temporary cellular components that may be synthesized by the cell or taken in from its surroundings. They are not essential to life or cellular activities and include such diverse materials as **pigment granules, glycogen, lipid droplets, crystals,** and **secretory granules**.

III. MOVEMENT OF MATERIALS ACROSS CELL MEMBRANES

- A. **Basic principles.** The maintenance of cell life depends on the continuous movement of materials into and out of the cell. Nutrients must get in, waste materials must get out, and ions must be moved in both directions. Movement across the plasma membrane occurs through **passive transport** and **active transport** mechanisms.
- B. **Passive transport mechanisms** are physical processes that do not require the expenditure of cellular or metabolic energy but utilize external sources of energy such as heat. They include **diffusion, dialysis, osmosis, facilitated diffusion,** and **filtration**.
  1. **Diffusion** is the random movement of particles (molecules or ions) under the influence of their own thermal energy, from an area of their **higher concentration to an area of their lower concentration**, or "downhill." Diffusion of molecules or ions may take place in a liquid, gas, or solid or through nonliving or living membranes that are permeable to them (Figures 3–4 and 3–5).
     a. **Diffusion in a liquid** is the movement of solute and solvent particles in all directions through a solution, or in both directions through a permeable membrane.

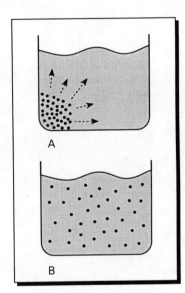

Figure 3–4. *Diffusion of sugar in a beaker of water. A,* As a result of the random movements of all molecules and ions, the molecules of dissolved sugar (solute) move in all directions through the solution down their concentration gradient from a region of high solute concentration to a region of low solute concentration. Similarly, the molecules of solvent (water) move down their own concentration gradient from high solvent concentration to lower solvent concentration. *B,* Eventually, the molecules of sugar and water are equally distributed and the solution is in a state of equilibrium.

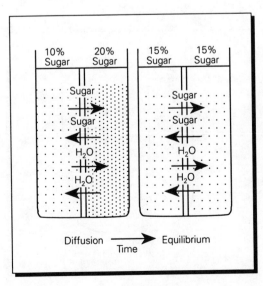

Figure 3–5. *Diffusion of sugar molecules through a permeable membrane. A,* A 10% sugar solution is separated from a 20% sugar solution by a membrane freely permeable to both sugar and water. Both substances diffuse rapidly through the membrane in both directions, but more sugar molecules move out of the 20% solution, where there are more of them, into the 10% solution, where there are fewer of them. At the same time, more water molecules move from their higher concentration in the 10% solution into their lower concentration in the 20% solution. *B,* Eventually, the result is an equilibrium of the two solutions on both sides of the membrane. Diffusion of sugar and water molecules still occurs, but equally in both directions.

b. **Net diffusion** is the movement of particles from an area of their own high concentration to an area of lower concentration; that is, along their own **concentration gradients**. Net diffusion means more particles are diffusing in one direction than in the other.
c. **The rate of net diffusion** of particles in a solution is increased by the following factors:
   (1) A **higher concentration gradient** because there are more particles
   (2) A **low molecular weight** because large particles are not as easily moved by colliding with each other
   (3) An **increase in temperature** because higher temperature increases random particle movement
d. **Equal diffusion** occurs after equilibrium is reached; that is, after net diffusion of the solute in one direction and the solvent in the opposite direction have caused the disappearance of the concentration gradients. Equal diffusion means that the number of solute and solvent particles diffusing in one direction equals the number diffusing in the other direction.
e. **Simple diffusion of substances across the plasma membrane** occurs through the lipid bilayer or through the protein channels.
   (1) Lipid-soluble substances that are small, nonpolar, and carry no charge, such as oxygen, carbon dioxide, fats, steroid hormones, urea, and alcohol, diffuse through the lipid bilayer.
   (2) Polar, electrically charged, or nonlipid-soluble substances diffuse through the pores of the channel (transmembrane) proteins.
f. **One example of diffusion in the body** is the exchange of oxygen and carbon dioxide at the lungs.
   (1) Inhaled oxygen diffuses through the lung cell membranes out of the lungs, where it is in high concentration, into the blood capillaries surrounding the lungs, where it is in lower concentration. The oxygen is then carried in the blood to diffuse into the body cells.
   (2) Carbon dioxide diffuses out of the body cells into the blood capillaries to be carried in the blood to the lungs. It diffuses from the blood capillaries surrounding the lungs, where it is in high concentration, into the lungs, where it is in lower concentration. Carbon dioxide is then exhaled into the air.

2. **Dialysis** is the separation of **crystalloid solute particles**, which have diameters that are smaller than 1 nm (e.g., ions, glucose, oxygen), by diffusion through a membrane permeable to them but impermeable to **colloid solute particles**, which have diameters from 1 nm to 10 nm (e.g., blood proteins). The principle of dialysis is used in the artificial kidney.

3. **Osmosis** is the net diffusion of water molecules through a **selectively permeable membrane**; that is, one that does not allow free passage of all the solutes present. The nondiffusible substance(s) will necessarily have a greater concentration on one side of the membrane than on the other side (Figure 3–6).

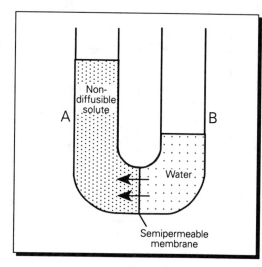

Figure 3–6. *Osmosis refers to the diffusion of water from an area of higher concentration of water to an area of lesser concentration of water.* The semipermeable membrane shown in this diagram separates two columns of fluid, one containing pure water and the other containing some solute that cannot penetrate the membrane. The more dilute solution (pure water in *B*) has a higher concentration of water molecules; the more concentrated solution with reference to the nondiffusible molecules (*A*) is less dilute because it has a lesser concentration of water molecules. Therefore, water moves down its own concentration gradient from chamber *B* to chamber *A*. Osmosis of water into chamber *A* causes an increase in the level of the fluid in chamber *A* and an increase in its pressure, which forces molecules and ions in the opposite direction and eventually is great enough to stop the osmosis of water.

a. In osmosis, water molecules move across the membrane from the area of higher water concentration to the area of lower water concentration.
b. The osmosis of water molecules into the more concentrated solution (lower water concentration) increases its volume and hydrostatic pressure. In a container of fixed volume, eventually the hydrostatic pressure of the water molecules is sufficient to balance the osmotic pressure moving the water molecules down their concentration gradient, and there is no further net movement of water.
c. The **osmotic pressure** of a solution is a potential pressure expressed in terms of the force or pressure it would take to stop further osmosis of water.
    (1) Osmotic pressure is the pressure that develops in a solution as a result of net osmosis into that solution. The greater the concentration of the solute, the greater its osmotic pressure. Thus, osmotic pressure is a measure of the **pulling power** of a solution for water molecules.
    (2) The osmotic pressure of a solution is dependent on the **number of solute particles** per unit volume of solution. Therefore, molar concentration (the number of molecules or ions in solution), rather than the percentage concentration of a solution, determines its potential osmotic pressure. (See Chapter 2, III A 2 d, on molarity.)
        (a) For a nonelectrolyte that does not dissociate in water (such as glucose), the number of solute particles in solution is the molar concentration of the solution.
        (b) For an electrolyte that does dissociate in water (such as sodium chloride), the number of solute particles in solution is determined by the molar concentration and the number of ions formed from each molecule.
    (3) The **osmolality** of a fluid is a measure of its solute particles.
        (a) One **osmol** is 1 mole (6.02×10^{23} molecules or mixture of ions and molecules) of a substance dissolved in 1,000 g of water. The **milliosmole (mOsm)**, which is 1/1,000th of an osmol, is used to describe the solute concentration of body fluids.
        (b) Most body fluids have an osmolality of 300 mOsm.
    (4) Two solutions with the same concentration of solute particles are **isosmotic** to each other.
        (a) A **hyperosmotic** solution has a greater concentration of solute particles and a greater osmotic pressure with respect to a more dilute solution (hyper = above).
        (b) A **hypoosmotic** solution is one that is more dilute, has less concentration of solute particles, and less osmotic pressure with respect to a more concentrated solution (hypo = below).
    (5) **Tonicity** is used to describe the effect solutions have on the shape or "tone" of cells according to the laws of osmosis (Figure 3–7).
        (a) A solution is **isotonic** to the cytoplasmic fluids of a cell if it has the **same concentration** of nondiffusible particles. Water will not osmose into or out of the cell. This is usually the case between extracellular fluid and intracellular fluid.
        (b) A solution (or the extracellular fluid) is **hypotonic** to the cell if it is **less concentrated** than the cellular contents. Net movement of water into the cell causes it to swell until it may rupture.
        (c) A solution is **hypertonic** to the cell if it is **more concentrated** than the cellular contents. Net movement of water out of the cell causes it to shrink, or **crenate**.
4. **Facilitated diffusion**, also known as **carrier-mediated** diffusion, is a mechanism whereby molecules that are both lipid-insoluble and too large to pass through the protein channels are assisted by **carriers**, which are special protein molecules in the external surface of the membrane. No energy is expended because the molecules move down their concentration gradients. Glucose and some amino acids are transported across the membrane by facilitated diffusion.

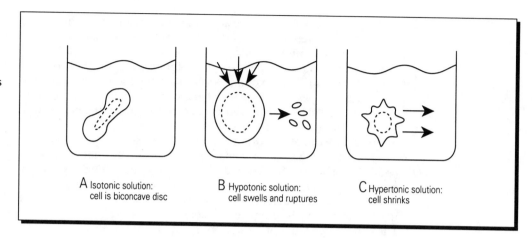

Figure 3–7. *Tonicity. A,* If a red blood cell is placed in a 0.85% NaCl solution, which is isotonic to the cell, it will neither shrink nor swell. *B,* If the cell is placed in a more dilute solution such as 0.3% NaCl, the cell swells and bursts as water diffuses into it. *C,* In a more concentrated solution, for example, 10% NaCl, water is drawn out of the cell and it shrinks.

   a. In facilitated diffusion, the carrier substance combines with the solute molecules to form a **solute-carrier complex**, which is soluble in the lipid bilayer, and thus transports the solute across the membrane. Once on the other side, the solute is released. The carrier breaks away from the complex, returns to the exterior of the membrane, and repeats the process.
      (1) The carriers exhibit **specificity**; i.e., they are highly selective in distinguishing between closely related molecules.
      (2) Facilitated diffusion can be inhibited by **competitive and noncompetitive inhibitor** molecules, which closely resemble the solute molecules.
      (3) The **rate** of passage of a solute through facilitated diffusion depends on its **concentration difference** on both sides of the membrane, the **number of carrier molecules** available, and how rapidly the **solute-carrier complex formation** takes place.
   5. **Filtration** is the forced movement of water and diffusible molecules across the plasma membrane as a result of high mechanical or fluid pressure, such as hydrostatic pressure or blood pressure. The blood pressure causes filtration across specialized blood vessels in the kidneys as a first step in the production of urine.
C. **Active transport** requires the use of metabolic energy derived from cellular chemical reactions and moves molecules or ions **against their concentration gradients**; i.e, uphill, from an area of lower concentration to an area of higher concentration. It allows movement of substances across cell membranes regardless of their extracellular or intracellular concentrations. Active transport includes **carrier-mediated mechanisms** and **bulk transport**.
   1. **Carrier-mediated active transport.** The carriers are integral proteins referred to as "pumps." Examples are the **sodium/potassium ion pump**, which is active in all living cells, and the **calcium pump**, which is important in muscle contraction.
      a. The sodium/potassium pump is an exchange pump that trades intracellular sodium for extracellular potassium. It maintains a gradient of these ions across the cell membrane and contributes to a difference in electrical voltage known as the **membrane potential**. (See Chapter 9, The Nervous System.)
      b. The breakdown of ATP at the interior surface of the membrane to adenosine diphosphate (ADP) releases the energy necessary to operate the pump.
      c. **Coupled transport (co-transport)** is a combination of diffusion and active transport.
         (1) In coupled transport, a specialized transport protein can couple active transport of one substance against its own concentration gradient with passive diffusion of a second substance.

(2) An example of a coupled transport system is **sodium-linked cotransport** in the kidney, in which glucose and amino acids are actively transported across kidney tubule cells while sodium diffuses passively (See Chapter 16, Urinary System.)
2. **Bulk transport** is an active process that transports large particles and macromolecules through the plasma membrane by enclosing them within portions or infoldings of the membrane to form a membrane-bounded sac or **vesicle (vacuole)**. Bulk transport includes endocytosis and exocytosis.
   a. **Endocytosis** (endo = inner) means taking into the cell. It includes phagocytosis and pinocytosis.
      (1) **Phagocytosis** (phago = to eat) is the engulfing of large solid substances by foldings of the plasma membrane to form a phagocytic vesicle.
         (a) The phagocytic vesicle fuses with a lysosome and the lysosomal enzymes destroy the contents.
         (b) Specialized **phagocytic cells** in the body remove disintegrating cells, foreign matter, and bacteria.
      (2) **Pinocytosis** (pino = to drink) is the engulfing of small drops of extracellular fluid, which may contain dissolved nutrients, and incorporating them into the cell.
      (3) **Receptor-mediated endocytosis** refers to the binding of receptor molecules on the cell surface with specific substances known as **ligands**. The receptor-ligand complex then undergoes endocytosis for transport into the cell.
   b. **Exocytosis** is the reverse of endocytosis. It is a method of ridding the cell of unwanted substances and one means for the release of useful cell products into the extracellular fluid.
      (1) The substance to be released is enclosed in a vesicle, which fuses with the cell membrane for exit.
      (2) Examples of exocytosis are the release of products from the secretory cells of the pancreas and the release of chemical transmitters from nerve cells at nerve endings.

IV. CELL METABOLISM

A. **Metabolism** refers to all the chemical reactions that occur within the cell. **Catabolism**, the breakdown of large organic macromolecules into smaller compounds, is the energy-generating part of metabolism. **Anabolism** is the energy-utilizing phase in which complex compounds are built up from simple building blocks. (Metabolism and the biochemical events of energy capture, storage, and release are described in detail in Chapter 15.)

B. Catabolism in the cell is performed by the mitochondria. The energy results from the chemical breakdown of glucose, amino acids, and fatty acids with glucose as the most important source.
   1. **Storage of energy as ATP.** Cellular energy derived from catabolism is not directly available for cell activities. Instead, it is stored in an energy-rich molecule, ATP, to be released when needed.
      a. ATP is formed from the nucleotide, adenosine monophosphate, by the attachment of two additional phosphate groups. These terminal phosphate radicals in ATP are connected to the rest of the molecule by **high-energy bonds.**
      b. When a terminal high-energy bond in ATP is broken, it yields a **phosphate group**, a molecule of **adenosine diphosphate (ADP)** and **energy**.
      c. ATP acts as an energy-donor in different parts of the cell by transferring one of its terminal phosphate groups to another molecule. ATP may also be reconstituted from ADP via the energy-producing systems of the cell.

2. The **chemical breakdown of glucose** occurs in three stages: glycolysis, the citric acid cycle, and electron and hydrogen transport. The latter two stages are known as **cellular respiration**.
   a. **Glycolysis** is an initial series of chemical reactions that result in the formation of pyruvic acid and minor amounts of ATP are formed. Glycolysis takes place in the cytoplasm and is **anaerobic** (not oxygen-consuming). The pyruvic acid, which crosses the double membrane of the mitochondrion to enter the matrix, is further broken down in the next stage.
   b. The **citric acid cycle (Krebs cycle)** occurs in the mitochondrial matrix. It yields a small amount of ATP, carbon dioxide, and electrons and hydrogen ions (protons).
   c. The electrons and protons are sent through a chain of coenzymes, the **hydrogen and electron transport system**. At each step along the chain, free energy is produced to phosphorylate ADP to ATP, a step called **oxidative phosphorylation**. The ATP diffuses out of the mitochondria into the surrounding cytoplasm to be used for cellular activities.
   d. **Lipid and protein catabolism** are described further in Chapter 15.
      (1) Lipid molecules are broken down into smaller compounds that enter the glycolytic pathway to form pyruvic acid or directly enter the citric acid cycle.
      (2) Proteins are degraded to amino acids, undergo the removal of amine groups, and enter the citric acid cycle.

V. CELL DIVISION

A. **DNA replication.** Before any cell divides, it must make a copy of the DNA molecule so that all the information it carries is passed on to its offspring.
   1. **DNA structure** is the basis for its ability to carry information. (See also Chapter 2, III C 4, Nucleic acids.)
      a. The DNA molecule is in the form of a ladder twisted into a **double helix**. The structural units of DNA are four different **nucleotides** assembled into the long DNA strands.
      b. Each nucleotide contains **phosphate**, the sugar **deoxyribose**, and a **nitrogenous base**, in that order.
      c. The four bases are **adenine (A), guanine (G), cytosine (C),** and **thymine (T)**.
      d. The side pieces of the DNA backbone (ladder) are made of joined phosphates and sugars. The cross connections (rungs) are formed by pairing of the bases through numerous weak hydrogen bonds.
      e. In **complementary base pairing**, adenine bonds only with thymine (A–T, T–A), and guanine only with cytosine (G–C, C–G).
      f. Although there are only four variations of the linkage, the linear sequence in which they occur can provide a nearly infinite variety of combinations.
         (1) The exact sequence of the nucleotide pairs is commonly referred to as the **genetic code**. It is the basis for the biological information carried by DNA.
         (2) A **gene** is a discrete unit of DNA, or coding region, with a specific hereditary (genetic) function. That is, a gene gives the instructions for the synthesis of a particular protein, which, in turn, is responsible for another activity in the cell.
         (3) The **genome** is the complete complement of an organism's genes.
      g. Because DNA is double-stranded, each strand carries a nucleotide sequence that is exactly complementary to the nucleotide sequence of its partner strand. For example, if the two strands are designated "1" and "1'," strand "1" can serve as a **template** for making a new strand "1'," and strand "1'" can serve as a template for making a new strand "1." (See Figure 3–8.)

2. **Stages of replication**
   a. The two strands of DNA are **unwound and separated** (unzipped) by unwinding enzymes, which cause the weak hydrogen bonds between the paired bases to break.
   b. The enzyme **DNA polymerase**, using the four kinds of complementary nucleotides freely present in the nucleus, matches and attaches the nucleotides to the exposed bases on each unzipped, single-stranded DNA.
   c. **Two complete DNA double helices are formed**, each identical in nucleotide sequence to the original DNA helix that served as the template. Thus, the genetic information is copied exactly.
   d. Such replication is termed **semiconservative** because it conserves each strand of the original DNA double helix while each also receives a newly synthesized matching partner strand.
3. **Errors in DNA replication**
   a. The replication machinery may make a mistake by skipping a base, adding one or more bases, or substituting the wrong base.
   b. A change in the DNA molecule also can occur as a result of exposure to potentially damaging physical and chemical agents such as **x-rays** or **carcinogens** in the environment.
   c. The resultant change in the sequence of nucleotides is a **mutation**, which will be faithfully copied in all future replications and can have harmful consequences on the cell.
4. **DNA repair** is a constant process, which can minimize accidental changes. A large variety of different **DNA repair enzymes** continuously scan the DNA molecule and remove damaged nucleotides.

B. **Chromosomes in human cells**
   1. **Chromosomes** are tightly coiled strands of DNA and associated protein. They are the condensed form of chromatin found in the nucleus and they become visible during cell division.
   2. All normal somatic (body) cells, with the exception of sex cells (ova and spermatozoa) possess **46 chromosomes**, or **23 pairs** of chromosomes. (Sperm and ova have only 23 chromosomes.)
   3. Of the 23 pairs, 22 pairs are **homologous** (matched) pairs called **autosomes**. Homologous chromosomes carry the genetic information covering the same traits.
   4. The twenty-third pair are known as the **sex chromosomes**, X and Y. They are homologous in females (X and X), but not in males (X and Y).
   5. A cell with both members of all pairs is termed **diploid (2n)**. A cell, such as the sperm or ovum, with only one of each pair is termed **haploid (n)**.
   6. **Mitosis** is the term for nuclear and cytoplasmic division of somatic cells.
      a. Mitosis maintains the diploid number of chromosomes. Through the process of chromosomal replication followed by nuclear and cytoplasmic division, mitotic cell division ensures that each daughter cell receives the same number and kinds of chromosomes as the parent cell.
      b. It thus preserves genetically identical chromosomal information, unchanged from generation to generation of cells.
   7. **Meiosis** is a special type of division that occurs in the formation of sperm and eggs. It reduces the diploid number of chromosomes to the **haploid** number.

C. The **cell cycle and mitosis**. The cell cycle, in cells that are capable of dividing, refers to the events in a cell's life span in the period between the time it was formed by cell division to the beginning of the next cell division. The greatest portion of the cycle (about 90%) is devoted to growth and synthesis, called **interphase**, with a smaller portion devoted to nuclear and cell division, or **mitosis**.

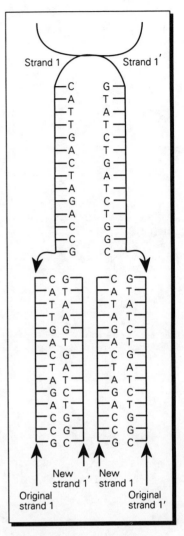

Figure 3–8. *Semiconservative replication of DNA.* Each replicated molecule contains either an original "1" strand or an original "1'" strand, thus conserving elements of the old coding.

Figure 3–9. *The cell cycle.*

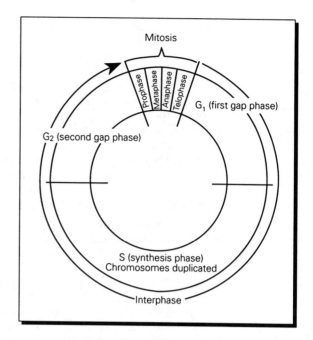

1. **Interphase** consists of the G_1 phase, S phase, and G_2 phase (Figure 3–9).
   a. In the G_1 **(gap 1) phase**, the cells metabolically are very active. All cell components are synthesized and the cells grow rapidly. In the nucleus, each chromosome is a single, unreplicated double helix of DNA with its associated histones and other chromosomal proteins. Nondividing cells generally remain in the G_1 phase for their entire life span.
   b. In the **S (synthesis) phase**, protein synthesis continues and DNA and chromosomal proteins (histones) are replicated. Each chromosome then consists of two identical DNA double helices called **chromatids**, joined at the **centromere**.
   c. The G_2 **(gap 2) phase** is an important period of cell metabolism and growth prior to mitosis.
      (1) The chromosomes are uncondensed and still in long strands.
      (2) The centriole divides and the mitotic spindle, produced by the microtubule fibers of the cell, begins to form in preparation for the coming nuclear division.
2. **Mitosis** includes the condensation and division of the chromosomes and **cytokinesis**, the actual division of the cytoplasm to form two offspring cells. Although it is a continuous process, it is divided into four subphases: **prophase, metaphase, anaphase,** and **telophase** (Figure 3–10).
   a. Prophase
      (1) The **chromosomes condense** into tight coils and supercoils, and become visible. Each consists of two **chromatids** held together by the **centromere**. The chromatids will be the chromosomes in the next generation of cells.
      (2) The **paired centrioles separate** and begin moving to opposite sides of the nucleus, propelled by the lengthening microtubules formed between them. Once there, they direct the assembly of the polar fibers of the mitotic spindle.
      (3) The **nucleoli disperse** and the **nuclear membrane** disappears, which allows the spindle to enter the nucleus. Short microtubules emanating from **kinetochores**, structures on the centromeres, now are able to interact with the polar fibers of the spindle to cause vigorous motion of the chromosomes.
      (4) Other microtubules radiate outward from the centrioles to form the **asters**.

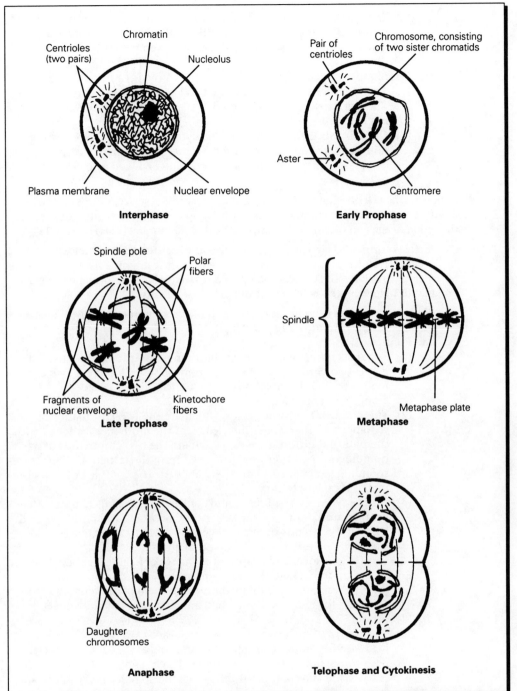

Figure 3–10. *Stages of mitosis.* Only four chromosomes are shown.

  b. Metaphase
     (1) The chromosomes (paired chromatids) line up at the **metaphase plate** or **equatorial plate** of the cell, so-called because it crosses from one side of the cell to the other at right angles to the spindle.
     (2) The **centromeres of all chromosomes are aligned** with one another.
     (3) The **kinetochores separate** and the chromatids are moved apart.
  c. Anaphase
     (1) As a result of changes in the length of the microtubules to which they are attached, the **paired chromatids (each now considered a chromosome) are moved** from the equatorial plate to their respective poles.

(2) The end of anaphase is characterized by two sets of complete chromosomes clumped together at either cell pole. The cytoplasmic organelles, previously replicated, also are evenly distributed to the poles.
d. **Telophase**
(1) **Two nuclei re-form** around the two aggregations of chromosomes. The chromosomes uncoil and disperse. The nuclear envelope and nucleoli re-form.
(2) **Cytokinesis** is the division of the cytoplasm. A **cleavage furrow** midway between the two chromosomal masses deeply constricts the cytoplasm, progresses around the cell, and divides it into two separate cells.

D. **Meiosis** is the cell division that occurs in the formation of sex cells (eggs and sperm). It reduces the number of chromosomes to the haploid number (23). At fertilization, the union of egg and sperm nuclei restores the diploid number (46) (Figure 3–11).
1. **Meiosis consists of two nuclear and cellular divisions**, termed Meiosis I and Meiosis II, which produce a total of four cells. During interphase before the first meiotic division, each chromosome replicates to result in the formation of chromatids held together by their centromeres, as in mitosis.
2. **Meiosis I separates each homologous pair** of chromosomes and distributes the members of the pair into daughter cells.
   a. **Prophase I**
   (1) **Synapsis** occurs while the chromosomes are still uncondensed. That is, the two chromatids of each chromosome seek out the two chromatids of their homologous partner and come to lie side by side along their entire length.
   (2) The **maternal chromosomes**, which are the two chromatids of each chromosome that were inherited from the mother, are aligned with the **paternal chromosomes**, which are the two chromatids of its homologous pair that were inherited from the father. All their corresponding genes are also precisely aligned. The complex of four chromatids is called a **tetrad**.
   (3) During synapsis, the four strands of the homologous chromosomes wrap around or **cross over** each other. Breakage and reuniting of the DNA in the strands occurs and genetic material is exchanged between the maternal and the paternal chromosomes.
      (a) The crossover and reciprocal exchange of fragments of DNA occurs at random. There are about ten for every tetrad in humans.
      (b) The result is a reshuffling or rearrangement of the genes that provides for genetic variation and uniqueness of the individual.
   b. **Metaphase I**
   (1) The pairs of homologous chromosomes, each with two sister chromatids held together at the centromere, are arranged at the equatorial plate.
   (2) Both chromatids of one chromosome of each homologous pair are oriented to the same pole of the cell so that the homologous chromosomes face opposite poles.
   (3) The spindle fibers from just one pole attach to the centromere of each chromosome.
   (4) The centromeres do not divide as they do in Metaphase I of mitosis.
   c. **Anaphase I**
   (1) Each chromosome (composed of two chromatids) is drawn to a pole.
   (2) Thus, a haploid group of chromosomes (23) is aggregated at each pole.
   d. **Telophase I**
   (1) As in mitosis, telophase reverses the events of prophase. The chromosomes disperse, the nuclear envelope re-forms, nucleoli reappear, and the spindle disassembles.
   (2) Cytokinesis occurs and the two cells are separated from each other.
   e. **Meiotic interphase** is short. There is no DNA replication.

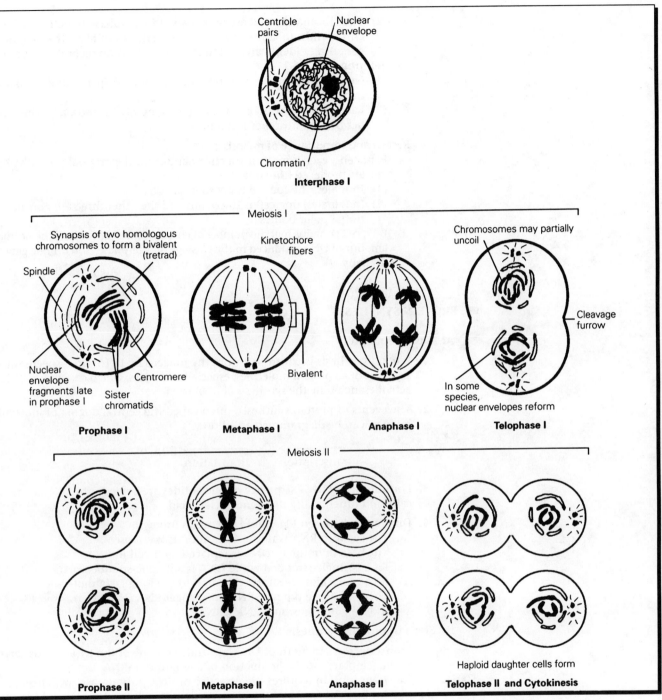

Figure 3–11. *Stages of meiosis in a hypothetical cell with a diploid number of four.*

3. **Meiosis II is similar to mitosis.**
   a. **Prophase II** events are the same as prophase in mitosis.
      (1) The centrioles separate and move to opposite poles.
      (2) The microtubules from each centromere attach to fibers from the centrioles at the opposite poles.
   b. **Metaphase II**
      (1) The chromatids line up on the equatorial plate of the cell.
      (2) The arrangement of chromatids in a pair, rather than in the tetrad of metaphase I, is called a **dyad**.
   c. **Anaphase II**
      (1) The centromeres divide, and the separating chromatids become chromosomes.

(2) The chromatids that are pulled apart in anaphase II are not sister chromatids. In contrast to the chromatids in mitosis, they are not genetically identical as a result of crossover and recombination.
    d. **Telophase II**
       (1) Nuclear membranes re-form, chromosomes disperse, and cytokinesis occurs.
       (2) Each new cell contains one of each kind of chromosome. The number of chromosomes is haploid.
4. **Results and importance of meiosis**
   a. Four cells, each containing a chromatid from the original prophase I tetrad, are produced from one parent cell.
     (1) In males, the four cells are spermatozoa.
     (2) In females, one cell is the ovum and the other three are non-functional polar bodies.
   b. Each cell contains half the number of chromosomes, one fourth the normal amount of DNA produced in the G_2 stage of interphase, and unique genetic diversity.

VI. PROTEIN SYNTHESIS

A. **Basic principles**
1. Protein synthesis is controlled from the nucleus by DNA. DNA specifies the structure of all protein molecules, especially enzymes, which catalyze all cellular activities including the synthesis of DNA itself.
2. The events of protein synthesis, often called the central dogma of molecular biology, can be diagrammed as follows:

$$\text{DNA} \xrightarrow{\text{transcription}} \text{RNA} \xrightarrow{\text{translation}} \text{protein}$$

3. Portions of the DNA sequence of nucleotides are copied into RNA, a chemically and functionally different nucleic acid.
4. Differences between RNA and DNA are as follows:
   a. The sugar in RNA is ribose instead of deoxyribose.
   b. The base thymine in DNA is replaced by uracil in RNA.
   c. RNA is single-stranded while DNA is a double-stranded helix.
   d. RNA molecules are short compared to the long DNA molecules.
   e. There are three types of RNA: messenger RNA (mRNA), transfer RNA (tRNA), and ribosomal RNA (rRNA).

B. **Steps in protein synthesis**
1. **Transcription.** A length of DNA unwinds and one of the two strands serves as the template for the production of a length of mRNA.
   a. Transcription requires the presence of **RNA polymerase**, which assembles the new bases, and **gene regulatory proteins**, which bind to special sequences of bases on the DNA molecule and determine the segments of DNA to be copied.
   b. As soon as the mRNA copy is completed, the original DNA double-stranded helix re-forms and releases the mRNA.
   c. The mRNA leaves the nucleus via the pores in the nuclear envelope and moves into the cytoplasm.
     (1) The message written on mRNA is in the **genetic code**. Each code word consists of three adjacent nucleotides, or triplet of bases, that forms a **codon**.
     (2) The triplet specifies one of the 20 common amino acids commonly found in proteins. For example, if the codon is GAG, it specifies the amino acid glutamic acid. (The nucleotide equivalent in DNA would be CTC.)

(3) Because there are four different nucleotides, there are 4^3, or 64, possible codon triplets and only 20 amino acids for which to code.
   (a) The code is called **degenerate** because many amino acids are designated by more than one codon.
   (b) The code is also **universal** in that the same codons specify the same amino acids in all living things.

2. **Translation** is the synthesis of proteins based on the translation of the base-sequence information found in the mRNA codons. It requires the participation of tRNA and rRNA.
   a. **tRNA** molecules are small, only 70 to 90 nucleotides long, and located in the cytoplasm.
      (1) Each tRNA molecule has a clover-leaf shaped three-dimensional conformation. One end of the clover-leaf contains an **anticodon**, a base triplet of nucleotides that is complementary to the codon on the mRNA.
      (2) The other end of the tRNA carries **one of the 20 amino acids** (found free in the cytoplasm), which has been enzymatically attached by a high-energy (ATP) linkage.
   b. **rRNA** molecules form the structural core of the **ribosome**, a complex composed of rRNA and almost 100 different proteins. A ribosome acts as a biochemical site on which tRNA molecules position themselves to read the message coded onto mRNA.

3. **Initiation of protein assembly**
   a. **A ribosome has a small subunit and a large subunit.** A newly transcribed strand of mRNA attaches to the smaller subunit and becomes positioned in a groove between it and the larger ribosomal subunit.
   b. The **anticodon of an initiator tRNA molecule**, carrying an amino acid, recognizes and **binds to a start codon on mRNA** to form an **initiation complex**.
      (1) The **start codon is always AUG**, which codes for the amino acid **methionine**. The initiator tRNA molecule has the **anticodon UAC** and carries methionine.
      (2) The anticodon/codon complex binds at the exact spot where the polypeptide chain is to start.
      (3) The binding groups the nucleotide bases into the **reading frame** that determines where the sequence of reading off nucleotide triplets begins.

4. **Elongation of the polypeptide chain** (Figure 3–12)
   a. In addition to the mRNA binding site, each larger ribosomal subunit has two tRNA binding sites.
      (1) The **P site** (for polypeptide) holds the tRNA with the growing polypeptide chain.
      (2) The **A site** (for amino acid) holds the tRNA with the next amino acid to be added to the chain.
   b. The **initiator tRNA molecule fits into the P site** on the ribosomal subunit. Its amino acid forms the front end of the polypeptide chain.
   c. Once initiation is complete, a **second tRNA** (an appropriate one that has an anticodon for the codon on the mRNA) **moves into the A site**. Its amino acid is linked to the initial amino acid by peptide bonds.
   d. **The tRNA in the P site moves out of the ribosome** away from the mRNA. It separates from its amino acid and becomes free to pick up another amino acid.
   e. As the ribosome moves three nucleotides to the right along the mRNA molecule, a process called **translocation**, the tRNA in the A site, carrying the growing polypeptide, moves into the P site and exposes the A site to an incoming third tRNA.
   f. The tRNA with its attached amino acid moves into the A site. Thus, one codon at a time is translated, using the appropriate tRNA molecules to add amino acids to the polypeptide.
   g. After each amino acid is bonded to the next, its tRNA is freed to move out into the cytoplasm and recycle; that is, pick up another amino acid.

Figure 3–12. *Diagram of protein synthesis. A,* In transcription, a messenger RNA (mRNA) complementary copy of a strand of DNA is formed. In translation, the nucleotides in the mRNA are matched, three at a time, to a complementary set of three nucleotides in the anticodon region of the transfer RNA (tRNA) molecule. The other end of the tRNA molecule holds an amino acid in a high-energy linkage. When matching takes place, the amino acid is added to the elongating polypeptide. *B,* Diagram of synthesis of proteins on ribosomes. The ribosomes serve as the sites on which the tRNAs position themselves to read the genetic message encoded on the mRNA. The ribosomes become attached to a start signal near one end of an mRNA molecule and move, three nucleotides at a time, toward the other end.

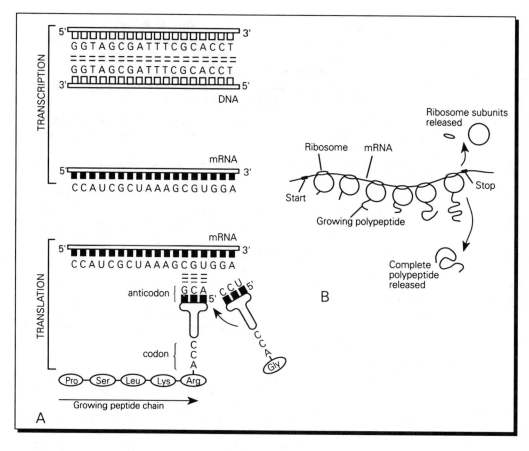

5. **Termination**
   a. When the ribosome runs into one of several mRNA **termination or stop codons** at the A site, a **release protein** binds with the stop codon to **terminate the translation process.**
   b. The polypeptide chain is then released from the ribosome.
   c. The release protein moves away from the A site and the ribosomal subunits separate and move into the cytoplasm to be available for another protein synthesis cycle.

Study Questions

Directions: Each question contains four suggested answers. Choose the **one best** response to each question.

1. All of the following statements about the plasma membrane are true EXCEPT
   (A) It regulates the passage of materials entering, but not exiting, the cell.
   (B) It is organized as a double layer of phospholipid molecules.
   (C) It is a boundary that surrounds and protects every cell.
   (D) It contains proteins that act as receptor sites for messages from endocrine glands.

2. Simple diffusion
   (A) occurs from regions of higher to lower concentrations by the interaction of solute molecules with specific membrane transport systems
   (B) moves a substance from an area of lower concentration to an area of higher concentration against its concentration gradient
   (C) is a process involving the unaided net movement of a substance from a region of higher to a region of lower concentration
   (D) can only take place across a living cell membrane

3. Which of the following statements is accurate concerning a cell placed in a hypertonic solution?
   (A) The size of the cell increases.
   (B) Water diffuses into the cell.
   (C) The number of solute particles is greater inside the cell than outside the cell.
   (D) The concentration of water outside the cell is lower than the concentration of water inside the cell.

4. Which of the following is an accurate description of ATP within the cell?
   (A) After ATP is formed, it is a very stable compound that cannot be broken down without an input of large amounts of energy.
   (B) ATP is an organic base found in the DNA of the nucleus of all cells.
   (C) The energy-donor ability of ATP is associated with the transfer of its third phosphate bond to other molecules.
   (D) The adenosine of ATP is a necessary component of all cellular chemical reactions.

5. The specific ability of a DNA molecule to carry coded hereditary information depends on
   (A) its amino acid sequence
   (B) the number of sugar-phosphate units in the molecule
   (C) the linear sequence of nitrogenous bases in the nucleotides
   (D) the total number of nitrogenous base pairs in a single strand of DNA

6. If a DNA strand has the nitrogenous base sequence T T A C G A, the complementary base sequence of an mRNA molecule would be
   (A) A A U G C U
   (B) A A T G C T
   (C) T T A C G A
   (D) U U T G C T

7. A sperm cell of an organism is found to contain eight chromosomes. What is the basic diploid (2n) number of chromosomes in this organism?
   (A) eight
   (B) 16
   (C) four
   (D) 24

8. In the synthesis of a protein, select the one factor likely to be most important in assuring that amino acid X will be positioned on mRNA where the mRNA directs it to be placed.
   (A) tRNA
   (B) rRNA
   (C) ATP
   (D) the concentration of amino acid X in the cytoplasm

9. An experimental compound X is found to interfere only with cytokinesis during the process of mitosis and has no effect on the other events. Which of the following is most likely to result if compound X were added to a culture of living cells undergoing cell division?
   (A) uncontrolled cell division
   (B) cells without any nuclei
   (C) multinucleated cells
   (D) identical daughter cells

Questions 10–13. For each description of cellular components, choose the structure with which it is most closely associated.
 (A) endoplasmic reticulum
 (B) mitochondria
 (C) Golgi apparatus
 (D) centrioles

10. Contain enzymes and are concerned with energy for cellular activities

11. Are found near the nucleus and are involved with cell division

12. Functions in the synthesis and transport of materials in the cytoplasm

13. Modifies and packages secretions for discharge from the cell

Questions 14–18. Match each stage of mitosis with the description of the events occurring in that stage. A stage may be used more than once.
 (A) prophase
 (B) metaphase
 (C) anaphase
 (D) telophase

14. The paired chromatids align along the equator of the cell.

15. The nuclear membrane disappears and the nucleolus disperses.

16. The chromatids separate at their centromeres.

17. The nuclei and nuclear membranes reform around each group of chromosomes.

18. Microtubules form and the paired centrioles separate and begin moving.

Answers and Explanations

1. **The answer is A.** (II B 1 a–c) The plasma membrane controls the passage of materials both into and out of the cell by the processes of simple diffusion, facilitated diffusion, active transport, and endocytosis/exocytosis. According to the fluid-mosaic model, the membrane is organized as a bilayer of lipid molecules interrupted by the presence of proteins, which exist as integral and peripheral proteins. Peripheral proteins are loosely bound to the bilayer surface and may be easily removed. The bulk of membrane proteins are integral proteins, which include most membrane-associated enzymes, receptors, and antigens.

2. **The answer is C.** (III B 1–4) Simple diffusion is the random movement of particles from areas of higher concentration to areas of lower concentration without the aid of energy or transport systems. It may take place in liquids, gases, or solids, and through nonliving or living membranes.

3. **The answer is D.** (III B 3 c (5)) A solution hypertonic to the cell is a more concentrated solution; that is, it has a greater concentration of solute particles, greater osmotic pressure, and a lower concentration of water than the solution inside the cell. Because water moves by osmosis from its area of higher concentration to lower concentration, the cell loses water and crenates.

4. **The answer is C.** (IV B 1 a–c) ATP stores energy in the form of high-energy bonds in the tail of its phosphate groups. Breaking the bond and the transfer of a terminal phosphate group to a variety of other compounds releases a large amount of energy that can be used to power other reactions. It is the phosphate bond of ATP, not the adenosine, that is integral to chemical reactions that require energy (endergonic reactions). The cleaving of a phosphate group from ATP and the release of energy is easily accomplished in the presence of the appropriate enzyme. ATP molecules are not found in DNA but are dissolved in the cytoplasm of cells where it is used to fuel cellular activities such as protein synthesis, active transport, cell division, cell secretion, and cell movement.

5. **The answer is C.** (V A 1) The genetic code depends on the linear sequence of nucleotides arranged into codons, which consist of three nucleotides. The joined sugar-phosphate units of a nucleotide form the ladder of the DNA molecule while the nitrogenous bases form the rungs. The DNA molecule is a double-stranded molecule arranged in the form of a helix.

6. **The answer is A.** (V A 1, VI A 4 b) In transcription of the coded message from DNA to mRNA, the base thymine in DNA is replaced by uracil.

7. **The answer is B.** (V B 1–7) Because meiosis, which is the type of cell division that occurs in the development of the gametes (ova and spermatozoa), results in halving the number of chromosomes found in all somatic cells, the diploid number of chromosomes in this organism would be twice the number found in the sperm cells.

8. **The answer is A.** (VI B 1, 2 a) Each clover-leaf–shaped tRNA molecule carries an anticodon, or base triplet complementary to the codon on mRNA, on one end of the molecule and one of 20 amino acids on the other end of the tRNA molecule. In translation of the mRNA message, an appropriate tRNA, carrying its specific amino acid, attaches by its anticodon to the appropriate mRNA codon. rRNA is contained in the ribosome subunits and functions as the site of protein synthesis. It performs a structural role in holding the tRNA and mRNA as the polypeptide chain elongates. The concentration of the amino acids is important to protein synthesis but not in the positioning of the appropriate amino acid on the mRNA.

9. **The answer is C.** (V C 2 a–d) Cytokinesis is the cytoplasmic division that occurs in the telophase stage of mitosis. A compound that interfered with cytokinesis, but not with DNA replication, spindle formation, or any of the other events of mitosis would be most likely to result in multinucleated cells.

10–13. **The answers are 3–B, 4–D, 5–A, 6–C.** (II A 1 a–g, 2 a–e) The most important function of mitochondria is the production of energy in the form of ATP. The centrioles function in cell division and are also the site of formation of cilia and flagella. The endoplasmic reticulum is the major site of the synthesis of cell products as well as their transport. The Golgi apparatus is the site of accumulation, concentration, packaging, and chemical modification of the products synthesized by the endoplasmic reticulum.

14–18. **The answers are 14–B, 15–A, 16–B, 17–D, 18–A.** (V C 2 a–d) In metaphase, the paired chromatids line up at the equatorial plate of the cell. The nuclear membrane and nucleolus disappear during prophase. Although the chromatids move to their respective poles during anaphase, separation of the chromatids at the centromeres occurs during metaphase. The nuclear membranes re-form around each aggregate of chromosomes during telophase. The paired centrioles begin to move to opposite poles of the cell during late prophase.

4 Human Genetics

I. INTRODUCTION

A. **Definition.** Genetics is the study of the fundamental units of inheritance (genes) and their transmission and expression from one generation to the next.

B. **Subdivisions of genetic science**

1. **Molecular genetics** is concerned with the chemical makeup of the gene. (See Chapter 3 IV A.)
2. **Developmental genetics** studies the way in which genes control embryonic development.
3. **Cytogenetics** deals with the role of the genes and the chromosomes in cellular activities.
4. **Behavioral genetics** involves the interactions between heredity and environmental components in human behavior. For example, it studies the genetic and environmental contributions to such aspects of behavior as personality, intelligence, mental illness, or sex roles.
5. **Population genetics** studies the principles of genetics as they affect whole populations of organisms.

II. TERMINOLOGY OF GENETICS. (See also Chapter 3, IV B–D.)

A. The **diploid** number of chromosomes found in normal somatic human cells is 46. Of these, 44 are **autosomes** (other than sex chromosomes). They occur in 22 **homologous (matched) pairs** in somatic cells and singly in gametes.

B. **The sex chromosomes are X and Y.**

1. In females, both of the sex chromosomes are X chromosomes.
   a. Both X chromosomes in females are not fully functional. At two weeks of embryonic life, either one of the X chromosomes (at random) becomes inactive.
   b. The condensed and inactivated X chromosome is present in female body cells and visible as the **Barr body** in the nuclei of certain epithelial cells.
2. In males, one sex chromosome is an X and the other is a Y.
   a. The X in the male appears identical to the two X's in the female.
   b. The Y is a short chromosome. It carries a gene that induces the development of testes and the male reproductive tract. In its absence the embryo develops as a female.

C. A **karyotype** is a systematized array of chromosomes, which allows their individual identification.
   1. The karyotype is prepared from a photomicrograph of metaphase chromosomes in a single cell.
   2. The chromosomes are arranged in pairs and assigned into groups based on their length and the position of their centromeres.
D. The **haploid number of chromosomes is 23**. It is produced by meiosis and occurs only in the gametes.
   1. Each gamete (sperm or ovum) contains a haploid set of chromosomes, one of each distinguishable kind. The sex chromosomes behave like homologous pairs of autosomes and separate during metaphase I of meiosis.
      a. In males, meiosis results in four viable haploid **spermatozoa**: two X-bearing and two Y-bearing.
      b. In females, meiosis results in one viable haploid **ovum** (egg cell) and three **polar bodies**, which fail to develop further (Figure 4–1).
   2. **Meiosis results in genetic variation.** The genes located on the maternal and paternal chromosomes are reshuffled and reorganized by **crossing-over** (the break and reunion of nonsister chromatids during metaphase I). The genetic consequence is the increase of genetic variation; i.e., the absolute uniqueness of each human being.

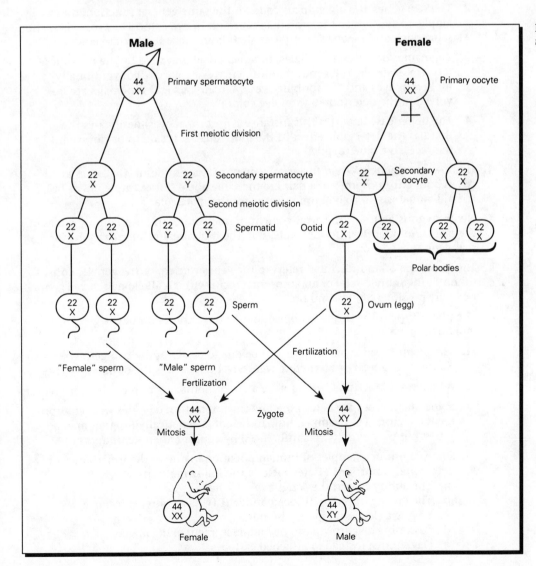

Figure 4–1. *Gametogenesis, fertilization, and sex determination.*

E. **Fertilization** is the merging of the egg nucleus with the sperm nucleus.
   1. Each gamete contains 23 chromosomes, one chromosome from each homologous pair in the parents.
   2. The **zygote** is the fertilized cell containing the restored diploid number of chromosomes. It contains all the information needed to divide by mitosis and develop from an embryo to an adult.
F. **Sex determination** occurs at conception.
   1. If an X-bearing sperm fertilizes an egg, the resulting zygote will be female (XX).
   2. If a Y-bearing sperm fertilizes an egg, the resulting zygote will be male (XY).
   3. Thus, the sex ratio at fertilization is theoretically ½ males and ½ females. Based on millions of births in the United States, however, there is a slight excess of males—1,026 (nonwhite) to 1,055 (white) males born for every 1,000 females.
G. The **gene** is the basic blueprint for the synthesis of all cellular proteins.
H. The **locus** is the physical place on a chromosome where a particular gene is located.
I. An **allele** is each member of the gene pair that occupies the same position on homologous chromosomes.
   1. Each chromosome exists in matching or homologous pairs; therefore, genes on the chromosomes also exist in pairs or **alleles**.
   2. The two genes at a given locus code for the same general function; for example, to synthesize the proteins that result in eye color. They may have the same nucleotide sequence or differ slightly in nucleotide arrangement.
   3. The names of the genes (alleles) in a pair are abbreviated by the use of letters. For example, the genes that influence eye color are conventionally "B" for brown eyes and "b" for blue eyes. (Actually, several gene pairs are involved in the determination of eye color.)
   4. The upper case letter (in this instance B for brown eyes) designates the gene that has the detectable effect in the individual. It is said to be **dominant** to the **recessive** blue (b) allele.
J. **Homozygous genes.** When the arrangement of nucleotides in the genes at a locus in both chromosomes of a pair are the same allele (for example, BB or bb), the individual is said to be **homozygous** for that particular gene.
K. **Heterozygous genes.** If the arrangement of nucleotides in both chromosomes of the pair is not identical (for example, Bb), the individual is **heterozygous** for that gene.
L. The **genotype** of an individual refers to the genes carried, or the genetic constitution of the individual. For an autosomal locus with an allelic pair, three genotypes are possible: BB, Bb, and bb.
M. The **phenotype** is the physical appearance or the **expression** of the gene in the individual.
   1. The phenotype for brown eyes can be due to either of the two genotypes—BB (homozygotes) or Bb (heterozygotes)—because B is the dominant gene.
   2. A blue-eyed individual would have a bb, or homozygous recessive genotype.
   3. Some phenotypes are determined by the interaction of genes with environmental factors. For example, human height is genetically programmed by genes, but proper growth is influenced by appropriate nutrition.
   4. Some common examples of human phenotypes due to the inheritance of **autosomal dominant alleles** include the following traits:
      a. The ability to curl the tongue in a U-shape
      b. The tasting of phenylthiocarbamide (PTC) as a bitter substance
      c. Having a cleft or dimple in the chin
      d. Possessing hair growth on the middle segment of the fingers
      e. Having free instead of attached earlobes.

III. MENDELIAN LAWS OF GENETICS

A. The Abbot Gregor Johann Mendel, a self-trained Augustine monk (1822–1884), developed the basic laws of heredity.

B. Mendel's laws provided the basis for modern genetics, which still analyzes the transmission of inherited traits by his methods.

1. **Mendel's Law of Segregation** states that the members of a pair of alleles segregate, or are separated, during the production of gametes. With random distribution, some gametes will contain the original maternal gene; others will contain the original paternal gene. The physical basis for the law is the separation of homologous chromosomes during anaphase I of meiosis.

2. **Mendel's Law of Independent Assortment** states that genes at different loci segregate independently of each other, i.e., when dealing with two or more pairs of genes, each pair separates to go into the gametes independently, provided that they are not on the same chromosome.

    a. When genes are positioned on the same chromosome they do not behave with complete independence but are inherited as a group and are said to be **linked**, especially if they are closely adjacent.

    b. Linkage groups of alleles can be broken and recombined in many different ways by crossing over between homologous chromosomes during meiosis. This adds to increased variation among offspring.

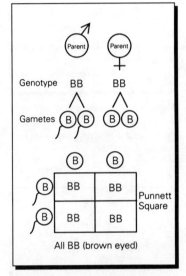

Figure 4–2. *Diagram of a simple Mendelian cross between two homozygous dominant parents. A Punnett square places the crosses in graphic form.*

IV. HUMAN INHERITANCE

A. The simplest pattern of inheritance is the one governed by a **single pair of genes**.

1. When there are two alleles at a locus, B and b, three genotypes will be present in the population: BB, Bb, and bb. The three genotypes can give rise to six different kinds of matings (for purposes of illustration, the examples use the trait for eye color—B for brown eyes and its alternative allele, b):

    BB x BB
    BB x Bb
    BB x bb
    Bb x Bb
    Bb x bb
    bb x bb

    a. A cross between two homozygous brown-eyed parents (BB x BB), designated as the **parental (P) generation**, gives rise solely to brown-eyed offspring, BB. The offspring are the **first filial (F_1) generation** (Figure 4–2). The Punnett Square, named after a British geneticist, is a common way of expressing genetic traits.

    b. A homozygous brown-eyed parent (BB) and a heterozygous brown-eyed parent would result in two kinds of progeny, all brown eyed, BB and Bb, in equal proportions (Figure 4–3).

    c. In a cross between a homozygous brown-eyed man (BB) and a homozygous blue-eyed woman (bb), all the sperm will carry B and the eggs will carry b. All offspring will be hybrids, genotypically heterozygous Bb and phenotypically brown eyed (Figure 4–4).

    d. A monohybrid cross is the mating between two heterozygous Bb individuals, called **monohybrids**. They produce a ratio of genotypes among the progeny of **one-fourth BB, one-half Bb, and one-fourth bb**, or 1:2:1. The phenotypes, however, are **three-fourths brown eyed (one-fourth BB plus one-half Bb) to one-fourth blue eyed**, or 3:1 (Figure 4–5).

        (1) The three-fourths to one-fourth, or 3:1, ratio of phenotypes is characteristic of inheritance in matings between two heterozygotes.

        (2) Such a ratio is an **expectation based on probability**, and does not represent a definite outcome.

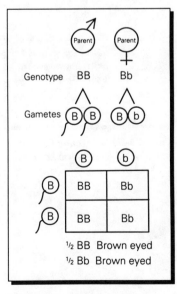

Figure 4–3. *Diagram of a cross between a homozygous dominant parent and a heterozygous parent.*

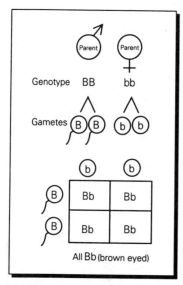

Figure 4-4. *Diagram of a cross between a homozygous dominant parent and a homozygous recessive parent.*

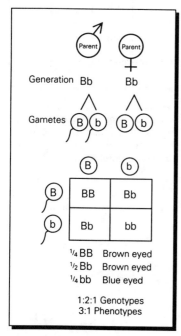

Figure 4-5. *Diagram of a simple monohybrid cross of two heterozygous parents.*

(a) It means that any child born to two heterozygotes has a 75%, 3 out of 4, or a 3:1 probability of having the dominant phenotype.

(b) Similarly, when both parents are **carriers** of a recessive gene, the probability of having a homozygous recessive child, therefore, is 25%, or 1 out of 4.

(c) If the first child is blue eyed, the odds that the second child will be blue eyed remain 1 out of 4.

(d) This holds for each subsequent child, regardless of the number of previously blue-eyed children.

e. If one parent has the genotype Bb and the other parent has the genotype bb (phenotype blue eyed), one half of the offspring would be brown eyed (Bb) and one half would be blue eyed (bb) (Figure 4–6).

f. If both parents have blue eyes (bb x bb), all the offspring will be blue eyed.

2. **Incomplete dominance** describes a type of inheritance in which neither allele is dominant to the other, and the heterozygote phenotype appears intermediate to the homozygous dominant or recessive phenotypes.

   a. For example, in Andalusion fowl (a type of chicken) a cross of a homozygous black fowl (dominant gene) with homozygous white fowl (recessive gene) produces a heterozygote F_1 that is blue-gray, indicating incomplete dominance of black to white.

      (1) A cross between the F_1 generation will produce a phenotypic ratio of 1 black : 2 blue-gray : 1 white, again indicating a case of incomplete dominance.

      (2) Note that in incomplete dominance the phenotypic ratio is also the genotypic ratio.

   b. In humans, there are some traits believed to be determined by incomplete dominance.

      (1) A curly-haired individual homozygous for a particular allele produces offspring with wavy hair when mated to a straight-haired individual.

      (2) Similarly, singing voice tone (bass, alto, tenor, and soprano) is thought to be determined by incompletely dominant alleles.

3. **Multiple alleles** refers to the occurrence of three or more alleles (gene forms) associated with a single locus although no individual carries more than two. The classic example is the inheritance of the ABO blood types which also involves **codominance**. The alleles are designated I^A, I^B, and I^O. I^A is dominant to I^O, and I^B is dominant to I^O, but they show no dominance with respect to one another.

   a. Individuals of genotype $I^A I^A$ or $I^A I^O$ have blood type A.
   b. Individuals of genotype $I^B I^B$ or $I^B I^O$ have blood type B.
   c. Individuals of genotype $I^O I^O$ have blood type O.
   d. Heterozygotes of genotype $I^A I^B$ have blood type AB and codominant alleles because the phenotypic characteristics are associated with both I^A and I^B.
   e. Based on the above information and using Mendelian genetics, the following situations are evident:

      (1) A type A individual mated with a blood type B individual can produce types A, AB, B, or O children.
      (2) A type A mated with a type A can produce type A or O children.
      (3) A type B mated with a type O can produce type B or O children.
      (4) A type AB mated with type O can produce type A or B children.

4. **Sex-linked genes** are genes (usually recessive) carried on the X chromosome. Inheritance of sex-linked (or X-linked) genes is unique because males receive their X chromosome only from their mothers and transmit it only to their daughters.

   a. **Red-green color blindness** is a well-known X-linked trait, which is recessive to normal vision. A female, who has two X chromosomes, must be a homozygous recessive to express the trait of color blindness. Males, with only one X chromosome, would be affected with only one recessive gene because they would lack a dominant allele to compensate for the recessive gene. The inheritance of color blindness is as follows:

(1) If C is the normal gene and c is the recessive color-blind gene, a woman of normal vision, whose father was color blind, must be a heterozygote and have the genotype $X^C X^c$. This is because all X chromosomes from her mother carried the normal allele and all X-bearing sperm from her father carried the c gene for color blindness.

(2) If she marries a normal man, his genotype must be $X^C Y$ because he has normal vision.

(3) A Punnett square (Figure 4–7) diagrams the expected genotypes. There would be no color-blind daughters (but half of the daughters would be a carrier), and half of the sons would be color blind and half would have normal vision.

b. **Hemophilia**, a bleeding disorder in which the blood fails to clot properly, is another X-linked condition in humans with its expression found predominantly in males.

5. **Sex-influenced traits** are characteristics due to **autosomal** genes rather than sex-linked genes. Their expression is limited to one sex or the other.
   a. Examples of sex-influenced traits are the growth of facial hair in males and the development of breasts in females. Both are influenced by male or female sex hormones at puberty.
   b. Common baldness in males also is an example of a sex-influenced trait, which shows a variable age of onset, a variation in its degree of expressivity in different men, and a variation in incidence among various populations (common among whites, less common in blacks, and rare among Asians).

6. **Polygenic inheritance** refers to traits governed in their expression by more than one pair of allelic genes. Human skin color appears to be transmitted by at least three pairs of genes, each pair located on a different chromosome.

B. A **dihybrid cross** involves parents differing in **two** traits carried on different nonhomologous chromosomes. If a man heterozygous for tongue curling (dominant) and free earlobes (dominant) marries a woman with the same genotype, the two pairs of alleles will segregate independently (according to Mendel's Law of Independent Assortment). The following symbols are used for the four alleles:

C = ability to curl tongue   E = free earlobes
c = inability to curl tongue   e = attached earlobes

The Punnett square that diagrams a cross between two heterozygote dihybrids (CcEe x CcEe) is shown in Figure 4–8. Each parent can produce four different gametes: CE, Ce, cE, ce.

C. **Simple inheritance and genetic disorders**

1. Many human traits with adverse effects (over 500 known and nearly 600 suspected) are caused by **recessive** genes that only are expressed homozygously. Among them are **cystic fibrosis**; **Tay-Sachs disease**, which is lethal; and **phenylketonuria (PKU)**, which causes mental retardation.
   a. The heterozygous carriers usually bear the recessive gene unknowingly. It becomes apparent only when two carrier parents happen to mate.
   b. Consanguinous marriages (between blood relatives) increases the risk that both parents have received the same detrimental gene through some common ancestor.

2. When defective genes are transmitted in a **dominant** pattern, each affected person has one affected parent. The normal children do not carry the harmful dominant gene and their offspring will be normal. Examples of adverse dominant defects are **achondroplasia**, a form of dwarfism, and familial **hypercholesteremia**, which causes elevated levels of blood cholesterol.

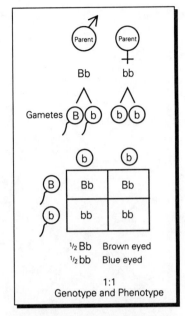

Figure 4–6. *Diagram of a cross between a heterozygous parent and a homozygous recessive parent.*

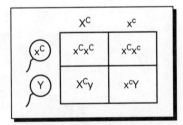

Figure 4–7. *Diagram of a cross between a man with normal vision and a woman carrier of the recessive gene for color blindness.*

Figure 4–8. *Diagram of a cross between parents, each of whom are heterozygous for two factors; tongue curling and free earlobes.*

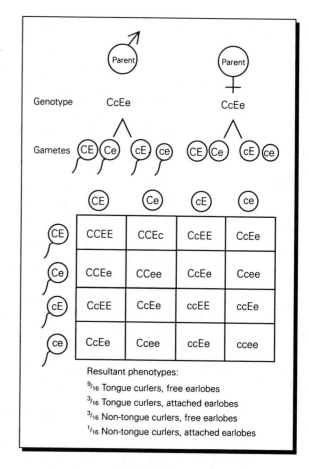

V. CHROMOSOMAL ABNORMALITIES

A. Chromosomal disorders may arise as a result of some aberration during the process of cell division. They can occur during the early stages of meiosis in development of gametes and they can also arise after fertilization from faulty mitotic division of the zygote.

1. An **abnormal number of chromosomes** may result from **nondisjunction**, which is the failure of homologous chromosomes to separate properly at anaphase I of meiosis. The result is a cell with too many or too few chromosomes.
    a. **Trisomy** indicates an extra chromosome so that the homologous set consists of three chromosomes instead of the normal two.
    b. **Monosomy** means the absence of a chromosome.
2. A **structural change in the chromosome** may occur as a result of environmental influences, such as drugs or infections, or part of a chromosome may be altered, lost, or moved during cell division.
    a. A **deletion** occurs if a piece of a chromosome breaks off and is lost.
    b. If it is not lost but improperly inserted on another chromosome, it is termed a **translocation**.

B. **Autosomal abnormalities**

1. In **trisomy 21** (Down's syndrome), there are three copies of the smallest autosome, number 21, for a total of 47 chromosomes instead of 46.

a. **Occurrence.** Trisomy 21 is the most frequent trisomy in live births, affecting about 1 in 700 children overall, although the risk of having an affected child increases with the age of the mother. It results in a child with mental and physical abnormality.
b. **Causes**
   (1) It arises from **meiotic nondisjunction of number 21 autosome,** most often during the development of the ovum, which will then have an extra chromosome 21. When such an ovum is fertilized by a normal sperm, the resulting zygote will have three copies of chromosome 21 instead of two.
   (2) A small percentage of Down's syndrome births (about 5%) results from an inherited chromosomal translocation, which can be identified in the carrier parent by chromosome analysis.
c. The risk of having an affected child statistically increases with the age of the mother.

2. **Edward's syndrome (trisomy 18),** with an incidence of 1 in 6,500 live births, usually is fatal within the first year. The same is true for **Patau's syndrome (trisomy 13).**
3. **Cri-du-chat or cat-cry syndrome** is a rare condition caused by a deletion of the shorter part of number 5 chromosome. Affected babies have physical deformation, mental retardation, and a laryngeal defect that results in a cry much like that of a mewing cat.

C. **Sex-chromosome abnormalities**
1. Trisomies and monosomies of sex chromosomes can occur during oogenesis and spermatogenesis and cause abnormal sperm or ova.
2. The result after fertilization can be an individual with an abnormal sex chromosome complement: XO, XXY, XXX, XYY. The presence of a Y chromosome results in a male, no matter how many X chromosomes are present.
3. At least one X chromosome must be present for survival. Table 4–1 shows the mating possibilities between abnormal and normal sperm and ova. Table 4–2 shows the most frequent clinical conditions and the major clinical symptoms.

D. **Prenatal determination of genetic defects**
1. **Amniocentesis** is the piercing of the abdomen and uterus of a pregnant woman during the fourteenth to sixteenth week of development to remove a sample of the fetal cells and their surrounding fluid for analysis. It makes possible the diagnosis of chromosomal abnormalities, certain enzymatic defects, and neural tube defects.

Table 4–1. Various mating possibilities between normal and abnormal sperm and ova and the genotypes produced.

Normal Ovum	Abnormal Sperm	Expected Genotype
X	O	XO
X	XX	XXX
X	XY	XXY
X	YY	XYY

Abnormal Ovum	Normal Sperm	Expected Genotype
XX	X	XXX
XX	Y	XXY
O	X	XO
O	Y	YO (inviable)

Reproduced by permission. *Biology of Women*, 3E, by Ethel Sloane. Delmar Publishers, Inc., Albany, New York; copyright © 1993.

Table 4–2. Frequencies and clinical characteristics of some sex-chromosome abnormalities.

Genotype	Name of Syndrome	Frequency	Symptoms
XO	Turner's syndrome (gonadal dysgenesis)	1/3,500 females	Female, is short in stature, has typical webbed neck, poorly developed breasts, and immature external genitalia. Ovaries may be absent. The mental capacity is normal.
XXY	Klinefelter's syndrome	1/800 males	Male, is above average in height, with long arms and legs relative to the rest of the frame; has small testes and penis and is usually sterile. The breasts may be somewhat enlarged.
XYY	Double Y-syndrome	1/700 males	Male, is taller than average, normally fertile, but possibly of somewhat lower intelligence.
XXX	Trisomy X	1/1,000 females	Female, is normal physically, fertile, but possibly with greater tendency toward mental retardation.

Reproduced by permission. *Biology of Women* 3E, by Ethel Sloane. Delmar Publishers, Inc., Albany, New York; copyright © 1993.

2. **Chorionic villi sampling** is another method of analysis, which can be performed in the first trimester of pregnancy, rather than at 14 to16 weeks. It involves the insertion of a catheter through the vagina and cervix into the uterus for withdrawal of a small amount of fetal tissue.

3. **Ultrasound scanning** during pregnancy is used to determine physical fetal abnormalities such as brain and spinal cord defects, heart, gastrointestinal or skeletal anomalies, or kidney or bladder problems. It is also used to date the age of the fetus or confirm a suspected multiple pregnancy.

Study Questions

Directions: Each question below contains four suggested answers. Choose the **one best** response to each question.

1. Alternative genes that code for a given trait are known as

    (A) chromatid pairs
    (B) alleles
    (C) gene partners
    (D) gametes

2. If both genes for a given trait are alike, the individual is said to be

    (A) homozygous for the trait
    (B) heterozygous for the trait
    (C) a hybrid
    (D) a twin

3. A medical technologist who is studying a human blood smear under the microscope notes that some of the white blood cells have a mass of chromatin, or Barr body, located close to the nuclear membrane. This indicates that the blood most likely came from

    (A) a color-blind individual
    (B) a normal woman
    (C) a hemophilic man
    (D) a diabetic

4. A student who is studying genetic crosses in the laboratory could best determine the phenotype for the eye color of a fruit fly by

    (A) examining the intact fruit fly and noting its characteristics
    (B) breeding it with another fruit fly to see what eye color occurs in the offspring
    (C) biochemical analysis of the fruit fly's DNA
    (D) microscopic analysis of the fruit fly's eye tissues

5. Cystic fibrosis is a condition caused by a recessive autosomal gene and may be lethal in early childhood if not treated. Suppose two normal people marry and have three normal children and one who has cystic fibrosis. If C represents the dominant gene and c the recessive gene, what are the possible genotypes of the parents and of each child?

    (A) father, CC; mother, Cc; normal child (1), CC; normal child (2), CC; normal child (3) CC; ill child, Cc.
    (B) father, cc; mother, Cc; normal child (1), cc; normal child (2), cc; normal child (3), cc; ill child, Cc.
    (C) father, Cc; mother, Cc; normal child (1), CC; normal child (2), Cc; normal child (3), Cc; ill child, cc.
    (D) father, CC; mother, CC; the genotypes of any of the children are impossible to determine from the information given.

6. The defective gene for red-green color blindness in humans is recessive and found on the X chromosome. A color-blind man marries a woman with normal vision who has no color-blind relatives. Which of the following statements would best describe their probable children?

    (A) Half of their sons will be colorblind.
    (B) Half of their daughters will be colorblind.
    (C) All of their children will be colorblind.
    (D) None of the children will be colorblind, but all of their daughters will be carriers.

7. A father has type AB blood and a mother has type A blood. All of the following blood types could occur in the offspring EXCEPT

    (A) type O
    (B) type A
    (C) type AB
    (D) type B

8. A child with Down's syndrome is found to have a normal chromosome number, but analysis of her karyotype indicates that chromosome 19 has a small chromosome 21 attached to it. Which of the following events could have produced such a structural abnormality?

    (A) deletion
    (B) duplication
    (C) translocation
    (D) nondisjunction

9. Chromosomal karyotype analysis of a sterile male of average intelligence, but who was somewhat above average in height, revealed that he had a Y chromosome. Tissue analysis indicated the presence of a single Barr body in the nuclei of somatic cells. The genotype of this individual with respect to the sex chromosomes was most likely

    (A) XXY
    (B) XYY
    (C) XY
    (D) YO

Answers and Explanations

1. **The answer is B.** (II A–I) Alleles are two or more different genes that occupy the same locus on specific paired chromosomes. Chromatid pairs are longitudinal strands, held together by a centromere, formed by the replication of a chromosome during mitosis or meiosis. Gametes are the spermatozoa and ova formed as a result of meiotic division.

2. **The answer is A.** (II A–J) When both members of an allelic pair are the same, the individual is said to be homozygous for that particular gene. If the nucleotide arrangement of one of the genes differs slightly, but both still code for the same general function, the individual is said to be heterozygous for that trait. A hybrid is an individual whose parents are different varieties of the same species or belong to different, but closely allied, species. For example, the offspring of a horse and a donkey is a hybrid, or cross-breed. A twin is one of two offspring that are born at the same time. Twins may be monozygotic (identical) and result from division of a single fertilized ovum, or may be dizygotic (fraternal) and derived from two separate zygotes.

3. **The answer is B.** (II B 1) Under ordinary light microscopy, the Barr body (the inactivated, condensed X chromosome) is visible as a mass of chromatin near the nucleus in certain female body cells such as white blood cells or cheek cells. There is no visible indication in a blood smear of color blindness, hemophilia, or diabetes.

4. **The answer is A.** (II M) The phenotype of an organism is the visible expression of the genetic constitution. It may be determined by the physical appearance of a particular trait, such as eye color. The genotype of an individual may be determined by breeding experiments. Neither biochemical tissue analysis nor microscopic analysis would be necessary for the determination of the phenotype of most physical characteristics.

5. **The answer is C.** (IV A 1 d) The ratio of three offspring with the dominant phenotype to one with the recessive phenotype is most likely to occur in a monohybrid cross between two heterozygous individuals. Both the father and the mother were carriers of the recessive gene for cystic fibrosis and had the genotype Cc. The probability is that of the four children produced by this marriage, one fourth would have the genotype CC and be normal, one half would have the genotype Cc and be normal, and one-fourth would have the genotype cc and develop cystic fibrosis.

6. **The answer is D.** (IV A 4 a) As a sex-linked trait, the gene for color blindness is not carried on the Y chromosome. A father (XY) who carries the defective gene on the X chromosome transmits the X to all daughters, all of whom, therefore, would be carriers. The normal mother (XX) transmits an X without the defective gene to any son and daughter. Subsequently, a carrier daughter married to a man with normal vision may pass the X bearing the defective gene to half of their sons.

7. **The answer is A.** (IV A 3) The father who has type AB blood has the genotype I^AI^B. The mother with type A blood has the genotype I^AI^A or I^AI^O. Thus, the only impossible blood type in a child would be I^OI^O, or type O.

8. **The answer is C.** (V A 1,2; B 1 a–c) Translocation of a chromosome means it has been improperly attached to another chromosome. A deletion occurs when a piece of a chromosome breaks off and is lost. Duplication of a chromosome is a normal event that occurs during interphase of cell division. Nondisjunction is the failure of homologous chromosomes to separate during anaphase I of meiosis. The result of nondisjunction is too many or too few copies of a chromosome.

9. **The answer is A.** (II B 1, V C 1–3) An individual with one Barr body in somatic cells must have at least two X chromosomes. The presence of both a Y and a Barr body indicates that the genotype must be XXY, a condition known as Kleinfelter's syndrome. A male with a genotype of XY would have no Barr bodies. The genotype of YO is a lethal combination; an individual needs at least one X chromosome for survival.

Tissues 5

I. **INTRODUCTION.** Tissues are groups of structurally similar cells (and their products) that are specialized to perform a particular function. Four basic tissue types are found in the human body: **epithelium, connective tissue, muscle tissue,** and **nervous tissue.** All body structures are composed of varying amounts of these tissues; most major organs incorporate all four tissue types.

II. **EPITHELIAL TISSUES**

   A. **Divisions.** Epithelial tissues can be divided into two classifications: **covering and lining epithelia** and **glandular epithelium.**
   1. **Covering and lining epithelia** are sheets of cells that **cover** the internal and external **body surfaces and organs** and line the body **cavities and hollow organs.**
      a. **Endothelium** is epithelium that lines blood vessels.
      b. **Mesothelium** is epithelium that lines several of the body cavities.
   2. **Glandular epithelium** is derived from covering and lining epithelia through downgrowths of cells into the underlying supporting tissue.
      a. **Exocrine glands** retain a duct or connection to the surface (e.g., salivary glands, digestive glands).
      b. **Endocrine glands** are ductless; they lose the connection to the surface and become a detached, solid mass (e.g., pituitary gland, adrenal glands).
   B. General characteristics
   1. Structure
      a. Typically, **one surface of epithelium is free** and faces either air or fluid.
      b. Epithelium has **no blood supply.** It is nourished by diffusion from blood vessels in underlying connective tissue, to which it is anchored by a nonliving **basement membrane (basal lamina).**
      c. Epithelial cells are **packed closely** with little intercellular material.
      d. Epithelial cells **reproduce rapidly** to replace damaged or lost cells.
   2. **Function.** Epithelial tissues perform diverse functions, which include the following:
      a. **Protection** from dehydration, trauma, mechanical irritation, and toxic substances

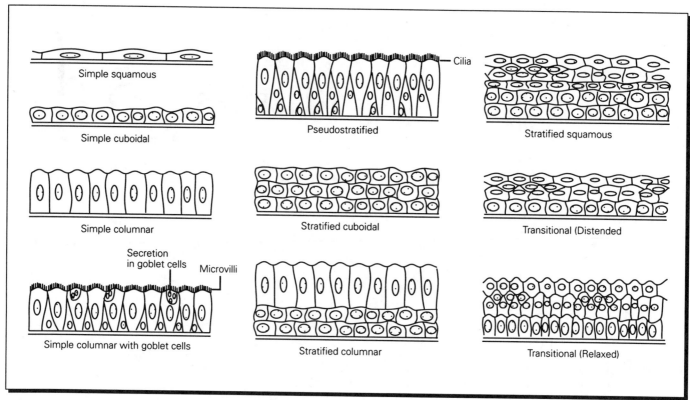

Figure 5-1. *Types of epithelial tissues.*

      b. **Absorption** of gases or nutrients, such as in the lung or digestive tract
      c. **Transport** of fluids, mucus, nutrients, or other particulate matter
      d. **Secretion** of synthesized products, such as hormones, enzymes, and perspiration, produced by glandular epithelium
      e. **Excretion** of wastes such as urine through filtration
      f. **Sensory reception** by specialized epithelial cells in the taste buds, nose, and ear
  C. **Covering and lining epithelia** (Figure 5–1)
    1. **Classification.** Covering and lining epithelia are classified according to the number of cell layers (**simple or stratified**) and the shape of the cells (**squamous, cuboidal, or columnar**). On the basis of the differences, the common classification is into the following eight types:
      a. **Simple epithelia** consist of a single layer of cells. They include the following:
        (1) **Simple squamous epithelium.** Squamous cells are thin and flattened.
        (2) **Simple cuboidal epithelium.** Cuboidal cells are cube shaped and are approximately equal in height and width.
        (3) **Simple columnar epithelium.** Columnar cells are taller than they are wide.
        (4) **Pseudostratified columnar epithelium.** In this type, all of the cells are in contact with the basement membrane, but the epithelium appears layered because of staggered cell heights.
      b. **Stratified epithelia** include membranes that are two or more cells thick. They are better able to withstand wear and tear than are simple epithelia. Types of stratified epithelia include the following:
        (1) **Stratified squamous epithelium**
        (2) **Stratified cuboidal epithelium**
        (3) **Stratified columnar epithelium**
        (4) **Transitional epithelium**

2. **Specializations at the surfaces of epithelial cells**
   a. **Lateral specializations** form intercellular junctions (**junctional complexes**), which serve as sites to keep cells together, act as seals to prevent the flow of materials through the intercellular spaces, and provide a means for intercellular communication. Types of intercellular junctions include the following:
      (1) **Zonula occludens** is a tight junction in which opposing plasma membranes fuse and obliterate the intercellular space, thus preventing any passage of substances via an intercellular route. It is situated just beneath the free surface of the epithelium.
      (2) **Zonula adherens** is located below the zonula occludens. It has an apparent intercellular space, usually 20 nm to 25 nm wide, with cytoplasmic filaments radiating outward.
      (3) **Desmosomes** are intercellular contacts in which an intercellular space of 30 nm separates the opposing cell membranes. Desmosomes are believed to be important for adhesion between cells.
      (4) **Gap junction** (nexus) is a contact that occurs not only between epithelial cells but also between the lateral surfaces of cardiac and smooth muscle cells. It consists of an intercellular gap of about 2 nm and may be a type of structural basis for cell-to-cell communication.
   b. **Apical specializations** may occur at the exposed surfaces of epithelial cells. They include the following:
      (1) **Microvilli** are cylindrical extensions, or folds, that project from the cell surface to increase the surface area. They are numerous in the cells that line the digestive tract and in the cells of the proximal convoluted tubules of the kidney.
      (2) **Cilia** are tiny, hair-like motile structures, much longer than microvilli, which are capable of rapid back-and-forth movement. They are found in the upper respiratory tract, the distal tubules of the kidney, portions of the female reproductive tract, and in the olfactory epithelium.
      (3) **Stereocilia** are long nonmotile processes, which actually are longer microvilli. They are located in the epididymis in the male reproductive tract.
3. **Characteristics and distribution of epithelial types**
   a. **Simple squamous epithelium** is a single layer of flattened cells with a central disc-like nucleus.
      (1) **Distribution.** Simple squamous epithelium occurs in the following areas:
         (a) Lining of blood and lymph vessels (endothelium)
         (b) Lining of body cavities (mesothelium)
         (c) Smallest ducts of many glands
         (d) Part of the kidney tubules
         (e) Terminal ducts and air sacs of the respiratory system
      (2) **Appearance** in a microscopic cross section is that of a sheet of flat, joined cells that look like a plate of fried eggs.
   b. **Simple cuboidal epithelium** consists of a single layer of six-sided cells with a central, ovoid nucleus.
      (1) **Distribution.** Simple cuboidal epithelium is present in the following locations:
         (a) Many glands and part of their ducts
         (b) Pigmented epithelium of the retina of the eye
         (c) Germinal (surface) layer of the ovary
         (d) Portions of the testis
         (e) Anterior surface of the lens of the eye
      (2) **Appearance** in a microscopic cross section is that of cubical blocks that look like a row of squares.
   c. **Simple columnar epithelium** consists of a single layer of rectangular-appearing cells with the nuclei located closer to the base of the cells.

(1) **Distribution.** Columnar cells are found in areas where absorption and secretion occur; for example,
  (a) Stomach, small and large intestine, and gall bladder
  (b) Oviduct and uterus
  (c) Portions of the respiratory tract
  (d) Many glands and parts of the ducts of some glands
(2) **Appearance** in a microscopic cross section is that of closely packed rectangles set on end.
(3) Unicellular glands, called **goblet cells**, occur in the columnar epithelium of the digestive tract. They secrete mucus, which causes the upper half of the cell to swell and resemble a goblet.

d. **Pseudostratified epithelium** resembles stratified epithelium. It is so named because the cells rest on the basement membrane, but many do not reach the free surface.
  (1) **Distribution.** Pseudostratified epithelial cells occur in the following:
    (a) Much of the respiratory tract (i.e., nasal cavity, pharynx, trachea, bronchi)
    (b) All of the male urethra and some of the female urethra
    (c) Most of the male reproductive ducts
  (2) **Appearance** in a microscopic cross section
    (a) Cells are cuboidal and columnar and thus vary in height; only the columnar cells reach the surface.
    (b) Cell surfaces have microscopic cilia.
    (c) Many mucus-secreting goblet cells are present.

e. **Stratified squamous epithelium** consists of many cell layers with columnar cells at the lowest levels, cuboidal cells at the intermediate level, and squamous cell layers at the free surface.
  (1) **Distribution.** Stratified squamous epithelium occurs wherever there is a need for protection from friction, drying, or mechanical injury.
    (a) The **epidermis** of the skin is "dry" stratified squamous epithelium; squamous layers are cornified (i.e., infiltrated with the synthesized protein keratin) to prevent water loss.
    (b) All **body orifices** (i.e., mouth, anal canal, urethral opening, vagina, and ear) are lined with "wet" stratified squamous epithelium, which serves as a protection against friction.
  (2) The **appearance** of stratified squamous epithelium in microscopic cross section is many layered, with thin squamous cells toward the surface.

f. **Stratified columnar epithelium** occurs rarely; generally, it is found where simple columnar or pseudostratified epithelium meets stratified squamous epithelium, in the pharynx and larynx, and in part of the urethra.

g. **Stratified cuboidal epithelium and transitional epithelium**
  (1) **Stratified cuboidal epithelium** occurs in testis tubules and ovarian follicles, ducts of sweat glands, and sebaceous glands.
  (2) **Transitional epithelium** is a specialized stratified cuboidal epithelium, the appearance of which varies with stretching. It is found in the urinary bladder and urinary tract.
    (a) In relaxed transitional epithelium, there are six to seven layers of cells.
    (b) In a distended urinary bladder, the epithelium is thinned to two to three layers of cells.

D. **Glandular epithelium.** Epithelial glands are the second subdivision of epithelial tissues. They are categorized broadly as **exocrine glands** or **endocrine glands**, depending on the relationship of the gland to the epithelial surface (see section II A 2 a, b). (Endocrine glands are discussed in Chapter 10.) Exocrine glands may be classified in several different ways.

1. **Number of cells**
  a. A **single cell** that acts as a glandular unit within a sheet of epithelial cells is a unicellular gland; for example, the **goblet cell** within the digestive epithelium.

b. A **multicellular gland** is a gland in which many cells act to produce a secretion. Normally, these cells are organized into tubes or sacs that open onto the epithelial surface. Examples of multicellular glands are sweat glands, salivary glands, mammary glands, and the exocrine part of the pancreas.

2. **Type of secretion**
    a. **Mucus-secreting glands** produce a viscous, slimy product rich in protein-polysaccharide complexes, which formerly were called mucopolysaccharides and mucoproteins and now are termed **glycosaminoglycans**. Mucus is produced by goblet cells in the intestine and other mucous cells in the stomach; salivary glands; and in the respiratory, urinary, and reproductive tracts.
    b. **Serous-secreting glands** produce a watery protein discharge, which frequently (but not always) contains enzymes. The exocrine portion of the pancreas is an example of an organ containing such cells.
    c. **Seromucous glands** produce a mixed product, owing to the presence of both serous and mucous cell types in the same gland, as in the parotid and submaxillary salivary glands.
3. **Method of release of the secretion**
    a. **Merocrine secretion** refers to release of the synthesized cell product through the cell membrane by exocytosis (fusion of the secretory vesicles with the membrane). The cell remains intact with no loss of cytoplasm. Merocrine secretion occurs in most of the exocrine glands, such as the pancreas, sweat glands, goblet cells, and salivary glands.
    b. **Holocrine secretion** involves the breakdown of the entire cell to form the secretion; thus, the whole cell is discharged. Holocrine secretion is seen exclusively in the sebaceous glands of the skin.
    c. **Apocrine secretion** refers to a type of secretion in which the product accumulates below the cell surface and can be released only by detachment of some of the apical cytoplasm of the cell. Observations with the electron microscope have not supported the existence of this type of secretion and the idea is largely outdated.
4. **Anatomical structure of the exocrine gland**
    a. If the shape of the secretory portion is a tube, the gland is referred to as **tubular**; if the shape is flask-like, the gland is **alveolar** or **acinar**.
    b. If the duct is **unbranched**, the gland is **simple**; if the ducts are **branched**, the glands are **compound**.

E. **Membranes.** In the study of tissues, a moistened epithelium and its underlying connective tissue are known as a membrane. Membranes are kept moist by mucus or serous fluid.
1. **Serous membranes** consist of mesothelium lying on a connective tissue layer. Serous membranes line closed body cavities and include the pericardial membranes (enclosing the heart), the pleural membranes (enclosing the lungs), and the peritoneal membranes (enclosing the abdominopelvic viscera).
    a. A **parietal** portion of the serous membrane lines the external wall of the cavities.
    b. A **visceral** portion covers the various organs enclosed within the cavities.
    c. The epithelial cells of the serous membranes secrete **serous fluid**, which lubricates the surfaces of the enveloped organs and allows their free movement within the thoracic and abdominopelvic cavities.
2. **Mucous membranes** are mucus-secreting sheets that line various hollow organs in the interior of the body. Mucous membranes typically consist of three layers.
    a. The **epithelium** may be simple, pseudostratified, or stratified, and is lubricated by specialized goblet cells or multicellular glands opening onto it.
    b. The **lamina propria** is connective tissue that supports the epithelium. It is rich in blood and lymphatic vessels and sometimes contains glands.
    c. The **muscularis mucosae** consists of a circular and longitudinal layer of smooth muscle.

Figure 5–2. *Various types of connective tissue.*

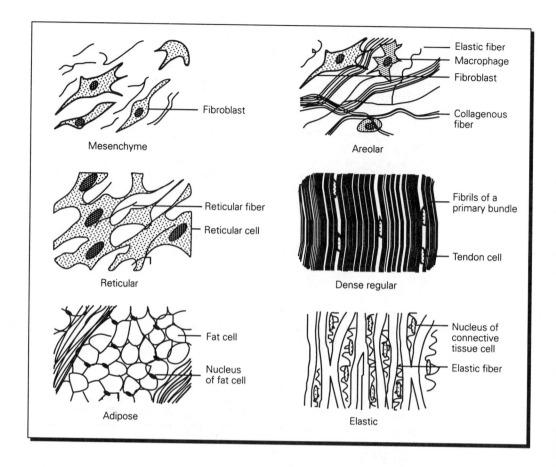

III. CONNECTIVE TISSUES give support to the body and its organs and connect other tissues together. It primarily consists of nonliving intercellular substances produced by certain connective tissue cells.

   A. **Classification**
      1. **Embryonic connective tissue** is found in the embryo and developing fetus. It includes two subtypes, mesenchyme and mucoid connective tissue.
         a. **Mesenchyme** is the unspecialized packing, wrapping, and supporting tissue of early embryonic life. All types of adult connective tissue cells derive from the star-shaped embryonic mesenchyme cell.
         b. **Mucoid tissue (Wharton's jelly)** appears temporarily in the normal, development of connective tissue and is also found in the umbilical cord.
      2. **Connective tissue proper** includes the following:
         a. **Loose (areolar)** connective tissue
         b. **Dense fibrous** connective tissue
         c. **Adipose** tissue
      3. **Specialized connective tissue** includes the following:
         a. Supporting connective tissue
         b. Cartilage
         c. Bone
         d. Vascular connective tissue (blood and lymph)
   B. **Components**
      1. All connective tissue is made up of living cells, usually located some distance apart.
      2. These cells are embedded in a **nonliving intercellular "ground" substance,** or **matrix,** which is semifluid to solid in consistency. Ground substance is composed of a mixture of glycosaminoglycans and proteins.

3. Three types of nonliving fibers, which are produced by cells called **fibroblasts**, are found within the matrix.
   a. **Collagenous fibers**
      (1) Collagenous fibers consist of bundles of variable numbers of parallel fibrils. They are composed chemically of the protein collagen.
      (2) Fresh fibers are white, wide, and strong. They yield gelatin or glue when boiled and leather when treated with tannic acid.
   b. **Elastic fibers**
      (1) Elastic fibers occur singly (not in bundles) and are chemically composed of the protein elastin.
      (2) They are yellow, much coarser but thinner than collagenous fibers, and not particularly strong but have great elasticity.
      (3) In the living body, they are stretched and under tension. Elasticity is diminished with aging.
   c. **Reticular fibers**
      (1) Reticular fibers are composed of collagen but differ in the number, diameter, and arrangement of the fibrils.
      (2) They are thin, inelastic, and branch to form a fine network, or **reticulum**, to support soft organs such as the liver and spleen.
      (3) Reticular fibers are the first connective tissue fibers to appear in development and are plentiful in the fetus and newborn.

C. **Functions**
   1. Connective tissue gives the body **form and support**; without the intercellular substances of connective tissue the body would be a jelly-like mass.
   2. Connective tissue **binds** various tissues together and provides a **packing material** between body parts, **stores fat**, and **aids in tissue repair**.
   3. The **ground substance** of loose connective tissue **provides a pathway** for blood vessels and nerves; nutrients, gases, and wastes are transported from capillaries to cells (and back) via the ground substance.
   4. **Ground substance** is a **barrier** to the spread of harmful bacteria and also it is an arena where the battle against bacteria takes place.

D. **Connective tissue proper** (Figure 5–2)
   1. **Areolar (loose) connective tissue** consists of several types of cells embedded in a matrix of loosely arranged collagenous and elastic fibers. It is delicate and flexible, has many blood vessels, and is somewhat resistant to stress.
      a. Cells (Figure 5–3)
         (1) **Fibroblasts** are the most common cell found in loose connective tissue.
            (a) Young fibroblasts have irregularly branched cytoplasmic processes and a large oval-shaped nucleus.
            (b) They are responsible for the synthesis of connective tissue fibers and ground substance.
         (2) **Macrophages** (histiocytes) are almost as common as fibroblasts. They are derived from white blood cells (monocytes) that circulate in the blood and migrate into connective tissue where they contribute to defense against infectious agents. They have the following characteristics:
            (a) They are large, irregularly shaped cells with an oval-shaped nucleus that is sometimes indented and smaller than the fibroblast nucleus.
            (b) They are phagocytic; that is, they have the ability to ingest bacteria, dead cells, and foreign material.
         (3) **Mast cells**, which are found in abundance along blood vessels, may develop from a type of white blood cell called a **basophil**.
            (a) Mast cells are large, oval-shaped cells filled with cytoplasmic granules.
            (b) They produce histamine, a substance that causes dilation of blood vessels, and heparin, an anticoagulant that prevents blood clotting.

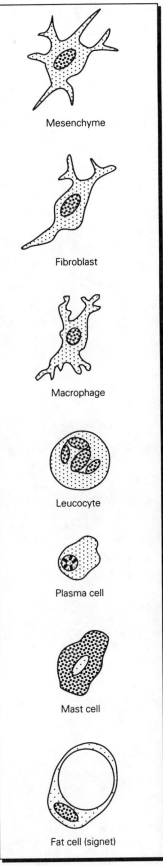

Figure 5–3. *Cells of loose areola connective tissue.*

(4) **Plasma cells** are relatively rare in connective tissue except in areas that are invaded by bacteria.
    (a) Plasma cells are round, with a nucleus that is often described as having a clock-face appearance.
    (b) They synthesize antibodies.
(5) **Adipose cells** are connective tissue cells that are specialized for the storage of fat.
(6) **Leukocytes** (white blood cells) frequently are found in connective tissue, having migrated from the blood vessels.

b. **Distribution**
    (1) Areolar connective tissue is very abundant in the body and is found under epithelial membranes and around glands and ducts.
    (2) It fills the spaces in epithelial organs and muscle and nerves, and ensheathes blood and lymph vessels.

2. **Dense connective tissue** has the same components as areolar connective tissue; however, the collagen and elastic fibers are more closely arranged. Dense connective tissue can be divided into two types: regular and irregular.
    a. **Dense regular connective tissue**
        (1) **Structure.** Completely collagenous (white) fibers are arranged in parallel bundles, which are oriented in one pattern to withstand tension exerted in the parallel direction.
        (2) **Distribution**
            (a) **Tendons** connect muscles to bones.
            (b) **Ligaments** attach bones to bones at a joint.
            (c) **Aponeuroses** are broad, flat tendons that serve to connect a broad muscle to bones.
    b. **Dense irregular connective tissue**
        (1) **Structure.** The predominantly collagenous fibers are arranged in irregular bundles; thus, the tissue can withstand tensions exerted from different directions.
        (2) **Distribution.** This tissue forms the coverings over **muscles** (deep fascia), **bone** (periosteum), **cartilage** (perichondrium), and **capsule wrappings** over organs (e.g., kidneys, liver).

3. **Elastic connective tissue**
    a. **Structure.** Elastic connective tissue contains freely branching elastic (yellow) fibers, which are arranged in parallel strands or in networks. Collagenous fibers and fibroblasts are present in the spaces between the fibers.
    b. **Distribution.** Elastic connective tissue is found in elastic ligaments (between adjacent vertebrae, suspensory ligament of the penis, true vocal cords) and in the walls of the largest arteries and airway passages.

4. **Adipose tissue** is a special kind of connective tissue in which the adipose cells (adipocytes) store fat in the form of large intracellular droplets.
    a. **General characteristics**
        (1) The fat droplet expands the cell so that the cytoplasm becomes reduced to a thin rim around the edge. The nucleus, pushed by the fat droplet, is also flattened and thin.
        (2) A microscopic cross section shows a fat cell with a nucleus that has a "signet-ring" appearance.
        (3) Fat cells are found scattered in loose connective tissue. When many fat cells become organized into a mass surrounded by a meshwork of reticular fibers, the mass is called adipose tissue.
    b. **Function**
        (1) Fat cells **synthesize fats, store them,** and **release them** when necessary as an energy reserve.
        (2) Adipose tissue is a packing around and between organs, bundles of muscle fibers, nerves, and blood vessels.
        (3) Adipose tissue is a poor heat conductor; therefore, it insulates the body from excessive temperature rise or loss.
    c. **Distribution.** Adipose tissue is located anywhere in conjunction with areolar connective tissue; for example,

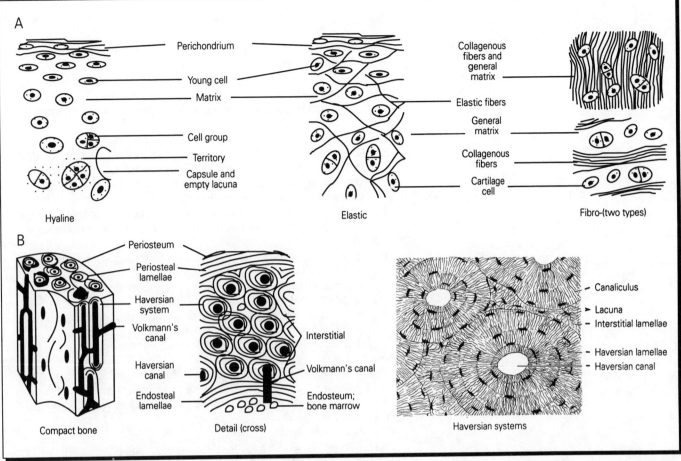

Figure 5–4. *Supporting connective tissue. A, cartilage. B. bone.*

- (1) Beneath the skin
- (2) In the mesenteries and the mediastinum
- (3) Around the kidneys and adrenal glands
- (4) On the surface of the heart
- (5) In bone marrow

5. **Reticular connective tissue** is composed of thin fibers that branch extensively and unite to form a delicate network that supports soft organs. In wound healing, reticular fibers develop first, then thicken to become typical collagenous fibers.

E. **Supporting connective tissues** (Figure 5–4). Cartilage and bone have the tensile strength provided by collagen fibers and additional materials in the ground substance, which provide rigidity and weight-bearing properties.

1. **Cartilage** contains a firm, rubber-like, mixture of glycosaminoglycans and proteins in the ground substance, which gives the tissue its plastic-like character. Most of the cartilage that develops in the body is replaced by bone. The persisting cartilage is divided into three types.
   a. **Hyaline cartilage**
      (1) **Distribution.** Hyaline cartilage occurs primarily in areas where strong support is necessary, but where flexibility is desirable; for example,
         (a) Ends of long bones (articulating surfaces)
         (b) Anterior ends of ribs
         (c) External ear
         (d) Fetal skeleton
         (e) Nose, larynx, trachea, and bronchi
      (2) **Structure**
         (a) **Chondrocytes** are mature cartilage cells. They occupy small spaces (**lacunae**) in a clear, glassy matrix.

i. **Chondroblasts**, which arise from mesenchyme, are immature chondrocytes. They proliferate and produce the matrix.
ii. As the intercellular matrix increases, the chondroblasts are isolated in the lacunae and become mature chondrocytes. Chondrocytes continue to divide and produce additional cartilage.
- (b) The **perichondrium** is a well-vascularized, dense connective tissue membrane that surrounds hyaline cartilage (except at the articular cartilage of bones.) The perichondrial cells adjacent to the cartilage can differentiate into chondroblasts and chondrocytes to form new cartilage.
- (c) Cartilage matrix lacks blood vessels; therefore, nutrients and gases must seep through to the chondrocytes from the perichondrium.
- (3) Growth
    - (a) **Interstitial growth** (an expansion of cartilage from within) occurs when the young chondrocytes divide, deposit matrix around themselves, and become progressively separated.
    - (b) **Appositional growth** (from the outside on top of previously existing layers) occurs when the innermost cells of the perichondrium differentiate first into **chondroblasts**, surround themselves with matrix, and become chondrocytes.
- b. **Fibrocartilage**
    - (1) **Distribution.** Fibrocartilage occurs in locations where tougher support or tensile strength than hyaline cartilage is necessary. It unites bones at joints where movement is limited; for example,
        - (a) Bones of the skull
        - (b) Pubic symphysis
        - (c) Intervertebral discs
    - (2) **Structure.** The chondrocytes often occur in groups or rows between numerous bundles of collagenous fibers.
- c. **Elastic cartilage** has a predominance of elastic fibers. This permits the stiffness of cartilage but the elasticity of movement.
    - (1) **Distribution.** Elastic cartilage occurs in the external ear, epiglottis, and some of the cartilages of the larynx.
    - (2) **Structure.** Elastic cartilage is similar in structure to hyaline cartilage, with the addition of a meshwork of branching, elastic fibers.

2. **Bone (osseous tissue)**, like cartilage, consists of cells, fibers, and matrix. However, it is much stronger than cartilage because the matrix contains inorganic calcium and phosphate salts that give it hardness and weight-bearing capacity. Unlike cartilage cells, bone cells have a rich blood supply through **canaliculi**, which are small channels that penetrate the calcified matrix.
    - a. **Types of cells**
        - (1) **Osteoblasts** synthesize the organic components of bone. They are responsible for forming new bone during growth, repair, and remodeling of bone.
        - (2) **Osteocytes** are mature cells that occupy lacunae within the matrix.
        - (3) **Osteoclasts** are the cells responsible for destroying and remodeling bone.
    - b. **Types of bone tissue.** Based on porosity, bone may be classified as compact bone or cancellous bone.
        - (1) **Compact (dense) bone** is solid, with the exception of the microscopic canaliculi. Compact bone is external in a long bone.
        - (2) **Cancellous bone** (also called spongy bone, or trabecular bone) has a lattice-like structure composed of thin, bony bars, or **trabeculae**, which enclose the marrow spaces. Cancellous bone is internal to compact bone.
    - c. **Structure of compact bone**
        - (1) The basic structure of adult compact bone is the **haversian system (osteon)**.
        - (2) Each haversian system has a **central haversian canal** that is surrounded by lamellae, which are concentric rings of intercellular substance.

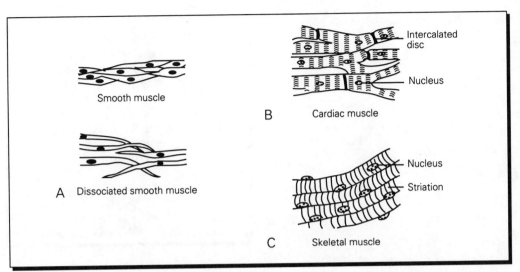

Figure 5-5. *A typical neuron. Structure of the three muscle types. A,* Smooth muscle cells are oval and elongate. *B,* Cardiac muscle is composed of irregular, branched cells. Intercalated discs are characteristic. *C,* Skeletal muscle consists of large, multinucleated, elongated fibers.

       (3) **Lacunae** containing **osteocytes** and **canaliculi** lie within the lamellae. Canaliculi branch from all surfaces of a lacuna to connect other canaliculi and with the haversian canals or the Volkmann's canals.
          (a) Volkmann's canals, which run at right angles to the haversian canals, pierce through lamellae to cross connect haversian systems.
          (b) Haversian canals conduct blood vessels, lymphatics, and nerves through compact bone.
       (4) **Periosteum and endosteum.** Bone is covered externally and internally by layers of bone-forming cells and dense connective tissue called **periosteum** and **endosteum**, respectively.
       (5) Bone growth can only be appositional on previously existing layers of bone.
   F. **Blood** is a specialized connective tissue in which the cells do not make the fluid matrix (plasma) in which they are suspended. (Blood, lymph, and blood-forming tissue will be described in Chapter 11.)

IV. **MUSCLE TISSUE** is the "flesh" of the body and comprises much of the walls of hollow organs and vessels of the body. The cells of muscle tissue, which are called fibers, are highly specialized for contractility. (For further discussion of muscle tissue, see Chapter 8.)

   A. **General characteristics**
      1. The elongated cells (fibers) contain many myofibrils, which are composed of contractile myofilaments.
      2. The nuclei of muscle cells are well defined.
      3. The cytoplasm is called **sarcoplasm**, the cell membrane is called the **sarcolemma**, and the smooth endoplasmic reticulum is called the **sarcoplasmic reticulum**.
      4. Muscle fibers can enlarge; however, except for the limited ability of smooth muscle cells in certain locations, they do not divide to proliferate after birth.
   B. **Classification**
      1. Functionally, muscle is classified as **voluntary** (controlled by the will) or **involuntary** (not under conscious control).
      2. Structurally, muscle is classified as **striated** (with cross-striping) or **unstriated** (unstriped). Striations are seen when the fibers are viewed in a microscopic section.
      3. On the basis of structure and function, muscle tissue is classified into the following types: smooth, skeletal, and cardiac (Figure 5–5).

Figure 5-6. *A typical neuron.*

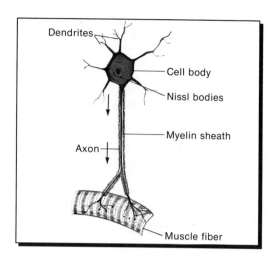

   a. **Smooth muscle** is **involuntary** and **nonstriated**.
      (1) **Distribution.** Smooth muscle occurs in the following locations:
         (a) Walls of hollow organs
         (b) Walls of ducts and vessels
         (c) Organs such as the skin, spleen, and penis.
      (2) **Structure**
         (a) Spindle-shaped cells vary greatly in length but are shorter in length and smaller in diameter than skeletal muscle cells.
         (b) Each cell contains a **centrally located nucleus.**
         (c) The fibers are bundled together in units or sheets (layers).
   b. **Skeletal muscle** is **voluntary** and **striated**.
      (1) **Distribution.** Individual fibers are aggregated into bundles to form functional groups called muscles, which are attached to the skeleton and are responsible for movement.
      (2) **Structure**
         (a) Individual fibers may be from 1 mm to 40 mm in length.
         (b) A fiber has **many nuclei** found under the sarcolemma at the periphery of the cell.
      (3) Cross-striations are composed of alternating light **I-bands** (isotropic) and darker **A-bands** (anisotropic).
   c. **Cardiac muscle** is **involuntary** and **striated**.
      (1) **Distribution.** Cardiac muscle occurs only in the heart.
      (2) **Structure**
         (a) Cardiac muscle fibers **branch** and form networks.
         (b) **Nuclei are single** and centrally located.
         (c) The cross-striations are closer together and not as clearly visible as in skeletal muscle.
         (d) **Intercalated discs**, which are heavier cross-bands visible under the microscope, are a characteristic feature of cardiac muscle. The disks are special **junctions** between cardiac muscle cells and represent areas of low electrical resistance to the spread of contraction.

V. **NERVOUS TISSUE.** As a communications network, nervous tissue is specialized to receive stimuli and conduct impulses to all parts of the body. (See Chapter 9 for a complete discussion of the nervous system.)

   A. **Structure.** Nervous tissue consists of two types of cells: neurons and neuroglia cells. Neurons are the structural and functional units of the nervous system (Figure 5-6).
      1. **Neurons**, or nerve cells, contain numerous processes called nerve fibers. Neurons consist of the following components:

a. The cell body of a neuron is the **perikaryon**, which contains the nucleus.
   b. Most neurons have many **dendrites**, which carry impulses to the perikaryon.
   c. Each neuron has only one **axon**, which carries an impulse away from the perikaryon.
  2. **Neuroglia** cells support nervous tissue and supply nutrients to neurons by connecting them to blood vessels.

B. **Divisions.** Anatomically, the nervous tissue consists of the **central nervous system** (i.e., the brain and spinal cord) and the **peripheral nervous system** (i.e., nerve fibers and groups of nerve cells called ganglia).

Study Questions

Directions: Each question below contains four suggested answers. Choose the **one best** response to each question.

1. Which of the following statements is most characteristic of epithelium?
    (A) It has abundant intercellular material.
    (B) It has no regenerative ability.
    (C) It gives support to other tissues and organs.
    (D) It contains tightly adhering cells.

2. All of the following specializations may be found on the apical surface of epithelial cells EXCEPT
    (A) microvilli
    (B) desmosomes
    (C) cilia
    (D) stereocilia

3. Epithelium that functions in a protective role in the body is likely to be
    (A) stratified
    (B) simple
    (C) phagocytic
    (D) highly vascularized

4. The most likely role of tight intercellular junctions such as zonula occludens in intestinal epithelium is to
    (A) prevent passage of unwanted substances into the body by way of an intercellular route
    (B) provide structural support for the epithelial cells
    (C) enhance the absorption of small molecules and ions
    (D) assist the movement of substances down the intestinal tract

5. Tendons and ligaments are composed primarily of
    (A) dense fibrous connective tissue
    (B) adipose tissue
    (C) muscle tissue
    (D) loose connective tissue

6. What type of epithelial tissue lines the urinary bladder and many of the urinary passages?
    (A) cuboidal
    (B) transitional
    (C) pseudostratified
    (D) columnar

7. All of the following histologic comparisons are true EXCEPT
    (A) Stratified squamous epithelium is more abundant in the body than stratified columnar epithelium.
    (B) There are more blood vessels penetrating epithelial tissue than connective tissue.
    (C) More heparin is produced by mast cells than by macrophages.
    (D) A skeletal muscle fiber is larger than a smooth muscle fiber.

Questions 8–15. Each question refers to characteristics of epithelial tissue and/or its cells and characteristics of connective tissue and/or its cells. Use the following key:

   (A) The characteristic applies to epithelium.
   (B) The characteristic applies to connective tissue.
   (C) The characteristic applies to **neither** epithelium nor connective tissue.
   (D) The characteristic applies to **both** epithelium and connective tissue.

8. has cilia and microvilli

9. contains intercalated discs

10. conducts impulses

11. has a protective function

12. contains abundant intercellular ground substance

13. contains phagocytic cells

14. stores fat

15. produces glandular secretions

Answers and Explanations

1. **The answer is D.** (II B 1 b–d) Characteristically, epithelium has no blood supply, has great regenerative ability, and contains closely packed cells with little intercellular material between them. By definition, sheets of epithelial cells cover body surfaces (e.g., skin) or line body cavities and epithelial tissues are part of various glands. Epithelium is supported by connective tissue.

2. **The answer is B.** (II C 2 a, b) Desmosomes are lateral specializations of epithelial cells; they are individual sites of cell adherence. Cilia, microvilli, and sterocilia are apical specializations. Cilia are associated with moving substances along the cell surface; microvilli with absorption, and stereocilia with absorption and possibly secretion.

3. **The answer is A.** (II C 1 b) Stratified epithelium is better able to withstand wear and tear than is simple epithelium. It has two or more layers of cells with the deepest layer touching the basal lamina. Simple epithelium occurs where secretion, absorption, and diffusion must take place. Epithelium is nourished by the blood vessels in underlying connective tissue; it has no blood supply of its own.

4. **The answer is A.** (II C 2 a (1), (2)) Tight intercellular junctions function to seal off any intercellular pathway to the passage of materials. This is of particular importance in absorptive epithelia such as the intestine because passage **across** the apical surface of epithelial cells allows selective absorption.

5. **The answer is A.** (III D 3 a) Tendons of muscles and ligaments of joints are composed of dense regular connective tissue, which primarily consists of parallel bundles of collagenous fibers. These fibers impart great strength to the connective tissue.

6. **The answer is B.** (II C 3 g) Transitional epithelium is a specialized stratified cuboidal epithelium, which is limited in distribution to the urinary tract. This tissue is found in the bladder, permitting the organ to expand. The cell shape changes from cuboidal to squamous, depending on whether the bladder is empty or full.

7. **The answer is B.** (I B 1 b; II C 3 e, f; III D 2 a (2), (3); IV B 3 a, b) Stratified epithelium is rare in the body, limited to epithelial junctions (e.g., anorectal junction), larynx, pharynx, and in large excretory ducts (e.g., urethra). In contrast, stratified squamous epithelium forms the epidermis of the skin and lines body orifices, serving a protective function.

    Epithelial tissue has no blood supply. It is nourished by underlying connective tissue, which serves as a pathway for blood vessels.

    Mast cells produce heparin, which prevents blood clotting. Macrophages are protective cells found in connective tissue. Among the function of macrophages is the ingestion of microorganisms.

    Individual skeletal muscle fibers may attain a length of 40 μm; the typical length of a smooth muscle fiber is 50 μm to 200 μm. Skeletal muscle fibers are also about 20 times wider than smooth muscle fibers.

8–15. **The answers are 8–a; 9–c; 10–c; 11–a; 12–a; 13–b; 14–b; 15–a.** (II C 2 b; IV B 3 c; V A, B; II C; II A; II C 4; II D 5; II D) Epithelial cells only may have cilia or microvilli on their apical surfaces. Intercalated discs are found in cariac muscle tissue, which is neither epithelial nor connective tissue. Epithelium has no blood supply. Nerve tissue and muscular tissue are specialized to conduct impulses. Epithelium is protective in function; connective tissue is supporting and connecting tissue. Only connective tissue has abundant intercellular ground substance; it has relatively few cells located some distance apart. Epithelial tissue consists of many cells packed closely together. Only connective tissue such as blood and areolar tissue contains phagocytic cells. Adipose tissue, which is a specialized form of connective tissue, stores fat. Glandular secretions are produced only by derivatives of epithelial tissue.

6 Integumentary System

I. **INTRODUCTION.** The integument forms the outer covering of the body. It consists of the skin and certain specialized skin derivatives, which include the hair, nails, and several kinds of glands.

   A. **Components of the integument**

   1. **Skin** (Figure 6–1) is the largest organ of the body. It weighs about 9 lb and covers an area of about 18 sq ft in a 150 lb male.
      a. **Epidermis** is the upper, or outer, layer. It consists of epithelial tissue.
      b. **Dermis** is the connective-tissue lower layer. It binds the epidermis to the underlying structures.
   2. **Fingernails and toenails** are skin specializations found only in humans and other primates.
   3. **Hair** is a skin specialization characteristic only of mammals.
   4. **Skin glands** in humans include **sebaceous glands**, **sweat glands**, and **mammary glands**, which are modified sweat glands.

   B. **Functions of the integument**

   1. **Protection.** The skin protects the body from microorganisms, fluid loss or gain, and from other mechanical and chemical irritants. Melanin pigment in the skin provides further protection against the sun's ultraviolet rays.
   2. **Body temperature regulation.** Both the blood vessels and the sweat glands in the skin function in the maintenance and regulation of the body temperature.
   3. **Excretion.** Fatty substances, water, and ions such as Na^+ are excreted through the glands of the skin.
   4. **Metabolism.** In the presence of sunlight or ultraviolet radiation, the synthesis of vitamin D, essential for bone growth and development, is initiated from a precursor molecule (7-dehydrocholesterol) found in the skin.
   5. **Communication.**
      a. All stimuli from the environment are received through the skin by means of numerous specialized receptors that detect sensations related to temperature, touch, pressure, and pain.
      b. Skin is the medium of facial expression and vascular reflexes important in communication.

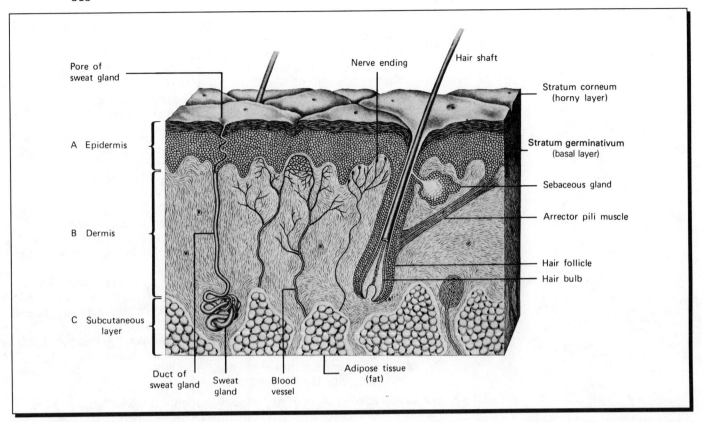

Figure 6–1. *Three-dimensional diagram of the skin in cross section.* Reproduced by Permission. *Biology of Women* 3E, by Ethel Sloane. Delmar Publishers, Inc., Albany, NY; Copyright 1993.

II. SKIN

A. Layers

1. The **epidermis** is the outer portion of the skin. It is stratified squamous, keratinized, epithelial tissue; it lacks blood vessels; and the cells are packed closely together. The epidermis is thickest on the palms of the hands and the soles of the feet, where it is stratified into the following five layers:
   a. **Stratum basale (germinativum)** is a single layer of cells resting on the connective tissue of the underlying skin layer, the dermis. Rapid cell division occurs in this layer, and the new cells are pushed up into the next layer.
   b. **Stratum spinosum** is the spiny or prickle-cell layer, so-called because the cells appear to be held together by spinelike projections. The spines are portions of intercellular contacts called desmosomes.
   c. **Stratum granulosum** consists of three to five cell layers or rows of cells filled with keratohyalin granules, precursors to the formation of **keratin**.
      (1) **Keratin** is a tough, resilient protein that waterproofs and protects exposed skin surfaces.
      (2) The keratin in epidermis is soft keratin, low in sulfur, in contrast to the hard keratin occurring in hair and nails.
      (3) As keratohyalin and keratin accumulate, the cell nuclei disintegrate, leading to cell death.
   d. **Stratum lucidum** is a clear, translucent layer of flattened, nonnucleated, dead or dying cells, four to seven cells deep.
   e. **Stratum corneum**, the uppermost layer of the epidermis, consists of 25 to 30 layers of heavily keratinized, nonliving scales, which become increasingly flattened as they approach the surface. (Thin epidermis, which covers all of the body except for the palms and soles, consists only of the basale and corneum layers.)

(1) The exposed surface of the stratum corneum undergoes constant wearing-away, or desquamation.
(2) There is constant renewal of the desquamated cells from cell division in the basale layer. The cells move upward to the surface, undergo keratinization, and die. Thus, the exposed surfaces of the body actually are enclosed in a dead husk of epidermal cells.
(3) The epidermis is totally replaced from the bottom up every 15 to 30 days.

2. The **dermis** is separated from the epidermis by the basement membrane, or lamina. It is composed of two layers of connective tissue.
   a. The **papillary layer** is loose, areolar connective tissue with fibroblasts, mast cells, and macrophages. It contains many blood vessels, which provide nourishment to the epidermis above.
      (1) Finger-like **dermal papillae**, which contain tactile **sensory receptors** and **blood vessels**, project upward into the epidermis.
      (2) In the palms and soles, papillae are numerous and tall with as many as 65,000/sq in.
      (3) The patterns of ridges and whorls visible on the palms and soles are unique to each individual and reflect the arrangement of these dermal papillae. The purpose of the ridges is to facilitate the grip by increasing friction.
   b. The reticular layer is deeper than the papillary layer. It is composed of irregular, dense connective tissue with collagen and elastic fibers. With age, normal deterioration of the bundles of collagen and elastic fibers results in skin wrinkling.

3. The **subcutaneous layer or hypodermis** (superficial fascia) binds the skin loosely to underlying organs. It contains varying numbers of fat cells, depending on the area of the body and the nutrition of the individual, and has many blood vessels and nerve endings.

B. **Color.** Differences in skin coloration are due to the following factors:

1. **Melanocytes**, located in the stratum basale, produce the pigment, **melanin**, which is responsible for brown-to-black tones.
   a. To a limited extent, melanin protects against the damaging ultraviolet rays of the sun. An increased production of melanin (tanning) occurs upon exposure to the sun.
   b. The number of melanocytes (about 1,000/mm^2 to 2,000/mm^2) does not vary among races, but genetic differences in greater melanin production and wider pigment dispersal result in racial differences.
   c. The nipples, areolae, circumanal region, scrotum, penis, and labia majora are areas of increased pigmentation; the palms and soles contain little pigment.

2. **Blood** in the dermal vessels underlying the epidermis shows through and results in pink tones. This is more apparent in Caucasian skin.

3. The presence and amount of the yellow pigment, **carotene**, mainly found in the stratum corneum and in the fat cells of the dermis and hypodermis, causes some skin tone differences.

III. **SKIN DERIVATIVES.** Nails, hair, and sebaceous and sweat glands are derived from the epidermis.

A. **Nails.** Fingernails and toenails are protective plates that arise from downgrowths of the epidermis into the dermis.

1. The nail is a curved plate of hard keratin that rests on the **nail bed,** which is supplied with blood vessels.

2. The **nail body** grows from the **nail root**, which is embedded in the skin. Nail growth is approximately 0.5 mm per week, faster in the summer than in the winter.

3. The **cuticle (eponychium)** is a curved fold of epidermis that covers the root. The **hyponychium** is thickened stratum corneum under the free border of the nail.
4. The **lunula** (half-moon) is a whitish, opaque area that appears closest to the cuticle.

B. **Hair.** Hairs, or pili, are present almost everywhere on the body, but most are tiny, colorless **vellus** hairs, or down. **Terminal** hairs are coarse and visible. They are confined to the scalp, eyebrows, and eyelashes until puberty, when they replace vellus hairs in the axillary and pubic areas (and on the face in males) as part of the secondary sex characteristics (see Chapter 18).

1. Hairs arise from hair **follicles**, which developed before birth by downgrowths from the epidermis into the dermis (Figure 6–2).
   a. The tubular hair follicle swells at its base as the **hair bulb**. The hair bulb is invaginated by a mass of loose connective tissue, blood vessels, and nerves called the **dermal papilla**, which provides nutrients to the growing hair.
   b. Cells of the hair bulb directly over the papilla are known as the **germinal matrix** of the hair, and are analogous to the stratum basale cells of the epidermis. Nourished by the blood vessels in the papilla, the cells of the germinal matrix divide and push up toward the surface of the skin to become the completely keratinized hair.
2. The hair consists of a **root**, the portion enclosed within the follicle, and the **shaft**, the portion above the surface. The root and shaft are made up of three layers of epithelium.
   a. **Cuticle** is the outer layer of scaly, dead cells.
   b. **Cortex**, the middle keratinized layer, forms the major portion of the shaft. It contains variable amounts of pigment, which determines hair color.
   c. A **medulla**, or central axis, consists of two to three layers of cells. It is poorly developed and frequently absent, especially in blond hair.

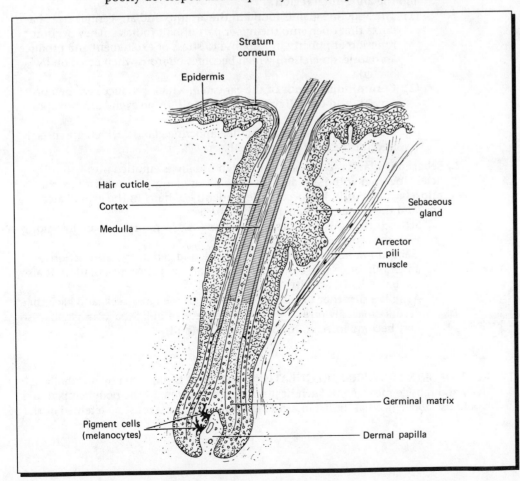

Figure 6–2. *Diagram of a hair within its follicle.* The follicle expands at its base to form the hair bulb, which is invaginated by the dermal papilla. The cells of the germinal matrix directly above the papilla give rise to the hair. Melanocytes in the germinal matrix are responsible for hair color. Reproduced by permission. *Biology of Women* 3E, by Ethel Sloane. Delmar Publishers, Inc., Albany, NY; Copyright © 1993.

3. The **arrector pili** muscle is a thin band of smooth muscle associated with the hair follicle. Contraction causes the hair to stand on end ("goose bumps") and results in secretion of the sebaceous gland. Each hair follicle contains one or more sebaceous glands (see section III C 2).
4. Hair growth is cyclic.
   a. There are definite periods of growth followed by a resting phase, when a hair reaches its limit in length.
      (1) During the resting period, the base of the hair becomes a keratinized club-shaped mass that remains attached to the follicle.
      (2) After a resting period, a new hair bulb forms below the old clubbed hair. The new hair pushes out old hair, causing it to be shed.
      (3) At any given time, 90% of the hairs on the head are actively growing and 10% are resting.
   b. Scalp hair grows from 2 to 6 years and rests 3 months before being shed.
   c. Body hair grows about 0.05 in./week. Scalp hair takes about 7 weeks to grow an inch.
   d. Baldness is a progressive deterioration of follicles. It is more prevalent in males, in whom it is a sex-influenced genetic trait that is expressed only when male hormone is present in the body.

C. **Glands of the skin**
   1. **Sweat (sudoriferous) glands** are divided into two types on the basis of structure and location.
      a. **Eccrine sweat glands** are simple, coiled, tubular glands that are not associated with hair follicles. They have widespread occurrence on the body, especially on the soles, palms, and forehead. The secretion (sweat) is watery and aids in evaporative cooling of the body to maintain body temperature.
      b. **Apocrine sweat glands** are specialized large, branched sweat glands with limited distribution. They are found in the axilla, mammary areola, and the anogenital region.
         (1) The apocrine glands found in the armpits and anogenital areas have ducts that open into the upper part of hair follicles. They begin to function at puberty in response to stress or excitement and produce an odorless secretion, which becomes odorous when acted on by bacteria.
         (2) **Ceruminous glands** of the ear canal, which produce cerumen or earwax, and the **ciliary glands of Moll** in the eyelid are also apocrine glands.
         (3) **Mammary glands** are modified apocrine glands, which are specialized for the production of milk.
   2. **Sebaceous glands** secrete **sebum**, which usually is emptied into hair follicles. Together with hairs and apocrine sweat glands, they constitute the **pilosebaceous unit**, but occur independently of hairs in the genital area, lips, and the nipples and areolae of the breasts.
      a. Sebaceous glands are holocrine glands (i.e., secretory cells are lost along with the sebum secretion).
      b. **Sebum** is a mixture of fats, waxes, oils, and cell debris. It functions as an emollient or skin softener and is a barrier against evaporation. It also has bactericidal activity.
      c. Acne is a **disorder of sebaceous glands** of the face, neck, and back that occurs especially during the second decade of life. Sebaceous glands also may become infected to cause furuncles (boils).

IV. **ROLE OF SKIN IN THERMOREGULATION.** Body heat is produced by metabolic activity and muscle movement. Such heat must be dissipated, or the body temperature would rise above normal limits; in a cold environment, heat must be retained or the body temperature would drop below normal limits.

- A. **Heat loss by the skin** occurs through **evaporation** of water secreted by the sweat glands and through **insensible perspiration** (i.e., the diffusion of water molecules through the skin).
    1. In hot, humid weather, sweating is more profuse, but the rate of evaporation is greatly reduced, with resulting discomfort. Thus, sweating as a cooling mechanism is efficient only in lower humidity.
    2. Sweating is controlled by the nervous system, which responds to overheating or cooling of the blood.
- B. **Heat retention** is a function of the skin and the adipose tissue in the subcutaneous layer. **Fat is a heat insulator** for the body and the degree of insulation is dependent on the amount of adipose tissue.
- C. The **blood vessels in the dermal papillae** are under control of the nervous system.
    1. When the blood vessels dilate, blood flow to the skin surface increases, permitting conduction of heat to the exterior.
    2. The blood vessels constrict to decrease blood flow to the skin surface in order to maintain central body heat.

Study Questions

Directions: Each question below contains four suggested answers. Choose the **one best** response to each question.

1. All of the following are true statements about the function of the integument EXCEPT
    (A) It prevents water loss.
    (B) It protects the body against invasion by disease organisms.
    (C) It synthesizes vitamin D necessary for nervous system functioning.
    (D) It helps maintain a constant body temperature.

2. Mitotic division occurs in which layer of the epidermis?
    (A) stratum granulosum
    (B) stratum basale
    (C) stratum corneum
    (D) stratum lucidum

3. A microscopic bit of material from the human body is analyzed and found to contain keratin with a low sulfur content. The material must be a particle from
    (A) the epidermis
    (B) a hair shaft
    (C) a fingernail or toenail
    (D) the hypodermis

4. The pink tones in Caucasian skin are due largely to the presence of
    (A) melanocytes in the epidermis
    (B) carotene
    (C) blood vessels in the dermis
    (D) keratin

5. Finger-like projections of the dermis that contain blood vessels and sensory receptors are called
    (A) arrector pili
    (B) follicles
    (C) germinal matrix
    (D) dermal papillae

6. Which of the following glands would secrete their product in response to a high environmental temperature?
    (A) eccrine sweat glands
    (B) apocrine sweat glands
    (C) sebaceous glands
    (D) ceruminous glands

7. Pimples and boils are associated with which of the following skin structures?
    (A) sweat glands
    (B) sebaceous glands
    (C) dermal papillae
    (D) melanocytes

Answers and Explanations

1. **The answer is C.** (I B 1–5) The integument synthesizes vitamin D in the presence of sunlight or ultraviolet light from a precursor molecule. The skin protects the body against entrance of microorganisms or mechanical and chemical irritants, and assists in body thermoregulation.

2. **The answer is B.** (II A 1 a–e) The epidermis grows as a result of cell division in the stratum basale. In thick skin the cells move upward to form the stratum spinosum, the stratum granulosum filled with keratohyalin granules, the stratum lucidum, and, finally, the nonliving stratum corneum.

3. **The answer is A.** (II A 1 c) Soft keratin with a low sulfur content is found in epidermis, in contrast to hard keratin with a high sulfur content found in hair and nails. The hypodermis contains no keratin.

4. **The answer is C.** (II B 1–3) Pink tones in Caucasian skin result from blood in the dermal vessels showing through the epidermis. Melanin, produced by melanocytes, is responsible for brown and black tones and carotene in the epidermis results in yellowish coloration. Keratin is not associated with skin color.

5. **The answer is D.** (III B 1 a) Dermal papillae, which contain capillary loops and sensory receptors, project upward into the epidermis. The arrector pili are smooth muscles associated with a hair follicle; they enclose a sebaceous gland. The germinal matrix refers to the cells of the hair bulb from which the hair grows.

6. **The answer is A.** (III C 1, 2; IV A) The secretion of the eccrine sweat glands is important in evaporative cooling of the body to maintain body temperature. Apocrine sweat glands, concentrated in the armpits and anogenital area, increase their activity in response to sexual excitement or emotional stress. Ceruminous glands, located in the external ear canal, produce earwax. Sebaceous glands produce sebum, which lubricates the surface of the skin and hair.

7. **The answer is B.** (III C 2 c) Acne and boils are associated with dysfunction or infection of sebaceous glands. Sweat glands function in thermoregulation. Dermal papillae extend into the epidermis to form characteristic whorls and ridges; they provide the pathway for nutrition of the epidermis. Melanocytes function in skin pigmentation.

7 Skeletal System

I. GENERAL INTRODUCTION TO THE SKELETAL SYSTEM

A. **Organization of the skeletal system.** The adult human skeleton consists of the bones (about 206) that make up the solid framework of the body. Although it is composed mainly of bone, the skeleton is completed in certain areas by cartilage. For purposes of study, the skeleton is organized into the **axial skeleton**, the **appendicular skeleton**, and the **joints** between bones. (See Color Plates 1 and 2.)

1. The **axial skeleton** is composed of the 80 bones that make up the long axis of the body and protect the organs of the head, neck, and torso.
   a. The **vertebral column** consists of 26 vertebrae separated by intervertebral discs.
   b. The **skull** is balanced on the vertebral column.
      (1) **Cranial bones** enclose and protect the brain and the special sense organs.
      (2) **Facial bones** give shape to the face and contain the teeth.
      (3) Six **auditory (ear) ossicles** are involved in transmission of sound.
      (4) The **hyoid bone**, which supports the tongue and larynx and assists in swallowing, is a separate bone of the skull.
   c. The **thoracic cage** (rib cage) includes the **ribs** and **sternum** and encloses and protects the thoracic organs.
2. The **appendicular skeleton** is composed of the 126 bones that make up the arms, legs, and the bony pectoral and pelvic girdles that anchor the arms and legs to the axial skeleton.
3. **Joints** are the articulations between two or more bones.

B. **Functions of the skeletal system**

1. **Support and shape** are provided to the body by the bones.
2. **Movement.** Bones articulate with other bones at joints and act as levers. As the muscles (which are anchored to bones) contract, the force applied to the levers results in movement.
3. **Protection.** The skeletal system protects the delicate soft organs of the body.
4. **Blood cell formation (hematopoiesis).** Red bone marrow, which is found in the adult in the sternum, ribs, bodies of the vertebrae, flat bones of the cranium, and the end portions of the long bones, is the site of production of red blood cells, white blood cells, and platelets of the blood.

5. **Storage reservoir for minerals.** Bone matrix is composed of approximately 62% inorganic salts, primarily calcium phosphate and calcium carbonate with lesser amounts of magnesium, chloride, fluoride, and citrate. The skeleton contains 99% of the body's calcium. The calcium and phosphorus in bone are stored to be withdrawn and utilized for body functions; they are replenished through nutrition.

C. **Composition of bone tissue**
1. Bone consists of cells and an extracellular matrix. The cells are **osteocytes, osteoblasts,** and **osteoclasts.** (See Chapter 5.)
2. **Bone matrix** is composed of organic **collagen fibers** imbedded in a ground substance and inorganic **calcium and phosphorus bone salts.**
    a. **Ground substance** of bone consists of a type of proteoglycans composed primarily of chondroitin sulfate and small amounts of hyaluronic acid associated with protein.
    b. **Bone salts** are in the form of calcium phosphate crystals called **hydroxyapatite** with the molecular formula $3Ca_3(PO_4)_2 \cdot Ca(OH)_2$.
    c. The association of collagen and hydroxyapatite crystals is responsible for the great tensile and compressional strength of bone. Bones are constructed the same way as reinforced concrete: the collagen fibers are like the rods of steel; the bone salts are like the cement, sand, and rock in concrete.
3. The two types of bone tissue, **cancellous (spongy) bone** and **compact bone,** were described in Chapter 5. Both compact and cancellous bone are the same in composition, but they differ in their porousness.
    a. **Compact bone** is tightly packed tissue that primarily is found as a layer over cancellous bone tissue. Its porosity depends on microscopic channels (**canaliculi**) containing blood vessels, which are linked to **haversian canals.**
    b. **Cancellous bone** consists of delicate, irregular bony bars that branch and intersect to form a network of bony spicules with cavities that contain marrow.
    c. The relative amounts of cancellous and compact bone varies in different bones and in different parts of the same bone. For a description of the organization of compact bone into lamellae and haversian systems, see Chapter 5.

D. **Anatomy of a typical long bone.** A long bone such as the femur has the following features (Figure 7–1):
1. The **diaphysis** (shaft) consists of a thick compact bone cylinder that encloses a large central **medullary** or **marrow cavity.**
    a. The marrow cavity contains yellow (adipose) bone marrow or red marrow, depending on the age of the individual.
    b. **Endosteum** lines the marrow cavity. It is composed of vascular areolar connective tissue.
    c. **Periosteum** covers the diaphysis.
        (1) Periosteum is a connective tissue sheath composed of two layers: the **outer layer** is dense, fibrous, connective tissue; the **inner layer** is osteogenic (bone forming) and consists of a single layer of osteoblasts.
        (2) **Sharpey's fibers** (connective tissue fibers) bind the periosteum down to the bone.
        (3) Periosteum envelops all bone except sesamoid bones, on articular surfaces and around tendon and ligament insertions.
        (4) It serves the following functions:
            (a) **Growth** of bone in width by means of the deeper, more cellular osteogenic layers
            (b) **Nutrition** of bone because the periosteum is highly vascularized and is the pathway for blood vessels to penetrate bone
            (c) **Regeneration** of bone following fractures
            (d) **Means of attachment** for tendons and ligaments

Figure 7–1. *Diagram of the parts of a typical long bone.*

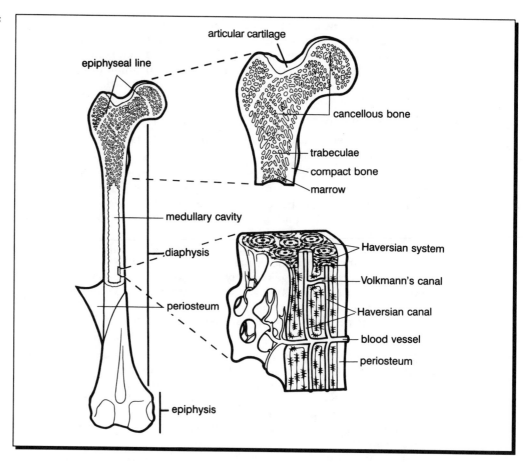

2. The **epiphyses** are the expanded ends of the bone with which the marrow cavity freely communicates.
    a. An epiphysis consists of internal cancellous bone, which is surrounded by compact bone and covered with **articular cartilage** (hyaline cartilage).
    b. The articular cartilages, which are located at the ends of articulating surfaces of bones, are lubricated by synovial fluid from the joint cavity. These cartilages allow for smooth movement at joints.

E. **Development of bone.** Osteogenesis (the growth and development of bone) is the process by which bone forms in the body. Because of the hard matrix in bone, interstitial growth (from within), such as occurs in cartilage, is impossible and bone forms by replacement of a preexisting tissue. The two types of bone formation are **intramembranous** ossification and **endochondral (intracartilaginous)** ossification.

  1. **Intramembranous ossification** occurs directly within the mesenchyme tissue of the fetus and involves replacement of preexisting membranous (mesenchyme) tissue. It gives rise to most of the flat bones of the skull, the so-called "membrane bones."
      a. In areas where bone is to form, groups of star-shaped mesenchyme cells differentiate into **osteoblasts** and form **ossification centers** (the earliest centers are in the 8th week of fetal life).
      b. The osteoblasts secrete an organic not-yet-calcified matrix called **osteoid**.
      c. Calcification of the osteoid mass by deposition of bone salts follows and traps the osteoblasts and their cell processes.
          (1) Once encapsulated by calcified matrix, the osteoblasts become **osteocytes**, which become isolated in **lacunae** and no longer secrete intercellular substance.
          (2) The channels left by the osteoblast processes become the **canaliculi**.

d. The islands of developing bone, or **spicules**, fuse together and branch to form the spongy, or **trabecular**, network of cancellous bone.
e. The result of early intramembranous ossification is the formation of vascular, primitive bone, surrounded by condensed mesenchyme that later will become periosteum. Because the collagen fibers run in all directions, the new bone is often called woven bone.
    (1) In areas of primitive spongy bone where compact bone is destined to form, the trabeculae become thicker and gradually obliterate the intervening connective tissue.
    (2) Where bone is to remain cancellous, the connective tissue spaces are replaced by bone marrow.
2. **Endochondral ossification** occurs by replacement of a preformed cartilaginous model. Most of the bones of the skeleton are formed in this process, which occurs within a small hyaline cartilage model in the fetus (Figure 7–2).
    a. The embryonic skeleton is formed by hyaline cartilage bones enclosed by perichondrium.
    b. A **primary ossification center** forms in the center of the shaft (diaphysis) of the cartilage model of a long bone.
    c. The **cartilage cells (chondrocytes)** in the area of the ossification center **increase in number** (proliferate) and **enlarge in size** (hypertrophy).
    d. The surrounding cartilage **matrix becomes calcified** through deposition of calcium phosphate.
    e. The perichondrium surrounding the diaphysis at the site of the ossification center changes to a **periosteum**. The inner osteogenic layer forms the **bone collar**, which surrounds the calcifying cartilage.
    f. The chondrocytes, which are cut off from nutrition by the bony collar and the calcified matrix, **degenerate** and lose their ability to maintain the cartilage matrix.
    g. A **periosteal bud** containing blood vessels and osteoblasts invades the calcified cartilage spicules through spaces made by the osteoclasts in the bony collar.
    h. When the bud arrives at the center, the osteoblasts lay down bone on the calcified cartilage spicules, using them as a framework. Bone growth spreads in both directions toward the epiphyses.
    i. After birth, **secondary ossification centers** arise in the cartilage of the epiphyses at both ends of the long bone.
    j. Two areas of cartilage are **not** replaced by bone.
        (1) The **articular cartilage** remains over the end of the bone.
        (2) An **epiphyseal plate** of cartilage is spared between the epiphysis and the diaphysis.

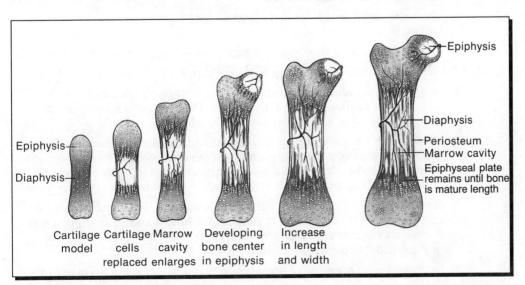

Figure 7–2. *Development of a long (endochondral) bone.*

k. All future elongation of a bone takes place as a result of division of cartilage cells (through interstitial growth) in the epiphyseal plate of cartilage.
  (1) Because bone can only grow appositionally, the interstitial growth of cartilage in the epiphyseal plate and in the above described processes of cartilage proliferation, enlargement, calcification, and replacement by bone are the means of bone elongation.
  (2) When full growth of the individual is complete, the cartilage in the epiphyseal plates is replaced totally by bone. Further growth in length is impossible and ceases.
  (3) Bone growth in thickness occurs as a result of appositional growth from the periosteum, coupled with osteoclastic remodeling from within.

F. **Remodeling of bone**
  1. Bones retain their external shape throughout their growth period as a result of constant remodeling, with bone deposition (by osteoblasts) and resorption (by osteoclasts) occurring on the surface of the bone and within the bone.
  2. Bone is a living, plastic tissue. It adapts its shape and architecture to stresses, activity, use, disuse, and disease through a balance of osteoblastic and osteoclastic action, which is controlled by hormones and nutritional factors.
     a. The **hormones** that affect development as well as the reorganization throughout life include **growth hormone, thyroid hormone, calcitonin, parathyroid hormone,** and **sex hormones (androgens and estrogens).**
     b. The **nutritional factors** necessary for proper bone growth and development include **calcium, phosphorus,** and **vitamins A and D.**

G. **Repair of fractures.** Bone cells and matrix are unable to repair themselves directly without the aid of associated tissues. Repair begins almost at the time of injury.
  1. When a bone is fractured, the first reaction is **hematoma** (large clot) formation. The blood vessels at the site **hemorrhage** and **clotting** takes place.
  2. The hematoma is then invaded by regenerating blood vessels, osteoblasts, and osteoclasts from the periosteum and endosteum.
     a. Macrophages in the blood remove the clot and debris.
     b. Osteoclasts remove damaged bone matrix.
  3. Rapidly dividing cells from the periosteum and endosteum fill in and surround the fracture and form a hyaline cartilage **external callus** (encircling the injury) and an **internal callus** (within the marrow cavity).
  4. The fracture is repaired by both endochondral ossification and intramembranous ossification that takes place on small cartilage fragments in the external and internal callus.
  5. The **bone callus** that is formed is remodeled and replaced by compact lamellar bone.
  6. Thus, the fracture is healed and restored to the original bone structure.

H. **Classification of bones according to their shapes**
  1. **Long bones** are found only in the **limbs.** They are elongate and cylindrical and consist of a **diaphysis** and **epiphyses.** Their function is to sustain the weight of the body and act in locomotion.
  2. **Short bones** are the wrist bones (**carpals**) and ankle bones (**tarsals**). They are cuboidal or oblong and usually are found grouped together to provide strength and compactness in areas of limited movement. Short bones are mostly cancellous bone that is surrounded by a thin layer of compact bone.
  3. **Flat bones** occur in the **skull, ribs,** and **breastbone.** Their plate-like structure provides a broad surface for muscle attachment and affords protection. Two plates of compact bone (known as outer and inner tables in the cranium) enclose a spongy layer (**diploe**).
  4. **Irregular bones** are those of irregular shape not belonging in the above cat-

egories and include **vertebrae** and **ear ossicles**. They are structurally similar to short bones—they are cancellous bone covered by a thin layer of compact bone.

5. **Sesamoid bones** are small, rounded bones that enter into the formation of joints or are associated with cartilage, ligaments, or another bone. An example is the **patella** (kneecap), the largest of the sesamoid bones.

I. **Surface markings of a bone.** Various markings on the surfaces of bones are visible because they are the site of muscle, ligament, or tendon attachments, or serve as passages for blood vessels or nerves. All markings are likely to be more pronounced in a muscular individual than in children or most women (Table 7–1).

II. **SKELETAL ANATOMY: AXIAL SKELETON.** The axial skeleton is composed of bony and cartilaginous parts that protect and support the organs of the head, neck, and trunk. Parts of the axial skeleton include the skull, hyoid bone, auditory ossicles, vertebral column, sternum, and ribs.

Table 7–1. Surface markings of bones.

General Term	Description	Example
Projections		
Tuberosity (trochanter)	Large, rough pronounced projection	Radial tuberosity; trochanter of femur
Tubercle	Blunt, slight, rounded	Greater tubercle of humerus
Spine	Slender, pointed	Scapular spine
Condyle	Rounded enlargement of an articulating surface	Occipital condyle
Epicondyle	Projection located above a condyle	Medial epicondyle of femur
Head	Large rounded articular end of long bone	Head of femur
Facet	Flat articular surface	Rib facets of thoracic vertebrae
Line	Slight ridge on the shaft of a bone	Linea aspera of femur
Crest	Pronounced ridge	Tibial crest
Depressions		
Fovea	Shallow pit	Fovea capitus of femur
Fossa	Deeper than a fovea	Mandibular fossa of the temporal bone
Sulcus	Groove to accommodate passage of blood vessels or nerve	Intertubular sulcus of humerus
Perforations		
Foramen	Large opening for passage of blood vessels, nerves, or ligaments	Foramen magnum of occipital bone
Canal	Tubular passage	Infraorbital canal
Fissure	Narrow, cleft-like passage lies between two bones	Sphenomaxillary fissure

98 Skeletal System

A. The **skull** consists of 22 bones: 8 cranial bones and 14 facial bones (Figures 7–3 and 7–4 and Plate 5).
   1. The **cranium** encloses and protects the brain.
      a. The **frontal bone** forms the forehead, roof of the nasal cavity, and roofs of the **orbits** (eye sockets).
         (1) The frontal bone embryologically develops in two halves, which fuse completely during early childhood.
         (2) The **frontal tuberosities** are two eminences that vary in size and are usually larger in young skulls.
         (3) The **superciliary arches** are two curved prominences that are joined medially by a smooth elevation called the **glabella**.
         (4) **Supraorbital margins**, which are located below the superciliary arches, form the upper borders of the orbits. The **supraorbital foramen** (or notch, in some skulls) is a passageway for an artery and nerve.
      b. The **parietal bones** form the sides and the roof of the cranium.
         (1) The **sagittal suture**, which joins the right and left parietal bones, is an immovable joint united by fibrocartilage.
         (2) The **coronal suture** joins the parietal bones to the frontal bone.
         (3) The **lambdoidal suture** joins the parietal bones to the occipital bone.
      c. The **occipital bone** forms the base and back of the cranium.
         (1) The **foramen magnum** is a large oval opening encircled by the occipital bone. It connects the cranial cavity with the spinal cavity.
         (2) The **external occipital protuberance** is a prominent projection above the foramen magnum.
         (3) The **occipital condyles** are two oval processes on the occipital bone, which articulate with the first cervical vertebra, the **atlas**.
      d. The **temporal bones** form the sides and the base of the cranium. Each irregularly shaped temporal bone consists of four parts.
         (1) The **squamous portion**, the largest, is a thin, flat plate that forms the temple. The **zygomatic process** projects from the squamous portion of each temporal bone. It meets the temporal portion of each **zygomatic bone** to form the **zygomatic arches**.

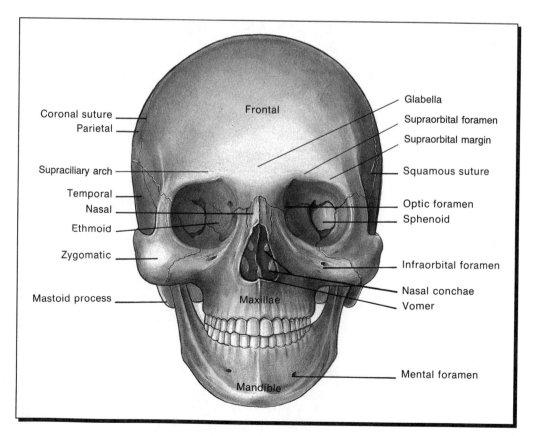

Figure 7–3. *Anterior view of the skull.*

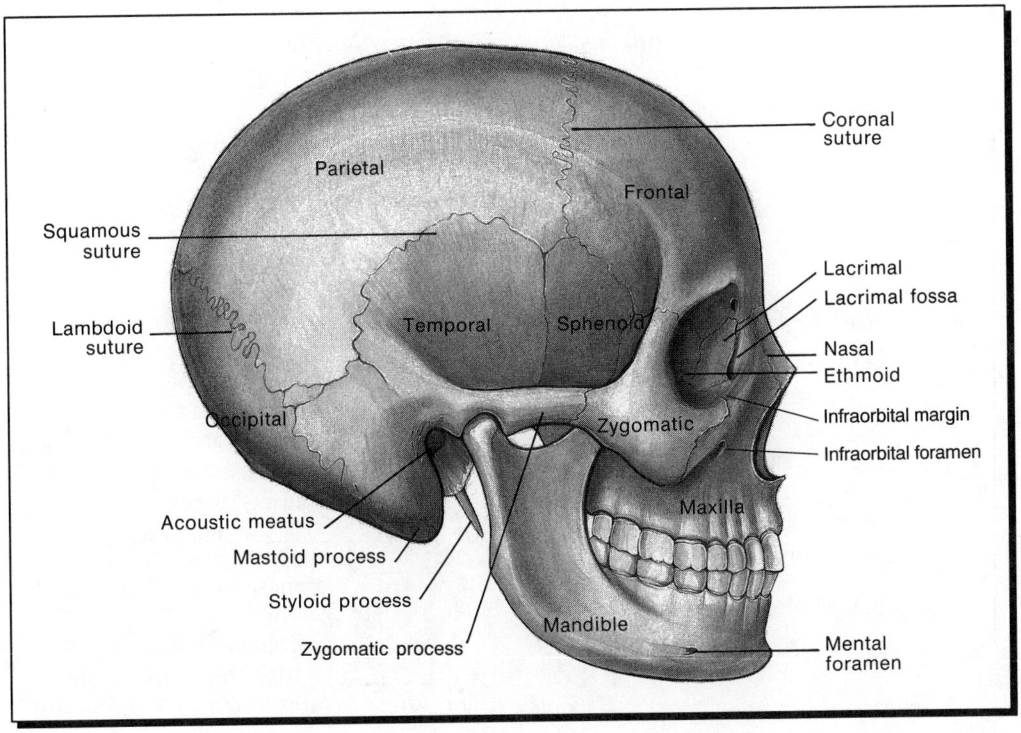

Figure 7–4. *Lateral view of the skull.*

(2) The **petrous portion** is deep within the base of the skull and cannot be seen from the side. It contains the structures of the middle and inner ear.

(3) The **mastoid portion** is behind and below the opening to the ear. The **mastoid process** is the rounded projection easily felt behind the earlobe.
   (a) In the adult, the mastoid contains air spaces, which are called **mastoid air cells** (sinuses), separated from the brain only by thin, bony partitions.
   (b) Inflammation of the mastoid air cells (**mastoiditis**) can occur from untreated middle ear infections.

(4) The **tympanic portion** is inferior to the squamous portion and anterior to the mastoid portion. It contains the **ear canal** (external auditory meatus) and bears a slender **styloid process** for attachment to the stylohyoid ligament.

e. The **ethmoid bone** is the principal, supporting structure of the nasal cavity and contributes to the formation of the orbits of the eyes (Figure 7–5A). It has four parts.

(1) The **cribriform plate** forms a part of the roof of the nasal cavity and is perforated for passage of olfactory nerves. The **crista galli** (so-called because of its resemblance to a cock's comb) is a smooth, triangular process that projects into the cranial cavity upward from the cribriform plate and serves as attachment for coverings of the brain.

(2) The **perpendicular plate** projects downward at right angles from the cribriform plate and forms part of the nasal septum separating the two nasal chambers.

(3) The **lateral masses** contain air cells or **ethmoid sinuses** in which mucous secretions are produced.

(4) **Superior** and **middle nasal conchae,** or turbinates, project medially and serve to increase the surface area of the nasal chambers. (The inferior nasal conchus is an independent bone).

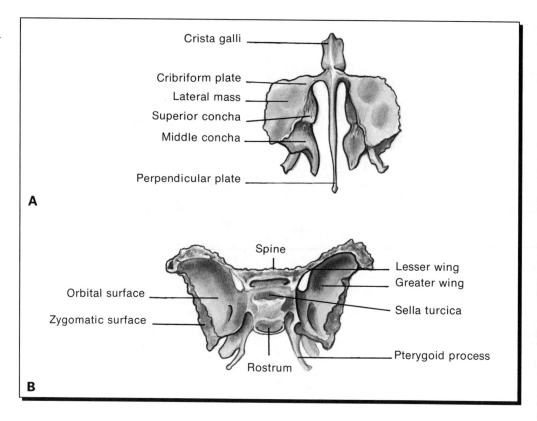

Figure 7–5. *A*, Posterior view of ethmoid bone. *B*, Superior aspect of sphenoid bone.

   f. The **sphenoid bone** is shaped like a bat with outstretched wings. It forms the anterior base of the cranium and articulates laterally with the temporals and anteriorly with the ethmoid and the frontals (Figure 7–5B).
      (1) The body of the sphenoid contains a depression, the **sella turcica** or "turkish saddle," which houses the pituitary gland.
      (2) The **greater wings** and the **lesser wings** project laterally from the body.
      (3) The **pterygoid processes** project inferiorly from the body and form part of the walls of the nasal chambers.
   g. The **auditory ossicles** consist of the **malleus, incus,** and **stapes.** Their function in hearing will be discussed in Chapter 9.
   h. The **wormian bones** are small, variable in number, and are located within sutures.
2. **Facial bones** are not in contact with the brain. They are united by immovable sutures except for the **mandible,** or **lower jaw.**
   a. The **nasal bones** form the bridge of the nose and articulate with the nasal septum.
   b. The **palatine bones** form the posterior portion of the roof of the mouth (hard palate), part of the orbits, and part of the nasal cavity.
   c. The **zygomatic (malar) bones** form the prominence of the cheekbones. The temporal process of each articulates with the zygomatic process of the temporal bone.
   d. The **maxillary bones** form the upper jaw.
      (1) The **alveolar processes** contain the sockets of the upper teeth.
      (2) The **zygomatic processes** extend out to meet the zygomatic bones, together to form the **infraorbital margins** of the orbits. The **infraorbital foramen** perforates the maxilla on each side to transmit nerves and blood vessels to the face.
      (3) The **palatine processes** form the anterior part of the hard palate.
      (4) The **maxillary sinuses,** which empty into the nasal cavity, are part of the four paranasal sinuses.

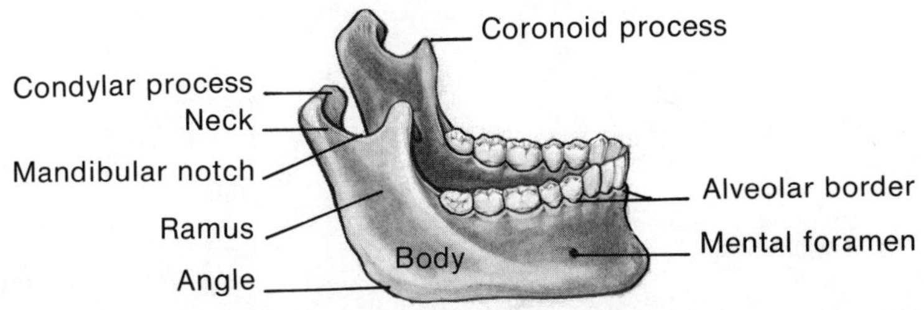

Figure 7–6. *Lateral view of disarticulated mandible.*

    e. The **lacrimal bones** are small and thin, and are located between the ethmoid and maxillary bones in the orbits. They contain a groove for passage of the lacrimal duct, which drains tears into the nasal cavity.
    f. The **vomer bone** forms the middle portion of the hard palate between the palatines and maxillae and contributes to the nasal septum.
    g. The **inferior nasal conchae (turbinates)**. See above with superior and middle conchae in section II A 1 e (4).
    h. The **mandible** is the lower jawbone (Figure 7–6).
       (1) The alveolar portion contains the sockets of the lower teeth.
       (2) The mandibular ramus, which is located on either side of the jawbone, has two processes.
          (a) The **condyloid process** serves for articulation with the temporal bone at the mandibular fossa.
          (b) The **coronoid process** serves as an attachment to the temporal muscle.
  3. The **hyoid bone** is a horseshoe-shaped bone that is unique because it has no articulations with other bones. It is suspended by muscles and ligaments from the styloid processes of the temporals.
  4. The **paranasal sinuses (frontal, ethmoidal, sphenoidal,** and **maxillary)**, consist of air spaces in the skull bones that communicate with each nasal cavity. They serve the following functions:
    a. To **lighten the bones** of the head
    b. To **provide resonance** to the voice and assist in speech production
    c. To **produce mucus**, which drains into the nasal chambers and assists in warming and humidifying the incoming air

B. **Vertebrae**
  1. The **vertebral column** bears the body weight and protects the spinal cord. It consists of vertebrae separated by intervertebral disks of fibrocartilage.
    a. There are **seven cervical vertebrae, 12 thoracic vertebrae, five lumbar vertebrae,** and **five sacral vertebrae** fused into the **sacrum,** and **three to five coccygeal vertebrae** fused into the coccyx.
    b. The 31 pairs of spinal nerves exit through intervertebral foramina between adjacent vertebrae.
  2. **Structure of the typical vertebra**
    a. The **body**, or **centrum**, bears most of the weight.
    b. The **neural (vertebral) arch**, which is formed by two **pedicles** and the **lamina**, encloses the neural cavity for passage of the spinal cord.
    c. A **spinous process** projects from the lamina posteriorly and inferiorly for muscle attachments.
    d. The **transverse processes** extend laterally.
    e. The **inferior articulating processes** and **superior articulating processes** bear facets for articulation with the vertebra above and the vertebra below.
  3. **Regional variations in vertebral characteristics** (Figure 7–7)
    a. All **cervical vertebrae** have transverse foramina for passage of the vertebral artery. The first and second cervical vertebrae are modified for support and movement of the head.

Figure 7–7. *Lateral and superior view* of *A*, the atlas, or first cervical vertebra; *B*, a cervical vertebra; *C*, a thoracic vertebra; and *D*, a lumbar vertebra.

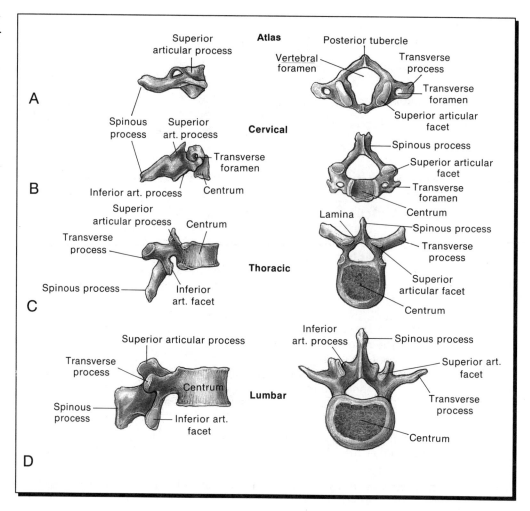

Figure 7–8. *A*, Superior view of a normal intervetebral disc. *B*, A herniated disc may press on the root of a spinal nerve, or *C*, may press on the spinal cord to result in back pain.

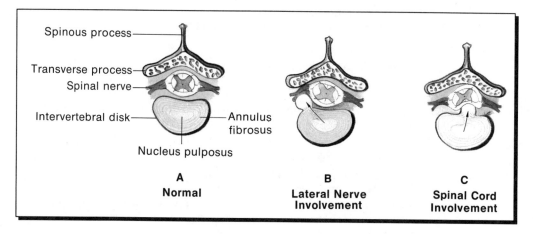

(1) The **atlas** is the first cervical vertebra. It lacks a body.
(2) The **axis** is the second cervical vertebra. It has odontoid process that projects upward to lie within the atlas.
(3) The seventh cervical vertebra has a long spinous process that can be felt and seen at the base of the neck. For this reason, it is sometimes called the **vertebra prominens**.
  b. The **thoracic vertebrae** have long spinous processes, which point downward, and articular facets on the transverse processes, which are used for rib articulations.
  c. The **lumbar vertebrae** are the largest and strongest. Their spinous processes are short and thick and project almost horizontally.
  d. The **sacrum** is a triangular bone. Its base articulates with the fifth lumbar vertebra.
(1) Laterally, there are many foramina (openings) in the sacrum for passage of arteries and nerves.
(2) The upper anterior border of the sacrum is the **sacral promontory**, an obstetrical landmark that is used as a guide in determining the size of the pelvis.
  e. The **coccyx** (tailbone) is fused and articulates with the tip of the sacrum, thereby forming a joint with slight movement. This is important during childbirth to allow for passage of the fetal head.

4. **Curvatures of the vertebral column**
   a. **Primary curvatures** are **concave** (C-shaped) and develop in the thoracic and pelvic areas during the fetal stage.
   b. **Secondary curvatures** are **convex** and develop in the cervical spine after birth when the infant holds up its head, and in the lumbar spine when the infant begins standing and walking.
   c. **Abnormal curvatures**
      (1) Scoliosis, which may appear during periods of rapid growth (as in adolescence), is a lateral curvature of the spine with rotation of the vertebrae.
      (2) Kyphosis, which may be congenital (present at birth) or the result of disease, is an exaggerated posterior curve in the thoracic region; it is commonly called **hunchback**.
      (3) Lordosis (swayback) is an exaggerated anterior curve in the lumbar area.

5. **Vertebral disorders**
   a. **Herniated (slipped) disc**
      (1) The intervertebral discs are located between the bodies of adjacent vertebrae and act to cushion stress between them.
      (2) Each disc contains a central mass, the **nucleus pulposus**, which is composed of pulpy, elastic cartilaginous tissue that is surrounded by an outer layer of fibrocartilage, the **annulus fibrosus**. The annulus fibrosus consists of concentric fibrous rings that hold the nucleus pulposus in place.
      (3) With age, or as a result of injury, the annulus fibrosus loses resiliency and may allow the nucleus pulposus to expand outward and press on the spinal cord or nerve roots, resulting in pain (Figure 7–8).
   b. **Spina bifida** is a congenital defect in which the two laminae of the vertebral arch fail to unite in the midline, thereby allowing tissues from the spinal cord to protrude. The defect is most common in the lumbar area.

C. **Sternum and Ribs** (Figure 7.9)

1. The **sternum** (breast bone) develops in three parts: upper **manubrium, body (gladiolus),** and **xiphoid process**.
   a. The articulation of the manubrium with the clavicle (collarbone) is the **jugular (suprasternal) notch**, an easily palpated bony landmark. Laterally, two **costal notches** articulate with the costal cartilages of ribs 1 and 2.

Figure 7-9. *A*, Anterior view of the thoracic cage and pectoral girdle. *B*, Posterior view of the thoracic cage and pectoral girdle.

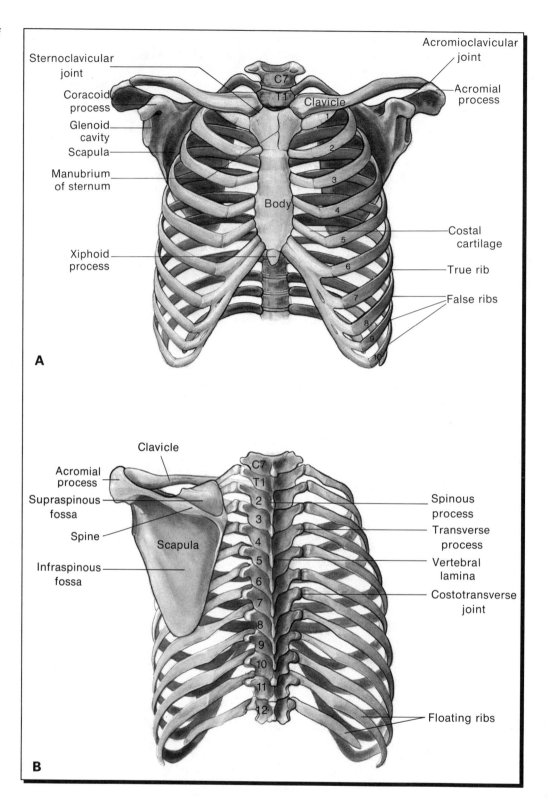

b. The body forms the main portion of the sternum. The lateral costal notches articulate directly with the costal cartilages of ribs 2 through 7 and indirectly with the costal cartilages of ribs 8 through 10.
c. The inferior xiphoid process is cartilaginous.

2. **Ribs.** All 12 pairs of ribs articulate posteriorly with rib facets on the transverse processes of the thoracic vertebrae.
   a. The first seven pairs (1 through 7) are **true ribs** and articulate with the sternum anteriorly.
   b. The next three pairs (8 through 10) are **false ribs**. They articulate indirectly with the sternum through a fusion of their cartilages with the rib above and then a fusion of the joined cartilages with the seventh costal cartilage.
   c. The 11th and 12th ribs are **floating ribs**. They have no anterior attachment.
   d. Although some ribs have particular markings, all share several common features.
      (1) The **head** and a **tubercle** articulate with the facets and the transverse processes of a vertebra.
      (2) The **neck** is roughened for ligament attachment.
      (3) The **shaft**, or body, of a rib has a convex external surface for muscle attachment and a costal groove to accommodate nerves and blood vessels on the internal surface.
      (4) Ribs contain red bone marrow, as does the sternum.

III. **SKELETAL ANATOMY: APPENDICULAR SKELETON.** The appendicular skeleton consists of the pectoral (shoulder) girdles, the pelvic girdles, and the bones of the arms and the legs.

   A. Each **pectoral girdle** has two bones—the **clavicle** and the **scapula**—and serves to attach the arm to the axial skeleton.
      1. The **scapula** (shoulder blade) is a triangular flat bone with three borders: a long **vertebral (medial) border** that runs parallel with the vertebral column; a short **superior border** that slopes out to the tip of the shoulder; and the **lateral border** (which completes the third side of the triangle) and is directed toward the arm.
         a. The **spine** of the scapula is a bony ridge that begins at the vertebral border and widens as it extends to the tip of the shoulder.
         b. The spine ends in the **acromion process**, which articulates with the clavicle; it overhangs the shoulder joint.
         c. The **corocoid process** is the hook-shaped projection of the superior border that serves as attachment for several muscles of the chest wall and arm.
         d. The **glenoid cavity** (glenoid fossa) is a shallow depression that is found at the junction of the superior and lateral borders. It receives the head of the humerus (arm bone).
      2. The **clavicle** (collarbone) is an S-shaped bone that articulates laterally with the acromion process of the scapula and medially with the manubrium at the clavicular notch to form the **sternoclavicular joint**.
         a. The medial two thirds of the clavicle is convex, or bowed forward.
         b. The lateral one third of the clavicle is concave, or bowed backward.
         c. The clavicle serves as attachment for certain muscles of the neck, thorax, back, and arm.

   B. The **upper limb** consists of the bones of the arm, the forearm, and the hand (Figures 7–10 through 7–12).
      1. The **humerus** is the single bone of the arm. It consists of a rounded **head**, which fits into the glenoid cavity, an anatomical **neck**, and a **shaft**, which extends distally.
         a. Two elevations, the **greater tubercle** and the **lesser tubercle**, are located at the upper end of the shaft and provide attachments for muscles.

Figure 7–10. *Right humerus: A,* posterior view; *B,* anterior view.

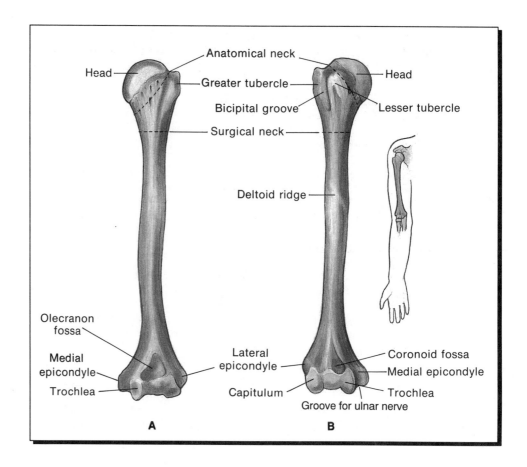

Figure 7–11. *Right radius and ulna: A,* posterior view; *B,* anterior view.

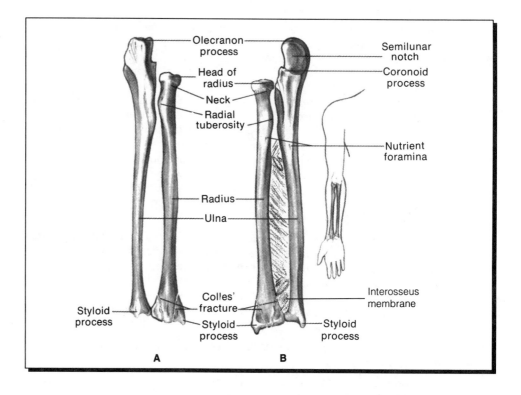

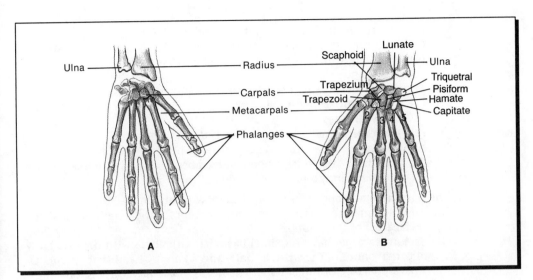

Figure 7–12. *Bones of the hand: A,* posterior aspect; *B,* palmar aspect.

    b. The shaft just below the tubercles narrows to a region that is referred to as the **surgical neck** because of the tendency of the humerus to fracture at this location.
    c. Midway down the shaft is the rough **deltoid tuberosity** that serves as the site for attachment of the deltoid muscle.
    d. The lower end of the humerus widens into the knobby **lateral** and **medial epicondyles** from which the muscles of the forearm and hand originate. The **ulnar nerve** winds behind the medial epicondyle and is responsive to blows or pressure, resulting in the "crazy bone" tingling.
    e. The articular surface of the humerus consists of the lateral **capitulum** (small head), which receives the radius of the forearm, and the **trochlea** (pulley), in which the ulna of the forearm moves.
    f. The **coronoid process** is above the trochlea on the anterior surface; the **olecranon process** is above the trochlea on the posterior surface. These indentations function to receive parts of the forearm bones when they move.
2. The bones of the forearm are the **ulna** on the medial side and the **radius** on the lateral (thumb side) connected by a flexible connective tissue, the **interosseus membrane**.
    a. **Ulna**
       (1) The proximal (upper) end of the ulna looks like an open wrench. The upper part of the wrench is the **olecranon process**, which fits into the olecranon fossa of the humerus when the forearm is fully extended. The lower part of the wrench is the **coronoid process**, which slides into the coronoid fossa of the humerus when the forearm is fully flexed. The **radial notch**, which is located below the coronoid process, accommodates the head of the radius.
       (2) The distal (lower) end of the ulna has a knobby expansion of the shaft called the **head**, which articulates with the ulnar process of the radius. The head continues onto the **styloid process** of the ulna.
    b. **Radius**
       (1) The proximal end of the radius is the disc-shaped **head** that articulates with the capitulum of the humerus and the radial notch of the ulna.
       (2) The **radial tuberosity** for attachment of the biceps muscle is located on the shaft of the radius just below the head.
       (3) The distal end of the radius has a concave **carpal surface** for articulation with the wrist bones, an **ulnar notch** on the medial surface for articulation with the ulna, and a **styloid process** on the lateral side.
3. **Bones of the wrist (carpus).** The wrist is made up of eight irregular **carpal bones** that are arranged in two rows of four bones each.
    a. The proximal row of carpals from the thumb side in anatomical position consists of the following bones:
       (1) **Navicular** (scaphoid), named because of its boat shape

(2) **Lunate**, named because of its half-moon shape
(3) **Triquetral** (triangular), named because of its three-cornered shape
(4) **Pisiform**, which means pea, named because of its approximate size and shape
   b. The distal row of carpals consists of the following bones:
      (1) **Trapezium**, formerly called the greater multangular because of its many surfaces
      (2) **Trapezoid**, smaller but also multisided
      (3) **Capitate**, named for its large, rounded head
      (4) **Hamate**, which means hook, named because of its prominent hook, which extends on the medial side of the wrist
4. The **hand** (metacarpus) consists of five **metacarpal** bones.
   a. All of the metacarpals are very similar except for the length of the first metacarpal of the thumb.
   b. Each metacarpal has a proximal **base** that articulates with the distal row of carpal bones of the wrist, a **shaft**, and a knobby **head** that articulates with a phalanx, or finger bone. The heads of the metacarpals form the prominent knuckles of the hand.
5. The finger bones are called **phalanges**; a singular bone is referred to as a **phalanx**.
   a. Each finger has three bones, the **proximal, middle,** and **distal** phalanges.
   b. The thumb has only proximal and middle phalanges.
C. The **pelvic girdle** transmits the weight of the trunk to the lower limbs and protects the abdominal and pelvic organs. It consists of two hip bones (also called **ossa coxae, innominate bones,** or **pelvic bones**) that meet anteriorly at the pubic symphysis and articulate posteriorly with the sacrum.
   1. Each **hip bone** resembles an electric fan with a hub and two blades (Figure 7–13).
      a. The hub is the cup-shaped socket called the **acetabulum**, which receives the head of the femur, or thighbone, at the hip joint.
      b. The **ilium** is a broad plate of bone that flares upward and outward from the acetabulum. It rises to a thick **iliac crest** that can be felt in the hands-on-hips position.

Figure 7–13. *Lateral view of the right pelvic bone.*

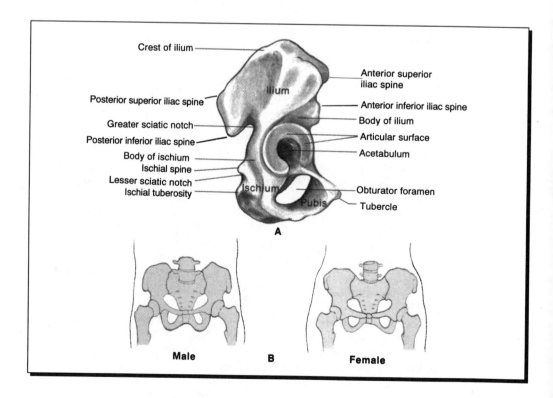

(1) The crest ends anteriorly in the **anterior superior iliac spine** and posteriorly in the **posterior superior iliac spine**. The spines provide attachments for muscles and ligaments.
(2) The **anterior inferior iliac spine** is a bulge below the anterior superior iliac spine. Just below the posterior superior iliac spine is the **posterior inferior iliac spine**.
(3) Below the posterior inferior iliac spine, the posterior border of the ilium is indented deeply by the **greater sciatic notch**.
c. The **ischium** is the posterior and inferior blade of the fan. Its medial border contributes to the greater sciatic notch.
(1) At the inferior limit of the greater sciatic notch is the projecting **ischial spine**, to which a ligament from the sacrum attaches.
(2) Inferior to the ischial spine is the **lesser sciatic notch**.
(3) The **ischial tuberosity** is the large projection of the ischium that supports the body when in a sitting position. It serves as an attachment for posterior thigh muscles.
(4) Anterior to the ischial tuberosity, a slender **ischial ramus** runs forward and upward to meet the **inferior pubic ramus**, which extends downward from the pubic bone.
d. The **pubis** completes the anterior and inferior blade of the hip bone. It mainly consists of two bars of bone: the superior and inferior pubic rami.
(1) The **superior pubic ramus** and the **inferior pubic ramus** meet their counterparts from the other side at the midline in the **symphysis pubis**.
(2) The **pubic arch** is the angle formed at the junction of the pubic bones below the symphysis.
(3) The **obturator foramen** is the large opening bordered by the ischial ramus, the inferior pubic ramus, and the superior pubic ramus. It is the largest foramen in the skeleton and is covered in life by the obturator membrane.

2. **Sex differences in the pelvis**
a. Based on measurement of the average dimensions of the pelves of men and women, about 50% of women possess a **gynecoid**, or true female pelvis, which is wider and roomier in all diameters than the pelvis of men, many of whom have the android, or true male pelvis.
b. Pelvic measurements show considerable variation; actually, there is as much variation in the size and shape of the pelvis among women as there is between women and men.

3. **Anatomical relationships of the pelvis**
a. The **false (greater) pelvis** is bounded by the upper flared parts of the two ilia and their concavities and by the two wings of the base of the sacrum.
b. The **true (lesser) pelvis** is formed by the rest of the sacrum and coccyx and by the ilium, pubis, and ischium on both sides.
(1) The boundaries of the opening to the true pelvis, or **pelvic inlet**, are called the pelvic brim. The diameters of the pelvic brim have obstetric significance in childbearing.
(2) The dimensions of the **pelvic outlet**, bounded by the ischial tuberosities, the lower rim of the symphysis pubis, and the tip of the coccyx, are also very important obstetrically.
(3) At birth, the ilium, ischium, and pubis are mainly cartilaginous, unfused, and separate. The ischium and pubis become bony and fuse at 7 to 8 years of age; total ossification of all cartilaginous portions is not complete until sometime between 17 and 25 years of age.

D. The **lower limb**. Anatomically, the proximal part of the lower limb between the pelvic girdle and the knee is the thigh; the portion between the knee and the ankle is the leg.
1. The **femur**, from the Latin word for thigh, is the longest, strongest, and heaviest bone in the skeleton (Figure 7–14).

Figure 7-14. Anterior aspect of the right femur.

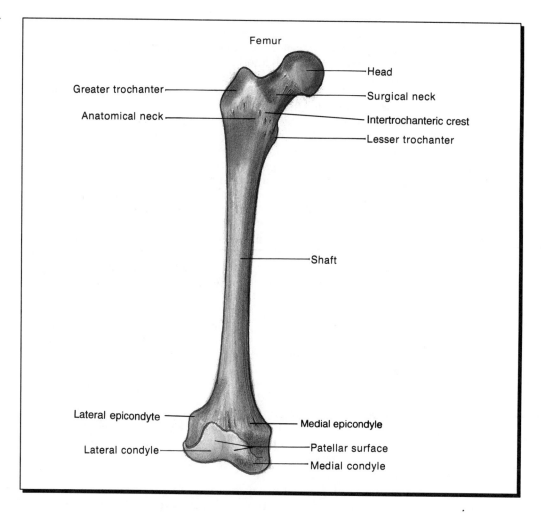

a. The proximal end of the femur bears a rounded **head** for articulation with the acetabulum. The smooth surface of the head is broken by a depression, the **fovea capitis**, for the attachment of a ligament that holds the head in place and carries blood vessels into the head.
   (1) The femur is not in a vertical line with the body. The head of the femur fits into the acetabulum to form an angle of approximately 125° at the neck of the femur; the shaft, therefore, can swing clear of the pelvis when the thigh moves.
   (2) The femoral angle generally is more oblique (less than 125°) in adult women because the pelvis is wider and the femurs are shorter.
b. The head tapers down into a thick **neck**, which continues as the **shaft**. The **intertrochanteric line** on the anterior surface and the **intertrochanteric crest** on the posterior surface delimit the neck from the shaft.
c. The upper end of the shaft bears two prominent processes, the **greater trochanter** and the **lesser trochanter**, for the attachment of muscles to move the hip joint.
d. The shaft is smooth and bears only one marking, the **linea aspera**, which is a rough strip that serves several muscle attachments.
e. The lower end of the shaft expands into the **medial condyle** and the **lateral condyle**.
   (1) On the posterior surface, the two condyles are raised with a deep **intercondylar fossa** between them. The triangular area above the intercondylar fossa is called the **popliteal surface**.

(2) On the anterior surface, the **medial and lateral epicondyles** are positioned above the two large condyles. The smooth articular surface between them is the **patellar surface**, which is concave to receive the **patella** (knee cap).
2. The **leg bones** are the medial **tibia** and the lateral **fibula** (Figure 7–15).
   a. The tibia is the larger medial bone; it transmits the weight of the body from the femur to the foot.
      (1) The head is expanded into the **medial** and **lateral condyles**, which are concave to articulate with the femoral condyles.
      (2) Flat wedges of cartilage, the **medial and lateral semilunar cartilages (menisci)**, are at the edges of the condyles to deepen the articular surfaces.
      (3) The **intercondylar eminence** lies between the tibial condyles.
      (4) The lateral condyle projects over a **fibular facet**, which receives the head of the fibula.
      (5) The **tibial tuberosity**, which serves for attachment of the ligament from the patella, projects on the anterior surface between the two condyles.
      (6) The **tibial (anterior) crest**, commonly known as the shin, is a sharp ridge that curves down the shaft on the anterior surface.
      (7) The lower end of the tibia is expanded to articulate with the **talus bone** of the ankle. The **medial malleolus** is the projection that forms the prominent bump at the medial side of the ankle.
   b. The fibula is the slenderest bone of the body, proportional to its length, and is not involved in weight-bearing. Its purpose is to increase the available area for muscle attachments in the leg.
      (1) The **head** of the fibula articulates with the fibular facet under the lateral condyle of the tibia.

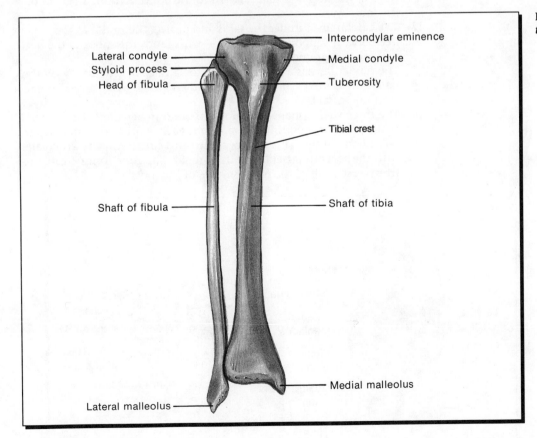

Figure 7–15. *Anterior aspect of the right tibia and fibula.*

Figure 7–16. Bones of the right foot: A, superior view; B, medial view, illustrating the longitudinal arch; C, lateral view.

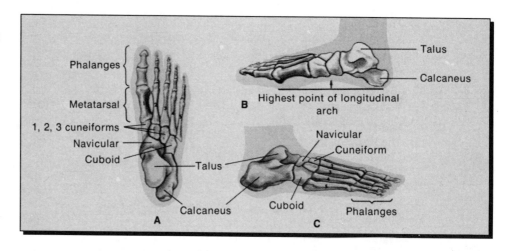

(2) The lower end of the shaft articulates medially with the fibular notch on the tibia and is prolonged laterally as the **lateral malleolus**, which, like the medial malleolus of the tibia, can be felt at the ankle.

3. The **ankle and foot** are comprised of 26 bones arranged into three sets. The **tarsal bones** resemble the carpals of the wrist but are larger; the **metatarsal bones** are similar to the metacarpals of the hand, and the phalanges of the toes are like the phalanges of the fingers (Figure 7–16).
   a. There are seven **tarsal bones**.
      (1) The **talus** articulates with the medial malleolus of the tibia and with the lateral malleolus of the fibula to form the ankle joint. It thus receives the entire weight of the limb, which it distributes, half downward to the heel and half forward to the bones forming the arch of the foot.
      (2) The **calcaneus** is under the talus and projects behind it as the heelbone. It supports the talus and withstands the shock of the heel striking the ground.
      (3) The **navicular** has a concave posterior surface to articulate with the talus and a convex anterior surface to articulate with three of the remaining tarsal bones.
      (4) The three wedge-shaped **cuneiforms** are numbered medial to lateral as first, second, and third cuneiforms. Each articulates with a metatarsal bone of the same number; the third cuneiform also articulates with the seventh tarsal bone, the cuboid. The cuneiforms create the **transverse arch** of the undersurface of the foot.

Figure 7–17. Structure of the knee joint: A, frontal section; B, coronal section.

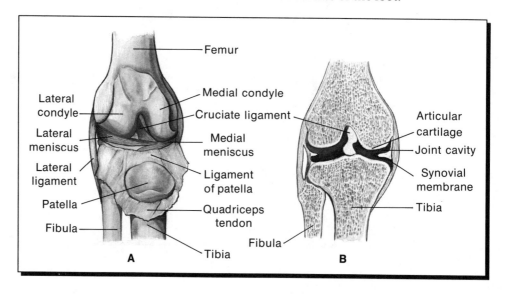

(5) The **cuboid** articulates anteriorly with the fourth and fifth metatarsals; posteriorly, it articulates with the calcaneous.
- b. The sole of the **foot** and the **longitudinal arch** are formed by the five slender **metatarsal** bones. Each metatarsal has a base, shaft, and head.
  (1) The metatarsals are referred to as numbers 1 through 5, starting from the medial (great) toe side.
  (2) The **bases** of the metatarsals articulate with the tarsal bones. The **heads** articulate with the phalanges.
  (3) The heads of the first two metatarsal bones form the ball of the foot.
  (4) The head of the first metatarsal has two **sesamoid bones** attached to its plantar surface.
- c. The **14 phalanges** of the toes, like the phalanges of the fingers, are arranged in a proximal, middle, and distal row. The great toe has only a proximal and distal phalanx.

IV. ARTICULATIONS

A. **General classification of joints.** An articulation, or joint, occurs when two surfaces of bone join together, whether or not movement is permitted at the junction. Joints may be classified according to their **structure** (based on the presence or absence of a joint cavity between the articulating bones and the kind of connective tissue associated with the joint); and according to their **function** (based on the amount of movement permitted at the joint).

B. **Structural classification of joints**
  1. **Fibrous joints** have no joint cavity and are held together by fibrous connective tissue.
  2. **Cartilaginous joints** have no joint cavity and are held together by cartilage.
  3. **Synovial joints** have a joint cavity and are held together by a surrounding articular capsule and ligaments.

C. **Functional classification of joints**
  1. **Synarthroses** are **immovable joints.** Structurally, they are held together by fibrous connective tissue or cartilage.
     a. **Sutures** are joints connected by dense fibrous connective tissue and are found only within the skull. Examples of sutures are the saggital suture or the parietal suture.
     b. **Synchondroses** are joints in which two bones are connected by hyaline cartilage. An example is the temporary epiphyseal plate between the epiphysis and the diaphysis in the long bones of a child. When the temporary synchondrosis ossifies, it is referred to as a synostosis.
  2. **Amphiarthroses** are **slightly movable joints** that allow very limited motion in response to torsion or compression.
     a. **Symphyses** are joints in which two bones are connected by a disc of fibrocartilage, which cushions the joint and allows some slight movement. Examples of symphyses are the **symphysis pubis** between the pubic bones and the **intervertebral discs** between the bodies of adjacent vertebrae.
     b. **Syndesmoses** occur when adjacent bones are connected by collagenous connective tissue fibers. Examples of syndesmoses are found in the side-to-side bones that are held together by an interosseus membrane, such as the radius and ulna, and the tibia and fibula.
     c. **Gomphoses** are joints in which a peg- or cone-shaped bone fits into a bony socket, as in teeth set into the alveoli (sockets) of the jawbones. The intervening fibrous connective tissue is the peridontal ligament.
  3. **Diarthroses** are **freely movable joints**, which are also known as **synovial joints** (from a Greek word meaning "with egg"). They have a joint cavity filled with synovial fluid, a joint (articular) capsule that holds the two bones together, and the ends of the bones in a synovial joint are covered with articular cartilage.

a. The **outer layer** of the joint capsule is dense, white, fibrous connective tissue that is continuous with the periosteum of the bones meeting at the joint.
   (1) Ligaments are thickenings of the capsule that serve to reinforce the joint capsule and provide stability.
   (2) Ligaments may be incorporated within the capsule or separated from it by outpouchings of the capsule.
b. The **inner layer** of the joint capsule is the **synovial membrane**, which lines the joint everywhere except over the articular cartilages.
   (1) Synovial membrane secretes **synovial fluid**, a clear, viscous material with the consistency of egg white. It is 95% water, has a pH of 7.4, and is a mixture of polysaccharides (mostly hyaluronic acid), proteins, and fats.
   (2) Synovial fluid functions to lubricate and nourish the articular cartilage surfaces. It also contains phagocytic cells to remove debris from injured or infected joint cavities.
   (3) In some synovial joints, such as the knee joint, there are **articular discs (menisci)** of fibrocartilage.
      (a) Articular discs modify the shape of the articulating surfaces of the bones to provide easier movement, greater stability, or shock absorption.
      (b) Injury to the articular disc in the knee is commonly called a torn cartilage.
   (4) **Bursae** are closed sacs lined with synovial membrane found outside the joint cavity. Bursae may be located under tendons or muscles. They are found when tendons or muscles glide over bony prominences or subcutaneously where the skin is exposed to friction, such as the elbow or kneecap.

D. **Classification of synovial joints** is based on the shape of the articulating surfaces.
   1. A **ball-and-socket joint** consists of the ball-shaped head of one bone that fits into a cup-shaped cavity of the other bone. This joint, which is known as a triaxial or multiaxial joint, permits the greatest range of movement, occurring in three planes. Examples of ball-and-socket joints are at the hip and the shoulder joints.
   2. In a **hinge joint**, a convex surface of one bone fits into the concave surface of the second bone. This joint permits movement in only one plane and is known as a uniaxial joint. Examples of hinge joints are at the knee and the elbow (Figure 7–17).
   3. A **pivot joint** is a conical or peg-shaped bone that fits into a depression in the second bone, which pivots around it. This joint, which is uniaxial, permits rotation about a central axis; for example, the joint formed as the atlas rotates around the odontoid process of the axis, and the joint between the proximal heads of the radius and the ulna.
   4. A **condyloid joint** consists of an oval condyle of one bone that fits into an elliptical cavity of the second bone. This joint is a biaxial joint and permits movement in two planes at right angles to each other. Examples of condyloid joints are the joints between the radius and the carpal bones and the joint between the occipital condyles of the skull and the atlas.
   5. In a **saddle joint**, the articulating surface of one bone is concave in one direction and convex in the other; thus, it fits into the second bone that is convex and concave in an opposite articulating surface, such as two saddles fitting together. This joint is a modified condyloid joint and permits the same movements. The only true saddle joint in the body is between the carpal and metacarpal of the thumb.

6. A **gliding joint** is one in which the articulating surfaces of both bones are flattened, enabling a sliding movement of one bone over another. Slight movement is allowed in all directions within the limitations of bony processes or ligaments surrounding the joint. Such joints are called nonaxial; for example, the intervertebral joints and the joints between the carpal bones and the tarsal bones.

E. **Movements at synovial joints** are produced by skeletal muscles attached to the bones forming the articulation. The muscles provide the force, the bones act as the levers, and joints are the fulcra.
  1. **Flexion** decreases the angle between two bones or body parts, as in bending the elbow (moving the arm forward), bending the knee (moving the leg backward), or bending the torso sideways.
     a. **Dorsoflexion** is pulling the foot upward at the ankle (elevating the dorsal part of the foot).
     b. **Plantar flexion** is bending the foot downward at the ankle.
  2. **Extension** increases the angle between two bones or body parts.
     a. Extension of a body part returns it to anatomical position, as in straightening the elbow or the knee joints after flexion.
     b. **Hyperextension** refers to increasing the angle of a body part beyond anatomical position to greater than 180°, as in bending the torso or the head backward.
  3. **Abduction** is movement of a body part away from the midline of the body, as in the abduction of the arm, or from the longitudinal axis of the limb, as in the fingers and toes.
  4. **Adduction**, the opposite of abduction, is the movement of a body part back to to the main axis of the body or to the longitudinal axis of the limb.
  5. **Rotation** is the movement of a bone around its own central axis without lateral displacement of the body part, as in shaking the head in a "no" movement.
     a. **Pronation** is a medial rotation of the forearm in the anatomical position, which results in the palm of the hand being directed backward.
     b. **Supination** is a lateral rotation of the forearm, which results in a palm-upward position.
  6. **Circumduction** is the combination of all of the above angular and rotary movements in succession to circumscribe a conical space, as in swinging the arms in a circle. Circumduction can take place at the hip, shoulder, trunk, wrist, and ankle joints.
  7. **Inversion** is the movement at the ankle joint that permits the sole of the foot to turn inward or medially.
  8. **Eversion** is the movement at the ankle joint that permits the sole of the foot to turn outward. Inversion and eversion of the foot are useful movements when walking over a rough and bumpy terrain.
  9. **Protraction** is moving a body part forward, as in jutting out the lower jaw or flexing the pectoral girdle forward.
  10. **Retraction** is pulling a body part back, as in retracting the mandible or retracting the pectoral girdle to pull back the shoulders.
  11. **Elevation** is moving a structure superiorly, as in closing the mouth (elevating the mandible) or shrugging the shoulders (elevating the scapulae).
  12. **Depression** is moving a structure inferiorly, as in opening the mouth.

F. **Joint disorders**
  1. **Sprains.** A sprain is an injury to a joint with stretching or possible tearing of some of the ligaments or tendons surrounding the joint. It is generally the result of sudden twisting or impact to the joint. Sprains are not uncommon in the knee, wrist, or ankle. A lesser injury with no rupture of tissue is a strain.

2. **Dislocation.** A dislocation, also called a luxation, refers to a displacement of the articulating surfaces of a joint. The knee and the shoulder joints are particularly vulnerable to dislocation.
3. **Bursitis,** an inflammation of a bursa associated with a joint, results from overexertion of the joint or from infection. It most commonly occurs in the subacromial bursa in the shoulder and results in painful and limited movement at the shoulder joint or in the bursa between the olecranon process and the skin ("tennis elbow"). Prepatellar bursitis (commonly known as "housemaid's knee") may result from frequent kneeling.
4. **Arthritis** is a general term for many kinds of joint disease, all characterized by pain, swelling, and inflammation, and all resulting in various degrees of crippling.
    a. **Rheumatoid arthritis** is a systemic disease affecting connective tissue with joint inflammation as the major manifestation.
        (1) It involves a thickening of the synovial membrane with subsequent damage to the articular cartilage.
        (2) Although periods of remission occur, the disease tends to be chronic and progressive.
    b. **Osteoarthritis** is a degenerative joint disease that appears to be associated with aging, obesity, or joint trauma.
        (1) Symptoms rarely occur before the age of 40, and eventually everyone acquires osteoarthritis to some extent.
        (2) The articular hyaline cartilage is destroyed and bone overgrowth at the articular margins with subsequent loss of the joint space results in pain, particularly after exercise. It is not a crippling disease unless the hip joint is involved.
    c. **Gouty arthritis,** which affects mostly mature males, is the result of an abnormality of nucleic acid metabolism, which causes uric acid crystals to be deposited in certain joints.
    d. **Infectious arthritis** results when bacteria or their products settle in the joint and produce inflammation.
        (1) Gonococcal arthritis is acutely painful and results from the invasion of the joints by the organism that causes gonorrhea.
        (2) Staphylococcal infections may also cause arthritic symptoms.

Study Questions

Directions: Each question below contains four suggested answers. Choose the **one best** response to each question.

1. A major difference between bone tissue and other connective tissues lies in which of the following?
   (A) the composition of the intercellular substance of bone
   (B) the inability of bone cells to metabolize nutrients for energy
   (C) the limited blood supply available to bone tissue
   (D) the presence of collagen fibers in bone

2. All of the following statements concerning the periosteum of a bone are true EXCEPT
   (A) The periosteum consists of an outer dense connective tissue layer and an inner osteogenic layer.
   (B) Periosteum is important for bone growth in width.
   (C) Blood vessels enter bone through periosteum.
   (D) All bone in the body is covered by periosteum.

3. Endochondral and intramembranous are terms that are used to describe
   (A) the function of adult bone
   (B) the development of bone
   (C) the porosity of bone
   (D) the chemical composition of bone

4. All of the following are steps in the formation of a long bone, such as the femur, EXCEPT
   (A) Hyaline cartilage forms a miniature model of the adult femur in the fetus.
   (B) A periosteal bone collar around the primary ossification center is produced by cells that have differentiated from perichondrium.
   (C) The femur becomes ossified after birth when the cartilage turns into bone.
   (D) Secondary ossification centers arise in the epiphyses of the femur.

5. Long bones differ from flat bones in that long bones
   (A) have an outer layer of compact bone
   (B) have a diaphysis
   (C) contain cancellous bone
   (D) contain marrow

6. Which of the following bones would most likely be involved in surgery to remove a tumor on the pituitary gland?
   (A) styloid process of the temporal
   (B) zygomatic process of the temporal
   (C) sella turcica of the sphenoid
   (D) cribriform plate of the ethmoid

7. Sinuses are found in all of the following bones EXCEPT
   (A) mastoid
   (B) maxillary
   (C) mandible
   (D) frontal

8. A vertebra has a relatively small body, a foramen in each transverse process, but no articular facets on the transverse processes, and a long and prominent spinous process. Which vertebra is it?
   (A) first cervical
   (B) second cervical
   (C) first thoracic
   (D) seventh cervical

9. The 11th and 12th ribs are called floating ribs because they lack
   (A) an anterior attachment to the sternum
   (B) a posterior attachment to thoracic vertebrae
   (C) an anterior attachment to lumbar vertebrae
   (D) any vertebral attachment either anteriorly or posteriorly

10. In the true female pelvis (gynecoid), as compared to the true male pelvis (android),
    (A) The dimensions of the pelvic outlet are lesser.
    (B) The angle formed at the junction of the pubic bones is narrower.
    (C) The distance between the anterior superior iliac spines is greater.
    (D) The ischium, ilium, and pubis remain unfused in adulthood.

11. A joint that permits flexion and extension only is known as a(n)
    (A) uniaxial joint
    (B) hinge joint
    (C) diarthrotic joint
    (D) all of the above

12. The combination of joint movements that allows turning a doorknob is
    (A) eversion-inversion
    (B) protraction-retraction
    (C) pronation-supination
    (D) abduction-adduction

Answers and Explanations

1. **The answer is A.** (I C 1, 2) The difference between bone tissue and other connective tissues is that the matrix of bone tissue contains calcium and phosphorus bone salts in the form of hydroxyapatite crystals. Like other connective tissues, the cells of bone metabolize nutrients for energy and have a good blood supply. The collagen fibers present in the bone matrix give bone its resiliency and flexibility, while the bone salts provide its strength and hardness.

2. **The answer is D.** (I D 1 c) Periosteum does not cover sesamoid bones, the articular surfaces of bones, or extend around tendon and ligament insertions on bone. The inner osteogenic layer of periosteum is essential for bone growth in width, and the pathway for blood vessels to penetrate bone is by way of the vascularized periosteum.

3. **The answer is B.** (I E 1, 2) Endochondral and intramembranous ossification refer to bone formation and development. Bone functions, such as support and protection, movement, storage of minerals, and blood cell formation, are independent of the type of bone formation. Compact and cancellous are terms used to describe the porosity of bone, while the chemical composition of all bone is the same.

4. **The answer is C.** (I E 1, 2 a–k) Osteogenesis, or the formation and development of bone, occurs through the replacement of the cartilage model by bone. The cartilage in the primary ossification center proliferates, hypertrophies, calcifies, and is subsequently invaded by osteoblasts, which lay down bone on the remains of the calcified cartilage. Secondary ossification centers arise in both epiphyses after birth.

5. **The answer is B.** (I H 1–5) The only difference between long bones and bones that are short, flat, or irregular is that long bones are elongate and have a diaphysis and epiphyses. All mature bone contains both compact and cancellous bone, and all contain either red or yellow bone marrow.

6. **The answer is C.** (II A 1 a–f). The sella turcica of the sphenoid bone surrounds the pituitary gland.

7. **The answer is C.** (II A 4) The frontal, maxillary, ethmoid, and sphenoid bones contain the paranasal sinuses. The mastoid bone also contains air cells or sinuses.

8. **The answer is D.** (II B 2 a–e) All cervical vertebrae have transverse foramina for passage of the vertebral artery and no rib facets, but only the seventh cervical vertebra is called the vertebra prominens because of its long spinous process. The first cervical vertebra has no body, the second cervical vertebra has an odontoid process, and all thoracic vertebrae possess articular facets on the transverse processes.

9. **The answer is A.** (II C 2 a–c) Floating ribs (11th and 12th ribs) have no anterior attachment to the sternum but articulate posteriorly with the thoracic vertebrae.

10. **The answer is C.** (III C 1 b, (3), (4)). Because the gynecoid pelvis is wider and roomier in all diameters than the android pelvis, the distance between the anterior superior iliac spines is greater, the dimensions of the pelvic outlet are greater, and the pubic angle is greater. The three bones of the pelvis fuse and totally ossify in both males and females.

11. **The answer is D.** (IV C 3, D 2) A hinge joint is a freely movable diarthrotic joint. It permits movement in only one plane and is also known as a uniaxial joint.

12. **The answer is C.** (IV E 1–12) Pronation is medial rotation of the forearm and supination is lateral rotation of the forearm, which would enable the turning of a doorknob. Abduction is movement of a body part away from the midline and adduction is the movement of a body part back to the midline. Eversion is the movement that permits turning the sole of the foot outward; inversion is the turning of the sole of the foot inward. Protraction is the moving of a body part forward from the main axis of the body; retraction is the pulling of a body part backward.

Muscular System 8

PART I: GENERAL STRUCTURE AND PHYSIOLOGY

I. **INTRODUCTION.** Muscle tissue, which constitutes about 40% to 50% of the the body weight, predominately consists of contractile cells called muscle fibers. Through contraction, muscle cells produce movement and do work.

   A. **Functions of the muscular system**
      1. **Movement.** Muscles produce motion of the bones to which they are attached and movements within the body's internal organ parts.
      2. **Body support and maintenance of posture.** Muscle supports the skeleton and maintains the body in a standing or seated position against the force of gravity.
      3. **Production of heat.** Muscle contractions metabolically produce heat to maintain normal body temperature.

   B. **Properties of muscle**
      1. **Contractility.** Muscle fibers contract and develop tension, which may or may not include shortening. Fibers are elongated because contraction in any diameter of a rounded or square-shaped cell would produce only limited shortening.
      2. **Excitability.** Muscle fibers forcefully respond when stimulated by nerve impulses.
      3. **Extensibility.** Muscle fibers have the capacity to stretch beyond their relaxed length.
      4. **Elasticity.** Muscle fibers can return to their original length after contraction or stretching.

   C. **Classification of muscle tissue.** Muscle is classified **structurally** on the basis of the presence or absence of **cross-striations**, **functionally** on the basis of whether the muscle contraction is under **voluntary** (conscious) or **involuntary** control, and also by **location**, as in cardiac muscle, which is found only within the heart. (See Chapter 5.)

   D. **Special terminology** in muscle tissue. The usual cellular organelles are present in a muscle fiber, but some have different names.
      1. The cytoplasm is called **sarcoplasm**.
      2. The endoplasmic reticulum is called **sarcoplasmic reticulum**.
      3. The plasma membrane is called the **sarcolemma**.
         a. **T-tubules** are sets of transverse tubules in skeletal and cardiac muscle formed by finger-like invaginations of the sarcolemma.

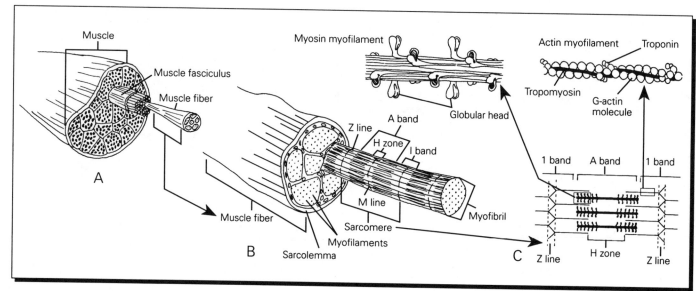

Figure 8-1. *Organization of a muscle.* A, A muscle is formed by thousands of muscle cells (fibers) surrounded by connective tissue sheaths and bundled into groups called fasciculi. B, Each muscle fiber, on which the banding patterns are visible, contains myofibrils. C, Each myofibril is composed of thick myosin myofilaments and thinner actin myofilaments.

      b. **Terminal cisternae** are sac-like structures of the sarcoplasmic reticulum on either side of a T-tubule.
      c. A T-tubule invagination and the adjacent terminal cisternae on either side constitute a **triad**.
  E. Types of muscle
    1. **Skeletal muscle** is striated, voluntary, and attached to the skeleton.
      a. Fibers are very long, up to 30 cm, and cylindrical, with widths ranging from 10 microns to 100 microns.
      b. Each fiber has many nuclei, which are arranged at the periphery.
      c. Contractions are rapid and forceful.
    2. **Smooth muscle** is nonstriated and involuntary. It is found in the walls of hollow organs, such as the urinary bladder and uterus, and in the walls of tubes, such as those in the respiratory, digestive, reproductive, urinary, and circulatory systems.
      a. Muscle fibers are spindle-shaped with an elongated central nucleus.
      b. Fibers are small, ranging in size from about 20 microns (encircling blood vessels) to 0.5 mm in the pregnant uterus.
      c. Contractions are strong and slow.
    3. **Cardiac muscle** is striated, involuntary, and found only in the heart.
      a. Fibers are elongated and branched with a single central nucleus.
      b. They are about 85 microns to 100 microns in length and 15 microns in diameter.
      c. **Intercalated discs** are special tight junctions at the sites of end-to-end contact of adjacent cardiac muscle cells.
      d. Contractions of cardiac muscle are strong and rhythmic.

II. SKELETAL MUSCLE

  A. **Organization** (Figure 8-1)
    1. Skeletal muscle consists of masses of fibers arranged in bundles called **fascicles**. The larger the muscle, the greater the number of fibers.
      a. The biceps brachii in the upper arm is a large muscle and may have 260,000 fibers.
      b. A small muscle, such as the stapedius in the middle ear, contains only about 1,500 fibers.
    2. **Fibrous connective tissue sheaths** surround each muscle and extend into it to envelop fascicles and individual fibers. They carry nerve impulses and

blood vessels into the muscle and mechanically transmit the force of contraction from one end of the muscle to the other.
- a. **Epimysium** is dense connective tissue that surrounds the entire muscle and is continuous with deep fascia.
- b. **Perimysium** refers to extensions of the epimysium that penetrate inward into the muscle to surround fascicle bundles.
- c. **Endomysium** is the delicate layer of connective tissue that surrounds each individual muscle fiber.

B. **Microscopic organization of a skeletal muscle fiber**
1. The **myofibrils**, specialized contractive units, constitute 80% of the volume of the fiber.
2. Each cylindric myofibril consists of **thick myofilaments** and **thin myofilaments**.
   - a. Thick myofilaments are primarily composed of the protein **myosin**. Myosin molecules are arranged to form a rod-like tail with two globular heads, similar to a double-headed golf club.
   - b. Thin myofilaments are composed of the protein **actin**. Two additional thin filament proteins, **tropomyosin** and **troponin**, are associated with actin.
3. **Banding is determined by the arrangement of the myofilaments**
   - a. The darker **A band** (anisotropic, or able to polarize light) consists of a vertical stack of thick myofilaments and overlapping thin myofilaments.
   - b. The lighter **I band** (isotropic, or nonpolarizing) is formed by thin actin myofilaments, which extend in both directions from a Z line to run part way into a stack of thick filaments.
   - c. The **Z line** is formed by a supporting protein that holds the thin myofilaments together along the length of a myofibril.
   - d. The **H zone** is a lighter area in the A band of myosin myofilaments that is not penetrated by thin filaments.
   - e. The **M line** bisects the center of the H zone. It is the result of another supporting protein that holds the thick myofilaments together within a stack.
   - f. A **sarcomere** is the distance from one Z line to another Z line.

C. **Mechanism of actin and myosin interaction**
1. **The sliding filament hypothesis**
   - a. During contraction, the actin and myosin myofilaments remain the same length but slide past each other, increasing the amount of overlap between the filaments.
   - b. The actin filaments slide to extend farther into the A band, narrowing and obliterating the H band.
   - c. The length of the sarcomere (from Z line to Z line) shortens in contraction.
   - d. The shortening of the sarcomeres shortens individual muscle fibers and the entire muscle.
2. **Molecular basis for contraction**
   - a. The **myosin molecule** is made up of two identical heavy protein chains and two pairs of light chains.
     - (1) The tails of the heavy chains are twisted around each other with the two globular protein heads, or **crossbridges**, projecting out at one end.
     - (2) Crossbridges link the thick to the thin filaments. Each crossbridge has an **actin binding site**, an **ATP binding site**, and **ATPase** (enzyme that hydrolyzes ATP) activity.
     - (3) Several hundred myosin molecules are arranged in each thick filament with their rod-like tails overlapping and their globular heads directed toward either end.
   - b. **The actin molecule** consists of three proteins.
     - (1) Fibrous **F-actin** is formed from two strands of globular G-actin twisted around each other.
     - (2) **Tropomyosin** molecules form filaments that extend over actin subunits and cover the sites that bind to the myosin crossbridges.

(3) **Troponin** molecules are bound to tropomysin molecules and stabilize the blocking position of the tropomyosin molecules. Troponin is a complex consisting of the following:
 (a) A polypeptide that binds to tropomyosin
 (b) A polypeptide that binds to actin
 (c) A polypeptide that binds to calcium ions

c. **In the absence of calcium** (Ca^{++}), tropomyosin and troponin prevent linkage between actin and myosin.

d. **In the presence of calcium**, a repositioning of troponin-tropomyosin allows contact between actin and myosin.

D. **Chemistry of contraction**

1. At the beginning of a contraction cycle, ATP is bound to the myosin head at the site of the hydrolyzing enzyme, ATPase.
2. ATPase breaks down ATP into ADP and inorganic phosphate. Both remain attached to the myosin head (ATP ———> ADP + P + energy).
3. The energy released by the hydrolysis activates the myosin head into a cocked position, ready to bind to actin.
4. Calcium ions, which have been released from the sarcoplasmic reticulum, bind to the troponin attached to tropomyosin and actin.
5. The troponin-calcium ions complex undergoes a conformational change that allows tropomyosin to move away from its actin-blocking position.
6. The myosin-binding site on the actin is then exposed to allow attachment to the actin-binding site on the myosin head.
7. At binding, the ADP and inorganic phosphate are released from the myosin head, and the myosin head moves and swivels in the opposite direction to pull the attached actin filament in toward the H band. This is referred to as the **power stroke** of the myosin head.
8. The myosin head remains strongly bound to actin until a new ATP molecule binds to it and weakens the link between actin and myosin.
9. The myosin head disconnects from actin, recocks, and is ready to attach to actin at a new site, swivel, and pull again to repeat the cycle.
10. This cycle occurs in thousands of myosin heads as long as there is nerve stimulation, sufficient calcium ions, and ATP.
11. **Muscle relaxation** occurs when nerve stimulation stops and calcium ions are no longer released. Calcium ions are transferred back to the sarcoplasmic reticulum by a calcium pump in the sarcoplasmic reticulum membrane.
12. **Rigor mortis.** ATP is required to release myosin from actin. Total depletion of ATP in a muscle and the inability to generate more ATP, as would occur after death, results in permanent actin and myosin attachment and muscle rigidity.

E. **Energy sources for contraction.** Because the ATP stored in a muscle is generally used up after about ten contractions, it must be regenerated for continued muscle activity through other sources.

1. **Creatine phosphate** (CP), another high-energy compound, is the immediate source of energy available to regenerate ATP from ADP (CP + ADP ———> ATP + creatine).
 a. CP allows muscle contractions to persis while additional ATP is formed by anaerobic and aerobic metabolism of glucose.
 b. CP provides energy for only about 100 contractions and must be resynthesized by production of more ATP (ATP + creatine ———> ADP + CP).
 c. Additional ATP is formed by the metabolism of glucose and fatty acids through anaerobic and aerobic reactions.

2. **Anaerobic reactions (the glycolytic pathway)**
 a. Muscle can contract for a brief period in the absence of oxygen by utilizing ATP generated by anaerobic glycolysis, the first step in cellular respiration.

b. Glycolysis takes place in the sarcoplasm, does not require oxygen, and involves the conversion of one molecule of glucose to two molecules of **pyruvic acid.**
c. Anaerobic glycolysis is rapid but inefficient, producing only two molecules of ATP per molecule of glucose. It can meet the ATP demands of contracting muscle for brief periods when the oxygen supply is inadequate.
d. **Lactic acid formation in anaerobic glycolysis**
   (1) In the absence of oxygen, pyruvic acid is converted to lactic acid.
   (2) If exercise is moderate and brief, adequate oxygen supply prevents accumulation of lactic acid.
   (3) Lactic acid diffuses out of the muscles and is carried to the liver for resynthesis back to glucose.

3. **Aerobic (oxygen-utilizing) reactions**
   a. During moderate activity, the pyruvic acid formed by anaerobic glycolysis drifts to the mitochondria of the sarcoplasm to enter the citric (tricarboxylic) acid cycle for oxidation.
   b. In the presence of oxygen, glucose is completely broken down to carbon dioxide, water, and energy (ATP).
   c. Aerobic reactions are slow but efficient, producing an energy yield of 36 moles of ATP per mole of glucose.

4. **Oxygen debt.** During short bursts of strenuous activity when the breakdown of ATP is very rapid, anaerobic energy stores become exhausted quickly. The respiratory system and the blood vessels cannot deliver enough oxygen to the muscles to regenerate ATP through aerobic reactions.
   a. Lactic acid accumulates, changes the pH, and causes muscle fatigue and soreness.
   b. The extra oxygen that must be inhaled after strenuous excercise is **oxygen debt.**
   c. The amount of oxygen respired remains above normal until the lactic acid is removed, either by oxidation back to pyruvic acid in the muscle or resynthesis to glucose in the liver.

F. **Nervous control of skeletal muscle contraction**
1. Each muscle fiber receives an ending from a **somatic motor neuron**, a nerve cell in the spinal cord that transmits impulses to a skeletal muscle.
2. The motor nerve ending, called an **axon** or nerve fiber, travels with numbers of similar fibers from other motor neurons in a **nerve.**
   a. A single axon fiber divides into a variable number of branches that form specialized **neuromuscular junctions** with skeletal muscle fibers.
   b. Each axonal terminal sits within a fluid-filled indentation (**synaptic cleft**) on the sarcolemma, which is thrown up into folds (Figure 8–2).
3. The **motor end plate** is the noncontiguous junction of a nerve axon branch and the skeletal muscle fiber.
4. A **motor unit** is one motor neuron (and its branches) and all the muscle fibers that it innervates.
   a. A motor unit may consist of only two or three muscle fibers or there may be more than a thousand muscle fibers in some large muscles.
   b. The fewer the number of fibers innervated by a neuron, the more precise the movement produced.
      (1) Muscles used in writing, for example, have few muscle fibers in a motor unit.
      (2) Large postural muscles that support the body may have as many as 800 muscle fibers/motor unit.
5. The **axon terminals (terminal boutons)** contain mitochondria and many small **synaptic vesicles**. When a nerve impulse reaches the axon terminal, the synaptic vesicles release the transmitter substance **acetylcholine** (ACh).

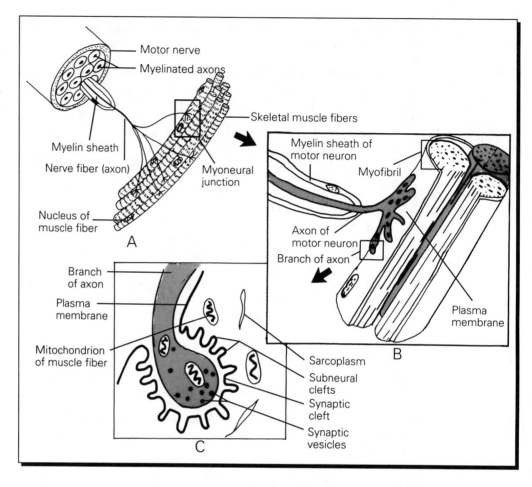

Figure 8–2. *Innervation of a skeletal muscle. A,* A single axon of a motor neuron branches to a number of muscle fibers. *B,* Each axonal branch forms a myoneural junction with a fiber. *C,* The muscle fiber membrane forms a motor end plate that surrounds the terminal branch of an axon. The axon is separated from the sarcolemma by the synaptic cleft across which the transmitter substance, acetylcholine, drifts.

      a. **ACh** diffuses across the synaptic cleft to bind to receptors on the folds of the sarcolemma. This causes a sudden change in the muscle membrane's permeability to sodium and potassium ions and results in a reverse in the polarization (electrical potential) of the membrane. (See Chapter 9 for discussion of membrane potential.)

      b. The message of the electrical impulse (depolarization) spreads inward to the muscle fiber by means of the T-tubules to the sarcoplasmic reticulum.

      c. The sarcoplasmic reticulum releases its store of calcium ions into the vicinity of the overlapping thick and thin filaments. This results in the interdigitation of actin and myosin and the shortening of the sarcomeres.

      d. The series of events is called **excitation-contraction coupling.**

  6. When the nerve impulse ceases, the membrane depolarization is over, the calcium ions are recaptured by the sarcoplasmic reticulum, and the contractile process stops.

  7. ACh is in contact with the sarcolemma for only a few milliseconds. It is almost immediately broken down by the enzyme **cholinesterase** released by the sarcolemma folds. Such ACh breakdown is necessary to limit the duration of contraction and allow for repetitive contractions.

  8. Skeletal muscle also contains many sensory nerve endings.

 G. **Characteristics of skeletal muscle contraction.** Much of the information concerning muscle contraction has come from muscle-nerve preparations in the laboratory, usually from the gastrocnemius muscle of a frog with its motor nerve attached. One end of the muscle is fixed and the other end is movable and attached to a recording device (**myogram**) that senses and displays changes in the muscle's length. Electodes are inserted directly into the muscle and stimuli are applied to illustrate fundamental characteristics of contraction.

1. **Threshold stimulus** is the minimum electrical voltage that causes a contraction in a single muscle fiber.
    a. **All-or-none response of a muscle fiber.** Once threshold stimulation is reached, a muscle fiber responds maximally or not at all as long as conditions in the fiber's environment do not change.
    b. Increasing the intensity of the stimulus beyond threshold produces no further increase in response of the single muscle fiber.
2. **Muscle twitch**
    a. When a muscle preparation is stimulated, each muscle fiber in the muscle obeys the all-or-none law, but different fibers have different thresholds.
    b. As the stimulus is increased in voltage, additional fibers are recruited to respond.
    c. A muscle twitch (maximum contraction of the entire muscle) will occur when the intensity of the stimulus is adequate for all the muscle fibers. Figure 8–3 is a muscle twitch recorded as a myogram.
        (1) **Latent period** is the time between the stimulus or electrical event and the mechanical event of contraction. During this time, the muscle fibers become depolarized, the calcium ions are released, and the chemical reactions get underway.
        (2) **Contraction period** is the time during which the muscle is shortening.
        (3) **Relaxation period** is the time during which the muscle is returning to its original length. The relaxation period is longer than the contraction period.
        (4) **Magnitude of response** is the height of the wave.
        (5) **Refractory period** is the very brief time after one stimulus during which a muscle is unresponsive to a second stimulus.
        (6) A muscle twitch is induced under laboratory conditions and does not usually occur in the body. Contraction of the ocular muscle (eye blink) with a contraction time of 10 milliseconds is the closest to a twitch response.
3. **Graded muscle responses.** Muscle twitches are of little practical use in body movements, which require even, controlled muscle contractions of varying strengths, depending on the task required. Whole muscles respond in a graded fashion to **frequency** and **intensity** of the nerve impulses to the motor units.

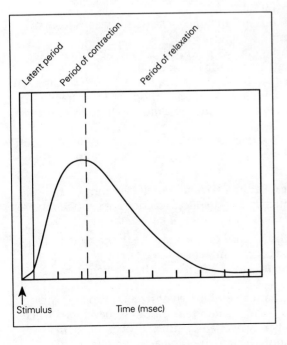

Figure 8–3. *A simple muscle twitch is composed of three periods: latent, contraction, and relaxation.*

a. **Wave summation** is a combination of twitches as a result of repetitive stimulation. When stimuli are applied in rapid succession so that the second contraction of the muscle begins before the first one is completely over, the two are added together to produce a greater and sustained contraction.
   (1) **Tetanic contraction.** If the frequency of stimuli increases beyond the ability of the muscle to relax at all, the contractions will fuse into a sustained and forceful contraction. Tetanic contractions are necessary and frequent in common muscle movements.
   (2) In the laboratory, continued stimulus to a muscle in tetany will produce **muscle fatigue** and the inability to sustain the contraction.
   (3) True muscle fatigue rarely, if ever, occurs in everyday muscle activities.
b. **Multiple motor unit summation** occurs when the different motor units within a muscle, each responding to different thresholds of stimulus, are activated. The more motor units responding, the greater the total strength of contraction.
c. Muscle activity in the body is graded, the result of asynchronous firing of motor nerve units at different frequencies, and utilizing both wave summation and multiple motor unit summation.

4. **Tonus.** Skeletal muscles in the body are always in a state of partial, sustained contraction called muscle tone. Nerve impulses from the spinal cord travel to the muscle fibers to keep roughly 10% of the fibers in a state of tetanic contraction on a rotating basis at all times.
   a. The degree of muscle tone depends on information from sensory receptors in muscle called muscle spindles, which sense the amount of contraction and relay the information to the spinal cord.
   b. Muscle tone is particularly important in postural muscles. It also generates body heat.

5. **Treppe.** If a resting muscle is stimulated at a moderate rate, the initial strength of contraction is much less than that of successive contractions and the myogram tracing looks like a staircase (Figure 8–4).
   a. The cause of treppe is unknown, but may be associated with an increase in the concentration, or perhaps the effectiveness, of calcium ions surrounding the myofibrils.
   b. The phenomenon of treppe is why all strenuous muscular activity should be preceded by a warm-up period of "working out" the muscles involved.

6. **Isometric and isotonic contractions**
   a. An **isometric contraction** is one in which a muscle develops force or tension without shortening to move a load.
      (1) Activation of crossbridges takes place, but no sliding of myofilaments occurs during an isometric contraction.
      (2) The tension developed in postural muscles acting to keep the head upright and the body standing are examples of isometric contractions.
   b. An **isotonic contraction** is one in which the muscle shortens to lift or move a load (do work).
   c. Muscles in the body can contract isometrically or isotonically. Most contractions are combinations of the two types. Walking or running, for example, utilizes both.

7. **Heat production by muscles.** Because skeletal muscles account for half the body weight, the heat liberated by the chemical reactions of contraction are a major source of body heat and maintenance of body temperature.

8. **Length-tension relationships in muscle.** Each muscle in the body has an optimum length at which maximal contraction force can be generated. Muscles contract most efficiently at the length they have when they are relaxed in the intact body.
   a. The sliding filament mechanism of muscle contraction explains the length-tension relationship. Maximum tension can be developed when the thin actin filaments barely overlap the thick myosin filaments so that sliding can occur along the length of the actin filaments.

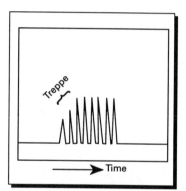

Figure 8–4. *Treppe.* Repeated stimuli of constant strength in a resting muscle produce an increase in the strength of the first few contractions.

- b. If a muscle is stretched beyond the optimum length, the thin filaments fail to overlap the thick filaments, leaving little availability of the myofilaments for actin-myosin interdigitation.
- c. If a muscle cell is shorter than the optimum length prior to contraction, less tension can be developed. The actin filaments soon overlap maximally, leaving fewer sites for interaction. The myosin filaments are compressed against the Z lines.

H. **Muscle fiber types.** Human skeletal muscles have three types of muscle fibers that differ in speed of contraction, resistance to fatigue, and ability to generate ATP.
1. **Slow-twitch red fibers** contain greater concentrations of the red respiratory pigment, **myoglobin**, which binds oxygen molecules to facilitate aerobic respiration.
    - a. The fibers are of thin diameter, surrounded by many capillaries to provide oxygen and nutrition, are slow to contract, and are resistant to fatigue.
    - b. Postural muscles of the back, adapted for constant activity, have a large proportion of slow red fibers.
2. **Fast-twitch white fibers** have no myoglobin, have fewer mitochondria and capillaries, but more stored glycogen and enzymes, which increase their capacity for anaerobic glycolysis.
    - a. The fibers are thicker, able to produce ATP at a high rate, but fatigue rapidly once their glycogen is depleted.
    - b. Fast white fibers are suited for quick bursts of muscular activity, such as running.
3. **Intermediate fibers** are red. They contain myoglobin and have properties and fatigue resistance intermediate to the other two types.
4. Most muscles contain a mixture of fiber types. It may be that genetically determined ratios of fast-to-slow twitch fibers are responsible for variations in athletic ability.

I. **Hypertrophy and atrophy**
1. **Muscle hypertrophy**, or enlargement of muscles, is the result of forceful, repetitive, muscular activity, rather than mild exercise. There is no increase in the number of fibers, but an increase in diameter and length of the fibers owing to an increase in fiber elements.
2. **Muscle atrophy** is the opposite of hypertrophy. If a muscle is not used, it will waste away. Eventually, the muscle fibers become infiltrated and replaced by fibrous and fatty tissue.

III. **SMOOTH MUSCLE.** Smooth muscle shares some mechanical and chemical properties with skeletal muscle but also has some unique characteristics.

A. **Myofilament differences**
1. Thick myosin filaments are longer than those in skeletal muscle.
2. Thin actin myofilaments lack troponin and tropomyosin.
3. Intermediate-sized myofilaments are present. They are not involved in the contractile process but are believed to serve as a cytoskeletal framework to support the cell.

B. **Contraction differences**
1. Although actin and myosin bind at the crossbridges in a smooth muscle cell, contraction in smooth muscle cells depends on the phosphorylation of myosin; i.e., when a phosphate group is attached to myosin.
2. In smooth muscle, the increased concentration of calcium ions binds with **calmodulin**, a protein that is structurally similar to troponin. The Ca^{++}/calmodulin complex activates **myosin kinase**, another intracellular protein, which phosphorylates myosin.

3. Some calcium ions are released from the sarcoplasmic reticulum, but most calcium enters through the plasma membrane upon opening of calcium ion channels.
4. When calcium ions are transported back into the sarcoplasmic reticulum and out across the plasma membrane, myosin is dephosophorylated, and the muscle relaxes.

C. **Smooth muscle types.** There are two major categories of smooth muscle based on how the muscle fibers are stimulated to contract.
1. **Multiunit smooth muscle** is found in the walls of large blood vessels, in the large airways of the respiratory tract, in the eye muscles that focus the lens and adjust pupil size, and in the arrector pili muscles of the hairs.
    a. Like skeletal muscle, multiunit smooth muscle is **neurogenic**; that is, it requires nerve stimulus to initiate contraction.
    b. Unlike skeletal muscle, there are no neuromuscular junctions. Neurotransmitter fluids are merely released into extracellular fluid surrounding the smooth muscle cells.
    c. Contractions of multiunit smooth muscle also can be influenced by certain hormones and drugs.
2. **Single-unit (visceral) smooth muscle** is found arranged in sheets in the walls of the hollow organs or viscera. All of the fibers in a sheet are able to contract as a single unit.
    a. Visceral smooth muscle is self-excitable or **myogenic** and does not require external nervous stimulation for contraction. It self-generates action potentials as a result of spontaneous electrical activity.
    b. The smooth muscle cells in a sheet are electrically linked together by communicating **gap junctions**, which quickly propagate an action potential throughout all the interconnected cells.
        (1) Autonomic nervous system endings on visceral, smooth muscle modify the rate and strength of contractions. Smooth muscle cells can be influenced by more than one type of neurotransmitter.
        (2) Other factors that influence visceral smooth muscle contractions are certain hormones, local metabolic intermediates produced in the vicinity of the muscle, mechanical stretch, and some types of drugs.

IV. **CARDIAC MUSCLE.** Cardiac muscle exhibits features of both skeletal and smooth muscle.

A. The myofilaments are organized into a regular banding pattern so cardiac muscle is striated.
1. Thin actin filaments contain troponin and tropomyosin. The mechanism of calcium ion action is the same as in skeletal muscle.
2. Cardiac muscle has T-tubules and a well-developed sarcoplasmic reticulum. It contracts according to the sliding filament mechanism.
3. Unlike skeletal muscle, part of the calcium ions released to initiate contraction come from extracellular fluids. Consequently, cardiac muscle is very sensitive to calcium imbalance in the body fluids.

B. Cardiac muscle is **myogenic** and initiates its own action potentials at a pacemaker without the necessity for nerve stimulation.
1. Gap junctions located at intercalated disks interconnect cardiac muscle cells and enhance the spread of depolarization throughout the heart.
2. Autonomic nerve endings on cardiac muscle, along with certain hormones, can modify the rate and strength of contraction.

PART II: MAJOR SKELETAL MUSCLES AND THEIR ACTIONS

I. GENERAL PRINCIPLES

A. **Movements are produced by skeletal muscles pulling on bones.** Most muscles in the body are attached to one bone, span at least one joint, and attach to another articulating bone.
  1. When the muscles contract, the shortening draws one of the bones toward the other one at the joint.
  2. Some muscles are not attached to bone at both ends. On the face, muscle is attached to the skin, which moves when the muscle contracts.

B. **Attachment and arrangement of skeletal muscles**
  1. **Origin** of a muscle is the point of attachment on the bone that is the more fixed and generally proximal end.
  2. **Insertion** of a muscle is the point of attachment that is the more moveable and generally more distal end.
      a. **Belly** of the muscle is the portion of the muscle between the origin and insertion.
      b. Functional reversal of origin and insertion may appear to occur, depending on the movement. For example, the pectoralis major, which originates on the clavicle, sternum, and ribs, inserts onto the humerus and adducts the arm. But when the muscle is used to pull up the body, as in climbing up a rope, the arms and hands are fixed and the body is the more movable part.
  3. **Muscle-tendon arrangements**
      a. In **parallel muscles**, the fascicles are organized parallel with the longitudinal axis of the muscle and end on a flat tendon to form a straplike muscle, as in the sartorius muscle.
      b. In **pennate muscles**, the fasciculi are arranged like barbs of a feather along the side of a tendon that runs the length of the muscle. Pennate muscles have a powerful contraction because all of the force of the myofibers is concentrated at the tendon.
          (1) A **unipennate** muscle has all fascicles on one side of the tendon, as in the semimembranosus muscle.
          (2) A **bipennate** muscle has fascicles converging at both sides of a tendon, as in the rectus femoris muscle.
          (3) A **multipennate** muscle has fascicles that converge at many tendons, as in the deltoid muscle.

C. **Muscles provide the force, bones serve as the levers, and joints serve as the fulcrums of levers.** A lever system consists of the **lever**, a rigid bar free to turn about a fixed point called its **fulcrum**; an object or **weight** to be moved; and a source of energy or **force** (power) that moves the weight. There are three types of lever systems in the body based on the position of the components (Figure 8–5).
  1. **Class I lever.** The fulcrum is between the force and the weight, as in a seesaw. This type of lever exists between the skull and the vertebral column. When the head is raised, the neck muscles provide the power, the facial bones are the weight, and the fulcrum is the joint between the skull and the backbone.
  2. **Class II lever.** The fulcrum is at one end, the force is at the opposite end, and the weight is between them, as in a wheelbarrow. This type of lever exists when the foot is raised on tiptoe. The posterior calf muscles provide the power, the leg bones are the weight, and the ball of the foot is the fulcrum.

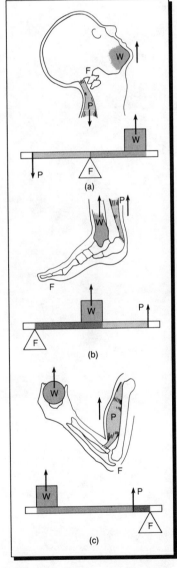

Figure 8–5. *Classes of levers: A, class I lever; B, class II lever; C, class III lever.* W = weight, P = power or force, and F = fulcrum.

3. **Class III lever.** The fulcrum is at one end, the weight is at the opposite end, and the force is between them, as in a forceps. This type of lever, the most common in the body, is exemplified by flexion of the forearm. The anterior upper arm muscles provide the power, the forearm bones and the hand are the weight, and the elbow joint is the fulcrum.

D. **Muscles that move a part generally do not lie over the part**, but lie proximal to the part moved. For example, the muscles that move the forearm lie on the upper arm; those that move the hand and wrist are on the forearm.

E. **Muscles act in groups rather than singly.** Rarely does a muscle act by itself. Some muscles in the group contract while others relax.

1. **Prime mover**, or agonist, is the muscle that performs most of the action in producing the movement.

2. **Synergists**, or fixators, are muscles that help the prime mover by contracting at the same time to assist in the movement or stabilize (fixate) a part so the movement is more effective.

3. **Antagonists** are muscles that relax when the prime mover and synergists are contracting. Antagonistic muscles have opposite actions and are located on opposite sides of the bone.

4. The same muscle can function as a prime mover, a synergist, or an antagonist, depending on the movement. For example, when the biceps brachii, the muscle that flexes the forearm, contracts, the triceps brachii, the muscle the extends the forearm, relaxes. When the triceps brachii contracts, the biceps brachii relaxes.

5. When the prime mover and synergists are stimulated by the nervous system to contract, the antagonists are reflexively inhibited.

F. **Muscles' names generally describe one or more of their features.** They may be named according to location, action, origin and insertion, direction of fibers, number of divisions, shape, points of attachment, or size, as in the following examples:

1. Action: **flexor** digitorum, **adductor** longus, **levator** anguli oris
2. Location: tibialis anterior, rectus **femoris**
3. Direction of fibers: **rectus** femoris, **transversus** abdominis
4. Shape: deltoid, trapezius
5. Number of divisions of the head (origin): biceps brachii, triceps brachii, quadriceps
6. Points of attachment: sternocleidomastoid
7. Size: maximus, minimus, brevis, longus

II. **MUSCLE TABLES** (Color Plates 3, 4, and 5)

Table 8-1. Muscles of the head and face: Muscles of facial expression.

Muscle	Description	Origin (O) and Insertion (I)	Action	Nerve Supply
Epicranius (Occipitofrontalis)	Broad muscular and tendinous layer over top and sides of skull; consists chiefly of the frontalis and occipitalis muscles			
Frontalis	Thin, broad muscle on forehead	O—galea aponeurotica* I—skin of eyebrow	Raises eyebrows; expresses surprise	Facial (VII) nerve
Occipitalis	Thin, broad muscle over base of skull	O—occipital bone mastoid process of temporal I—galea aponeurotica*	Draws scalp backward	Facial (VII) nerve
Orbicularis oculi	Broad, thin muscle that encircles the orbit and forms part of the eyelids	O—medial wall of orbit (frontal and maxillary bones) I—circumference of orbit, lower eyelid	Closes eye (sphincter muscle of eyelids); permits winking, blinking, and squinting	Facial (VII) nerve
Corrugator supercilii	Small muscle at medial end of eyebrow	O—on frontal bone of medial end of eyebrow I—skin of eyebrow	Pulls eyebrows together and downward; with orbicularis oculi produces wrinkles of forehead; involved in frowning	Facial (VII) nerve
Levator palpebrae superioris	Thin muscle located between the back of the bony orbit and upper eyelid	O—roof of orbit I—skin of upper eyelid	Raises upper eyelid	Oculomotor (III) nerve
Levator labii superioris	Broad muscle from orbit to upper lip	O—inferior margin of orbit I—muscle and skin of upper lip	Elevates upper lip; expresses sadness or seriousness, flares nostrils	Facial (VII) nerve
Depressor labii inferioris	Small muscle between the lower lip and the lower jaw	O—mandible I—lower lip	Depresses lower lip; expresses irony or a pout	Facial (VII) nerve
Levator anguli oris	Small muscle between side of nose and corner of the mouth	O—maxilla, below infraorbital foramen I—angle of mouth	Elevates angle of mouth; produces furrow between nose and upper lip	Facial (VII) nerve
Depressor anguli oris	Triangular muscle between corner of mouth and lower jaw	O—body of mandible, below mental foramen I—angle of mouth	Pulls angle of mouth downward (in opening mouth and expressing sadness)	Facial (VII) nerve
Zygomaticus major and minor	Muscle pair located between cheekbone and corner of mouth	O—zygomatic bone I—corners of mouth	Pulls angle of mouth up and outward	Facial (VII) nerve
Platysma	Flat, unpaired superficial muscle that extends from the front of the upper thorax, along the neck, to the corners of the mouth	O—fascia of chest and shoulder I—mandible; muscles and skin of lower part of the face	Draws lower lip and corners of mouth down; expresses surprise of horror; wrinkles skin of the neck	Facial (VII) nerve

* Galea aponeurotica is an extensive aponeurosis (broad, flat tendon) over the top and sides of skull that lies between the frontalis and occipitalis muscles.

Table 8-1. Muscles of the head and face: Muscles of facial expression.

Muscle	Description	Origin (O) and Insertion (I)	Action	Nerve Supply
Risorius	Narrow muscle extending diagonally from corner of mouth to in front of ear	O—fascia of cheek I—skin at angle of mouth	Draws angle of mouth laterally	Facial (VII) nerve
Orbicularis oris	Consists of muscle fibers having different directions that encircle the mouth	O—muscle fibers around mouth opening I—corners of mouth; superficially, skin of lips	Closes, purses, and protudes lips	Facial (VII) nerve
Mentalis	Small, conical chin muscle	O—anterior, medial mandible I—skin of chin	Raises and protrudes lower lip; wrinkles skin over chin	Facial (VII) nerve
Buccinator	Main cheek muscle	O—molar region of maxillae and mandible I—orbicularis oris muscle (below), which encircles mouth	Compresses cheek against teeth; sucks in cheeks, as when blowing	Facial (VII) nerve

Table 8-2. Muscles of the head: Muscles of mastication.

Muscle	Description	Origin (O) and Insertion (I)	Action	Nerve Supply
Temporalis	Broad, fan-shaped muscle at side of head (temple) and above ear	O—temporal fossa I—coronoid process and ramus of mandible	Elevates mandible (closes jaws); retracts mandible	Trigeminal (V) nerve
Masseter	Powerful multilayered muscle at corner of lower jaw	O—zygomatic arch; maxilla I—angle and lateral surface of ramus of mandible	Elevates mandible (closes jaws)	Trigeminal (V) nerve
Medial pterygoid	Thick two-headed muscle on inside of lower jaw; lies between corner of lower jaw and upper jaw	O—medial surface of lateral pterygoid plate of sphenoid bone; maxilla, palatine I—inner surface of mandible near its angle	Elevates mandible (closes jaws); moves mandible side to side (chewing movements)	Trigeminal (V) nerve
Lateral pterygoid	Short, two-headed muscle extending from mandibular condyle of lower jaw to bone behind eye	O—lateral surface, greater wing, and lateral pterygoid plate of sphenoid bone I—anterior side of mandibular condyle; capsule of temporomandibular joint	Protracts mandible (opens jaws); moves mandible side to side	Trigeminal (V) nerve

Table 8–3. Muscles that move the tongue.

Muscle	Description	Origin (O) and Insertion (I)	Action	Nerve Supply
Hypoglossus	Flat sheet of muscle on side of tongue	O—body and greater horn of hyoid bone I—side and inferior aspect of tongue	Depresses and retracts tongue	Hypoglossal (XII) nerve
Styloglossus	Slender muscle superior and at right angles to hypoglossus	O—styloid process I—side and inferior aspect of tongue	Retracts and elevates tongue	Hypoglossal (XII) nerve
Genioglossus	Vertical sheet of muscle fibers, which fans out into inferior and posterior aspect of tongue	O—geniol tubercle on mandible behind symphysis I—inferior aspect of tongue and body of hyoid bone	Protrudes tongue (posterior fibers); retracts tip of tongue (anterior fibers); depresses tongue	Hypoglossal (XII) nerve

Table 8–4. Muscles that move the eyeball.

Muscle	Description	Origin (O) and Insertion (I)	Action	Nerve Supply
Superior rectus	Muscle runs parallel to long axis of eye	O—tendon ring attached to orbital cavity I—superior, central part of sclera of eyeball	Rolls eyes upward	Oculomotor (III) nerve
Inferior rectus	Muscle runs parallel to long axis of eye	O—tendon ring attached to orbital cavity I—inferior, central part of sclera of eyeball	Rolls eyes downward	Oculomotor (III) nerve
Medial rectus	Muscle runs parallel to long axis of eye	O—tendon ring attached to orbital cavity I—midway on medial side of sclera of eyeball	Rolls eyes medially	Oculomotor (III) nerve
Lateral rectus	Muscle runs parallel to long axis of eye	O—tendon ring attached to orbital cavity I—midway on lateral side of sclera of eyeball	Rolls eyes laterally	Abducens (VI) nerve
Superior oblique	Muscle runs diagonally to long axis of eye	O—sphenoid bone (superior, medial part of orbital cavity) I—into sclera of eyeball between the superior and lateral recti	Rotates eyeball on its axis; directs cornea downward and laterally	Trochlear (IV) nerve
Inferior oblique	Muscle runs diagonally to long axis of eye	O—orbital plate of maxilla I—into sclera of eyeball between the superior and lateral recti	Rotates eyeball on its axis; directs cornea upward and laterally	Oculomotor (III) nerve

Table 8–5. Muscles of the throat and neck: Infrahyoid muscles that move the hyoid bone and laryngeal cartilages superiorly during swallowing and speech.

Muscle	Description	Origin (O) and Insertion (I)	Action	Nerve Supply
Superiorly				
Digastric	V-shaped muscle below lower jaw	O—anterior belly: inner surface of inferior border of mandible; posterior belly: mastoid process of temporal bone I—hyoid bone via intermediate tendon between anterior and posterior bellies	Elevates hyoid bone; assists in opening jaws (depresses mandible)	Anterior belly: trigeminal (V) (mandibular division); posterior belly: facial (VII) nerve
Stylohyoid	Slender muscle below angle of jaw	O—styloid process of temporal bone I—body of hyoid bone	Draws hyoid superiorly and posteriorly, elongating floor of the mouth during swallowing	Facial (VII) nerve
Mylohyoid	Flat, triangular muscle just under digastric: the two mylohyoids form a muscular floor beneath front of the mouth	O—internal surface of mandible I—body of hyoid bone	Elevates hyoid and floor of mouth enabling tongue to move food bolus into pharynx	Trigeminal (V) nerve (mandibular branch)
Geniohyoid	Narrow muscle, deep to medial side of mylohyoid	O—internal surface of mandibular symphysis I—body of hyoid bone	Draws hyoid anteriorly and superiorly, shortens floor of mouth and widens pharynx	Cervical nerve C1
Inferiorly				
Sternohyoid	Thin, narrow muscle between sternum and hyoid	O—medial end of clavicle; superior and posterior part of manubrium of sternum I—inferior border of hyoid bone	Draws hyoid inferiorly, depresses larynx	Cervical nerves C1, C2, C3
Sternothyroid	Shorter muscle lateral to sternohyoid; lies between sternum and larynx	O—manubrium of sternum; edge of first rib I—thyroid cartilage of larynx	Draws thyroid cartilage inferiorly	Cervical nerve C1, C2, C3
Thyrohyoid	Small muscle; appears as superior continuation of sternothyroid; lies between larynx and hyoid bone	O—thyroid cartilage of larynx I—hyoid bone	Draws hyoid inferiorly draws thyroid cartilage superiorly and elevates larynx	Cervical nerve C1 via the hypoglossal (XII) nerve
Omohyoid	Muscle with two bellies; extends from shoulder to hyoid bone	O—superior border of scapula I—hyoid bone	Depresses and retracts hyoid	Cervical nerves C1, C2, C3

Table 8-6. Muscles of the neck and vertebral column: Muscles that move the head and trunk.

Muscle	Description	Origin (O) and Insertion (I)	Action	Nerve Supply
Sternocleido-mastoid	Muscle passes obliquely along side of neck; prominent when contracted	O—two origins: sternum, clavicle I—mastoid process of temporal	Simultaneous contraction of both muscles flexes neck and head; contraction of one muscle rotates face toward opposite side	Accessory (XI) nerve; cervical nerves C2 and C3 of cervical plexus
Semispinalis capitis, cervicis, and thoracis	Composite muscle located along back from thoracis region to head	O—transverse processes of C7–T12 I—occipital bone (capitus) and spinous processes of C1–C7 (cervicis) and T1–T4 (thoracis)	Simultaneous contraction of both sides extends head and vertebral column; contraction of one side rotates head toward opposite side	Cervical and thoracic nerves
Splenius capitis and cervisis	Broad two-part muscle lies on back of neck between base of skull and upper thorax	O—ligamentum nuchae*; spines of C7–T6 I—mastoid process of temporal bone and occipital bone (capitus); transverse processes of C2–C4 vertebrae (cervicis)	Extends head and neck; contraction of one side rotates head and turns face toward same side	Cervical nerves
Scalenus (anterior posterior, medial)	Muscles on side of neck located deep to platysma and sternocleidomastoid	O—transverse processes of cervical vertebrae I—first two ribs	Flexes and rotates neck; elevates first and second rib (assists in inspiration)	Cervical nerves

* The **ligamentum nuchae** is a broad, fibrous septum in the back of the neck and it separates right and left sides.

Table 8–7. Posterior muscles that move the vertebral column.

Muscle	Description	Origin (O) and Insertion (I)	Action	Nerve Supply
Quadratus lumborum	Quadrilateral muscle in lower back between superior hip and last (12th) rib	O—iliac crest I—last rib; transverse processes of first lumbar vertebrae	Contraction on one side: lateral flexion of vertebral column; contraction on both sides: extend vertebral column; draw ribs inferiorly	Thoracic nerve T12, lumbar nerve L1
Erector spinae group (sacrospinalis)	The erector spinae group are posterior muscles of the back and the prime movers of back extension and erect posture. Each consists of three columns: (1) iliocostalis, (2) longissimus, and (3) spinalis, which subdivide.			
Lateral column				
Iliocostalis lumborum		O—crest of ilium I—last 6 to 7 ribs	Extends vertebral column and bends it to one side	Lumbar nerves
Iliocostalis thoracis		O—lower six ribs I—first six ribs; transverse process of seventh cervical vertebra	Extends vertebral column and bends it to one side	Thoracic nerves
Iliocostalis cervicis		O—angles of ribs 3, 4, 5, 6 I—transverse processes of cervical vertebrae 4 to 6	Extends vertebral column and bends it to one side	Cervical nerves
Intermediate column				
Longissimus thoracis		O—transverse processes of lumbar vertebrae I—transverse processes of all thoracic vertebrae and ribs 9 and 10	Extends thoracic part of vertebral column	Spinal nerves
Longissimus cervicis		O—trse processes of upper 4 to 5 thoracic vertebrae I—transverse processes of cervical vertebra 2 through 6	Extends cervical part of vertebral column	Spinal nerves
Longissimus capitis		O—transverse processes of upper 4 to 5 thoracic vertebrae; transverse processes of last 4 and 5 cervical vertebrae I—mastoid process of temporal bone	Contraction of both muscles: extends head; contraction of one muscle: bends head and rotates face toward same side	Cervical nerves
Medial column				
Spinalis thoracis (cervicis usually rudimentary, poorly developed)		O—spinous processes of first 2 lumbar and last 2 thoracic vertebrae I—spinous processes of upper thoracic vertebrae	Extends vertebral column	Spinal nerves

Table 8–8. Thoracic (respiratory) muscles.

Muscle	Description	Origin (O) and Insertion (I)	Action	Nerve Supply
Diaphragm	A broad, dome-shaped (in relaxed state) musculo-fibrous septum that separates the thoracic and abdominal cavities; forms the floor of the thoracic cavity and roof of the abdominal cavity; has openings for passage of aorta, inferior vena cava, and esophagus	O—xiphoid process; costal cartilages of last 6 ribs; lumbar vertebrae I—central tendon*	Contraction pulls diaphragm downward to increase vertical diameter of thorax on inspiration	Phrenic nerve
External intercostals	Eleven pairs of muscles lie between adjacent ribs on each side	O—inferior border of rib above the muscle I—superior border of rib below the muscle	Elevate ribs and increase volume of thoracic cavity; draw adjacent ribs together; synergist of diaphragm	Intercostal nerves
Internal intercostals	Eleven pairs of muscles pass between adjacent ribs; internal to and at right angles to external intercostals	O—inferior surface of rib and costal cartilages I—superior border of rib below	Lower (depress) ribs and decrease volume of thoracic cavity; draw adjacent ribs together	Intercostal nerves

* The **central tendon** is the large central tendinous part of the diaphragm encircled by peripheral muscle fibers.

Table 8–9. Muscles of the abdominal wall.

Muscle	Description	Origin (O) and Insertion (I)	Action	Nerve Supply
Rectus abdominis	Pair of medial long muscles separated from each other by the **linea alba*** (tendinous line) running vertically between them; tendons run horizontally across the long muscles (tendinous insertions) and divide them into four segments; muscles lie on the anterior body surface partially enclosed by a fibrous sheath	O—pubic crest; ligaments of pubic symphysis I—costal cartilages of ribs 5 through 7 and xiphoid process	Flexes and rotates vertebral column; draws sternum to pubis; compresses abdomen to increase intra-abdominal pressure	Thoracic nerves 17–T12
External oblique	Largest, most superficial of three muscles over lateral and anterior surfaces of abdomen; aponeurosis forms inguinal ligament between trunk and thigh; fibers run obliquely	O—lower 8 ribs I—linea alba*; iliac crest; pubis; by aponeurosis to xiphoid process	Contraction of both sides compresses abdomen, flexes vertebral column, and increases intra-abdominal pressure; contraction on one side bends column laterally	Thoracic nerves T7–T12; lumbar nerve L1
Internal oblique	Deep (internal to external) oblique; fibers run obliquely, but at right angles to external oblique	O—iliac crest; inguinal ligament; thoraco-lumbar fascia I—costal cartilages of last 3 to 4 ribs; linea alba	Same as above	Thoracic nerves T7–T12; lumbar nerve L1
Transversus abdominis	Innermost of three flat muscles of abdominal wall; fibers run transversely deep to internal oblique	O—iliac crest; inguinal ligament; thoraco-lumbar fascia; costal cartilages of last 6 ribs I—linea alba*; xiphoid process; pubis	Constricts abdomen; compresses contents of abdomen	Thoracic nerves T7–T12; lumbar nerve L1

***Linea alba** is the tendinous band that runs along the midline of the abdominal musculative.

Table 8-10. Muscles of the pelvis.

		Pelvic Diaphragm		
Muscle	Description	Origin (O) and Insertion (I)	Action	Nerve Supply
Levator ani	Broad, thin muscle that forms the sling-like floor of the pelvic cavity; consists of pubococcygeus and iliococcygeus muscles	O—anteriorly from superior ramus of pubis; posteriorly from spine of ischium I—end of coccyx; muscle and connective tissue of perineum	Supports pelvic viscera and slightly raises pelvic floor; constricts pelvic outlets	Sacral nerves S3 and S4; pudendal nerve
Coccygeus	Triangular flat muscle located posterior to the levator ani in floor of pelvic cavity	O—spine of ischium I—end of sacrum; coccyx	Same as levator ani; draws coccyx anteriorly; supports floor of pelvis.	Sacral nerves S3 and S4
		*Muscles of the Perineum**		
Superficial transverse perineus	Narrow muscular strip passing nearly transversely across front of anus	O—ischial tuberosity I—central tendon of perineum	Helps to stabilize the central tendon and support pelvic viscera	Pudendal nerve (sacral nerves S2, S3, S4)
Deep transverse perineus	Spans distance between ischia; surrounds vagina in females	O—ramus of ischium I—central tendon of perineum	Support of pelvic organs	Pudendal nerve (sacral nerves S2, S3, S4)
Bulbospongiosus	Muscle consists of two symmetric parts and lies along midline in front of anus; in female surrounds opening of vagina	O—central tendon of perineum I—urogenital diaphragm; corpus spongiosum of penis, and deep fascia on dorsum of penis in male; pubic arch plus root and dorsum of clitoris in female	Helps expel last drops of urine or semen from urethra and assists in erection of penis in male; constricts vagina in female	Pudendal nerve (sacral nerves S2, S3, S4)
Ischiocavernosus	Muscle lies between pelvis and base of penis or clitoris	O—ischial tuberosity; ischial and pubic rami I—corpus cavernosum of penis in male and clitoris in female	Erection of penis in male and clitoris in female	Pudenal nerve (sacral nerves S2, S3, S4)
Sphincter urethrae	Circular sphincter muscle surrounding urethra	O—ischial and pubic rami I—midline raphe	Constricts urethra; helps eject urine; helps eject semen in male	Pudendal nerve (sacral nerves S2, S3, S4)

*The **perineum** is the pelvic floor and other structures occupying the pelvic outlet. The boundaries of the perineum are the pubic symphysis anteriorly, the ischial tuberosities laterally, and the coccyx posteriorly.

Table 8–11. Muscles of the neck, thorax, and shoulder: Posterior muscles that move the pectoral girdle.

Muscle	Description	Origin (O) and Insertion (I)	Action	Nerve Supply
Trapezius	Superficial flat, triangular muscle that covers the back of the neck and shoulder	O—occipital bone ligamentum nuchae*; spine of seventh cervical and all thoracic vertebrae I—clavicle; acromion process and spine of scapula	Stabilizes, elevates, rotates, and retracts scapula; lower fibers depress scapula; with shoulder fixed, draws head backward and laterally	Accessory (XI) nerve; cervical nerves C3 and C4
Rhomboideus major	Large muscle located between vertebral column and scapula, deep to the trapezius	O—spines of T2 through 5; supraspinal ligament I—inferior angle and vertebral (medial) border of scapula	Steadies, retracts, and rotates scapula	Cervical nerve C5 (dorsal scapular nerve)
Rhomboideus minor	Muscle lies between vertebral column and scapula, superior to rhomboideus major	O—ligamentum nuchae*; spines of C7 and T1 I—vertebral (medial) border and superior angle of scapula	Steadies, retracts, and rotates scapula	Cervical nerve C5 (dorsal scapular nerve)
Levator scapulae	Thin, strap-like muscle between the scapula and the neck vertebrae; lies superior to rhomboideus minor and deep to trapezius	O—transverse processes of C1 through C4 I—vertebral (medial) border of scapula	Elevates, retracts, and steadies scapula; when scapula is fixed, pulls neck to the same side	Cervical nerves C3, C4, C5

*The ligamentum nuchae is a broad, fibrous septum in the back of the neck. It separates right and left sides.

Table 8–12. Muscles of the neck, thorax, and shoulder: Anterior muscles that move the pectoral girdle.

Muscle	Description	Origin (O) and Insertion (I)	Action	Nerve Supply
Pectoralis minor	Thin, triangular muscle between ribs and shoulder; deep to pectoralis major	O—anterior ends of ribs 3, 4, 5, and aponeurosis over intercostal muscles I—coracoid process of scapula	Draws glenoid end of scapula forward and downward; rotates scapula; when scapula is fixed, draws rib cage superiorly	Cervical nerve C8; thoracic nerve T1 (pectoral nerve)
Subclavius	Small, cylindrical muscle between lateral clavicle and first rib	O—junction of first rib and its costal cartilage I—underside of clavicle	Draws clavicle forward and down	Cervical nerves C5 and C6 (brachial plexus)
Serratus anterior	Large, broad muscle attaches to a series of lateral ribs and extends around the back beneath the scapula where it attached to vertebral (medial) border of scapula	O—superior borders of first 8 to 9 ribs; aponeurosis on intercostals I—anterior (deep) surfaces of superior angle, vertebral (medial) border, and inferior angle of scapula	Draws scapula forward; rotates scapula; holds scapula against chest wall	Cervical nerves C5, C6, C7 (long thoracic nerve of brachial plexus)

Table 8–13. Muscles of the neck, thorax, and shoulder: Muscles that move the arm.

Muscle	Description	Origin (O) and Insertion (I)	Action	Nerve Supply
Pectoralis major	Thick, fan-shaped muscle that covers front of chest	O—three origins: clavicle, sternum and costal cartilages of ribs 2 through 6; aponeurosis of external oblique muscle I—greater tubercle of humerus	Flexes, adducts, and rotates arm medially; raises ribs in forced inspiration	Lateral pectoral nerve (cervical nerves C5, C6, C7); medial pectoral nerve C8; thoracic nerve T1)
Latissimus dorsi	Broad, thick, triangular muscle that covers the lower thorax and back (lumbar region)	O—spines of last 6 thoracic, all lumbar and sacral vertebrae; crest of ilium; lower 3 to 4 ribs I—floor of intertubercular groove of humerus	Extends, adducts, and rotates arm medially; draws shoulder downward and backward	Thoracodorsal nerve (cervical nerves C6, C7, C8)
Deltoid	Large, thick, triangular muscle forming rounded mass over shoulder and upper humerus	O—lateral third of clavicle; acromion process and spine of scapula I—deltoid tuberosity of humerus	Abducts arm; anterior part flexes and rotates arm medially; posterior part extends and rotates arm laterally	Axillary nerve (cervical nerves C5 and C6)
Subscapularis	Large, triangular muscle occupying subscapular fossa on back of scapula	O—subscapular fossa of scapula I—lesser tubercle of humerus	Chief medial rotator of arm; helps hold head of humerus in glenoid cavity, stabilizing shoulder joint	Subscapular nerve (cervical nerves C5, C6, C7)
Supraspinatus	Muscle occupies supraspinous fossa on back of scapula	O—supraspinus fossa of scapula I—top of greater tubercle of humerus	Abducts arm; stabilizes shoulder joint	Suprascapular nerve (cervical nerves C4, C5, C6)
Infraspinatus	Muscle occupies infraspinous fossa on back of scapula	O—infraspinous fossa of scapula I—back of greater tubercle of humerus	Rotates arm laterally; stabilizes shoulder joint	Suprascapular nerve (cervical nerves C4, C5, C6)
Teres minor	Elongated muscle located between lower scapula and upper arm; lies inferior to infraspinatus	O—lower (axillary) border of scapula I—greater tubercle of humerus	Rotates arm laterally; stabilizes shoulder joint	Axillary nerve (cervical nerves C5 and C6)
Teres major	Thick, rounded muscle located inferior to teres minor	O—inferior angle of scapula, posterior surface I—crest of lesser tubercle of anterior humerus	Adducts, extends and medially rotates arm	Lower subscapular nerve (cervical nerves C6 and C7)
Coracobrachialis	Small elongated muscle runs from scapula to arm	O—coracoid process of scapula I—middle of humerus, medial edge	Flexes and adducts arm	Musculocutaneous nerve (cervical nerves C5, C6, C7)

Table 8–14. Muscles of the Arm: Muscles that move the arm and forearm.

Muscle	Description	Origin (O) and Insertion (I)	Action	Nerve Supply
Biceps brachii	Two-headed muscle over front of arm; forms prominent bulge above bend of the elbow	O—short head on coracoid process of proximal scapula; long head on tuberosity above glenoid I—tuberosity of proximal radius via common tendon	Flexes forearm at elbow joint; supinates forearm; weakly flexes arm at shoulder	Musculocutaneous nerve (cervical nerves C5 and C6)
Brachialis	Muscle deep to biceps brachii covers lower half of front of arm	O—distal end of anterior side of humerus I—tuberosity and coronoid process of ulna, anterior side	Powerful flexor of forearm	Musculocutaneous nerve (cervical nerves C5 and C6); radial nerve (cervical nerve C7)
Brachioradialis	Superficial muscle on radial side (thumb side) of elbow and forearm	O—lateral supracondylar ridge of distal humerus I—distal radius, just above styloid process	Effective flexor of forearm when forearm is in a partially flexed position	Radial nerve (cervical nerves C5 and C6)
Triceps brachii	Three-headed fleshy muscle covering the back of the arm	O—long head on infraglenoid tuberosity of scapula; lateral head on posterior side of humerus above radial groove; medial head on posterior side of humerus below radial groove I—olecranon process of ulna, via common tendon	Extends forearm	Radial nerve (cervical nerves C6, C7, C8)
Anconeus	Small, triangular muscle distal to triceps brachii and on lateral surface of proximal ulna at elbow	O—lateral epicondyle of humerus I—olecranon process of ulna	Extends forearm; abducts pronated forearm	Radial nerve (cervical nerves C7 and C8)

Table 8–15. Muscles of the forearm: Muscles that move the wrist and hand.

Muscle	Description	Origin (O) and Insertion (I)	Action	Nerve Supply
Anterior Muscles: Superficial flexors				
Pronator teres	Two-headed muscle between medial side of distal humerus and middle of the radius	O—medial epicondyle of humerus; medial side of coronoid process of ulna I—middle of the radius on lateral surface via common tendon	Pronates forearm and hand; weakly flexes forearm	Median nerve (cervical nerves C6 and C7)
Flexor carpi radialis	Muscle lies obliquely between medial side of distal humerus and base of thumb	O—medial epicondyle of humerus I—second and third metacarpals of hand	Flexes wrist; aids abduction of hand	Median nerve (cervical nerves C6 and C7)
Palmaris longus	Small muscle lies obliquely between medial side of distal humerus and base of the palm; has long tendon of insertion; may be absent	O—medial epicondyle of humerus I—carpal (wrist) ligaments (called flexor ratinaculum); palmar aponeurosis	Flexes wrist; tenses palmar aponeurosis (fascia of palm) during hand movements	Median nerve (cervical nerves C7 and C8)
Flexor carpi ulnaris	Two-headed muscle lies along medial surface of forearm between medial (ulnar) side of distal humerus and medial wrist	O—medial epicondyle of humerus; medial margin of olecranon process and upper two thirds of dorsal side of ulna I—pisiform and hamate bones of wrist; fifth metacarpal	Flexes and adducts hand	Ulnar nerve (cervical nerves C7 and C8)
Flexor digitorum superficialis	Two-headed muscle lies on front of forearm between distal humerus and the four fingers deep to flexor carpi radialis, palmaris longus, and flexor carpi ulnaris (above), but above flexor digitorium profundus and flexor pollicus longus (below)	O—medial epicondyle of humerus; medial side of coronoid process of ulna; anterior surface at radius from radial tuberosity to insertion of pronator teres I—second (middle) phalanges of four fingers	Flexes the four fingers and wrist; important in forceful flexion of fingers against resistance	Median nerve (cervical nerves C7 and C8; thoracic nerve T1)
Extensor digiti minimi	Superficial muscle on back of forearm lying between distal humerus and little finger; medial to and often connected with extensor digitorum	O—lateral epicondyle of distal humerus I—posterior extensor aponeurosis on little finger	Extends little finger; gives little finger independent movement; extends hand in conjunction with extensor digitorum	Radial nerve (cervical nerves C7 and C8)
Extensor carpi ulnaris	Long muscle on backside of forearm along ulnar (little finger) side between distal humerus and hand	O—lateral epicondyle of distal humerus I—base of fifth metacarpal ulnar side	Extends and adducts hand	Radial nerve (cervical nerves C7 and C8)

Table 8–15 (cont'd). Muscles of the forearm: Muscles that move the wrist and hand.

Muscle	Description	Origin (O) and Insertion (I)	Action	Nerve Supply
Posterior Muscles: Deep				
Supinator	Deep muscle on back of elbow; lies between distal humerus and upper third of radius; almost completely surrounds radius	O—lateral epicondyle of distal humerus, ligament of elbow joint; proximal ulna I—lateral surface of proximal third of radius	Supinates forearm; biceps brachii aids it in forceful supination	Radial nerve (cervical nerves C5 and C6)
Abductor pollicis longus	Lies on back of forearm between middle of forearm and the thumb immediately distal to supinator; lateral to extensor pollicis longus and brevis; becomes superficial in distal quarter of forearm on thumb side	O—posterior surface of middle third of ulna and radius; interosseous membrane I—base of first metacarpal lateral side	Abducts and extends thumb	Radial nerve (cervical nerves C7 and C8)
Extensor pollicis brevis	Lies posterior on forearm between middle of radius and thumb; medial to abductor pollicis longus	O—middle of radius, posterior surface; interosseous membrane I—base of first phalanx of thumb, posterior surface	Extends first phalanx of thumb	Radial nerve (cervical nerves C7 and C8)
Extensor pollicus longus	Lies posterior on forearm between middle of ulna and thumb; medial to and larger than extensor pollicis brevis; partly covers it	O—lateral side of middle of ulna, posterior surface; interosseous membrane I—base of last phalanx of thumb	Extends last phalanx of thumb	Radial nerve (cervical nerves C7 and C8)
Anterior Muscles: Deep flexors				
Flexor digitorum profundus	Large muscle deep to flexor digitorum superficialis	O—upper two thirds of ulna adjoining interosseous membrane* I—distal row of phalanges of fingers 2 through 5	Flexes fingers and assists in flexing wrist; involved in less forceful, unresisted flexion of fingers	Ulnar nerve and median nerve (cervical nerve C8; thoracic nerve
Flexor pollicis longus	Muscle lies on radial (thumb) side of forearm lateral to flexor digitorum profundus; tendon crosses wrist and continues to the lower thumb	O—anterior surface of upper radius; adjacent part of interosseous membrane* I—distal phalanx of thumb	Flexes joints of thumb	Median nerve (cervical nerve C8; thoracic nerve T1)
Pronator quadratus	Flat, square muscle lying across front of distal radius and ulna	O—distal part of anterior surface of ulna I—distal end of anterior surface of radius	Pronates forearm and hand; also holds distal radius and ulna together	Median nerve (cervical nerve C8; thoracic nerve T1)

*The interosseus membrane is fibrous connective tissue between the radius and ulna.

Table 8–15 (cont'd). Muscles of the forearm: Muscles that move the wrist and hand.

Muscle	Description	Origin (O) and Insertion (I)	Action	Nerve Supply
Posterior Muscles: Superficial extensors listed from lateral to medial aspect of arm				
Extensor carpi radialis longus	Muscle on radial (thumb) side of forearm; parallels brachioradialis	O—lateral supracondylar ridge of distal humerus I—posterior side of base of second metacarpal, radial side	Extends and abducts hand	Radial nerve (cervical nerves C6 and C7)
Extensor carpi radialis brevis	Slightly shorter than, and covered by, extensor carpi radialis longus	O—lateral epicondyle of distal humerus I—posterior side of base of third metacarpal, radial side	Extends and abducts hand	Radial nerve (cervical nerves C7 and C8)
Extensor digitorum	Superficial muscle on back of forearm lying between distal humerus and fingers	O—lateral epicondyle of distal humerus I—by 4 tendons to posterior phalanges of fingers 2 through 5	Extends fingers and hand	Radial nerve (cervical nerves C7 and C8)
Extensor indicis	Narrow muscle on posterior forearm; Located between distal ulna and index finger; runs parallel and medial to extensor policis longus	O—distal part of ulna on posterior surface, below extensor pollicis longus; interosseous membrane I—extensor expansion* of index finger; joins tendon of extensor digitorum to index finger	Extends index finger; gives index finger independent movement, as in pointing	Radial nerve (cervical nerves C7 and C8)

* The **extensor expansion** is a tendinous hood over the back of the proximal phalanx of the finger.

Table 8-16. Muscles of the hip: Muscles that move the thigh.

Muscle	Description	Origin (O) and Insertion (I)	Action	Nerve Supply
Muscles of the Pelvic Girdle				
Iliopsoas consists of three muscles: psoas major, iliacus, and psoas minor. The latter is a weak flexor of the vertebral column.				
Psoas major	Long, large muscle located lateral to lumbar vertebrae and passing through pelvis to medial, superior thigh (femur)	O—transverse processes, body, and discs of all lumbar vertebrae I—lesser trochanter of distal femur	Flexes thigh; flexes lumbar vertebrae laterally; flexes trunk when thigh is fixed	Lumbar nerves L2 and L3
Iliacus	Flat, triangular muscle located on inner surface of hip lateral to psoas major	O—iliac fossa of pelvis I—lesser trochanter of distal femur, along with psoas major	Flexes thigh when the pelvis is fixed, flexes trunk when thigh is fixed	Femoral nerve (lumbar nerves L2 and L3)
Psoas minor	Small muscle located anterior to psoas major. It is frequently absent.	O—anterior lateral surface of T12 and L1 vertebrae I—lesser trochanter of femur	Flexes pelvis on vertebral column; assists psoas major	Lumbar nerves L1 and L2
Anterior Muscles				
Pectineus	Flat, quadrangular muscle located between pubic bone of pelvis and medial superior thigh; inferior to psoas major and superior to adductor longus (below)	O—superior ramus of pubis I—line from lesser trochanter to linea aspera of femur	Flexes and abducts thigh	Femoral nerve (lumbar nerves L2 and L3)
Adductor longus	Triangular muscle located between pubic bone of pelvis and middle of the thigh; medial muscle is the medial border of femoral triangle*; most anterior of three adductors	O—front of pubis near symphysis pubis I—linea aspera of femur (middle third of femur)	Adducts, laterally rotates, flexes thigh; stabilizer during flexion and extension of thigh	Obturator nerve (lumbar nerves L2, L3, L4)
Adductor brevis	Triangular muscle located between pubic bone of pelvis and medial superior thigh; deep to pectineus and adductor longus	O—inferior ramus of pubis I—line from lesser trochanter to linea aspera of femur	Adducts and laterally rotates thigh; stabilizer during flexion and extension of thigh	Obturator nerve (lumbar nerves L2, L3, L4)
Adductor magnus	Large triangular muscle located on medial thigh between pubis and ischium of pelvis and medial, upper femur	O—inferior ramus of pubis; inferior ramus of ischium and ischial tuberosity I—tubercle on medial condyle of femur and linea aspera of femur	Adducts, laterally rotates, and flexes thigh; inferior part extends thigh (synergist for hamstrings); stabilizer during flexion and extension	Obturator nerve (lumbar nerves L2, L3, L4)
Gracilis	Superficial, long thin muscle on medial thigh between lower pelvis and knee	O—lower medial edge of pubis and ischium I—medial surface of upper tibia	Adducts thigh; flexes and medially rotates leg	Obturator nerve (lumbar nerves L2 and L3)
Tensor fascia lata	Superficial lateral hip muscle	O—anterior end of iliac crest; anterior superior iliac spine of pelvis I—iliotibial tract to tibia	Tenses iliotibial tract**; aids in flexion, abduction, and medial rotation of thigh	Superior gluteal nerve (lumbar nerves L4 and L5; sacral nerve S1)

***Femoral triangle** is in the upper third of the front of thigh. The borders of the triangle are formed laterally by the medial border of the sartorius muscle, medially by the medial border of the adductor longus, and superiorly by the inguinal ligament (between anterior superior iliac spine and pubis of pelvis). The femoral triangle contains the femoral vessels and nerve.

Table 8-16 (cont'd). Muscles of the hip: Muscles that move the thigh.

Muscle	Description	Origin (O) and Insertion (I)	Action	Nerve Supply
Posterior Muscles				
Gluteus maximus	Large, superficial muscle on back of hip forming bulk of buttock; lies between lower vertebral column and superior thigh	O—posterior end of crest of ilium; dorsal surface of sacrum and coccyx and associated ligaments I—iliotibial tract of of fascia lata**; gluteal tuberosity of proximal femur	Extends and laterally rotates thigh; extends hip against resistance, as when raising the trunk after bending over; through the iliotibial tract it braces the knee joint	Inferior gluteal nerve (lumbar nerve L5, sacral nerves S1 and S2)
Gluteus medius	Partially superficial muscle on posterior hip (pelvis); muscle lies between upper, lateral pelvis and superior femur; lower part of muscle covered by gluteus maximus	O—outer surface of ilium: upper, lateral part I—greater trochanter of femur, lateral side	Abducts and medially rotates thigh; during walking, acts to stabilize the pelvis on the femurs	Superior gluteal nerve (lumbar nerve L5, sacral nerve S1)
Gluteus minimus	Posterior fan-shaped muscle between upper, lateral pelvis and superior femur; located on back of hip deep to gluteus medius; smallest of gluteal muscles	O—outer surface of ilium: upper, lateral part I—greater trochanter of femur, anterior surface	Adducts thigh; medially rotates thigh; during walking, acts to stabilize pelvis on femurs	Superior gluteal nerve (lumbar nerve L5, sacral nerve S1)
Piriformis	Pyramidal muscle between sacrum and superior femur; lies alongside gluteus medius inferior to gluteus minimus	O—anterior sacrum and adjoining edge of ilium I—greater trochanter of femur, superior border	Rotates thigh laterally; abducts flexed thigh; stabilizes hip joint	Sacral nerves S1 and S2
Obturator internus	Surrounds most of obturator foramen	O—margin of obturator foramen; inner surface of obturator membrane† I—medial surface of greater trochanter of femur	Rotates thigh laterally; abducts flexed thigh; stabilizes hip joint	Sacral plexus (lumbar nerve L5, sacral nerve S1)
Obturator externus	Flat, triangular muscle over external surface of pelvis and obturator foramen	O—rami of pubis and ischium at margin of obturator foramen; outer surface of obturator membrane† I—trochanteric fossa of posterior femur (between greater trochanter and head)	Rotates thigh laterally; abducts thigh	Obturator nerve (lumbar nerves L3 and L4)
Quadratus femoris	Short, quadri-lateral muscle between lower pelvis and superior femur	O—tuberosity of ischium I—crest below greater trochanter of femur	Rotates thigh laterally nerve S1)	Nerve to femoris (lumbar nerve L5, sacral

Fascia lata is the tough, thick, deep fascia completely surrounding the thigh. A specialized, ligamentous band within this fascia and located along the lateral thigh is the **iliotibial tract.

†The obturator membrane is fascia covering the obturator foramen.

Table 8–17. Muscles of the thigh: Muscles that move the leg and knee joint.

Muscle	Description	Origin (O) and Insertion (I)	Action	Nerve Supply
Anterior Muscles				

Quadriceps femoris is a very large, fleshy muscle that covers the front and sides of the thigh and is composed of four separate muscles (rectus femoris, vastus lateralis, vastus medialis, and vastus intermedius) that share a common tendon of insertion, the patellar tendon, which contains the patella (kneecap), and extends to the tibia. The quadriceps femoris is the great extensor muscle of the leg and is important in climbing and running.

Muscle	Description	Origin (O) and Insertion (I)	Action	Nerve Supply
Rectus femoris	Middle part of the front of the anterior thigh; extends from the lower pelvis across hip joint straight down femur to anterior tibia; referred to as the kicking muscle	O—two tendons on ilium of pelvis: attached to the anterior inferior iliac spine and above brim of the acetabulum I—base of patella and anterior tibia	Extends leg at knee and flexes thigh at hip	Femoral nerve (lumbar nerves L2, L3, L4)
Vastus lateralis	Largest of the four muscles, located on lateral side of thigh; extends from proximal thigh to superior tibia	O—lateral edge of linea aspera, greater trochanter, and gluteal tuberosity of proximal femur I—lateral border of patella and anterior tibia	Extends leg at knee	Femoral nerve (lumbar nerves L2, L3, L4)
Vastus medialis	Thick muscle located on medial surface of thigh; forms bulge in inferior medial thigh	O—medial edge of linea aspera of femur along with ridges above and below I—medial border of patella and medial tibia (medial condyle)	Extends leg at knee	Femoral nerve (lumbar nerves L2, L3, L4)
Vastus intermedius	Located on anterior femur between vastus lateralis and vastus medialis, deep to rectus femoris	O—anterior surface of shaft of femur, upper two thirds I—lateral border of patella and lateral tibia (lateral condyle)	Extends leg at knee	Femoral nerve (lumbar nerves L2, L3, L4)
Sartorius	Long, strap-like, superficial muscle that originates on upper, lateral pelvis, crosses thigh obliquely, and descends to inside of knee; longest muscle in the body	O—anterior superior iliac spine of pelvis I—medial aspect of proximal tibia	Flexes leg on thigh; flexes thigh on pelvis	Femoral nerve (lumbar nerves L2 and L3)
Gracilis	Long, thin, superficial muscle of inner thigh; lies between lower, medial pelvis and upper, medial (tibia)	O—inferior medial edge of pubis and ischium I—proximal tibia, medial surface	Flexes leg and rotates it medially; adducts thigh	Obturator nerve (lumbar nerves L2 and L3)

Table 8-17 (cont'd). Muscles of the thigh: Muscles that move the leg and knee joint.

Muscle	Description	Origin (O) and Insertion (1)	Action	Nerve Supply
Posterior Muscles				
Hamstrings are three muscles on back of thigh: biceps femoris, semitendinosus, and semimembranosus. They span the hip and knee joints.				
Biceps femoris	Two-headed muscle covers posterior and lateral side of thigh between inferior pelvis and superior tibia and fibula	O—long head: ischial tuberosity (on pelvis); short head: linea aspera of femur I—proximal fibula, lateral surface; lateral condyle of tibia	Flexes and laterally rotates leg at knee; long head extends thigh	Sciatic nerve (lumbar nerve L5, sacral nerves S1 and S2), tibial branch to long head and common peroneal branch to short head
Semitendinosu	Lies on back of thigh between lower pelvis and upper leg (tibia)	O—ischial tuberosity (on pelvis) I—proximal tibia, medial surface	Flexes and medially rotates leg at knee; extends thigh at hip	Sciatic nerve, tibial branch (lumbar nerve L5, sacral nerves S1 and S2)
Semimembranosus	Muscle with membranous tendon of origin lies deep to semitendinosus	O—ischial tuberosity (on pelvis) I—proximal tibia medial surface	Flexes and medially rotates leg at knee; extends thigh at hip	Sciatic nerve, tibial branch (lumbar nerve L5, and sacral nerves S1 and S2)

Table 8–18. Muscles of the leg: Muscles that move the ankle and foot.

Muscle	Description	Origin (O) and Insertion (I)	Action	Nerve Supply
Anterior Superficial Muscles				
Tibialis anterior	Superficial, large, thick muscle lateral to superficial margin of tibia (shin)	O—proximal half of tibia, lateral surface including lateral condyle I—medial cuneiform and base of first metatarsal of foot, medial surface	Dorsiflexes foot and inverts foot (sole of foot is turned medially)	Deep peroneal nerve (lumbar nerves L4 and L5)
Extensor hallucis longus	Muscle on anterior of leg between middle of the leg and the big toe	O—anterior surface of middle section of fibula; interosseous membrane I—distal phalanx of big toe, superior surface	Extends big toe; dorsiflexes foot and assists foot inversion	Deep peroneal nerve (lumbar nerve L5, sacral nerve S1)
Extensor digitorum longus	Anterior lateral leg, immediately lateral to tibialis anterior	O—proximal three quarters of fibula, medial surface; lateral condyle of tibia; upper interosseous membrane* I—second and third phalanges of lateral four toes (two to five), superior surface	Extends lateral four toes; dorsiflexes foot	Same
Peroneus tertius	Small muscle located between inferior, lateral fibula and foot; is the lower, lateral part of extensor digitorum longus	O—distal one third of fibula, medial surface, and adjacent interosseous membrane I—base of fifth metatarsal (little toe side), posterior surface	Everts and plantar flexes foot	Superficial peroneal nerve (lumbar nerves L4 and L5, sacral nerve S1)
Lateral Superficial Muscles				
Peroneus longus	Superficial muscle on lateral leg between superior leg and foot	O—proximal two thirds of fibula, lateral surface I—base of first metatarsal and medial cuneiform bone; tendon crosses bottom of foot lateral to medial side	Everts foot; plantar flexes foot	Superficial peroneal nerve (lumbar nerves L4 and L5, sacral nerve S1)
Peroneus brevis	Short muscle on inferior, lateral leg deep to peroneus longus; tendon of insertion winds around lateral malleolus and passes to foot	O—distal two thirds of fibula, lateral surface I—base of fifth metatarsal, lateral side	Everts foot; plantar flexes foot	Superficial peroneal nerve (lumbar nerve L5, sacral nerves S1 and S2)
Posterior Superficial Muscles				
The *triceps surae* consists of three muscles formed by the two heads of the gastrocnemius and the soleus.				
Gastrocnemius	Two-headed superficial calf muscle; lies between lower thigh and heel; crosses two joints; forms prominent bulge in upper calf	O—posterior femur; medial head: medial condyle of femur; lateral head: lateral condyle of femur I—by calcaneal (Achilles) tendon to the calcaneus bone	Plantar flexes foot; flexes leg at knee; important in locomotion	Tibial nerve, (lumbar nerves L4 and L5, sacral nerves S1 and S2)

*The **interosseous membrane** is tough, fibrous connective tissue between the fibula and tibia.

Table 8–18 (cont'd). Muscles of the leg: Muscles that move the ankle and foot.

Muscle	Description	Origin (O) and Insertion (I)	Action	Nerve Supply
Soleus	Large, broad calf muscle deep to gastrocnemius; lies between superior leg and heel; crosses only the ankle joint	O—posterior upper one quarter of fibula; middle one third of tibia, medial border I—joins tendon of gastrocnemius to form calcaneal (Achilles) tendon attached to calcaneus bone	Plantar flexes foot; important in posture	Tibial nerve
Plantaris	Calf muscle with a small belly near the two heads of the gastrocnemius; long slender tendon extending to the heel; may be absent	O—ridge above lateral condyle of femur I—slender tendon joins calcaneal (Achilles) tendon to calcaneus bone	Assists gastrocnemius in plantar flexion of foot and flexion of leg	Tibial nerve

Posterior Deep Muscles

Muscle	Description	Origin (O) and Insertion (I)	Action	Nerve Supply
Popliteus	Thin, flat, triangular muscle on back of knee; deep to heads of gastrocnemius	O—lateral condyle of femur, plus lateral meniscus of knee I—posterior, superior tibia below medial condyle	Rotates tibia medially on femur with foot off the ground; rotates femur laterally with foot fixed	Tibial nerve
Tibialis posterior	Long muscle deep to soleus lies alongside lateral surface of tibia behind tibialis anterior	O—proximal tibia and fibula; interosseous membrane between tibia and fibula I—tendon passes in back of medial malleolus of tibia to several tarsal bones and metatarsals 2, 3, and 4 on underside of foot	Inverts foot; assists in plantar flexion of foot	Tibial nerve
Flexor hallucis longus	Lateral, deep muscle along lower fibula; tendon crosses back of ankle, winds behind medial malleolus and runs across the bottom of the sole of the foot to the end of the big toe	O—posterior, lower fibula; interosseous membrane I—distal phalanx of great toe, inferior surface	Flexes great toe; plantar flexes foot; active in tip-toe movements	Tibial nerve
Flexor digitorum longus	Medial, thin muscle along tibia; tendon of insertion passes behind the medial malleolus, crosses the bottom of the sole of the foot obliquely, and divides into four parts—one to each of the lateral four toes	O—posterior middle tibia I—distal phalanges of four lateral toes, on underside of toes	Flexes four lateral toes; plantar flexion of foot	Tibial nerve

Study Questions

Directions: Each question below contains five suggested answers. Choose the **one best** response to each question.

1. All of the following are functions of muscle EXCEPT
   (A) movement
   (B) protection
   (C) body support
   (D) posture maintenence
   (E) production of body heat

2. Which of the following statements about smooth muscle is true?
   (A) Smooth muscle is striated and involuntary.
   (B) Nuclei are peripherally located in the fibers.
   (C) Fibers are small and spindle shaped.
   (D) Branching fibers are a characteristic.
   (E) Contractions are rapid and forceful.

3. In a skeletal muscle fiber, which of the following best describes the composition of the structure known as a triad?
   (A) actin, troponin, and tropomyosin
   (B) terminal cisterna, transverse tubule, and terminal cisterna
   (C) A band, I band, and H band
   (D) sarcolemma, sarcoplasm, and sarcoplasmic reticuum
   (E) ATP, CP, and glycogen

4. In the sliding filament model of muscle contraction, the _____ myofilaments move toward each other.
   (A) myosin
   (B) thick
   (C) actin
   (D) actinomyosin
   (E) none of the myofilaments moves

5. An entire skeletal muscle is surrounded by
   (A) sarcolemma
   (B) epimysium
   (C) perimysium
   (D) epidermis
   (E) endomysium

6. Which of the following substances increases in quantity during muscle contraction?
   (A) adenosine trophosphate (ATP)
   (B) lactic acid
   (C) creatine phosphate
   (D) glucose
   (E) oxygen

7. The role of calcium ions in the contraction of skeletal muscle is
   (A) Calcium ions directly activate ATPase in the myosin head.
   (B) Calcium ions bind to lactic acid to remove it from the contracting muscle.
   (C) Calcium ions bind to the troponin-tropomyosin complex and remove their inhibitory action on actin/myosin interaction.
   (D) The release of calcium ions triggers the immediate regeneration of creatine phosphate to power the contraction.
   (E) Calcium ions diffuse across the synaptic cleft to result in depolarization of the muscle membrane.

8. The all-or-none response of a muscle fiber means
   (A) All of the muscles in a region of the body contract together to perform a movement.
   (B) All of the muscle fibers in a single muscle contract together.
   (C) A muscle fiber responds maximally to threshold stimulation or not at all.
   (D) All of the force generated by contraction of a muscle has no relationship to its muscle fiber elements.
   (E) When a muscle fiber contracts, it uses all of its stored ATP within the first few seconds.

9. All of the following are likely to result in muscle contraction in the laboratory EXCEPT
   (A) Calcium ions are injected into the muscle.
   (B) Acetylcholine is placed on the region of the motor end plate of the muscle.
   (C) ATP is injected into the muscle.
   (D) An electrode delivers a threshold stimulus to the muscle.
   (E) A high voltage stimulus is delivered to the muscle.

10. The term that means a continued mild or partial contraction is muscle
    (A) tone
    (B) twitch
    (C) stimulation
    (D) tetanus
    (E) summation

Questions 11–15. Choose the letter from the figure that most appropriately corresponds to the structure or statement. A letter may be used once, more than once, or not at all.

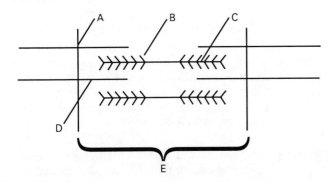

11. Z-line

12. sarcomere

13. location of troponin and tropomyosin

14. myosin head

15. actin myofilament

Questions 16–20. Select the most appropriate muscle from the list below to match each of the following facial expressions,

(A) zygomaticus
(B) orbicularis oculi
(C) depressor anguli oris
(D) frontalis
(E) orbicularis oris

16. kissing, pursing the lips

17. raising the eyebrows

18. smiling

19. frowning

20. squinting

Answers and Explanations

1. **The answer is B.** (I A) Protection is not a function of the muscular system.

2. **The answer is C.** (I E 2) Smooth muscle fibers are small and spindle shaped. Cardiac muscle has branching fibers and is striated and involuntary. Skeletal muscle has peripherally located nuclei and has rapid and forceful contractions.

3. **The answer is B.** (I D 3c) The triad refers to a set of three membranous channels, and is formed by a single transverse tubule and two terminal cisternae, or elements of the sarcoplasmic reticulum.

4. **The answer is C.** (II C 1) In the sliding filament hypothesis, during contraction, the actin filaments slide toward each other to extend further into the A band.

5. **The answer is B.** (II A 2) Epimysium surrounds the entire muscle, perimysium surrounds bundles of fibers (fasciculi), and endomysium surrounds each fiber.

6. **The answer is B.** (II E 4) Lactic acid increases in quantity during muscular contraction. ATP, CP, glucose, and oxygen all decrease in quantity.

7. **The answer is C.** (II D 4) Calcium ions bind to the troponin-tropomyosin complex and remove its inhibitory action on actin-myosin interdigitation. Acetylcholine diffuses across the synaptic cleft to bind with receptors on the muscle cell membrane, which causes a change in electrical potential and depolarization of the membrane. Lactic acid diffuses out of the muscle during brief and moderate exercise.

8. **The answer is E.** (II G 1) The all-or-none response of a muscle fiber means a muscle fiber responds maximally to threshold stimulation or not at all.

9. **The answer is C.** (II D 4; F 5; G) ATP injected into the muscle will not result in muscle contraction. An electrical stimulus to the muscle, acetylcholine on the motor end plate, and injection of calcium ions all would be likely to cause muscle contraction.

10. **The answer is A.** (II G 1–7) Muscle tone is a state of partial, sustained contraction in a muscle that results from tetanic contraction of about 10% of the muscle fibers on a rotating basis. Tetanic contraction is sustained and forceful contraction that results from high-frequency stimulation from the nervous system. A muscle twitch is a brief, maximum contraction of a muscle in response to a stimulus.

11–15. **The answers are 11–A, 12–E, 13–D, 14–B, 15–D.** (II B 3; C 1,2)

16–20. **The answers are 16–E, 17–D, 18–A, 19–C, 20–B.** (Muscle Tables)

The Nervous System 9

PART I: ORGANIZATION, CELLS, AND NERVE IMPULSES

I. INTRODUCTION

A. The nervous system is a complex and continuous series of organs composed primarily of nervous tissue. It is the mechanism by which the internal environment and external stimuli are monitored and regulated. The specialized qualities of **irritability**, or sensitivity to stimuli, and **conductivity**, or the ability to transmit a response to stimulation, are utilized by the nervous system in three broad ways:

1. **Sensory input.** The nervous system receives sensations or stimuli by way of **receptors**, which are located externally in the body proper (**somatic receptors**) or internally (**visceral receptors**).
2. **Integrative activities.** The receptors transform the stimuli into electrical impulses that pass along nerves to the brain and spinal cord, which interpret and integrate the stimuli so that a response to the information can occur.
3. **Motor output.** Impulses from the brain and spinal cord elicit appropriate responses from muscles and glands of the body, which are known as **effectors**.

B. **Structural organization of the nervous system**

1. The **central nervous system (CNS)** consists of the brain and spinal cord protected, respectively, by the bone of the cranium and the vertebral canal.
2. The **peripheral nervous system (PNS)** includes all other nervous tissue in the body. It consists of the cranial nerves and the spinal nerves that link the brain and the spinal cord with the receptors and effectors. Functionally, the PNS is subdivided into an afferent system and an efferent system.
   a. **Afferent (sensory) nerves** transmit information from sensory receptors to the CNS.
   b. **Efferent (motor) nerves** transmit information from the CNS to muscles and glands. The efferent system of the PNS has two subdivisions.
      (1) The **somatic (voluntary) division** is concerned with changes in the external environment and the formation of voluntary motor responses in the skeletal muscles.
      (2) The **autonomic (involuntary) division** controls all internal involuntary responses in smooth muscles, cardiac muscle, and glands by transmitting nerve impulses along two pathways.
         (a) **Sympathetic nerves** emerge from the thoracic and lumbar areas of the spinal cord.

(b) **Parasympathetic nerves** emerge from the brain and the sacral areas of the spinal cord.
(c) Most internal organs under autonomic control have both sympathetic and parasympathetic innervation.

II. CELLS OF THE NERVOUS SYSTEM

A. The **neuron** is the functional unit of the nervous system and consists of a cell body and its cytoplasmic extensions (Figure 9–1).
   1. The **cell body, or perikaryon**, of a neuron controls the metabolism of the entire neuron. It consists of the following components:
      a. A single **nucleus**, a prominent **nucleolus**, and other organelles, such as the Golgi complex and mitochondria, but it lacks centrioles and cannot replicate
      b. **Nissl bodies**, which are composed of rough endoplasmic reticulum and free ribosomes and play a role in protein synthesis
      c. **Neurofibrils**, which are neurofilaments and neurotubules visible by light microscopy when stained with silver
   2. **Dendrites** are cytoplasmic extensions that are usually multiple and short, and conduct impulses **to** the cell body.
      a. The surfaces of dendrites are studded with **dendritic spines**, which are specialized for contact with other neurons.
      b. Neurofibrils and Nissl bodies extend into dendrites.
   3. The **axon** is a single process, which is thinner and longer than the dendrites. It conducts impulses **away** from the cell body to another neuron, to another cell (muscle or gland cell), or to the cell body of the neuron that gives rise to the axon.
      a. **Axon origin.** An axon arises from the cell body at the axon hillock, a region devoid of Nissl bodies.
      b. **Axon size.** Axons may be less than 1 mm to more than 1 m in length (1 mm = 0.04 in.; 1 m = 3.28 ft). At its termination, an axon branches extensively.
         (1) The branching twigs have swellings called **synaptic knobs, presynaptic terminals, or terminal boutons.**
         (2) **Sidebranches (collaterals)**, which terminate in similar twigs with swellings, may occur distally.
      c. **Axon coverings**
         (1) All axons in the PNS are enveloped by a **sheath of Schwann**, also called a **neurilemma**, which is produced by Schwann cells.
            (a) Larger axons (which measure more than 2 μm in diameter) have an inner sheath of **myelin**, a lipoprotein complex produced by the plasma membrane of Schwann cells. These axons, which appear white, are called **myelinated fibers.**
            (b) In the PNS, Schwann cells myelinate the axons by encircling them in jelly-roll fashion.
            (c) **Myelin** functions as an electrical insulator and speeds conduction of nerve impulses.
            (d) **Nodes of Ranvier** represent gaps between adjacent Schwann cells. They are sites on the axon where the myelin and the sheath of Schwann are interrupted, thus partially uncovering the axon.
            (e) Axons of small diameter usually are unmyelinated and embedded in Schwann cell cytoplasm.
         (2) Axons in the CNS lack a neurilemma sheath.
            (a) **Myelinated fibers** without a neurilemma occur in the white matter of the brain and spinal cord.
               i. In the CNS, myelin is produced by oligodendrocytes rather than by Schwann cells.

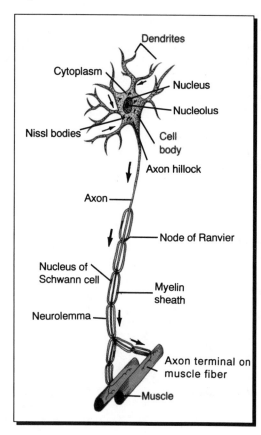

Figure 9–1. *Structure of a typical multipolar neuron.* The arrows show the direction of the nerve impulse.

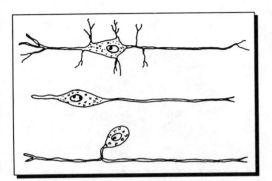

Figure 9-2. *The three major types of neurons based on polarity.* **Top,** Multipolar (CNS and motor ganglia of PNS). **Middle,** Bipolar (retina, olfactory mucosa, inner ear). **Bottom,** Unipolor (dorsal root ganglia).

      ii. Myelin is responsible for the white appearance of white matter.
    (b) **Unmyelinated fibers** without a neurilemma occur in the gray substance of the brain and spinal cord.
  (3) **Final terminations** of all nerve fibers lack both neurilemma and myelin.
  (4) **Regeneration** of injured neurons requires a neurilemma.
    (a) Neurons cannot divide mitotically, but fibers can regenerate if the cell body is intact.
    (b) If an axon is severed, the surrounding neurilemma (sheath of Schwann cells) divides mitotically to bridge the cut ends.
    (c) The axon distal to the cut degenerates; the portion of the axon closest to the cell body sprouts branches.
    (d) The empty neurilemma sheath provides a cellular tube to guide the regenerating axon; any additional axon branches that enter the hollow sheath disintegrate.
  (5) Neurons in the CNS lack a neurilemma and do not regenerate.

B. **Classification of neurons**
  1. **Function.** Neurons are classified functionally based on the direction in which they transmit impulses.
    a. **Sensory (afferent) neurons** conduct electrical impulses from receptors in the skin, sense organs, or an internal organ to the CNS.
    b. **Motor (efferent) neurons** relay impulses from the CNS to the effectors.
    c. **Interneurons (association neurons)** are found entirely within the CNS. They link sensory and motor neurons or relay information to other interneurons.
  2. **Structure.** Neurons are classified structurally according to the number of their processes (Figure 9-2).
    a. **Multipolar** neurons have one axon and two or more dendrites. Most motor neurons, which are found in the brain and spinal cord, are of this type.
    b. **Bipolar** neurons have one axon and one dendrite. They are found in sense organs, such as the eyes, ears, and nose.
    c. **Unipolar (pseudounipolar)** neurons appear to have a single process, but they originally were bipolar.
      (1) The two processes (axon and dendrite) fused during development into a single stem that branches to form a Y shape.
      (2) All sensory (afferent) neurons of the spinal ganglia are pseudounipolar.
      (3) The process of the pseudounipolar neuron that carries the sensation message to the cell body appears structurally as an axon but acts functionally like a dendrite.
      (4) True **unipolar** neurons have a single process. They occur in the embryo and in the photoreceptors of the eye.

C. **Neuroglial cells.** Commonly called **glia**, neuroglial cells are the auxiliary supporting cells of the CNS that function as connective tissue. Unlike neurons, glial cells can undergo mitosis throughout life and are responsible for nervous system tumors (Figure 9-3).

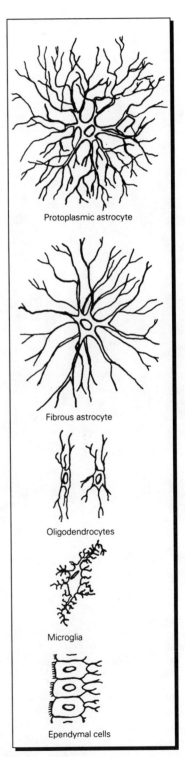

Figure 9-3. *Types of neuroglia.*

Figure 9–4. *Structure of nerves.*

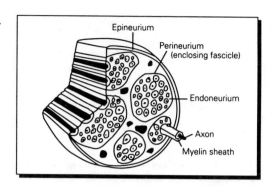

1. **Astrocytes** are star-shaped cells that have numerous long processes, many of which attach to the walls of blood capillaries by pedicles or "vascular feet."
   a. They provide structural support and may regulate the transport of materials between the blood and neurons.
   b. Vascular feet are believed to contribute to the **blood-brain barrier,** or the difficulty in passage of certain macromolecules from the blood plasma to the brain tissue.
   c. **Fibrous astrocytes** are located in the white matter of the brain and spinal cord; **protoplasmic astrocytes** are found in the gray matter.
2. **Oligodendroglia (oligodendrocytes)** resemble astrocytes, but the cell bodies are small and the processes are fewer and shorter.
   a. Oligodendrocytes in the CNS are analogous to Schwann cells found in the PNS.
   b. They form the myelin sheaths that are wrapped around axons in the CNS.
3. **Microglia** are found near neurons and blood vessels and are believed to have a phagocytic role. They are small cells and have fewer processes than other types of glial cells.
4. **Ependymal cells** form the epithelial membrane that lines the cerebral (brain) cavities and the cavity of the spinal cord.

D. Neuron groups
   1. A **nucleus** is a collection of neuron cell bodies located inside the CNS.
   2. A **ganglion** is a collection of neuron cell bodies located outside the CNS in the PNS.
   3. A **nerve** is a collection of nerve cell processes (fibers) located **outside** the CNS. The fibers are held together and supported by connective tissue, which carries the blood vessels and lymphatics (Figure 9–4).
      a. **Endoneurium** surrounds individual nerve fibers.
      b. **Perineurium** surrounds a group of fibers that is bundled together into a **fascicle.**
      c. **Epineurium,** the outermost covering, surrounds groups of fascicles, which form the nerve or nerve trunk.
   4. **Mixed nerves.** Most peripheral nerves are mixed; they contain both unmyelinated and myelinated afferent and efferent fibers.
   5. A **tract** is a collection of nerve fibers inside the brain or spinal cord that have a common origin and destination.
   6. A **commissure** is a band of nerve fibers that joins corresponding opposite sides of the brain or spinal cord.

III. THE NERVE IMPULSE

A. **Resting potential (membrane potential).** A resting nerve cell, as does any cell of the body, maintains a difference in electrical potential (voltage) across the cell membrane between the inside of the cell and the extracellular fluid around the cell. The inside of a resting nerve cell is about -50 millivolts (mV) to -80 mV rela-

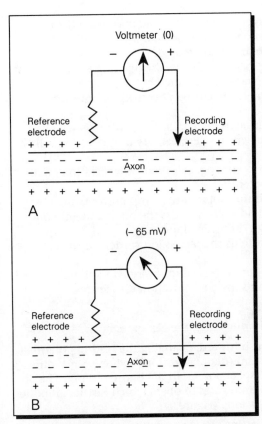

Figure 9–5. *Polarization of the axon membrane.* In *B*, the intracellular resting potential is measured by placing a recording electrode inside the cell and connecting it by a voltmeter to a reference electrode outside the cell while the cell is immersed in a saline solution. The voltmeter registers -65 mV, illustrating that the inside of the cell is negative with respect to the outside (-60 mV to -80mV, depending on the cell). In *A*, both electrodes are outside the cell and the electrical potential measures zero.

tive to the outside, depending on the neuron and the extracellular conditions that surround it.

1. The cell membrane in a resting state is said to be electrically charged, or **polarized**. The polarized state can be verified experimentally by minute electrodes placed inside and outside the membrane (Figure 9–5).
2. **Polarization (resting potential)** is caused by unequal concentrations of sodium (Na^+) and potassium (K^+) ions inside and outside the cell and the differential permeability of the membrane to these and other ions.
    a. Neuron membranes are freely permeable to K^+ and chloride (Cl^-) and relatively impermeable to Na^+ ions.
    b. They are impermeable to large, negatively charged intracellular protein molecules.
    c. The **concentration of K^+ ions inside** the cell membrane is higher than outside; the **concentration of Na^+ ions outside** the cell membrane is higher than inside.
    d. Because the membrane is about 75 times more permeable to K^+ ions than it is to Na^+ ions, the K^+ ions diffuse out of the cell much more rapidly than Na^+ ions diffuse into the cell.
    e. As the positively charged K^+ ions leak out of the cell, they leave behind negatively charged protein molecules that are too large to diffuse through the membrane. This results in the electronegativity on the inside.
3. **Diffusion** and **active transport (sodium-potassium pump)** are responsible for ion movements across the plasma membrane.
    a. **Diffusion occurs through channels** in the cell membrane down the concentration gradient of the individual ion.
        (1) Some channels are passive and always open to allow free passage of some ions.
        (2) Other channels are active (gated) channels, controlled by **ion gates**, specific to each kind of ion. Gated channels open and close in response to various stimuli.

(3) A gate consists of a charged protein molecule that spans the thickness of the membrane and undergoes a change in conformation (shape) when the membrane is stimulated.
(4) **Ion gates are voltage regulated**; that is, the opening and closing of the gates depends on a change in the membrane potential.
(5) All **voltage-gated channels are closed at resting membrane potential.**
(6) The leaking of K⁺ ions through permanently open nongated channels is responsible for the greater permeability of the resting cell membrane to K⁺.

b. **Active transport** of Na⁺ and K⁺ ions against their concentration gradients maintains the resting potential.
(1) The **ATP-dependent sodium-potassium pump** prevents the eventual equilibrium of Na⁺ and K⁺ ions across the plasma membrane that would occur through diffusion alone.
(2) The pump consists of proteins that act as ion carriers in the cell membrane.
(3) It carries three Na⁺ ions out of the cell for every two K⁺ ions pumped back in and thus maintains the concentration differences.

B. **Action potential**
1. When a nerve fiber is stimulated sufficiently, the Na⁺ gates open.
2. Positively charged sodium ions rush inward, changing the resting potential (polarization) to an **action potential (depolarization)** as the -65 mV differential shifts to an electrical peak (spike potential) of almost +40 mV. The depolarization causes even more sodium gates to open, which further amplifies the response in a positive feedback loop (Figure 9-6).
3. The action potential is very brief, lasting less than one thousandth of a second.

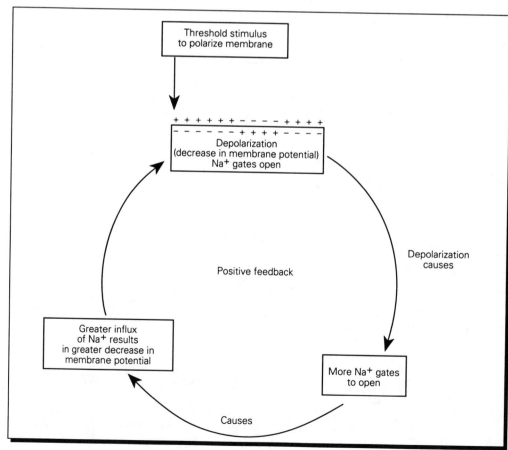

Figure 9-6. *Positive feedback loop responsible for opening additional Na⁺ gates after depolarization to a threshold level.*

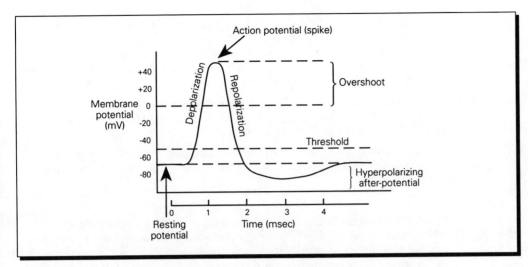

Figure 9–7. *The action potential.* It consists of a rapid depolarization, repolarization (slightly slower), followed by a hyperpolarization and a return of the membrane to the resting potential. The brief positive phase is called the overshoot. Not all cells show the hyperpolarizing after-potential.

    4. The **sodium gates then close**, stopping the inrush of Na^+. The **potassium gates open**, causing the K^+ to flow out rapidly.

    5. **Repolarization** (reverse polarity) is the restoration of the potential back to the resting state (Figure 9–7).
        a. The sodium-potassium pump assists in restoration of the original ionic concentration gradients across the cell membrane.
        b. The energy-dependent pump eliminates the excess Na^+ that has entered the cell and recovers the K^+ that has diffused out of the cell.

    6. **All-or-none response**
        a. Threshold stimulus for depolarization typically occurs with a 15 mV to 20 mV change from the resting potential state.
        b. Once the threshold for depolarization has been attained, an action potential is generated. This is the all-or-none response: The neuron responds completely or it does not respond at all.

    7. **Refractory periods**
        a. **Absolute refractory period** is the time during which the Na^+ ion gates are closing, the K^+ gates are still open, and the nerve fiber is absolutely unresponsive to another stimulus of any strength. It lasts less than 1 msec.
        b. **Relative refractory period** follows the end of the absolute refractory period. It lasts 2 msec, and is the time during which a stimulus of higher strength may trigger a second action potential.

C. **Propagation of the nerve impulse**
    1. After initiation, the **action potential is propagated** along the length of the nerve fiber with a constant speed and amplitude.
    2. Local electrical currents spread to adjacent areas of the membrane, which causes the **sodium gates to open** and which results in a **wave of depolarization** that moves along the fiber.
    3. In this way, the **nerve signal**, or **impulse**, is **transmitted** from one site in the nervous system to another.

D. **Velocity of the nerve impulse.** The speed of conduction is proportional to the axon diameter.
    1. **Unmyelinated (Group C) fibers** conduct impulses slowly. The thinner the fiber, the slower the nerve impulse conduction.
    2. **Myelinated fibers** allow for impulse conduction with a relatively small increase in diameter for the following two reasons:
        a. **Insulation and reduction of electrical resistance** is provided by the concentric wrapping of the myelin.

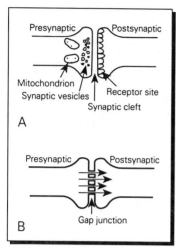

Figure 9-8. *Diagram of two types of synapses. A,* Chemical synapse. Neurotransmitter substance is released in synaptic vesicles from the presynaptic terminal. The chemical diffuses across the synaptic cleft and interacts with receptor sites on the postsynaptic membrane. *B,* Electrical synapse. Transmission occurs through gap junctions that provide the pathway for ionic flow (electrical current).

    b. **Saltatory conduction.** The action potential leaps from one exposed membrane section to another (from one node of Ranvier to another node of Ranvier) in a process called **saltatory conduction.** Saltatory conduction requires relatively little energy expenditure because depolarization occurs only at the nodes.

3. **Group A myelinated fibers** have the greatest diameter and conduct impulses with the greatest velocity. They are found in both the sensory and motor nerves of the PNS.

4. **Group B myelinated fibers** are medium sized and conduct impulses more slowly than type A fibers. They are found in the autonomic nervous system (ANS).

E. **Synapses**

1. **Synaptic transmission.** A synapse is the site (noncontiguous junction) where transference of the impulse occurs from the axon termination of one neuron to another neuron or to muscles or glands.

    a. In transmission from neuron to neuron, the connection may be from the **axon on one neuron to the dendrites**, the **cell body**, or the **axon of the second neuron.**

    b. The **presynaptic neuron** carries the impulse **toward** the synapse. The **postsynaptic neuron** carries the impulse **away** from the synapse. A single neuron can be postsynaptic at its dendrites or cell body and presynaptic at its axon ending.

2. **Chemical synapses**

    a. In a chemical synapse, a neurotransmitter (chemical) is released from the terminals of the presynaptic axon, drifts across a **synaptic cleft**, and attaches to receptors on the postsynaptic membrane (Figure 9-8).

        (1) Presynaptic axon endings are called **terminal boutons.** They release the neurotransmitter from **synaptic vesicles** when the action potential reaches the terminal, **calcium ion channels open**, and calcium ions enter the terminal bouton.

        (2) The calcium ions facilitate the passage of neurotransmitter across the synaptic cleft and binding to postsynaptic receptors.

        (3) Chemical transmission is **unidirectional** because neurotransmitter is released only from presynaptic neurons.

    b. **Synaptic delay** is the time required to cross a chemical synapse. It takes considerably more time for release, diffusion, reception, and effect of the neurotransmitter at a synapse than for the propagation of an action potential along a nerve fiber.

    c. **Excitatory synapses.** Some neurotransmitters excite the postsynaptic neuron, lead to depolarization, and cause **excitatory postsynaptic potentials (EPSPs).**

    d. **Inhibitory synapses.** Neurotransmitters that cause an **increase** in the resting potential of the postsynaptic neuron are inhibitory; that is, they make it more negative by decreasing permeability to Na^+ inflow and increasing permeability to K^+ outflow. This increase in internal negativity is called **hyperpolarization** and results in **inhibitory postsynaptic potentials (IPSPs).**

    e. **Summation.** The effect of chemical transmission on the postsynaptic neuron is a summation of the amount and type of neurotransmitter reaching the postsynaptic membrane.

        (1) **Temporal summation** is the increase in the amount of neurotransmitter by an increase in the **frequency** of stimulation by one or more presynaptic neurons.

        (2) **Spatial summation** is stimulation of increased **numbers** of excitatory presynaptic terminals to produce increased amounts of neurotransmitter.

        (3) When both EPSPs and IPSPs impinge on a postsynaptic membrane, the final result, excitation or inhibition, is determined by the algebraic sum of the excitatory and inhibitory effects, temporal summation, and spatial summation.

f. **Inactivation.** The neurotransmitter molecules released into the synaptic cleft must be rapidly **inactivated** to allow repolarization of the postsynaptic neuron for passage of the next impulse.
    (1) The neurotransmitter may be inactivated by enzymatic action.
    (2) The neurotransmitter molecules may be taken back up into the neuron that released them and recycled for additional use.
    (3) Neurotransmitter may passively diffuse away from the synaptic cleft.
g. **Synaptic fatigue.** A synapse is subject to fatigue after continued repetitive stimulation at a rapid rate. After a few milliseconds, the discharge rate of the postsynaptic neuron diminishes, although the presynaptic neuron keeps firing.
    (1) In the brain, synaptic fatigue serves as a protective mechanism against excessive neuronal excitability.
    (2) Exhaustion of stored transmitter in the presynaptic neuron is the main reason behind synaptic fatigue, but inactivation of the membrane receptors of the postsynaptic neuron may be a factor.
h. **Synapses are highly susceptible to changes in physiological conditions.**
    (1) **Alkalosis** above the normal pH 7.4 increases neuronal excitability. At pH 7.8, convulsions can occur because the neurons are so easily excited that they initiate spontaneous discharge.
    (2) **Acidosis** below the normal pH 7.4 greatly depresses neuronal discharge. A fall below pH 7.0 results in coma.
    (3) **Anoxia**, or oxygen deprivation, depresses neuronal excitability after only a few seconds.
    (4) **Drugs** may either increase or decrease neuronal excitability.
        (a) **Caffeine** reduces the threshold for transmission and facilitates the passage of impulses.
        (b) **Local anesthetics** (e.g., novocaine and procaine) that "freeze" an area increase the membrane threshold for excitation (hyperpolarization) of the nerve endings.
        (c) **General anesthetics** decrease neuronal activity throughout the body.
i. **Neuromodulation.** Chemicals such as hormones, which can enhance or dampen the synaptic response, are called neuromodulators. They may act at either presynaptic or postsynaptic sites.

3. **Electrical synapses.** When two excitable cells communicate by the direct flow of electric current through an area of low electrical resistance, the synapse is said to be an electrical synapse.
    a. **Gap junctions** link the electrically coupled cells. They are believed to be low resistance pathways between the two cells.
    b. Electrical synapses **lack synaptic delay**, which is present in chemical synapses. They are found in smooth and cardiac muscle and in the brain.
    c. Generally, electrical synapses allow **transmission in both directions** rather than unidirectionally as in chemical synapses.

4. **Neurotransmitters**
    a. **Acetylcholine (ACh)** is released by motor neurons that end on skeletal muscles (neuromuscular junctions). It also is released by parasympathetic neurons in the ANS and by certain neurons in the brain.
        (1) Most **ACh is synthesized** in the bodies of motor neurons from **choline** and **acetyl coenzyme A**; then it is transported to the axon terminals and stored in the synaptic vesicles.
        (2) After release, ACh is **broken down** into acetate and choline by an enzyme called **acetylcholinesterase**. The choline is taken up by the axon terminal and recycled.
        (3) **Acetylcholinesterases**, such as **esterine** and **prostigmine**, are used therapeutically in **myasthenia gravis**, a disease characterized by muscle weakness in which there is a decreased responsiveness of skeletal muscle cells to ACh.

b. **Catecholamines** include **norepinephrine (NE)**, **epinephrine (E)**, and **dopamine (DA)**. They contain a catechol nucleus and are all derived from the amino acid tyrosine.
    (1) Catecholamines are classified also as **monoamines** because they all possess a single amine group.
    (2) **All three are neurotransmitters in the CNS**; NE and E also function as hormones secreted by the adrenal gland.
    (3) Catecholamines are inactivated after release by
        (a) **Re-uptake** by axon terminals
        (b) Enzymatic degradation by **monoamine oxidase (MAO)** present in the presynaptic neuron endings
        (c) Enzymatic degradation by **catecholamin-O-methyl-transferase (COMT)** present in the postsynaptic neuron
c. **Seratonin** is a monoamine, but it does not contain a catechol nucleus. It is derived from the amino acid tryptophan and is found in the CNS and in certain cells in the blood and digestive system.
d. Some amino acids, such as **glycine, glutamic acid, aspartic acid**, and **gamma aminobutyric acid (GABA)**, function as neurotransmitters.
e. A number of **neuropeptides**, ranging from about two amino acids to about 40 amino acids in chain length, have been identified in body organs. Compounds such as **Substance P, enkephalins, bradykinin**, and **cholecystokinin** may serve as true neurotransmitters or as neuromodulators to affect the release of, or the response to, an actual transmitter. All have nonneural as well as neural effects.

IV. NEURONAL POOLS AND CIRCUITS

A. **Neurons** in the CNS are arranged in thousands of **neuronal pools**. Some groups contain only a few neurons, while other pools consist of enormous numbers of neurons.

B. Neuronal pools are organized into patterns called **circuits** over which signals are processed (Figure 9-9).

1. A **simple series circuit** is one in which a single presynaptic neuron stimulates a single postsynaptic neuron, which, in turn, stimulates another in a series, and so on.

2. A **diverging circuit** is one in which impulses from a single branching presynaptic neuron are transmitted to many postsynaptic neurons along one tract or along separate tracts going in different directions. This kind of circuit permits one stimulus to result in the contraction of several skeletal muscles. It is common along sensory pathways.

3. A **converging circuit** is one in which a single postsynaptic neuron receives information from several fibers branching off a single presynaptic neuron or from several fibers of different presynaptic neurons. This kind of circuit allows the possibility of reacting in the same way to many different stimuli.

4. In a **reverberating (oscillating) circuit**, once the presynaptic neuron is stimulated, the impulse is transmitted along a succession of postsynaptic neurons, some of which send the impulse back through the circuit by way of collateral axons. Reverberating circuits can continue for several seconds or for several hours. They are believed to occur in arousal responses and in the ability to remain awake, as well as in the control of respiration.

5. A **parallel circuit** is constructed so that a single presynaptic neuron stimulates a group of postsynaptic neurons, each of which synapses with a single postsynaptic cell. Such a circuit allows a single signal from the presynaptic neuron to be converted into a stream of continued impulses to the final output neuron.

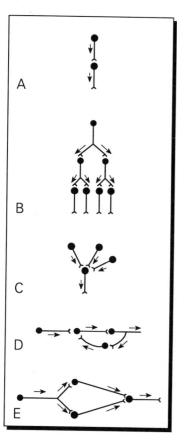

Figure 9-9. *Transmission along circuits of neuronal pools: A, simple series circuit; B, diverging circuit; C, converging circuit; D, reverberating circuit; E, parallel circuit.*

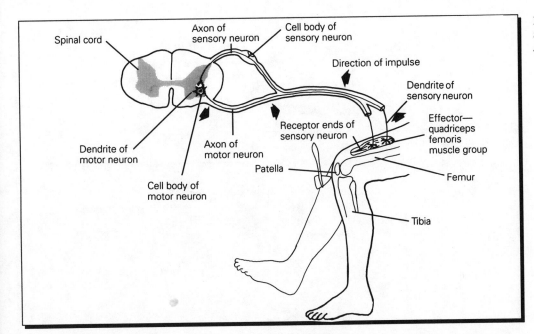

Figure 9-10. *The knee-jerk reflex is a simple reflex arc involving only a sensory neuron and a motor neuron.*

V. REFLEXES

A. A reflex is an automatic response to a specific stimulus that travels over a route known as the **reflex arc**. Most of the involuntary body processes (e.g., heart rate, breathing, digestive activities, and temperature regulation) and automatic responses (e.g., withdrawal from a painful stimulus or the knee-jerk) are reflex actions.

B. **All reflex arcs (pathways) contain the same components.**
   1. The **receptor** is the distal end of a dendrite, which receives the stimulus.
   2. The **afferent pathway** travels along a **sensory neuron** to the brain or spinal cord.
   3. The **center** is the site of the synapse, which takes place in the gray matter of the CNS. The impulse may be transmitted, rerouted, or inhibited at the center.
   4. The **efferent** pathway is taken along the axon of a **motor neuron** to the effector, which will respond to the efferent impulses with a characteristic action.
   5. The **effector** is the responding skeletal, cardiac, or smooth muscle or gland.

C. The most simple reflex is a **two-neuron, or monosynaptic ipsilateral reflex arc**. It is also called a **stretch reflex**.
   1. **Monosynaptic** means that only one synapse occurs between a sensory and a motor neuron.
   2. The term **ipsilateral** means that both neurons terminate on the same side of the body.
   3. The **patellar reflex**, or **knee-jerk**, is an example of a stretch reflex that is used for neurological examination (Figure 9-10).
      a. When the patellar tendon is tapped, muscle spindles (sensory receptors) of the quadriceps femoris muscle send impulses through the cell body of a sensory neuron (located in dorsal root ganglia) to the gray matter of the spinal cord.
      b. The sensory neuron synapses with a motor neuron, which transmits impulses to the quadriceps femoris, resulting in muscle contraction and extension of the leg at the knee.
      c. Stretch reflexes, also known as **myotatic, tendon,** or **proprioceptive** reflexes, are important in maintaining posture.

Figure 9–11. *The withdrawal reflex, a polysynaptic reflex arc.*

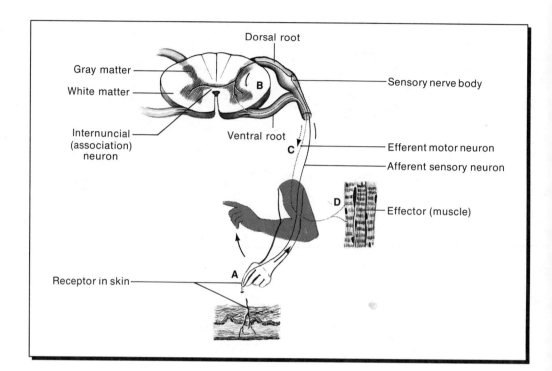

Figure 9–12. *The crossed extensor reflex.* When the flexor reflex occurs on one side, the extensor muscle on the opposite limb is stimulated to contract.

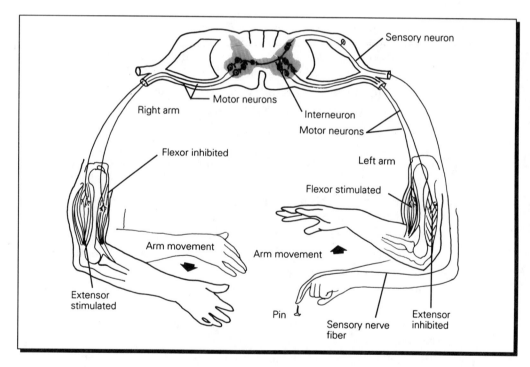

D. Most reflexes (other than stretch reflexes) are **polysynaptic** or multisynaptic. They contain at least **three neurons** and **two synapses** with an **interneuron** (association or internuncial neuron) between the sensory and the motor neurons.

1. The **flexor or withdrawal reflex**, which is elicited by a painful stimulus, is protective and takes place in the body as frequently as the stretch reflex. (Figure 9–11).

2. The **crossed extensor reflex**, which is closely associated with the flexor reflex, is an extension of the contralateral limb that occurs as a result of flexion of the limb on the ipsilateral side (Figure 9–12).

E. In more **complex reflexes**, sensory signals that are received from the eyes, ears, skin, or other sensory receptors are involved with many more integrative elements and motor elements. Complex reflexes also involve memories stored from previous experiences.

PART II: CENTRAL NERVOUS SYSTEM AND PERIPHERAL NERVOUS SYSTEM

I. **BRAIN.** The human brain comprises 2% of the total body weight, consumes 25% of its oxygen, and receives 15% of cardiac output.

  A. **Embryonic development**
     1. A primitive nervous system begins to form in the third week of embryonic life.
     2. A thickening, the **neural plate**, appears on the dorsal midline axis of the embryo. All neural tissues will eventually arise from the neural plate.
     3. The neural plate invaginates to form the **neural groove** and two **neural folds**. By the fourth week of pregnancy, the superior edges of the folds fuse in the midline to form the **neural tube**.
     4. **The neural tube gives rise to the brain and spinal cord.**
        a. Cell layers of the neural tube form the **ependyma**, the **gray matter**, and the **white matter**.
        b. Neural cells (**neural crest**) on either side of the neural tube become nerves, ganglia, and related cells, such as those of the adrenal medulla (Figure 9–13).
     5. The cranial part of the neural tube develops three swellings (vesicles) as it differentiates to form the brain: the forebrain, the midbrain, and the hindbrain (Figure 9–14).
        a. The **forebrain (prosencephalon)** becomes subdivided into the telencephalon and the diencephalon.
           (1) The **telencephalon** gives rise to the cerebral hemispheres or cerebrum, and the basal ganglia and corpus striatum (gray matter) of the cerebrum.
           (2) The **diencephalon** becomes the thalamus, hypothalamus, and epithalamus.
        b. The **midbrain (mesencephalon)** continues to be called the midbrain in the adult. It consists of the cerebral peduncles and the corpora quadrigemmina.
        c. The **hindbrain (rhombencephalon)** becomes subdivided into the metencephalon and the myelencephalon.
           (1) The **metencephalon** becomes the pons and cerebellum.
           (2) The **myelencephalon** becomes the medulla oblongata.
        d. The **cavity in the neural tube** remains and develops into the ventricles of the brain and the central canal of the spinal cord.

  B. **Protective coverings** of the brain consist of the outer bony skull and three connective tissue coverings called the **meninges**. Meningeal layers are the pia mater, the arachnoid layer, and the dura mater (Figure 9–15).
     1. The **pia mater** is the thin and delicate innermost covering, which is closely adhered to the brain. It contains numerous blood vessels that supply the nervous tissue.
     2. The **arachnoid layer** (middle) is external to the pia mater and contains few blood vessels.

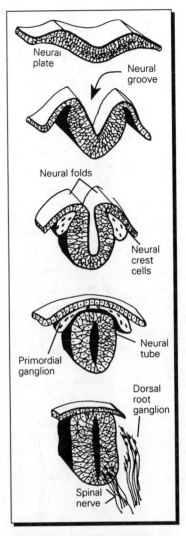

Figure 9–13. *Formation of the neural tube that gives rise to the brain and spinal cord.*

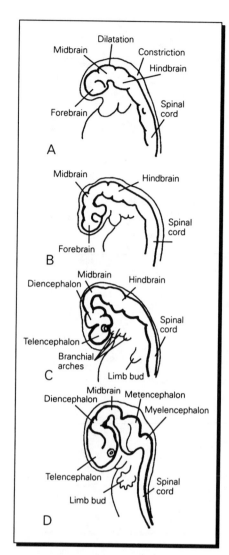

Figure 9-14. *Development of the brain and spinal cord: a and b, the three primary vesicles; c and d, secondary expansions of the walls to form five secondary vesicles from which the major divisions of the brain develop.*

Figure 9-15. *Meninges of the brain.*

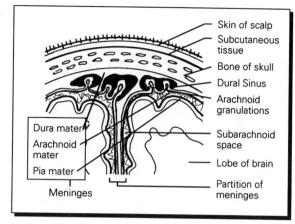

   a. The **subarachnoid space** separates the arachnoid from the pia mater and contains cerebrospinal fluid, blood vessels, and web-like bridges of tissue that secure the arachnoid to the pia mater below.
   b. Tiny tufts of arachnoid tissue, the **arachnoid villi**, project into venous (dural) sinuses of the dura mater.
3. **Dura mater**, the outermost layer, is thick and comprised of two layers. They generally are continuous with each other but are separated in a few specific sites.
   a. The **outer periosteal layer** of the dura mater adheres to the inner surface of the cranium and serves as the inner periosteum of the skull.
   b. The **inner meningeal layer** of the dura mater extends deeply into the fissures of the brain and folds back onto itself to form the following:
      (1) The **falx cerebri** is located in the longitudinal fissure between the cerebral hemispheres. It attaches to the crista galli of the ethmoid bone.
      (2) The **falx cerebelli** forms a midline partition between the cerebellar hemispheres.
      (3) The **tentorium cerebelli** separates the cerebrum from the cerebellum.
      (4) The **diaphragma sellae** extends over the sella turcia, the bony enclosure for the pituitary gland.
   c. In some regions, the two layers are separated by large blood vessels, the **venous sinuses**, which drain blood from the brain.
   d. The **subdural space** separates the dura mater from the arachnoid in the cranial and spinal cord regions.
   e. The **epidural space** is a potential space between the outer periosteal and the inner meningeal layers of the dura mater in the region of the spinal cord.

C. **Cerebrospinal fluid** surrounds the subarachnoid spaces around the brain and spinal cord and fills the ventricles within the brain.
   1. **Composition.** Cerebrospinal fluid resembles blood plasma and interstitial fluid, but it lacks the protein content.
   2. **Production.** Cerebrospinal fluid is produced by
      a. The **choroid plexuses**, which are cauliflower-shaped networks of capillaries that project from the pia mater into the two ventricles of the brain
      b. **Secretion** by the ependymal cells, which surround the cerebral blood vessels and line the central canal of the spinal cord
   3. **Circulation** of cerebrospinal fluid is as follows (Figure 9–16):
      a. The fluid moves from the **lateral ventricles** via the **interventricular foramen (of Munro)** to the **third ventricle** of the brain, where more fluid is added from the choroid plexus of the third ventricle.
      b. From the third ventricle, the fluid flows through the **cerebral aqueduct (of Sylvius)** to the **fourth ventricle**, where there are again contributions from a choroid plexus.

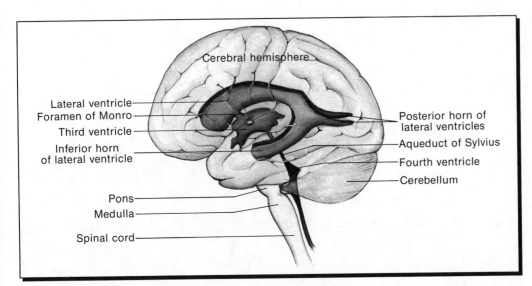

Figure 9–16. *Lateral view of the brain ventricles.*

      c. The fluid flows through three openings in the roof of the fourth ventricle to circulate through the **subarachnoid space** around the brain and spinal cord.
      d. It is reabsorbed at the **arachnoid villi (granulations)** into the venous sinuses of the dura mater and returned to the bloodstream from which it was produced.
      e. Cerebrospinal fluid is reabsorbed as rapidly as it is produced, leaving about 125 ml circulating. It is normally under slight pressure (10 mm Hg to 20 mm Hg), but any obstruction to reabsorption results in a buildup in fluid and an increase in intracranial pressure.
   4. The **function** of cerebrospinal fluid is to cushion the soft tissue of the brain and spinal cord and serve as a medium for exchange of nutrients and wastes between the blood and brain and spinal cord.
   5. Clinically, a sample of cerebrospinal fluid can be withdrawn for examination by a **lumbar puncture (spinal tap)**, which is the insertion of a hollow needle into the subarachnoid space between neural arches of the third and fourth lumbar vertebrae.
D. **White matter and gray matter** are contained in the brain and spinal cord.
   1. **Gray matter** makes up the **outer portion of the brain** (called the **cortex**) and the **inner portion of the spinal cord**. It contains neuron cell bodies, unmyelinated and myelinated fibers, protoplasmic astrocytes, oligodendrocytes, and microglia.
   2. **White matter** makes up the **inner portion of the brain** and the **outer portion of the spinal cord**. It contains predominantly myelinated (but also unmyelinated) fibers, oligodendrocytes, fibrous astrocytes, and microglia.
E. **Structure of the cerebrum.** The **cerebrum** is composed of two cerebral hemispheres, which comprise the largest part of the brain (Figure 9–17).
   1. The **cerebral cortex** consists of six layers of nerve cells and fibers. The thickness of each of these layers differs in various areas of the cerebrum.
   2. **Ventricles I and II (lateral ventricles)** are contained within the cerebral hemispheres.
   3. The **corpus callosum**, which is composed of myelinated fibers, joins the two hemispheres.
   4. **Fissures and sulci.** Each hemisphere is subdivided by **fissures** (deep grooves) and **sulci** (shallow grooves) into four lobes (**frontal, parietal, occipital,** and **temporal**) named according to the bones under which they lie.
      a. The **longitudinal fissure** divides the cerebrum into a left and right hemisphere.

Figure 9–17. *Lateral view of the right cerebral hemisphere.*

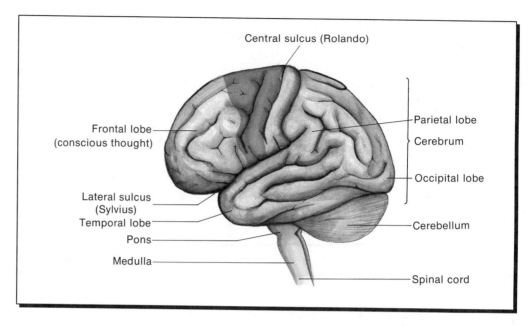

      b. The **transverse fissure** separates the cerebral hemispheres from the cerebellum.
      c. The **central sulcus** (fissure of Rolando) separates the frontal and parietal lobes.
      d. The **lateral sulcus** (fissure of Sylvius) separates the frontal and temporal lobes.
      e. The **parieto-occipital sulcus** separates the parietal and occipital lobes.
   5. **Gyri.** The surface of the cerebral hemispheres has convolutions called **gyri**. Gyri of importance include the following:
      a. the **precentral gyrus** in each hemisphere is located in the frontal lobe immediately in front of the central fissure. It contains neurons responsible for voluntary motor activities.
      b. The **postcentral gyrus** is located immediately behind the central fissure. It contains the neurons involved in sensory activities.
 F. **Functional areas of the cerebral cortex** include **primary motor areas, primary sensory areas,** and the **association or secondary areas** that are adjacent to the primary areas and function at a higher level of integration and interpretation.
   1. **Primary motor areas** of the cortex
      a. The **primary motor area** is in the precentral gyrus. Here, the neurons (**pyramidal**) control the voluntary contractions of skeletal muscles. Their axons travel in the pyramidal tracts.
      b. The **premotor area** of the cortex is located just anterior to the precentral gyrus. The neurons (**extrapyramidal**) control skilled learned motor activities of a repetitive nature, such as typing.
      c. **Broca's area** is located anterior to the premotor area at its lower margin. It may be present in only one hemisphere (usually the left) and is associated with speech abilities.
   2. **Primary sensory areas** of the cortex
      a. The **primary sensory area** is in the postcentral gyrus. Here, neurons receive general sensory information concerning pain, pressure, temperature, touch, and proprioception from the body.
      b. The **primary visual area** is located in the occipital lobe and receives information from the retinas of the eyes.
      c. The **primary auditory area**, located on the upper margin of the temporal lobes, receives nerve impulses concerned with hearing.
      d. The **primary olfactory area**, located on the medial surface of the temporal lobes, is concerned with the sense of smell.

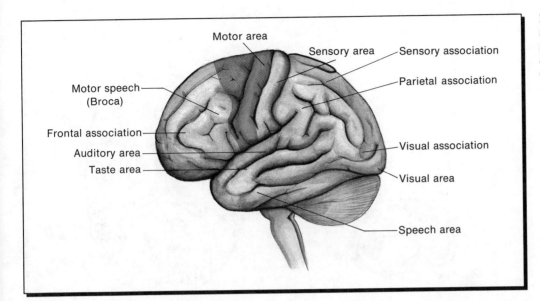

Figure 9–18. *Lateral surface of the right cerebral hemisphere illustrating some functional areas and some of the Brodmann association areas.*

    e. The **primary taste (gustatory) area**, located in the parietal lobe near the inferior part of the postcentral gyrus, is involved in perception of taste.
3. **Association areas** have been mapped out in a system called the Brodmann classification (Figure 9–18).
    a. The **frontal association area**, which is located on the frontal lobe, is the site of higher intellectual and psychic functions.
    b. The **somatic (somesthetic) association** area, which is located on the parietal lobe, is concerned with interpretation of shape and texture of objects and positional relationship of body parts.
    c. The **visual association area**, which is located on the occipital lobe, and the **auditory association area**, which is located on the temporal lobe, contribute to the interpretation of visual and auditory experiences.
    d. **Wernicke's speech area**, which is located in the superior part of the temporal lobe, is concerned with language comprehension and formulation of speech. It is connected to Broca's speech area.
4. **Brain lateralization and cerebral dominance.** The two cerebral hemispheres are symmetrical in structure but asymmetrical in some functions.
    a. The dominant hemisphere is concerned with language, speech, analysis, and calculation.
    b. The nondominant hemisphere is responsible for spatial perception and nonverbal ideation, or thought.
5. **Cerebral tracts.** White matter of the cerebrum is comprised of three types of fibers.
    a. **Long and short association tracts** interconnect neurons within the same hemisphere.
    b. **Commissure fibers** connect one hemisphere to the corresponding area of the other hemisphere; for example, the corpus callosum.
    c. **Projection fibers** are part of ascending and descending pathways coming and going from neurons located in other parts of the brain.
6. **Basal ganglia** are islands of gray matter (neurons) that lie deep within the white matter of the cerebrum. They are paired nuclei that are associated with gross body movement and connect with neurons in the precentral gyrus. Disorders of the basal nuclei result in diseases related to motor activities such as Parkinsonism, chorea, and athetosis. Included in the basal ganglia are the following structures (Figure 9–19):
    a. **Caudate nucleus**, named for its tail-like shape, is associated with unconscious skeletal muscle movements.

Figure 9-19. *A*, Basal ganglia. *B*, Transverse section of the cerebral hemisphere at the level of A—A.

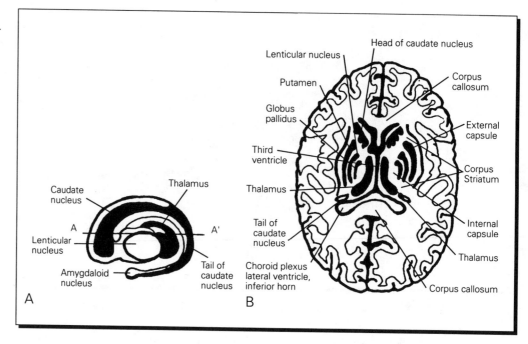

  b. **Amygdaloid nucleus** is the tail of the caudate nucleus.
  c. **Lenticular (lentiform) nucleus** consists of two parts, the **putamen** and the **globus pallidus**, which are collectively called the corpus striatum because of its striped myelinated and nonmyelinated fiber appearance. The globus pallidus regulates muscle tone and precise muscle movement.
  d. **Claustrum** is a thin layer of gray matter found between the putamen and the insular lobe of the cerebrum, which is deep within the lateral sulcus.
 G. **Diencephalon**, meaning "between brain," lies between the cerebrum and the midbrain and is hidden by the cerebral hemispheres with the exception of a basal view. It consists of all of the structures surrounding the third ventricle.
  1. The **thalamus** consists of two oval masses (½ in. wide by 1½ in. long) of gray matter partially covered by white matter. Each bulges out to form the side wall of the third ventricle.
   a. Many important **sensory and motor nuclei** are located in the thalamus; e.g., geniculate nuclei, ventral nuclei, and ventrolateral nuclei.
   b. Thalamus is the main sensory relay station for afferent fibers from the spinal cord to the cerebrum.
    (1) Axons of sensory neurons coming in from the body synapse with thalamic nuclei for the perception of the crude awareness of sensation.
    (2) **Thalamic fibers** extend in the thalamocortical tracts to the sensory area of the cerebrum for finer localization, discrimination, and interpretation of sensation.
    (3) Some of the **efferent (motor) tracts** that leave the cerebrum also synapse with thalamic neurons.
  2. The **hypothalamus** lies inferior to the thalamus and forms the floor and lower part of the side wall of the third ventricle.
   a. **Structure**
    (1) The anterior part of the hypothalamus is the gray matter surrounding the **optic chiasma**, which is a crossing of the optic nerves.

(2) The midportion of the hypothalamus consists of the **infundibulum** (stalk) of the posterior pituitary gland to which the anterior pituitary gland is attached.
  b. **Function**
    (1) The hypothalamus plays an important role in the **control of the ANS activities** that make up the vegetative functions necessary for living, such as regulation of heart rate, blood pressure, body temperature, water balance, appetite, digestive tract, and sexual activity.
    (2) The hypothalamus also functions as a **brain center for emotions** such as pleasure, pain, excitement, and rage.
    (3) The hypothalamus **produces hormones** that regulate the release or inhibition of pituitary gland hormones, thus affecting the entire endocrine system.
 3. The **epithalamus** forms the thin roof of the third ventricle. A small mass, the **pineal body**, which has possible neuroendocrine function, extends out from the posterior end of the epithalamus.
H. The **limbic system** consists of a group of structures in the cerebrum and diencephalon involved with emotional and primarily subconscious activities of behavior.
 1. The **cingulate gyrus**, the **hippocampal gyrus**, and the **pyriform lobe** are portions of the limbic system in the cerebral cortex.
 2. The **fornix** and the **septal area** in the frontal part of the brain near the root of the olfactory bulbs are subcortical portions of the limbic system.
 3. Parts of the hypothalamus, the mammillary bodies, the amygdaloid nucleus, and certain of the anterior thalamic nuclei are also included in the limbic system.
I. The **midbrain** is a short, constricted portion of the brain that connects the pons and cerebellum with the cerebrum and functions as a conduction pathway and a reflex center. The midbrain, pons, and medulla oblongata are called the brain stem (Figure 9–20).
 1. The **corpora quadrigemmina** are four rounded eminences known as colliculi that comprise the roof of the midbrain.

Figure 9–20. *Midsagittal section of the brain.*

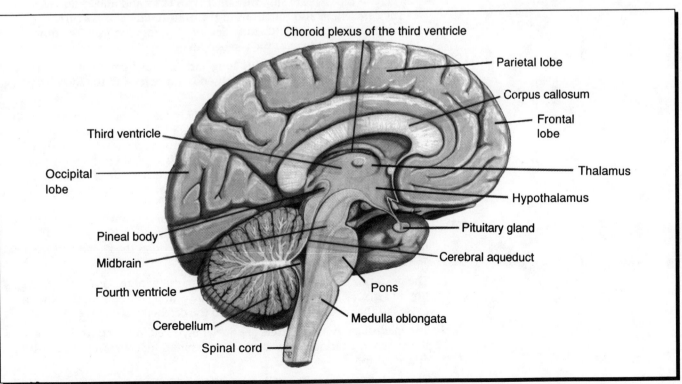

a. Two **superior colliculi** are concerned with visual reflexes.
b. Two **inferior colliculi** are concerned with auditory reflexes.
2. The **cerebral peduncles** are two cylindrical fiber bundles formed by ascending and descending tracts that make up the floor of the midbrain.
3. The midbrain contains the **aqueduct of Sylvius**, the canal that connects the third ventricle with the fourth ventricle.
4. The **nuclei of cranial nerves (CN) III, IV, and part of V** are contained within the midbrain.
5. The **substantia nigra** is an area of pigmented neurons that are important in motor function.
6. The **red nucleus** is a pinkish oval mass of neurons concerned with muscle tone and posture.

J. The **pons** (the name means bridge) consists almost entirely of white matter. It links the medulla, with which it is continuous, with various parts of the brain by way of the cerebral peduncles.
1. A **respiratory center** is located in the pons and regulates the rate and depth of breathing.
2. The **nuclei of CN V, VI, and VII** are located in the pons, which also receives information from **CN VIII**.

K. The **cerebellum** lies inferior to the pons and is the second largest portion of the brain.
1. Structure. The cerebellum consists of a central constricted part, the **vermis**, and two lateral masses, the **cerebellar hemispheres**.
   a. As in the cerebrum, the gray matter forms a cortex on the surface, which is pushed up into folds (folia) that are separated by fissures.
   b. A cross section of the cerebellum with the gray matter outside and white matter inside looks like a tree and is referred to as the **arbor vitae**, or tree of life.
2. Functions. The cerebellum is responsible for the **fine coordination and precise control of muscle movements**. It ensures that movements initiated elsewhere in the CNS are smooth rather than jerky and uncoordinated.
   a. The cerebellum also functions in the **maintenance of posture**.
   b. It helps maintain **equilibrium**. Sensory information from the inner ear is delivered to a lobe of the cerebellum.

L. The **medulla oblongata** is about 1 in. long and extends from the pons to the spinal cord with which it is continuous. It ends at the level of the foramen magnum of the skull.
1. The **anterior or ventral medulla** consists of bulges of white matter called the **pyramids**, which are continuations of the axons in the cerebral peduncles.
   a. **Decussation of the pyramids.** Just superior to the level of the spinal cord, the fissure between the two pyramids smooths out because about 85% of the pyramid fibers cross over to the other side of the cord.
   b. These **pyramidal tracts** (lateral corticospinal tracts) are the major **motor** pathways from the cerebrum to the spinal cord.
   c. Decussation results in the right side of the brain controlling the left side of the body and conversely.
   d. The remaining 15% of the axons continue on as the lateral corticospinal tracts and cross within the spinal cord.
2. The **posterior or dorsal medulla** consists partially of the upward continuation of **sensory** tracts. Nuclei serve as relay centers for information that is passed to the higher brain centers or to the cerebellum.
3. The **medullary (vital) centers** are nuclei involved in the control of functions such as heart rate, blood pressure, respiration, coughing, swallowing, and vomiting.
4. The nuclei that give rise to **CN IX, X, XI, and XII** are located in the medulla.

M. The **reticular formation**, or **reticular activating system** (RAS), is a network of nerve fibers and scattered cell bodies found throughout the medulla oblongata, pons, and midbrain. It is important in the **initiation and maintenance of alertness and wakefulness**.

II. **SPINAL CORD.** The cord of nervous tissue enclosed within the vertebral column that extends from the medulla of the brain stem to the level of the first lumbar vertebra is the spinal cord.

A. **Functions of the spinal cord**
1. The spinal cord controls many of the **reflex activities** of the body.
2. It transmits impulses to and from the brain via ascending and descending tracts.

B. **General structure of the spinal cord** (Figure 9–21).

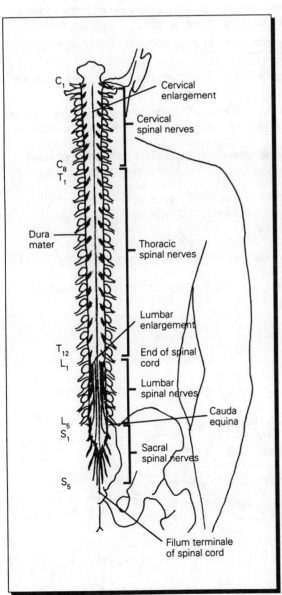

Figure 9–21. *Posterior view of the spinal cord showing roots of the 31 pairs of spinal nerves.* The spinal nerves are named for the general region of the cord from which they emerge and they are numbered in sequence.

176 The Nervous System

1. The spinal cord is a slightly flattened, hollow cylinder. Although the diameter of the spinal cord varies, it is usually about the size of the little finger. Its average length is 42 cm.
2. Two swellings, the **cervical and the lumbar enlargements**, mark the site of exit of the large spinal nerves that supply the arms and the legs.
3. **Thirty-one pairs of spinal nerves** exit from successive levels of the cord through intervertebral foramina.
4. The cord terminates at the lower level of the first or second lumbar vertebra. The lower spinal nerves that arise before the end of the cord course downward as the **cauda equina** to emerge from the spinal column at appropriate lumbar and sacral intervertebral foramina.
   a. The **conus medullaris (terminalis)** is the caudal tip of the cord.
   b. The **filum terminale** is the fibrous extension of the pia mater that attaches from the conus medullaris to the end of the vertebral column.
5. The meninges (dura mater, arachnoid, and pia mater) that cover the brain also cover the cord.
6. A deep **anterior (ventral) median fissure** and the shallower **posterior (dorsal) fissure** travel the length of the cord and divide it into right and left halves.

C. **The internal structure of the spinal cord** consists of a core of gray matter surrounded by white matter (Figure 9–22).
   1. A small **central canal** is surrounded by gray matter, which is shaped like the letter H.
   2. The upper and lower bars of the H are called **horns**, or columns, and contain **cell bodies, dendrites of association and efferent neurons**, and **unmyelinated axons**.
      a. The **posterior (dorsal) gray horns** are the vertical upper bars of gray matter. They contain the cell bodies that receive signals through spinal nerves from sensory neurons.
      b. The **anterior (ventral) gray horns** are the vertical lower bars. They contain the motor neurons whose axons send impulses through spinal nerves to muscles and glands.
      c. The **lateral horns** are protrusions between the posterior and anterior horns in the thoracic and lumbar areas. They contain the cell bodies of neurons in the ANS.
      d. The **gray commissure** connects the gray matter on the left and right sides of the cord.

Figure 9–22. Section of the spinal cord with spinal nerves.

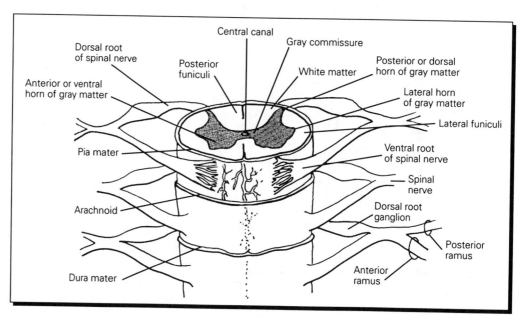

D. Each **spinal nerve** has a dorsal root and a ventral root. The **dorsal root** consists of groups of sensory fibers entering the cord. The **ventral root** is the ventral connection and carries motor fibers from the cord.
   1. Each root enters or leaves the cord by way of seven to ten rootlets.
   2. The dorsal and ventral roots on each side of each spinal cord segment unite to form the spinal nerve.
   3. **Dorsal root ganglia** are enlargements on the dorsal root that contain the cell bodies of sensory neurons.
E. **Spinal tracts.** The white matter of the cord, which is composed of myelinated axons, is divided into the **anterior, posterior, and lateral funiculi**. Within the funiculi are **fasciculi**, or **tracts**. Tracts are named for their location, origin, and destination.
   1. **Sensory or ascending tracts** carry information from the body to the brain. Ascending tracts of importance include the following:
      a. **Fasciculus gracilis and fasciculus cuneatus**
         (1) **Origin and destination.** Impulses from touch and position-sense receptors enter the spinal cord by way of the dorsal root (neuron I). The axons enter the cord, ascend to synapse with the gracilis and cuneatus nuclei in the lower medulla (neuron II). The axons cross over to the opposite side and synapse in the lateral thalamus (neuron III). Termination is in the somesthetic area of the cerebral cortex.
         (2) **Function.** These tracts convey information regarding touch, pressure, vibration, body position, and joint movement from the skin, joints, and muscle tendons.
      b. **Ventral (anterior) spinocerebellar tracts** (paired)
         (1) **Origin and destination.** Impulses from kinesthetic (awareness of body position) receptors in muscles and tendons enter the spinal cord by way of the dorsal root (neuron I) and synapse in the posterior horn (neuron II). Axons ascend on either the same or the opposite side to terminate in the cerebellar cortex.
         (2) **Function.** The ventral spinocerebellar tracts convey information regarding movement and position of entire limbs.
      c. **Dorsal (posterior) spinocerebellar tracts**
         (1) **Origin and destination.** Impulses from the dorsal spinocerebellar tracts originate and terminate the same as ventral spinocerebellar tracts; however, the axons of neuron II in the posterior horn ascend on the same side to the cerebellar cortex.
         (2) **Function.** The dorsal spinocerebellar tracts convey information on subconscious proprioception (awareness of body position, balance, and direction of movement).
      d. **Ventral (anterior) spinothalamic tracts**
         (1) **Origin and destination.** Impulses from tactile receptors in the skin enter the spinal cord via the dorsal root (neuron I) and synapse in the posterior horn on the same side (neuron II). Axons cross over to the opposite side and ascend to synapse in the thalamus (neuron III). Axons terminate in the somesthetic area of the cerebral cortex.
         (2) **Function.** The ventral spinothalamic tracts convey information regarding touch, temperature, and pain.
   2. **Motor (descending) tracts** convey motor impulses from the brain to the spinal cord and spinal nerves to the body. Motor tracts of importance include the following:
      a. **Lateral corticospinal (pyramidal) tracts**
         (1) **Origin and destination.** Neuron I originates in the motor area of the cerebral cortex. The axon descends to the medulla where most of the fibers (85%) decussate and continue down to the posterior horn to synapse directly or through interneurons with lower motor neurons (neuron II) in the anterior horn. Axons terminate on the motor end plates of skeletal muscle.

(2) **Function.** The lateral corticospinal tracts conduct impulses for coordinated and precise voluntary movements.
  b. **Ventral (anterior) corticospinal (pyramidal) tracts.**
    (1) **Origin and destination.** Neuron I originates in the pyramidal cells in the motor area of the cerebral cortex and descends to the spinal cord. Here, the axon crosses over to the opposite side just before synapsing either directly or through interneurons with neuron II in the anterior horn.
    (2) **Function.** The ventral corticospinal tracts have the same function as lateral corticospinal tracts; that is, they conduct impulses for coordinated and precise voluntary movements.
  c. **Extrapyramidal tracts.** Fibers in this system arise from other centers; e.g., motor nuclei in the cerebral cortex and subcortical areas in the brain.
    (1) **Reticulospinal tracts** originate in the reticular formation (neuron I) and terminate (neuron II) on the same side on the lower motor neurons in the anterior horn of the spinal cord. Impulses exert a facilitating influence on leg extensors and arm flexors and an inhibiting influence related to posture and muscle tone.
    (2) **Lateral vestibulospinal tracts** originate in the lateral vestibular nucleus of the medulla (neuron I) and descend on the same side to terminate (neuron II) in the anterior horn of the spinal cord. Impulses reinforce muscle tone in reflex activity.
    (3) **Medial vestibulospinal tracts** originate in the medial vestibular nucleus in the medulla and cross over to the opposite side to end in the anterior horn. This tract does not descend below the cervical level. It is concerned with control of the head and neck muscles.
    (4) The **rubrospinal tract**, which originates from the red nucleus of the midbrain, the **olivospinal tract**, which originates from the inferior olive of the medulla, and the **tectospinal tract**, which originates from the tectum of the midbrain, are also extrapyramidal tracts involved with posture and muscle tone.

III. **PERIPHERAL NERVOUS SYSTEM.** The PNS consists of all nervous tissue found outside the brain and spinal cord. It includes the cranial nerves, which arise from the brain; the spinal nerves, which arise from the spinal cord; and their associated ganglia and sensory receptors.

  A. **Cranial nerves** (Figure 9–23). **Twelve pairs of CN,** designated by Roman numerals as well as by name, emerge from various parts of the brain stem. Some cranial nerves contain only sensory fibers, but most contain both sensory and motor fibers. They are classified as follows:
    1. **Olfactory nerves (CN I)** are **sensory nerves.** They originate in the olfactory epithelium of the nasal mucosa. Bundles of sensory fibers lead to the olfactory bulb and travel via the olfactory tract to the tip of the temporal lobe (olfactory gyrus), where the perception of the sense of smell is located.
    2. **Optic nerves (CN II)** are **sensory nerves.**
      a. Impulses from the rods and cones of the retina of the eye are conveyed to cell bodies the axons of which form the optic nerves. Each optic nerve leaves the eyeball at the blind spot and enters the cranial cavity through the optic foramen.
      b. The fibers from the nasal half of each eye cross on the anterior part of the hypothalamus to form the **optic chiasma;** the fibers from the temporal half of each eye pass uncrossed.
      c. All fibers continue along as the **optic tract,** synapse at the lateral geniculate nuclei of the thalamus, and project up to the visual areas of the occipital lobes for the perception of the **sense of sight.**
    3. **Oculomotor nerves (CN III)** are **mixed but are mostly motor nerves.**

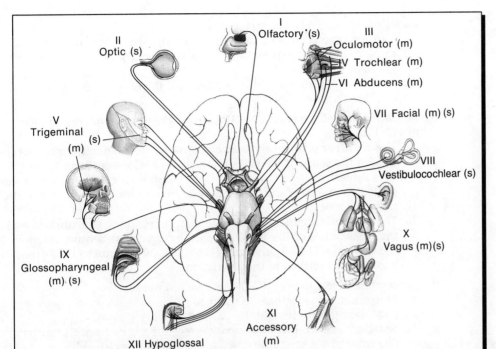

Figure 9–23. *Cranial nerves: origin and destination.*

    a. **Motor neurons** originate in the midbrain and carry impulses to **all eyeball muscles** (except the superior oblique and lateral rectus muscles), to the **muscle that raises the eyelid**, and to certain **smooth muscles of the eye**.
    b. **Sensory fibers** carry muscle sense information (proprioceptive awareness) from the **eye muscles innervated** to the brain.
4. **Trochlear nerves (CN IV)** are **mixed but are mostly motor nerves**. They are the smallest of the cranial nerves.
    a. **Motor neurons** originate on the roof of the midbrain and carry impulses to the **superior oblique muscles** of the eyeball.
    b. **Sensory fibers** from muscle spindles convey muscle sense information from the superior oblique muscles to the brain.
5. **Trigeminal nerves (CN V)**, the largest of the cranial nerves, are **mixed but are mostly sensory nerves**. They form the major sensory nerves from the face and nasal and oral cavities.
    a. **Motor neurons** originate in the pons and innervate the **muscles of mastication**, except for the buccinator muscle.
    b. **Sensory neuron** cell bodies are located in the **trigeminal (semilunar) ganglia**. The fibers branch distally into three divisions.
        (1) The **ophthalmic branch** carries information from the **eyelid, eyeball, tear glands, side of the nose, nasal cavity, and skin of the forehead and scalp**.
        (2) The **maxillary branch** carries information from the **skin of the face**, the **oral cavity** (upper teeth, gum, and lip), and the **palate**.
        (3) The **mandibular branch** carries information from the **lower teeth, gum, lip, skin of the jaw, and temporal area of the scalp**. The motor root of the trigeminal nerve travels with the mandibular branch.
6. **Abducens nerves (CN VI)** are **mixed but are mainly motor nerves**.
    a. **Motor neurons** originate from a nucleus in the pons to innervate the **lateral rectus muscles** of the eyes.
    b. **Sensory fibers** carry proprioceptive messages from the lateral rectus muscles to the pons.

7. **Facial nerves (CN VII)** are **mixed nerves**.
   a. **Motor neurons** are located in nuclei in the pons. They innervate the **muscles of facial expression** as well as the **tear glands** and the **salivary glands**.
   b. **Sensory neurons** carry information from the taste receptors on the anterior two thirds of the tongue.
8. **Vestibulocochlear nerves (CN VIII)** are **purely sensory nerves** and have two divisions.
   a. The **cochlear** or **auditory branch** conveys information from the receptors for the **sense of hearing** in the Organ of Corti of the inner ear to the cochlear nuclei in the medulla, to the inferior colliculi, to the medial geniculate nuclei of the thalamus, and then to the auditory area in the temporal lobe.
   b. The **vestibular** branch carries information regarding **equilibrium** and the **orientation of the head in space** received from sensory receptors in the inner ear. The impulses travel to the vestibular nuclei in the medulla and are then relayed to the cerebellum.
9. **Glossopharyngeal nerves (CN IX)** are **mixed nerves**.
   a. **Motor neurons** originate in the medulla and innervate the **speech and swallowing muscles** and the **parotid salivary glands**.
   b. **Sensory neurons** carry information regarding **taste** from the posterior one third of the tongue and general sensation from the pharynx and larynx; also, they convey information about blood pressure from sensory receptors within certain blood vessels.
10. **Vagus nerves (CN X)** are **mixed**.
    a. **Motor neurons** originate in the medulla and innervate almost **all of the thoracic and abdominal organs**.
    b. **Sensory neurons** carry information from the **pharynx, larynx, trachea, esophagus, heart**, and **abdominal viscera** to the medulla and pons.
11. **Spinal accessory nerves (CN XI)** are **mixed but contain primarily motor fibers**.
    a. **Motor neurons** originate in two areas.
       (1) The **cranial portion** begins in the **medulla** and innervates **voluntary muscles of the pharynx and larynx**.
       (2) The **spinal portion** arises from the **cervical spinal cord** and innervates the **trapezius** and the **sternocleidomastoid** muscles
    b. **Sensory neurons** convey information from the same muscles innervated by the motor portion; i.e., the pharynx and larynx, trapezius, and sternocleidomastoid muscles.
12. **Hypoglossal nerves (CN XII)** are **mixed but primarily motor nerves**.
    a. **Motor neurons** originate in the medulla and supply the **muscles of the tongue**.
    b. **Sensory neurons** carry information from muscle spindles in the tongue.

B. **Spinal nerves.** Thirty-one pairs of spinal nerves emerge from the cord via a dorsal (posterior) and ventral (anterior) root. Distal to the dorsal root ganglion, the two roots combine to form a spinal nerve. All are mixed nerves (motor and sensory), carrying information to the cord via afferent neurons and away from the cord via efferent neurons.
   1. Spinal nerves are named and numbered according to the region of the vertebral column from which they emerge.
      a. **Cervical nerves: eight pairs** numbered C1 through C8
      b. **Thoracic nerves: 12 pairs** numbered T1 through T12
      c. **Lumbar nerves: five pairs** numbered L1 through L5
      d. **Sacral nerves: five pairs** numbered S1 through S5
      e. **Coccygeal nerves: one pair**
   2. **Divisions.** After the spinal nerve leaves the cord through the intervertebral foramen, it branches into four divisions.

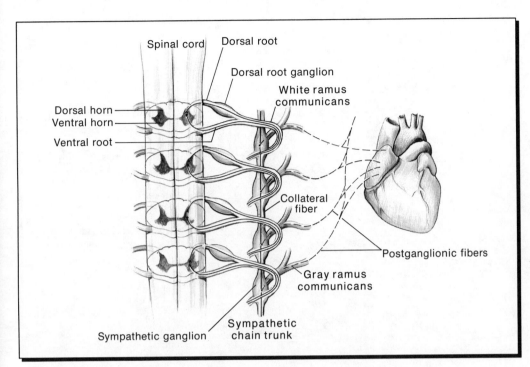

Figure 9-24. *Pathway for sympathetic fibers from the spinal cord to the sympathetic ganglia and the viscera.*

    a. The small **meningeal branch** re-enters the spinal cord through the same foramen from which the nerve exited and then supplies the meninges and blood vessels of the spinal cord and the intervertebral ligaments.
    b. The **dorsal (posterior) ramus** consists of fibers distributed posteriorly to supply the muscles and skin of the back of the head, neck, and trunk in the region of that spinal nerve.
    c. The **ventral (anterior) branch** consists of fibers that supply the anterior and lateral body trunk and the limbs.
    d. The **visceral branch** is part of the ANS. It has a **white ramus communicans** and a **gray ramus communicans** that establish a connection between the spinal cord and the ganglia of the sympathetic trunk of the ANS (Figure 9-24).

3. **Plexuses** are networks of nerve fibers formed by the ventral rami of all spinal nerves with the exception of those of T1 to T11, which give rise to the intercostal nerves.
    a. **Cervical plexus** is formed by the ventral rami of the first four cervical nerves—C1, C2, C3, C4—and a portion of C5. It innervates neck muscles and the skin of the head, neck, and chest. The most important nerve arising from this plexus is the **phrenic nerve**, which supplies the diaphragm.
    b. **Brachial plexus** is formed by the ventral rami of cervical nerves C5, C6, C7, C8, and the first thoracic nerve T1, with contributions from C4 and T2. The nerves from the brachial plexus supply the upper limb and some muscles of the neck and shoulder.
    c. **Lumbar plexus** is formed by the ventral rami of lumbar nerves L1, L2, L3, and L4, with a contribution from T12. The nerves from this plexus innervate the skin and muscles of the abdominal wall, the thigh, and the external genitalia. The largest nerve is the **femoral** nerve, which supplies the flexor muscles of the thigh and the skin of the anterior thigh, hip region, and lower leg.
    d. **Sacral plexus** is formed by the ventral rami of sacral nerves S1, S2, and S3, with contributions from L4, L5, and S4. The nerves from this plexus innervate the lower limbs, buttocks, and perineal region; the largest nerve is the **sciatic nerve**.

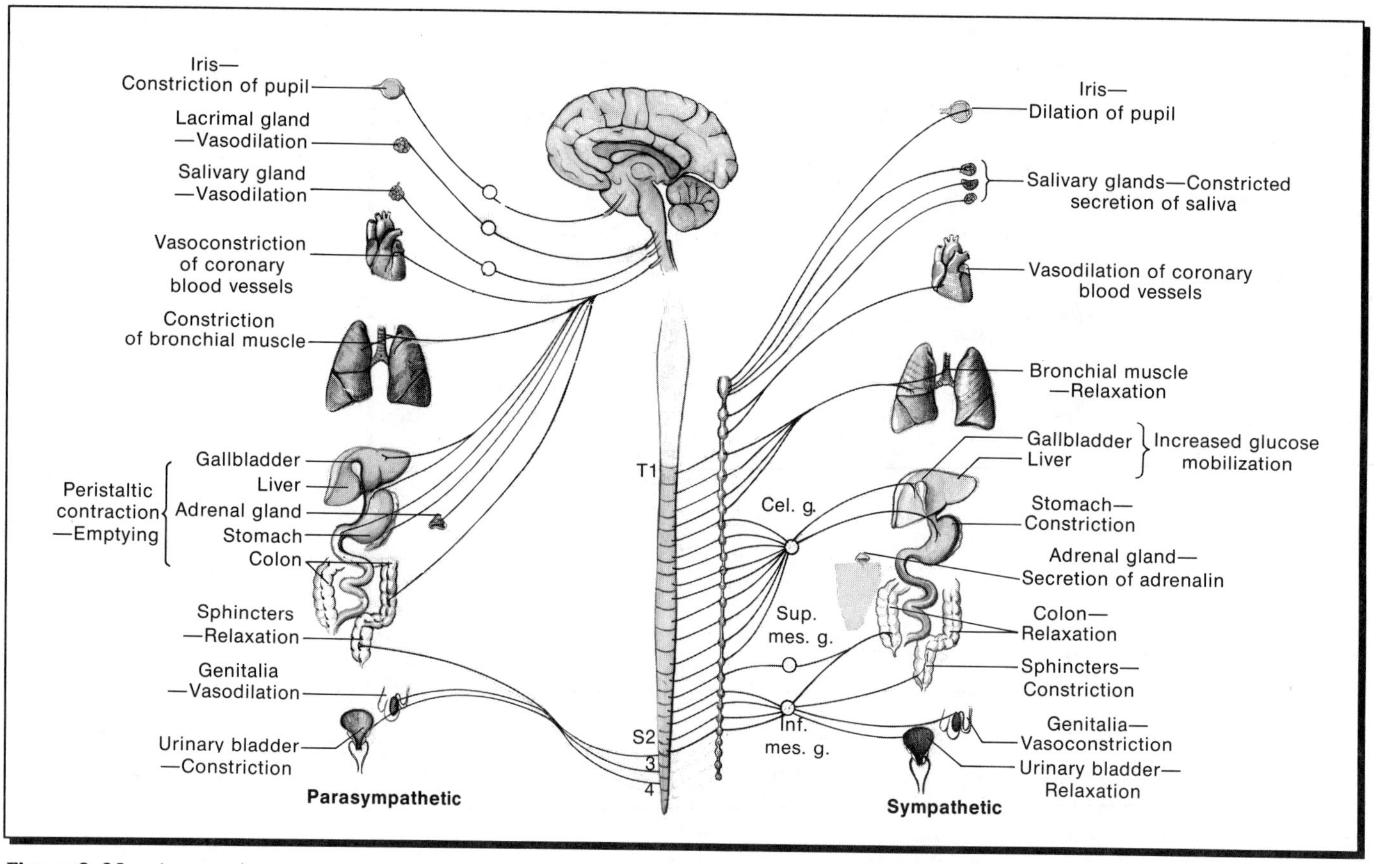

Figure 9-25. *Autonomic nervous system.* The parasympathetic division is on the left and the sympathetic division is on the right.

e. **Coccygeal plexus** is formed by the ventral rami of S5 and the coccygeal spinal nerve, with a contribution from the ramus of S4. It gives rise to the **coccygeal nerve** that supplies the region of the coccyx.

C. Autonomic nervous system

1. **Description**
   a. The ANS is a **visceral efferent motor system.** It innervates the heart; all smooth muscle, such as that of blood vessels and viscera; and glands.
   b. The ANS is without voluntary input; however, it is controlled by centers in the hypothalamus, medulla, and cerebral cortex and by additional centers in the reticular formation of the brain stem.
   c. **Visceral afferent (sensory)** fibers conveying sensations of pain or fullness and messages concerning heart rate, blood pressure, and respiration, are carried to the CNS along the same pathways as the visceral motor fibers of the ANS.

2. **Divisions**
   a. The ANS has two divisions: the **sympathetic system** and the **parasympathetic system** (Figure 9-25).
   b. Most organs innervated by the ANS receive dual innervation by nerves originating from each of the two divisions.
   c. The sympathetic and the parasympathetic divisions of the ANS are **anatomically different** and **functionally antagonistic.**
   d. Both divisions have **two neurons** between the CNS and the effector. The first, or **preganglionic neuron,** lies inside the CNS. The second, or **postganglionic neuron,** lies outside the CNS.

3. The **sympathetic division,** also known as the **thoracolumbar division,** has a short preganglionic neuron and a long postganglionic neuron.
   a. The preganglionic neuron cell bodies are located in the lateral horn of gray matter in the **thoracic and upper lumbar** segments of the spinal cord.

b. The **myelinated axons**, called **preganglionic fibers**, exit through the ventral roots along with somatic efferent fibers.
c. The preganglionic fibers travel as the **white rami communicantes** to the nearest ganglion of the **paravertebral sympathetic ganglion chain**, which is located along the length of the vertebral column on both sides. When the preganglionic fibers reach the ganglion, they follow one of three pathways.
   (1) Preganglionic fibers may synapse with the postganglionic neurons in the sympathetic ganglion at the level of entry.
      (a) The unmyelinated postganglionic axons (after synapse) form the **gray rami communicantes** and travel back to the spinal nerve.
      (b) Then these axons pass in the dorsal and ventral rami to smooth muscle effectors.
   (2) Preganglionic axon fibers may pass up or down the sympathetic chain and synapse in a ganglion at a higher or lower level.
      (a) The postganglionic fibers travel back along gray rami communicantes into a spinal nerve at that level.
      (b) The fibers innervate the effectors in the region supplied by those nerves.
   (3) Preganglionic fibers in the **thoracic region** may pass right through the sympathetic trunk (without synapsing) to form nerves called **greater and lesser splanchnic nerves** that lead to **collateral ganglia**, where synapse occurs.
      (a) The **celiac, superior mesenteric,** and **inferior mesenteric collateral ganglia** contain the postganglionic neurons, which are adjacent to the organs innervated.
      (b) The postganglionic axon fibers leave the ganglia and supply the abdominal and pelvic viscera.
d. The only **exception** to the two-neuron system is the innervation of the **adrenal medulla gland**. The sympathetic preganglionic fibers that travel to the adrenal medulla do not synapse with postganglionic neurons before the gland. Special medulla cells take the place of the sympathetic ganglion cells.

4. The **parasympathetic division**, or **craniosacral division**, has long preganglionic neurons that extend close to the organs being innervated and has short postganglionic fibers.
   a. The preganglionic neuron cell bodies are located in the **nuclei of the brain stem** and leave via CN III, CN VII, CN IX, CN X, and CN XI, and in the lateral gray matter of the **second, third, and fourth sacral segments** of the spinal cord and leave via the ventral roots.
   b. The **postganglionic neurons** are located in the **terminal ganglia** found just outside or in the wall of the organ supplied.
   c. Parasympathetic fibers originating from the **cranial region** of the cord innervate the eye, structures in the head, and the abdominal and pelvic viscera.
   d. Parasympathetic fibers originating from the **sacral region** of the cord form the **pelvic splanchnic nerves** and innervate the urinary system, the reproductive system, and parts of the lower large intestine.
   e. Parasympathetic fibers do not travel in the dorsal and ventral rami of spinal nerves. As a result, effectors in the skin (sweat glands, arrector pili muscles, and cutaneous blood vessels) receive no parasympathetic innvervation.

5. Neurotransmitters of the ANS
   a. **Acetylcholine** is released by sympathetic preganglionic fibers and parasympathetic preganglionic and postganglionic fibers, which are called **cholinergic** fibers.
   b. **Norepinephrine (noradrenaline)** is released by sympathetic postganglionic fibers, which are called **adrenergic** fibers. Norepinephrine and the related substance, epinephrine, are also released by the adrenal medulla.

Table 9–1. Effect of autonomic stimulation on various organs.

Organ/System	Sympathetic Effect	Parasympathetic Effect
Eye		
Pupil	Dilation	Constriction
Ciliary muscle	No effect, relaxation for far vision	Contraction for near vision
Glands		
Lacrimal (tear)	No effect	Stimulates secretion
Sweat	Copious secretion	No effect
Salivary	Thick, viscous secretion	Serous, watery secretion
Gastrointestinal	Inhibition or no effect	Watery secretion and enzymes
Arrector pili muscles	Contraction, erects hairs and causes "goosebumps"	No effect
Heart		
Muscle	Increase in rate and force of contraction	Decrease in rate and force of contraction
Coronary vessels	Vasodilation	Vasoconstriction
Lungs	Dilates bronchi, constricts blood vessels	Constricts bronchi
Gastrointestinal tract muscles	Inhibits peristalsis, stimulates sphincters	Stimulates peristalsis, inhibits sphincters
Liver	Glycogen hydrolysis to release glucose	No effect
Gallbladder	Inhibition, relaxation	Stimulation, release of bile
Kidney	Vasoconstriction, decreases urine output, promotes formation of renin	No effect
Blood vessels	Constricts most, increases blood pressure; dilates in skeletal muscles during exercise	Little or no effect
Penis/clitoris	Penile ejaculation	Erection of penis and clitoris
Cellular metabolism	Increases rate	No effect
Adipose cells	Stimulates fat breakdown for energy	No effect

(1) Norepinephrine and epinephrine can combine with four different receptors named alpha$_1$, alpha$_2$, beta$_1$, and beta$_2$, which are located on the membranes of effector cells.

(2) The predominance of the different receptors on the effector cells and their combination with norepinephrine and epinephrine determine the various actions of these two neurotransmitters.

6. Physiological effects of the sympathetic and the parasympathetic systems (Table 9–1)

  a. In general, the function of the sympathetic system is to **mobilize energy** during stressful situations through an increase in heart rate, blood pressure, blood sugar concentration, and blood flow to the skeletal muscles.

b. The parasympathetic system works opposite the sympathetic system: It **conserves and restores energy** through a decrease in heart rate and blood pressure and stimulation of the digestive tract to process food.
c. The two systems function together below a conscious level to maintain the internal environment, or homeostasis.

PART III: SENSORY RECEPTORS

I. **CLASSIFICATION OF SENSORY RECEPTORS.** Sensory receptors serve to transduce environmental stimuli into nervous impulses. They may be classified on the basis of the source of the stimuli that affect the receptor endings, the type of sensation detected by the receptors, distribution of the receptors, or the presence or absence of a covering on the receptor ending.

A. **Source (location) of the sensation**
1. **Exteroceptors** are sensitive to stimuli external to the body and are located at or near the body surface; for example, touch, pressure, cutaneous pain and temperature, smell, sight, and hearing.
2. **Proprioceptors** are located within the body in muscles, tendons, and joints, and include the equilibrium receptors of the inner ear. When stimulated, they convey the sense of position of body parts, amount of muscle tone, and equilibrium.
3. **Interoceptors** (visceroceptors) are affected by stimuli arising within the visceral organs and blood vessels that have motor innervation from the ANS. Examples are stimuli arising from changes during digestion, excretion, and circulation.

B. **Type of sensation detected**
1. **Mechanoreceptors** are sensitive to stretch, vibration, pressure, proprioception, hearing, equilibrium, and blood pressure.
2. **Thermoreceptors** are sensitive to change in temperature.
3. **Pain receptors (nociceptors)** are sensitive to tissue damage. All sensory receptors may function as nociceptors if the stimulus is strong enough.
4. **Photoreceptors** detect light energy.
5. **Chemoreceptors** are sensitive to changes in ion concentration, pH, levels of blood gases, and blood glucose. They include receptors for the sense of taste and smell.

C. **Distribution of the receptors**
1. **General senses** refer to information from the body as a whole.
2. **Special senses** refer to the sense organs located in the head.

D. **Endings of sensory receptors** are usually of two types.
1. **Free nerve endings** lack any cellular coverings and are located in skin, connective tissue, and blood vessels. They sense pain, light touch, and temperature.
2. **Encapsulated nerve endings** are enclosed in various types of capsules and are located in skin, muscles, tendons, joints, and body organs. The following receptors are encapsulated:
    a. **Pacinian corpuscles** detect vibratory stimuli and pressure. They are numerous in the fingers, external genitalia, and breasts.
    b. **Meissner's corpuscles** and **Merkle's disks** detect touch.
    c. **Ruffini's corpuscles** are responsive to tension in surrounding connective tissue and monitor pressure. They are found extensively on the plantar surface of the feet.
    d. **Krause's end bulbs** are lightly encapsulated and believed to contribute to touch pressure, position sense, and movement sense.
    e. **Neuromuscular spindles** monitor the muscle tone (stretch and tension) in muscle and **Golgi tendon organs** monitor the tension in tendons.

II. **EYE AND THE SENSE OF VISION.** The eye is an optical system that focuses light rays on photoreceptors, which transduce (change) light energy to nerve impulses.

   A. **Accessory structures of the eye**
   1. The **orbit** is the bony depression that contains the eyeball.
      a. Only one fifth of the orbit cavity is occupied by the eyeball; the remainder of the cavity contains connective and adipose tissue and the extrinsic eye muscles, which originate on the orbit and insert on the eyeball.
      b. There are two **openings** in the orbit: the **optic foramen** serves for passage of the optic nerve and ophthalmic artery, and the **superior orbital fissure** serves for passage of the nerves and arteries associated with the eye muscles.
   2. Three pairs of eye muscles (two pairs of rectus muscles and one pair of oblique muscles) allow the eye to turn freely vertically, horizontally, and obliquely.
   3. The **eyebrows** protect the eye from perspiration; the upper and lower **eyelids (palpebrae)** protect the eye from dessication and dust.
   4. The **palpebral fissure**, or space between the upper and lower lids, varies in size among individuals and determines the apparent size of the eye.
   5. The **medial canthus** is formed by the medial junction of the upper and lower lids; the **lateral canthus** is formed by the lateral junction of the upper and lower lids.
   6. The **caruncle** is the small elevation at the medial junction. It contains sebaceous and sweat glands.
   7. The **conjunctiva** is a thin protective layer of epithelium that lines each lid (palpebral conjunctiva) and folds back over the anterior surface of the eyeball (bulbar, or ocular, conjunctiva).
   8. The **tarsal plate** in each eyelid is a ridge of dense connective tissue. Meibomian glands, which are enlarged sebaceous glands in the tarsal plate, secrete an oily barrier that prevents tears from overflowing the lower lids.
   9. **Lacrimal apparatus** is important in the production and draining of tears.
      a. **Tears** contain salts, mucous, and lysozyme, a bacteriocide. They bathe the surface of the eye and keep it moist.
      b. Blinking compresses the **lacrimal glands** to cause the production of tears.
      c. Tears pass into the **punctum** of the **lacrimal papilla**, which is connected to a **lacrimal sac**. The sac opens into the **nasolacrimal duct**, which, in turn, leads into the nasal chamber.
   B. **Structures of the eye** (Figures 9–26 and 9–27)
   1. The tough **outermost** coat of the eyeball is the **fibrous tunic**. The posterior portion of the fibrous tunic is the opaque **sclera**, which consists of white fibrous connective tissue.

Figure 9–26. External anatomy of the eye showing lacrimal apparatus and other accessory structures.

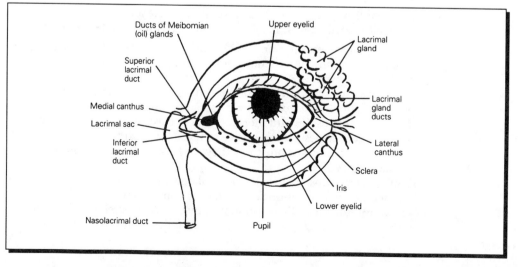

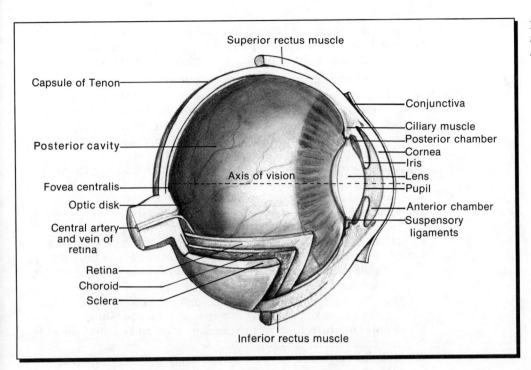

Figure 9–27. *Midsagittal section through the eye. See also Color Plate 16.*

   a. The sclera gives the eyeball form and shape and provides an attachment for the extrinsic muscles.
   b. The **cornea** is the transparent anterior continuation of the sclera over the front of the eye. It transmits light and focuses light rays.
 2. The **middle layer** of the eyeball is called the **vascular tunic (uvea)** and it consists of the **choroid**, the **ciliary body**, and the **iris**.
    a. The **choroid layer** is heavily pigmented to prevent internal reflection of light rays, is heavily vascularized to nourish the eye, and is elastic in order to pull on the suspensory ligaments.
    b. The **ciliary body**, an anterior thickening of the choroid layer, contains blood vessels and **ciliary muscles**. The muscles are attached to **suspensory ligaments** to which the **lens** is attached. The muscles are important in visual accommodation, or the ability to change focus from objects in the distance to close objects in front of the eye.
    c. The **iris**, the anterior continuation of the choroid, is the visible colored part of the eye. It consists of connective tissue and circular and radial muscles, which function to control the diameter of the pupil.
    d. The **pupil** is a rounded opening in the iris through which light enters the interior of the eye.
 3. The **lens** is a clear, biconvex structure located immediately behind the pupil. It has great elasticity, a feature that is lost with the aging process.
 4. **Cavities of the eye.** The lens separates the interior of the eye into two cavities: the anterior cavity and the posterior cavity.
    a. The **anterior cavity** is divided into two chambers.
       (1) The **anterior chamber** is located behind the cornea and in front of the iris; the **posterior chamber** is located in front of the lens and behind the iris.
       (2) The chambers contain **aqueous humor**, a clear fluid produced by the ciliary process and which nourishes the lens and cornea. Aqueous humor drains into the **canal of Schlemm** to enter the venous blood circulation.
       (3) The **intraocular pressure** of the aqueous humor is important in maintaining the shape of the eyeball. If drainage of aqueous humor is obstructed, the pressure may rise and result in damage to vision, a condition known as **glaucoma**.

b. The **posterior cavity** lies between the lens and the retina and is filled with **vitreous humor**, a transparent gel that also contributes to maintaining the shape of the eyeball and in holding the retina against the cornea.
5. The **retina**, the **innermost layer** of the eye, is thin and transparent. It consists of an outer pigmented layer and inner nervous tissue layers.
   a. **Outer pigmented layer** of the retina adheres to the choroid layer. It is a single layer of cuboidal epithelial cells containing melanin pigment and functions to absorb excess light and prevent internal reflection of light rays through the eyeball. It also stores vitamin A.
   b. **The inner nervous tissue (optical) layer**, which is adjacent to the pigmented layer, is a complex structure consisting of different types of neurons arranged in at least ten separate layers.
      (1) **Rods** and **cones** are the photosensitive receptors that are located adjacent to the pigmented layer.
         (a) **Rods** are bipolar, cylindrical neurons, which have modified light-sensitive dendrites. Each eye contains about 120 million rods located mainly at the periphery of the retina. Rods are not sensitive to color and are responsible for night vision.
         (b) **Cones** are involved in color perception. They function at high intensities of light and are responsible for day vision.
      (2) **Bipolar neurons** form an intermediate layer and connect the rods and cones to the ganglion cells.
      (3) **Ganglion cells** contain axons that come together at a specific region of the retina to form the optic nerve.
      (4) **Horizontal cells** and **amacrine cells** link lateral synapses and are other cell types found in the retina.
      (5) **Light passes** through the **ganglion layer**, the **bipolar layer**, and the **cell bodies of the rods and cones** to stimulate their dendritic processes and initiate nerve impulses. Then the nerve impulses pass in the **reverse** direction through the two nerve cell layers.
   c. The **blind spot (optic disc)** is the exit point of the optic nerve. Because no photoreceptors are present at this area, no sensation of vision occurs when light falls on it.
   d. The **macula lutea** is a yellowish area slightly lateral to the center.
   e. The **fovea** is a central depression in the macula lutea that is devoid of rod cells and contains only cone cells. It is the visual center of the eye; images focused here are interpreted clearly and sharply by the brain.
   f. **Visual pathways to the brain** (Figure 9–28)
      (1) The optic nerves formed by the axons of the ganglion cells leave the eyes and meet just superior to the pituitary gland as the **optic chiasma**.
      (2) At the optic chiasma, the neuron fibers originating from the temporal (lateral) half of each retina remain on the same side while the neuron fibers originating from the nasal (medial) half of each retina cross over to the opposite side.
      (3) After the optic chiasma, the axon fibers form the **optic tracts**, which continue to synapse with neurons in the lateral geniculate nuclei of the thalamus. Their axons pass to the occipital lobe cortex.
      (4) Some of the axons connect to the superior colliculi, the oculomotor, and the pretectal nuclei to participate in pupillary and ciliary reflexes.
C. Optical properties of the eye
   1. **Refraction** is the deflection, or bending, of light rays as they pass from one medium to another medium of a different optical density. The more convex a surface, the more the refractive, or bending, power.
      a. The **cornea** is responsible for about 70% of the refractive power and is the "coarse adjustment" of the eye.
      b. The **lens** is responsible for most of the remainder of refraction and is the "fine adjustment" of the eye.
      c. The **aqueous** and the **vitreous fluids** are responsible for minimal refraction.

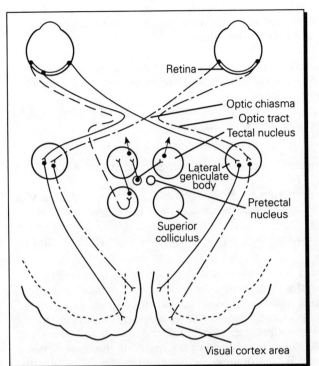

Figure 9–28. *Nerve pathways to the brain.* The right hemisphere of the brain receives information from the lateral field of vision on its own (right) side and from the nasal field of vision on the opposite side; this situation is reversed for the left hemisphere of the brain. Thus, if the right tract is damaged, the right half of each retina will be blinded; if the left tract is damaged, the left half of each retina is cut off from the brain.

2. **Accommodation** is the automatic adjustment process of the lens to focus objects clearly at varying distances.
   a. A **convex lens** (thickest in the center and thinner at the periphery) **converges** light rays; a **concave lens** (thinnest at the center and thicker at the periphery) **diverges** light rays.
   b. A convex (more rounded) lens focuses on near objects; a concave (flattened) lens focuses on distant objects.
   c. In **emmetropia**, or normal accommodation, **contraction** of the **ciliary muscles** reduces the pull of the suspensory ligament on the lens, which then bulges forward to become more convex, or rounded, for near vision. **Relaxation** of the **ciliary muscles** again increases the pull of the suspensory ligaments on the lens, which assumes a flattened shape for distant vision.
   d. The power of accommodation, an unconscious reflex, decreases with age as a result of the decreased elasticity of the lens, which can no longer bulge forward as much as it did in youth. The condition is **presbyopia** and is corrected with bifocal lenses.
   e. **Convergence** of the eyeballs when viewing close objects assists accommodation by ensuring that the images in both eyes fall on corresponding portions of the retina.
   f. **Constriction of the pupils** also reflexly occurs during accommodation to screen out the most divergent light rays and allow the formation of sharp images on the retina.
3. **Visual defects**
   a. **Myopia (nearsightedness)** (Figure 9–29)
      (1) An eyeball that is too long for the refractive power of the eye, or a lens system that is too strong, results in the image focusing at a point in front of the retina.
      (2) The result is nearsightedness, so-called because the eye can focus on near objects.
      (3) Myopia is corrected with a concave lens placed in front of the eye to cause enough refraction to focus distant objects on the retina.
   b. **Hyperopia (farsightedness)** (Figure 9–30)
      (1) An eyeball that is too short or a weak lens system results in the image focusing behind the retina, which results in blurred vision for near objects.

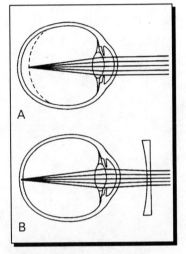

Figure 9–29. *A*, Myopia, or nearsightedness. The light rays from distant objects focus in front of the retina. *B*, Correction of myopia by a concave lens that diverges light rays before they enter the eye.

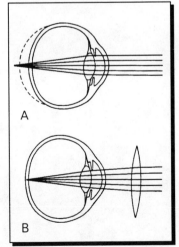

Figure 9–30. *A*, Hyperopia, or farsightedness. The light rays from distant objects focus behind the retina. *B*, Correction of hyperopia by a convex lens that converges light rays before they enter the eye.

(2) Hyperopia is corrected with a convex lens in front of the eye to bring objects to focus on the retina.
   c. **Astigmatism**
      (1) When the curvature of the cornea or the lens is uneven, the light rays passing through will be unevenly refracted and the image will be blurred in one plane.
      (2) Astigmatism is corrected by a special lens in front of the eye that has different correcting curvatures in the appropriate planes.
D. **Physiology of vision**
   1. **Rhodopsin (visual purple)** is a pigment contained in rods that has two sub-units.
      a. **Retinal**, also called **retinene** or **retinaldehyde**, is synthesized from vitamin A. It exists in two isomeric forms: a curved 11-cis-retinal and a straight **all-trans retinal**.
      b. **Opsin**, also called **scotopsin**, is a protein in loose chemical combination with 11-cis-retinal.
   2. **Bleaching of rhodopsin** from purple to pink occurs when light strikes the retina. The light causes the 11-cis-retinal bound to opsin to change shape to the all-trans form, which results in its detachment from opsin.
      a. The separation of opsin and retinal initiates a nerve potential in the rod cells (receptors), which causes stimulation of the bipolar and ganglion cells of the retina. This is transmitted to the brain via the optic nerve.
      b. Unlike other nerve cell membranes, Na^+ channels in the rod cell membrane are open in the absence of stimulation (light). Thus, in the dark, the inward passage of Na^+ results in depolarization and the release of inhibitory transmitter. The bipolar neurons and ganglion cells are not stimulated.
      c. When rods are stimulated by light, the release of Ca^{++} from within the rod cell causes the Na^+ channels to close. Because the Na^+ conductance is decreased, the inside of the cell becomes even more negative, or **hyperpolarized**. Inhibitory transmitter release is decreased and the bipolar cell depolarizes.
      d. The action potential is generated by a **hyperpolarized** rather than depolarized membrane.
   3. **Resynthesis of rhodopsin** occurs in the dark, when all-trans retinal is converted back to 11-cis-retinal and rejoined to opsin. This reaction requires energy and enzymes.
   4. Rods function at low light intensities because the bleaching reaction requires little illumination.
   5. **Dark and light adaptation** is the automatic adjustment of vision to the intensity of light falling on the retina when movement from darkness into bright light or the reverse occurs.
      a. The time required for dark adaptation (ability to see in dim light) is partly determined by the time needed to resynthesize and accumulate stores of rhodopsin.
      b. In bright light all available rhodopsin is broken down rapidly leaving little available for generating action potentials in the rods, and the eye is said to be light-adapted. About 20 minutes is required in dim light for dark-adaptation.
      c. The synthesis of rhodopsin and iodopsin (the pigment in cone cells) requires vitamin A, a precursor to retinal.
      d. Lack of sufficient dietary vitamin A can result in visual abnormalities as a result of rod and cone degeneration.
         (1) **Night-blindness**, a condition in which light sensitivity is decreased, occurs early in vitamin deficiency. It is particularly noticeable at night when little light is available for adequate vision.
         (2) Prolonged vitamin A deficiency also affects cones. Treatment with vitamin A restores retinal function if the rods and cones have not been destroyed.

(3) The B vitamins also are necessary for proper functioning of the retina and all neural tissues.
            e. Dark and light adaptation also involves pupillary reflexes to allow more light or less light to enter the interior of the eye.
        6. **Color vision**
            a. Each eye contains 6 to 7 million bipolar cone cells responsible for visual acuity and color vision.
            b. Cones contain **iodopsin**, which is retinal combined with a different opsin than that found in rods.
            c. Iodopsin may be blue-sensitive, red-sensitive, or green-sensitive, resulting in different cones being selectively sensitive to differences in colors.
            d. The same process of pigment decomposition that occurs in rods to create action potentials also occurs in cones. Because iodopsin pigments do not respond in dim light, cones function only in bright light.

III. **EAR: SENSES OF HEARING AND EQUILIBRIUM**

   A. **Structure of the ear.** The ear is divided into external, middle, and internal parts (Figure 9–31).
      1. The **external ear** consists of the **pinna**, or **auricle**, which is the cartilaginous flap that collects sound waves to funnel them into the **external auditory canal (meatus)**, a narrow passageway about 2.5 cm long that leads from the auricle to the tympanic membrane.
      2. The **tympanic membrane (eardrum)** is the boundary to the middle ear.
         a. The tympanic membrane is cone-shaped and covered with skin on its external surface and mucous membrane on its internal surface.
         b. It separates the external ear from the middle ear and is of suitable tension, size, and thickness to mechanically vibrate with sound waves.
      3. The **middle ear** lies in an air-filled cavity within the petrous portion of the temporal bone.
         a. The **eustachian (auditory) tube** connects the middle ear to the pharynx.
         b. The usually closed tube can be opened by yawning, swallowing, or chewing. It functions to equalize the air pressure on both sides of the tympanic membrane.

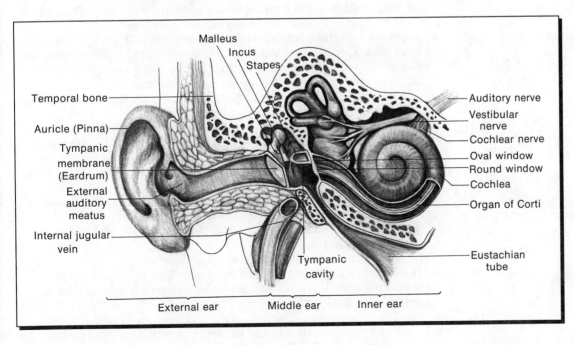

Figure 9–31. *Structure of the ear.*

Figure 9–32. *Diagrammatic view of the ear.*

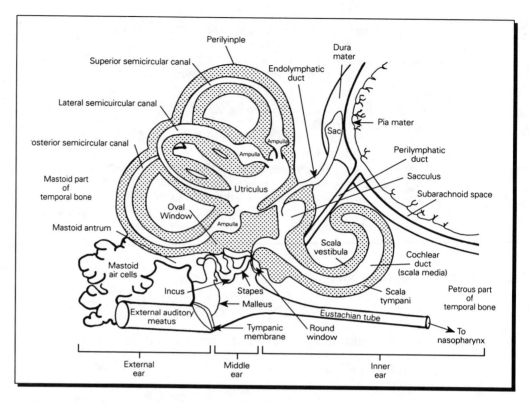

4. Three **auditory ossicles**, named for their shape, are the **malleus** (hammer), **incus** (anvil), and **stapes** (stirrup). They conduct the vibrations from the typanic membrane to the **oval window**, which separates the middle ear from the inner ear.
   a. The **stapedius muscle** is attached to the stapes, which fits into the oval window, and pulls it outward. The **tensor tympani muscle** is attached to the handle of the malleus, which rests against the tympanic membrane, and pulls the oval window inward.
   b. Loud sounds result in a reflex that causes contraction of **both** muscles, protectively damping the sounds.
5. The inner (internal) ear is filled with fluid and located in the temporal bone medial to the middle ear. It consists of two parts: the bony labyrinth and the membranous labyrinth inside the bony labyrinth (Figure 9–32).
   a. The **bony labyrinth** is a maze of chambers filled with **perilymph**, a fluid similar to cerebrospinal fluid. It is hollowed out of the petrous portion of the temporal bone and is divided into three portions: the vestibule, the semicircular canals, and the snail-shaped cochlea.
      (1) The **vestibule** is the central portion of the bony labyrinth that connects the semicircular canals with the cochlea.
         (a) The lateral wall of the vestibule contains the **oval window** and the **round window**, which communicate with the middle ear.
         (b) Membranes cover the windows to prevent loss of perilymph.
      (2) The bony cavities of the **semicircular canals** project from the posterior part of the vestibule.
         (a) The **anterior and posterior** semicircular canals are oriented in the vertical plane at right angles to each other.
         (b) The **lateral** semicircular canal lies horizontally and at a right angle to the other two.
      (3) The **cochlea** contains the receptor for hearing.
   b. The **membranous labyrinth** is a series of hollow tubes and sacs that lie within the bony labyrinth and follow its contours. It is filled with **endolymph**, a fluid similar to intercellular fluid.
      (1) The membranous labyrinth in the region of the vestibule gives rise to two sacs, the **utricle** and the **saccule**, which are connected by the short, narrow **endolymphatic duct**.

(2) The **semicircular ducts** filled with endolymph lie within the semicircular canals of the bony labyrinth filled with perilymph.
(3) Each semicircular duct, the utricle, and the saccule contain the receptors for **static equilibrium** (how the head is oriented in space relative to gravitational forces) and **dynamic equilibrium** (whether the head is moving and what is its rate and direction).
(4) The utricle is connected to the semicircular ducts; the saccule is connected to the cochlear duct within the cochlea.

B. Cochlea and the physiology of hearing
  1. The **cochlea** makes two and one half turns around a central bony core, the **modiolus**, which contains blood vessels and nerve fibers of the cochlear branch of the vestibulocochlear nerve (VIII). The cochlea is divided by partitions into three separate channels.
     a. The **cochlear duct** or **scala media**, which is part of the membranous labyrinth and is connected to the saccule, is the middle channel and is filled with **endolymph.**
     b. The two parts of the bony labyrinth above and below the scala media are the **scala vestibuli** and the **scala tympani**. The two scalae are filled with perilymph and are continuous via an opening at the apex of the cochlea, which is called the **helicotrema.**
        (1) **Reissner's membrane** (vestibular membrane) separates the scala media from the scala vestibuli, which connects with the oval window.
        (2) The **basilar membrane** separates the scala media from the scala tympani, which connects with the round window.
     c. The scala media contains the **organ of Corti**, which rests on the basilar membrane (Figure 9–33).
        (1) The organ of Corti consists of **receptors**, called **hair cells**, and **supporting cells**, which cup the lower ends of the hair cells and rest on the basilar membrane.
        (2) The **tectorial membrane** is a ribbon-like gelatinous structure that extends over the hair cells.
        (3) The basal ends of the hair cells contact branches of the cochlear portion of the vestibulocochlear nerve. The hair cells have no axons and synapse directly with the cochlear nerve endings.
  2. **Sound waves** (vibrations) enter the external auditory meatus and create vibrations in the tympanic membrane. These vibrations are carried along the ear ossicles to the oval window, pushing it inward and creating pressure waves in the noncompressible perilymph of the scala vestibuli.
  3. The **pressure waves** in the scala vestibuli are carried through to the scala tympani and cause the round window to bulge outward.
  4. The fluid-conducted vibrations also cause upward and downward vibrations in the basilar membrane, with the extent of the movement varying with the **frequency** and **amplitude** (strength) of the vibrations.
     a. The basilar membrane gradually widens as it extends from the stapes to the helicotrema. The narrow end of the membrane moves in response to all sound frequencies; the wider end moves only in response to low frequencies.
     b. **Pitch** of the sound is a function of the **frequency** of vibrations (cycles) of sound waves per second. Humans hear sounds between 20 and 20,000 cycles per second.
     c. **Intensity** of the sound is a function of the amplitude of the wave. The greater the amplitude, the louder the sound and the greater the vibration of the basilar membrane.
  5. The **hair cells** are bent by the movement of the basilar membrane; this, then, initiates the nerve impulses.
  6. **Neural pathway.** The cochlear nerve fibers synapse in the medulla and in the midbrain to ascend to the auditory cortex, which is located deep in the lateral fissures of the cerebral hemispheres.

Figure 9–33. *A,* Organ of Corti. *B,* Close-up view. Vibration of the basilar membrane causes the hair cells to rub against the tectorial membrane and results in their stimulation. Nerve impulses are sent along the cochlear nerve to the brain.

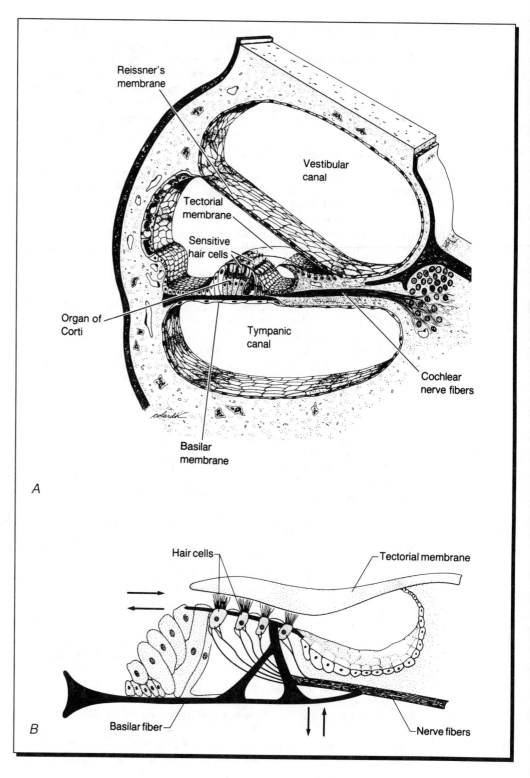

C. **Equilibrium and the vestibular apparatus.** The vestibular apparatus is the term used for the utricle, saccule, and the semicircular ducts, which contain the receptors for equilibrium and balance (Figure 9–34).

1. **Static equilibrium** is the sense of the position of the head with respect to gravity when the body is not moving. It is also the sense that responds to changes in **linear acceleration**; that is, the speed and direction of the head and body movement in a straight line.

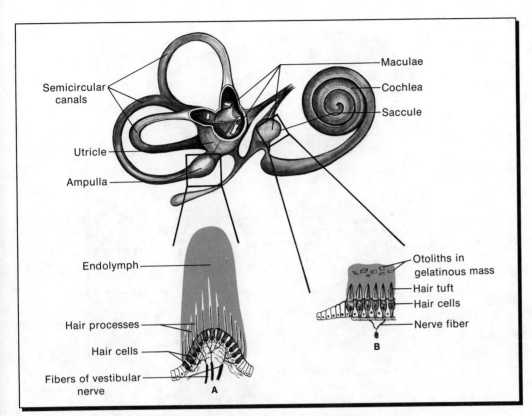

Figure 9-34. Diagram of *A*, a crista, located within each semicircular canal; *B*, a macula, located in the wall of the utricle and saccule.

   a. **Maculae** are the receptors for static equilibrium. One macula is located in the wall of the utricle and one in the wall of the saccule.
   b. Each macula consists of a patch of supporting cells and specialized receptor cells called **hair cells**. The hairs (actually cilia surrounded by microvilli) project up into a gelatinous mass, which contains calcium deposits called **otoliths** (otoconia, statoconia).
   c. When the head is in an upright position, the otoliths sit on top of the hair cells. When the head is tilted, the pull of gravity on the otoliths shifts their direction, bends the hair cells, and results in activation of the receptor cells.
   d. The hair cells also detect linear acceleration or deceleration, as in a car or on an elevator. The changes in the weight of the otoliths stimulate the hair cells.
   e. Receptor activity is transmitted to endings of the vestibular nerve (CN VIII) coiled around the bases of the hair cells.
2. **Dynamic equilibrium** is the sense of position of the head in response to angular or rotational movement.
   a. **Ampullae** are the receptors for dynamic equilibrium. Each of the semicircular ducts within the semicircular canals contains a swollen region, the ampulla, which contains an elevation or **crista**.
   b. The crista consists of supporting cells and hair cells that project up into a gelatinous layer called the **cupula**.
   c. Movement of the head causes the endolymph with the semicircular canals to cause movement of the cupula. The bending of the hair cells generates the nerve potential.
   d. Each semicircular duct—anterior, lateral, or posterior—responds to a particular rotational movement determined by the orientation of the duct.
3. **Neural pathway for the sense of equilibrium**
   a. The receptor endings form the vestibular branch of CN VIII. The cell bodies of the sensory neurons are located in the superior and inferior vestibular ganglia near the membranous labyrinth.

Figure 9-35. *A,* Regions of the tongue sensitive to various tastes. *B,* types of papillae that contain *C,* tastebuds in their furrows.

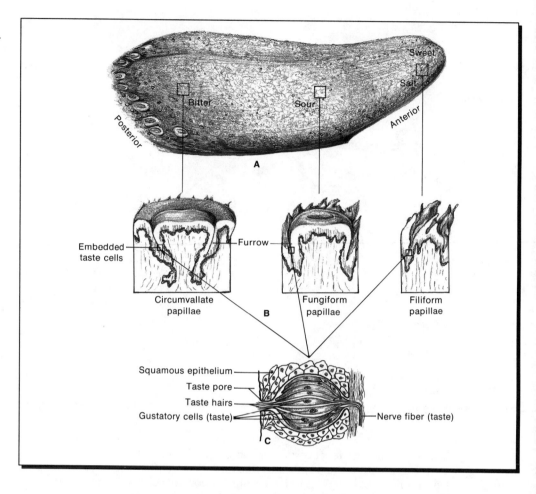

b. The impulses are transmitted from the vestibular ganglia to vestibular nuclei located at the junction of the medulla and pons. From here, the sensory information is integrated and relayed to the cerebellum.
c. The vestibular nuclei also receive information from visual receptors and proprioceptive receptors in the neck and limbs. This information is relayed through the medulla to the cerebellum, the reticular formation, and several nuclei to control reflexly the muscles of the eyes, head, and neck.

IV. **GUSTATION: THE SENSE OF TASTE**

A. **Structure of taste buds** (Figure 9-35)
  1. Receptors for taste are the **taste buds,** chemoreceptors that are located primarily on the tongue, but also are present on the soft palate and epiglottis.
  2. Taste buds are found in projections of the tongue mucosa called papillae.
  3. Each taste bud is a cluster of **supporting cells** and **sensory cells** with hairs that protrude into a central taste pore where they are bathed with saliva.

B. **Function of taste buds**
  1. The substance to be tasted must be in fluid or dissolved in the saliva.
  2. It combines with a receptor on the taste hairs. This stimulates sensory dendrites coiled around the sensory cells and results in nerve impulses, which are transmitted along the facial (CN VII) and glossopharyngeal (CN IX) nerves via the taste pathways to the insula of the cerebral cortex.

C. Taste sensations
1. Taste buds sensitive to **sweetness** are localized on the tip of the tongue.
2. **Sour substances** are tasted mainly on the **sides of the tongue.**
3. **Salty substances** can be tasted over most of the tongue, but reception primarily is concentrated on the sides.
4. **Bitter substances** excite taste buds on the **back of the tongue.**

V. OLFACTION: THE SENSE OF SMELL

A. Olfactory chemoreceptors are specialized neurons that are located in the **olfactory epithelium** on the roof of the nasal cavity.

B. The olfactory epithelium contains **supporting cells, basal cells,** and the **olfactory cells,** which are bipolar neurons the dendrites of which end in fine olfactory hairs that project into the mucus that coats the nasal cavity (Figure 9–36).

C. The mechanism by which the olfactory cells are stimulated by odors is not completely known. **Depolarization** occurs and results in an action potential that is conducted along olfactory nerve fibers to the olfactory bulbs and the olfactory areas in the cerebral cortex.

D. Olfactory receptors adapt to odors rapidly: It is possible to become unaware of strong smells after one minute.

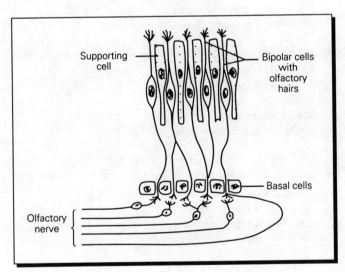

Figure 9–36. *Olfactory epithelium.*

Study Questions

Directions: Each question below contains four suggested answers. Choose the **one best** response to each question.

1. The cell in the central nervous system functionally equivalent to a Schwann cell is the
   (A) astrocyte
   (B) neuron
   (C) oligodendrocyte
   (D) microglial cell

2. During depolarization of a nerve fiber membrane
   (A) sodium ions pass from the intracellular fluid to the extracellular fluid
   (B) potassium ions pass from the extracellular fluid to the intracellular fluid
   (C) the inner aspect of the membrane becomes positively charged and the outer aspect becomes negatively charged
   (D) all of the above are correct

3. Which of the following is responsible for regeneration of injured nerve fibers in the peripheral nervous system?
   (A) dorsal roots
   (B) myelin sheath
   (C) neurilemma
   (D) neuroglia

4. Which is the correct sequence of the following events after threshold potential is reached?
   (1) Depolarization of the membrane
   (2) Sodium voltage-regulated channels open and sodium ions diffuse inward
   (3) Repolarization of the membrane
   (4) Sodium channels close; potassium ions diffuse outward
   (A) 1, 2, 3, 4
   (B) 2, 1, 4, 3
   (C) 1, 2, 4, 3
   (D) 4, 1, 3, 2

5. Which of the following best describes spatial summation in the transmission of a postsynaptic potential?
   (A) One presynaptic neuron discharges excitatory neurotransmitter at the postsynaptic membrane several times in rapid succession.
   (B) Many different presynaptic neurons discharge excitatory neurotransmitter at the postsynaptic membrane simultaneously.
   (C) A presynaptic neuron releases inhibitory neurotransmitter at the postsynaptic membrane to cause hyperpolarization.
   (D) Many presynaptic neurons discharge inhibitory neurotransmitter and one presynaptic neuron discharges excitatory neurotransmitter at the postsynaptic membrane.

6. Saltatory conduction
   (A) occurs only in a saline solution
   (B) occurs only in the absence of nodes of Ranvier
   (C) is more rapid than conduction in an unmyelinated fiber
   (D) is slower than conduction in an unmyelinated fiber

7. If the threshold stimulus for a single neuron is 100 mV, what response would be expected from a stimulus of 200 mV?
   (A) no response
   (B) half that of 100 mV
   (C) the same as that of 100 mV
   (D) twice as strong as that of 100 mV

8. An efferent impulse from the spinal cord to a skeletal muscle is conducted by
   (A) one neuron
   (B) at least two neurons
   (C) at least three neurons
   (D) generally more than three neurons

9. An efferent impulse from the spinal cord to a smooth muscle is conducted by
   (A) one neuron
   (B) at least two neurons
   (C) at least three neurons
   (D) more than three neurons

10. The afferent impulse that results in the perception of a light touch on the ankle of the right leg is conducted by
    (A) one neuron
    (B) at least two neurons
    (C) at least three neurons
    (D) more than three neurons

11. Which of the following is found within the subarachnoid space?
    (A) air
    (B) blood
    (C) cerebrospinal fluid
    (D) mucus

12. A blow to which of the following lobes of the brain could cause a sensation of "seeing stars?"
    (A) parietal
    (B) occipital
    (C) frontal
    (D) temporal

13. If the cerebral hemispheres of a laboratory rat were destroyed experimentally, it could not
    (A) breathe
    (B) see
    (C) reflexly move its legs
    (D) urinate involuntarily

14. Atropine is a drug that causes relaxation of the smooth muscle of the digestive tract, a dry mouth sensation, and dilated pupils. Therefore, atropine is a(n) _____ drug and mimics the effect of the _____ nervous system.
    (A) adrenergic, parasympathetic
    (B) adrenergic, sympathetic
    (C) cholinergic, parasympathetic
    (D) cholinergic, sympathetic

15. Which of the following fibers secrete acetylcholine as their transmitter substance?
    (A) all somatic motor neuron axons
    (B) all axons ending on skeletal muscle fibers
    (C) all postganglionic parasympathetic fibers
    (D) all of the above

Questions 16–23. Match the injury on the left with the cranial nerve on the right. A letter may be used once, more than once, or not at all.

16. difficulty in swallowing   (A) glossopharyngeal
17. difficulty in retracting shoulders (B) hypoglossal
18. dizziness, loss of equilibrium   (C) spinal accessory
19. deafness   (D) vestibulocochlear
20. inability to smile or frown   (A) trigeminal
21. loss of sensation on the face   (B) facial
22. inability to taste   (C) oculomotor
23. inability to chew   (D) abducens

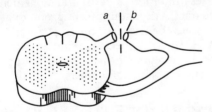

Questions 24 and 25 refer to the above diagram of the spinal cord.

24. What would be the result of the cut shown in the diagram between *a* and *b*?
    (A) death of the organism
    (B) total paralysis
    (C) the ventral root takes over the function of the dorsal root
    (D) sensation is lost in the structures innervated by that nerve

25. If a stimulus were applied at the site of letter *b*,
    (A) there would be a reflex motor response in the muscles supplied by this nerve
    (B) there would be no response in the muscle supplied by this nerve
    (C) the organism would feel pain in the structures innervated by this nerve
    (D) an impulse would travel backward to enter the cord at the ventral root

26. The iris of the eye is a continuation of the
    (A) retina
    (B) sclera
    (C) choroid
    (D) conjunctiva

27. All of the following are refractive media of the eye EXCEPT
    (A) lens
    (B) aqueous humor
    (C) cornea
    (D) retina

28. In nearsightedness the image falls
    (A) in front of the retina
    (B) in back of the retina
    (C) on the exit of the optic nerve
    (D) on the retina

29. In their pathway from the retina of the eyes to the brain, the nerve fibers
    (A) from the temporal half of each retina cross over to the opposite side
    (B) from the nasal half of each retina cross over to the opposite side
    (C) remain on the same side as the eye from which they emerge
    (D) all cross over to the opposite side

30. All of the following would be likely to cause hearing loss or total deafness EXCEPT
    (A) cochlear nerve damage
    (B) vestibular nerve damage
    (C) fusion of the ear ossicles
    (D) ruptured tympanic membrane

Answers and Explanations

1. **The answer is C.** (Part I, II C) Oligodendrocytes in the CNS are analogous to Schwann cells in the PNS and form the myelin sheaths around axons in the CNS. Astrocytes are supporting cells and may function in the transport of materials from blood vessels to neurons. Microglia are believed to play a phagocytic role in the CNS.

2. **The answer is C.** (Part I, III B 1, 2, 3) During depolarization, the positively charged sodium ions on the surface of the neuron cell membrane rush inward, causing the inside of the membrane to become positively charged and resulting in the action potential.

3. **The answer is C.** (Part I, II A 3 c) A neurilemma is required for regeneration of injured fibers. If an axon is severed, the surrounding Schwann cells divide mitotically to form a cellular tube, which bridges the cut ends and serves to guide the regenerating portion of the axon nearest the cell body.

4. **The answer is B.** (Part I, III A, B) The generation of an action potential requires, in sequence, the opening of the sodium channels to allow inward passage of sodium ions, depolarization, the closing of the sodium channels, and the opening of the potassium channels to allow outward passage of potassium ions, repolarization.

5. **The answer is B.** (Part I, III E 2) Summation is a function of the amount of neurotransmitter reaching the postsynaptic membrane. Spacial summation is accomplished by stimulation of increased numbers of excitatory presynaptic terminals ending on the postsynaptic membrane, thus resulting in an increased amount of neurotransmitter. Temporal summation results from an increased frequency of stimulation by one presynaptic neuron ending on a postsynaptic neuron membrane.

6. **The answer is C.** (Part I, III D 2 b) Saltatory conduction occurs in myelinated fibers and is more rapid because the action potential "leaps" from one exposed node of Ranvier to another.

7. **The answer is C.** (Part I, III B 6) Once the minimum stimulus necessary to initiate an action potential is attained, an increase in the intensity of the stimulus does not increase the strength of the response. This is the all-or-none response.

8. **The answer is A.** (Part II, II c 2 b) From the spinal cord, the efferent pathway to a skeletal muscle is taken along the axon of a motor neuron located in the posterior lateral horn of the cord.

9. **The answer is B.** (Part II, III C 3, 4) The ANS innervates smooth muscle. Both the sympathetic and the parasympathetic divisions are two neuron systems, with the first (preganglionic) neuron located in the brain or spinal cord, and the second (postganglionic) neuron located in a chain of ganglia outside the CNS or near the organ innervated.

10. **The answer is C.** (Part II, II E 1 a) The fasciculus gracilis and fasciculus cuneatus sensory tracts in the spinal cord convey sensory information concerning light touch and position sense to the brain. At least three neurons are required to carry the information to the somesthetic area of the cerebral cortex.

11. **The answer is C.** (Part II, I C 3) Cerebrospinal fluid circulates through the subarachnoid space around the brain and spinal cord and is reabsorbed at the arachnoid villi into the dural sinuses.

12. **The answer is B.** (Part II, I F 2 b; Part III, II B 5 f) The neurons of the primary visual cortex and the secondary visual association area are located in the occipital lobe of each cerebral hemisphere and are responsible for sight perception. Stimulus to these neurons by a blow to the occipital lobe may result in a visual sensation. Damage to the area could result in blindness.

13. **The answer is B.** (Part II, I F) The cerebrum is necessary for visual perception, which occurs in the occipital lobe. Respiration, reflex movement, and involuntary urination all can occur when the organism is decerebrate ("brain dead").

14. **The answer is B.** (Part II, III C, also Table 9–1) A drug that causes inhibition of peristaltic movement of the digestive tract, a dry mouth feeling, and dilation of the pupils mimics the effects of the sympathetic division of the ANS. Thus, its effects are similar to those caused by release of norepinephrine (noradrenaline) from sympathetic postganglionic fibers.

15. **The answer is D.** (Part II, III C 5 a, b; also refer to Chapter 8, Part I, II F 5) Acetylcholine is released by sympathetic preganglionic fibers, parasympathetic pre- and postganglionic fibers, and the axons on motor neurons ending on skeletal muscle.

16–23. **The answers are 16–A, 17–C, 18–D, 19–D, 20–B, 21–A, 22–B, 23–A.** (Part II, III A) The motor fibers of the glossopharyngeal nerves (CN IX) innervate the swallowing muscles. The spinal accessory nerves (CN XI) innervate the trapezius muscle. The vestibular branch of CN VIII carries information regarding dynamic and static equilibrium from the inner ear to the cerebellum. The cochlear branch of CN VIII carries information from the receptor for hearing in the inner ear. The facial nerve (CN VII) innervates the muscles of facial expression. The maxillary branch of the sensory division of the trigeminal nerves (CN V) conveys sensory information from the skin of the face. The facial nerve carries information from the taste receptors on the anterior two thirds of the tongue. The muscles of mastication are innervated by the motor division of the trigeminal nerves.

24. **The answer is D.** (Part II, II D) The dorsal root of a spinal nerve contains sensory fibers entering the cord; the ventral root carries motor fibers away from the cord. Interruption of the dorsal root results in the loss of sensation from the structures innervated at that level of the spinal nerve.

25. **The answer is B.** (Part II, II D) A stimulus applied at *B* would have neither a motor nor a sensory effect. A stimulus applied at *A*, however, would result in a motor response in the muscle innervated by that spinal nerve.

26. **The answer is C.** (Part III, II B 2 C) The iris is the anterior continuation of the choroid layer, which is a vascularized, pigmented, elastic coat. The sclera is the tough, outermost coat of the eyeball; its anterior continuation is the cornea. The retina is the innermost layer and the visual portion of the eye. The conjunctiva is a thin, protective layer of epithelium that lines the upper and lower lids and extends over the anterior surface of the eyeball.

27. **The answer is D.** (Part III, II C 1) The cornea and the lens are responsible for most of the refractive power of the eye; the aqueous and vitreous fluids also are refracting media. The retina contains the receptors for vision.

28. **The answer is A.** (Part III, II C 3) In myopia (nearsightedness), an eyeball that is too long for the refractive power of the eye results in the image focusing at a point in front of the retina. In hyperopia (farsightedness), a short eyeball or a weak lens system results in the image focusing behind the retina. Only images brought to focus at the macula lutea on the retina are perceived most sharply. The exit of the optic nerve contains no photoreceptors; thus, no sensation of vision occurs at this site.

29. **The answer is B.** (Part III, II B 5 f) In the visual pathway to the brain, the neuron fibers originating from the nasal half of each retina cross over to the opposite side at the optic chiasma. The neuron fibers originating from the temporal half of each retina remain on the same side without crossing.

30. **The answer is B.** (Part III, III B 1–4) Hearing would be unaffected by damage to the vestibular portion of the vestibulocochlear nerve. Any damage to organs of sound wave transmission (ear ossicles, tympanic membrane) or the cochlear nerve would be likely to cause hearing loss or deafness.

10 Endocrine System

I. **INTRODUCTION**

A. **General description of the endocrine system**
1. The endocrine system interacts with the nervous system to **regulate and co-ordinate** body activities.
2. Endocrine control is mediated via chemical messengers, or **hormones**, which are released by the endocrine glands into the body fluids, absorbed into the bloodstream, and transported by the circulatory system to **target tissues** (cells).
3. Hormones affect their target cells via hormone receptors, which are protein molecules with binding sites for specific hormones.
4. Hormonal responses of the body generally are **slower**, of **longer duration**, and of **wider distribution** than the immediate response of muscles and glands to nervous system stimuli.

B. **Characteristics of endocrine glands**
1. **Endocrine glands have no ducts.** They secrete hormones directly into the tissue fluid that surrounds their cells. In contrast, exocrine glands, such as the salivary glands, secrete their products into ducts.
2. **Endocrine glands usually secrete more than one hormone.** (The parathyroid glands, which secrete only parathyroid hormone, are an exception.)
   a. Approximately 40 to 50 hormones have been identified in humans.
   b. New hormones are being found in a variety of body sites including the gastrointestinal (GI) tract, the CNS, and the peripheral nerves.
3. **Hormone concentration in the circulation is low.**
   a. Hormones circulate in the bloodstream in extremely small quantities when compared with other biologically active substances, such as glucose or cholesterol.
   b. Although hormones reach most body cells, only the particular target cells with receptors specific for that hormone are affected.
4. **Endocrine glands are well-supplied with blood vessels.** Microscopically, they consist of cords or clumps of secretory cells surrounded by many capillaries and supported by connective tissue.

C. **Glands of the endocrine system** (Figure 10–1) include the following:
1. Anterior and posterior pituitary gland
2. Thyroid gland

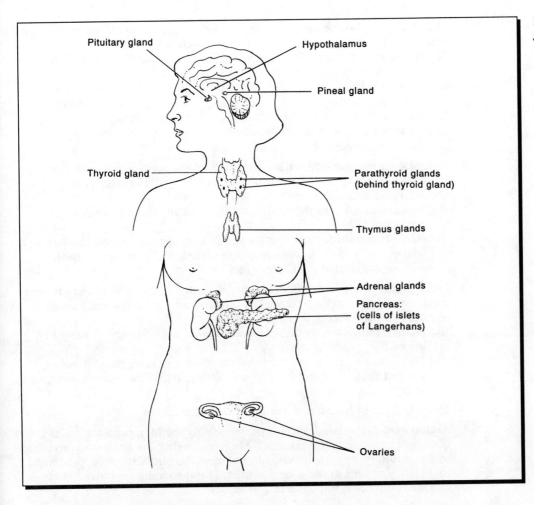

Figure 10–1. *The major endocrine glands.*

    3. Four parathyroid glands
    4. Two adrenal glands
    5. Islets of Langerhans of the endocrine pancreas
    6. The two ovaries
    7. The two testes
    8. Pineal gland
    9. Thymus gland

D. Activities regulated or influenced by the endocrine system include the following:
    1. Reproduction and lactation
    2. Immune system processes
    3. Acid-base balance
    4. Fluid intake, intracellular and extracellular fluid volume balance
    5. Carbohydrate, protein, lipid, and nucleic acid metabolism
    6. Digestion, absorption, and nutrient distribution
    7. Blood pressure
    8. Stress resistance
    9. Adaptation to environmental change

E. Types of hormones

1. The **endocrine hormones** are those secreted by the major organs or tissue of what is classically considered part of the endocrine system.
   a. Hormones do not act locally; they travel in the bloodstream for long distances to influence target tissue.
   b. Endocrine hormones may be secreted by individual cells or by groups of cells found within nonendocrine tissue (for example, insulin and glucagon produced by the islet cells of the exocrine pancreas).
   c. Some hormones, such as the placental hormones found during pregnancy, are produced temporarily.
2. **Neurohormones** are synthesized within neurosecretory nerve cells. They function and are secreted like hormones, but generally operate over a shorter, more defined distance.
   a. An example of neurohormones are the neuropeptides produced by neurons in the CNS.
   b. **Neurotransmitters** that operate across a synapse or **neuromodulators** that enhance or inhibit the response of neurons to neurotransmitters also are considered neurohormones.
3. **Prostaglandins** are hormone-like substances that are derived from fatty acid arachidonic acid. They are formed in tiny amounts in body tissues under normal and pathological conditions.
   a. Prostaglandins are synthesized and released to act locally on neighboring cells.
   b. They affect a variety of functions including effects on blood pressure, smooth muscle contraction, blood clotting, digestion, reproduction, and inflammatory responses.

F. The **biochemistry of hormones** includes two major classes.
   1. **Amino acid derivatives**, such as proteins, polypeptides, peptides, amines, or conjugated protein complexes such as glycoproteins, are hormones produced by the pituitary, hypothalamus, adrenal medulla, pineal, thyroid, pancreatic islet cells, and cells in the digestive tract. They generally are water soluble and are transported unbound in the blood.
   2. **Steroids** are fat-soluble lipid compounds synthesized from cholesterol. They are produced by the ovaries, testes, placenta, and the outer part of the adrenal glands and include testosterone, estrogen, progesterone, aldosterone, and cortisol. They circulate in the plasma bound to transport proteins.

G. **Mechanisms of hormonal action.** There are two major mechanisms by which hormones and related molecules produce their effects. One is by stimulation of enzyme activity within the cell and the other involves gene activation via transcription and translation.
   1. **Enzyme activation** involves **membrane-bound receptor (second messenger) systems** (Figure 10–2).
      a. The molecules of many protein and polypeptide hormones (first messenger) bind to a **fixed receptor on the cell surface** specific to that hormone.
      b. The **hormone-receptor complex** stimulates the formation of **cyclic adenosine 3',5'-monophosphate (cAMP)** as the second messenger, which can carry the first message from a variety of hormones.
         (1) The synthesis of cAMP involves one or more membrane-bound **G-proteins**, a family of guanine nucleotide-binding regulatory proteins.
         (2) The G-protein is conformationally altered, allowing the inactive guanosine diphosphate (GDP) to be exchanged for an activating enzyme, guanosine triphosphate (GTP).
         (3) The G-protein-GTP complex activates the enzyme **adenylate cyclase**, which leads to the production of cAMP.
      c. Each molecule of cAMP activates many molecules of an appropriate **cAMP-dependent protein kinase**.
         (1) Protein kinase enzymes catalyze specific phosphorylation reactions (transfer of phosphate groups) to key enzymes in the cytoplasm.

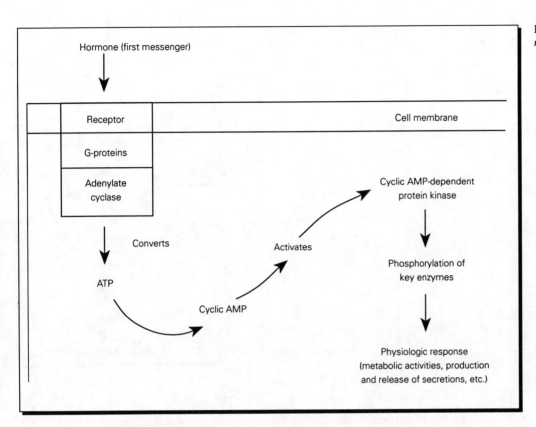

Figure 10–2. *The cAMP-mediated mechanism of hormone action.*

        (2) Each protein kinase molecule activates many molecules of its appropriate enzyme. Thus, a small concentration of circulating hormone can be amplified to result in major intracellular enzyme activity.
    d. Enzyme activation by protein kinase results in characteristic **chemical reactions and physiological** effects, depending on the inherent nature of the cell.
    e. cAMP is rapidly broken down by the intracellular enzyme **phosphodiesterase.** This limits the duration of effect of cAMP.
  2. Compounds other than cAMP that act as second messengers for certain hormones have been discovered. These include **inositol triphosphate** (IP_3), **cyclic quanosine monophosphate** (GMP), and a complex of calcium bound to **calmodulin,** which is an intracellular regulatory protein.
  3. Gene activation involves **intracellular receptor systems** (Figure 10–3).
    a. Steroid hormones, thyroid hormones, and some polypeptide hormones pass through the cell membrane to enter the cell. They bind to **mobile internal receptors** located in the cytoplasm or the cell nucleus.
    b. The **hormone-receptor complex** moves to the DNA at sites on or near the genes whose transcription is stimulated by the hormone. There the complex binds to a **DNA receptor specific for the hormone.**
    c. The genes are activated by the complex to form the **transcription of mRNAs,** which diffuse into the cytoplasm.
    d. The mRNAs are **translated into the proteins and enzymes** that constitute the cellular response to the hormone

H. Regulation of the rate and quantity of hormone secretion
  1. Hormonal secretion by an endocrine gland may be stimulated or inhibited by blood levels of a hormone (produced by the gland itself or by another endocrine gland) or a nonhormone (for example, glucose or calcium).
  2. Feedback control mechanisms are involved in the stimulation or inhibition of hormonal secretion.

Figure 10-3. *The gene activation mechanism of hormone action.*

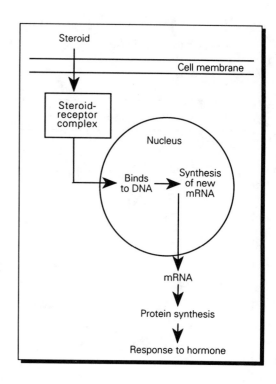

        a. **Negative feedback.** If the rising blood levels of a hormone or the nonhormone substance result in the **inhibition** of further secretion of the hormone, the mechanism is known as negative feedback system.

        b. **Positive feedback.** If the blood levels of the hormone or nonhormone substance result in the **enhancing** of the secretion of the endocrine gland, the mechanism is said to be a positive feedback system.

    3. Release of hormone from an endocrine gland also may be stimulated by **nerve impulses carried along nerve fibers** that end directly on the gland cells, or, as in the posterior pituitary gland, by **neurosecretions** that are stored in the gland as hormones.

II. HYPOPHYSIS (PITUITARY GLAND)

  A. Morphology

    1. Size and location
        a. The pituitary gland is an oval organ approximately the size of a pea. It weighs about 0.5 g.
        b. It is attached to the floor of the hypothalamus of the brain by a stalk called the **infundibulum** (hypothalamic stalk).
        c. The pituitary rests in a saddle-shaped depression of the sphenoid bone (sella turcica) and is enclosed by an extension of the dura mater.

    2. Divisions of the gland
        a. **The anterior lobe (adenohypophysis)** of the gland consists of the pars distalis, the pars tuberalis, and the pars intermedia.
            (1) The **pars distalis** constitutes the bulk of the anterior lobe.
            (2) The **pars tuberalis** in humans is reduced to a thin plate of epithelial cells on the superior portion of the pars distalis. It has no known endocrine function but is the most vascular portion of the anterior lobe.
            (3) The **pars intermedia**, adjacent to the pars distalis, is conspicuous in infants but greatly reduced in adults.
        b. **The posterior lobe of the pituitary (neurohypophysis)** consists of the pars nervosa and the infundibulum.

(1) The **pars nervosa** is connected with the hypothalamus of the brain. It contains the axonal endings from neurosecretory neurons of the hypothalamus and neuroglial-like cells (pituicytes), which are not believed to have a secretory function.

(2) The **infundibulum (neural stalk)** connects the neurohypophysis with the brain.

3. **Embryological origin of the lobes**
   a. The **adenohypophysis** arises from an upgrowth, or outpocketing (Rathke's pouch), of the epithelium in the roof of the primitive mouth cavity.
   b. The **neurohypophysis** develops from a downward projection of the neural tube at the base of the hypothalamus. It retains direct neural connections to the brain via the infundibulum.

4. **Pituitary-hypothalamus relationships.** Both vascular and neural connections between the hypothalamus and the hypophysis are essential to the functions of the pituitary gland (Figure 10–4).
   a. The hypothalamic-hypophyseal portal system
      (1) The blood supply to the **posterior lobe** (neurohypophysis) is via two inferior hypophyseal arteries, which branch off the internal carotid artery, enter the posterior lobe, and form a capillary network. Venous drainage is through hypophyseal veins into a dural sinus.

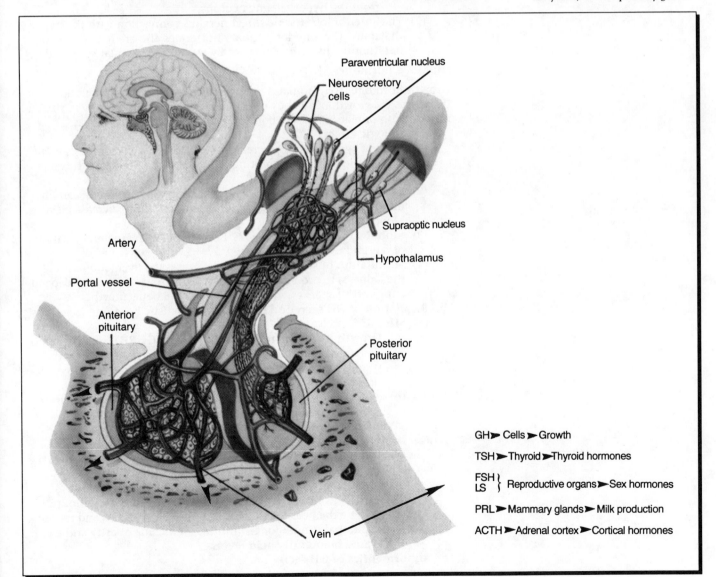

Figure 10–4. *The anatomic and vascular relationships between the hypothalamic neurosecretory cells, the hypothalamic–hypophyseal portal system, and the pituitary gland.*

(2) The blood supply to the **anterior lobe** (hypophysis) is **indirect**. The superior hypophyseal artery (a branch of the internal carotid artery) enters the median eminence of the hypothalamus and the infundibular stalk and forms a **first capillary network**.

(3) The first capillary bed is drained by the **hypophyseal portal veins**, which give rise to a **second capillary bed** in the lower portions of the anterior lobe.

(4) The **hypothalamic-hypophyseal portal system** refers to the two capillary beds described (one in the hypothalamus and one in the adenohypophysis) and the veins between them. Via this system, hormones produced in the hypothalamus are transported directly to the adenohypophysis without entering the general blood circulation.

b. Neural relationships

(1) The posterior lobe is supplied directly by neurons of the **supraoptic nucleus** and **paraventricular nucleus** in the hypothalamus. Their axons extend down the infundibular stem as the **hypothalamo-hypophyseal nerve tracts** to enter the neurohypophysis.

(a) The hypothalamic neurons secrete two neurohormones, **oxytocin** and **antidiuretic hormone** (ADH), which are transported along the axons and stored in the neurohypophysis.

(b) The hormones are released from the neurohypophysis on signal from the hypothalamic neurons.

(2) The anterior lobe has no direct nervous connection with the hypothalamus. The anterior pituitary hormones also are released on signal from the hypothalamus but by way of the vascular connection.

B. Anterior lobe hormones

1. Growth hormone (GH) (somatotropic hormone (STH)) is a protein hormone. It controls the growth of all body cells that are capable of increasing in size and number with primary effects on bone growth and skeletal muscle mass.

   a. Physiologic effects

   (1) **Protein synthesis.** GH increases the rate of protein synthesis in all cells of the body by promoting the uptake of amino acids through the cell membrane.

   (2) **Conservation of carbohydrates.** GH decreases the rate of carbohydrate utilization by body cells, thus promoting the increased level of blood glucose.

   (3) **Mobilization of stored fats.** GH causes the increased mobilization of fats and use of fats for energy.

   (4) **Stimulation of skeletal growth.** GH causes the liver (and possibly the kidneys) to produce **somatomedins**, a group of pituitary-dependent growth factors essential for cartilage and bone growth.

   b. Regulation of GH secretion is by two antagonistic hormone secretions.

   (1) Stimuli for release

   (a) **Growth-hormone-releasing-hormone (GHRH)** from the hypothalamus is carried via the hypothalamic-hypophyseal portal system to the anterior pituitary, where it stimulates GH synthesis and release.

   (b) Additional stimuli for GH release include **stress, malnutrition,** and **activities that lower blood sugar, such as fasting and exercise.**

   (2) Inhibition of release

   (a) The secretion of GHRH is inhibited by **increased levels of GH** in the blood via the negative feedback mechanism.

   (b) **Somatostatin, a growth-hormone-inhibiting-hormone (GHIH)** from the hypothalamus, is carried to the anterior pituitary by the portal system. It inhibits GH synthesis and release.

   (c) Additional stimuli for GH inhibition include **obesity** and **elevated blood fatty acid levels.**

   c. Abnormalities of GH secretion

(1) **Dwarfism.** Hyposecretion (**deficiency**) of GH during childhood results in stunting of growth. Human growth hormone is used therapeutically in cases of pituitary dwarfism.
(2) **Gigantism.** Hypersecretion (**excess secretion**) of GH during adolescence and prior to the closing of the epiphyseal plates results in excessive growth of the long bones (pituitary gigantism). This type of overproduction of GH generally results from a pituitary tumor and is very rare.
(3) **Acromegaly.** Hypersecretion of GH after closing of epiphyseal plates cannot cause increased length of long bones but does cause a disproportionate enlargement of soft tissues, growth in thickness of flat bones of the face, and increase in hand and foot size. It also is relatively uncommon.

2. **Thyroid-stimulating hormone (TSH)**
   a. **Physiologic effects of TSH.** TSH, also known as **thyrotropin**, controls the amount of thyroxine and triiodothyronine hormones secreted by the thyroid gland. TSH increases the growth and development of thyroid gland cells, their rate of hormone production, and the hormone's effect on cell metabolism.
   b. **Regulation of TSH secretion.** The synthesis and release of TSH is controlled by hypothalamic **thyrotropin-releasing hormone (TRH)** carried by the hypothalamic-hypophyseal portal system.
      (1) In turn, TRH secretion is regulated by the blood level of circulating thyroid hormone (negative feedback) and by the metabolic rate of the body.
         (a) If the circulating level of thyroid hormone increases and the body's rate of metabolism rises, TRH will be inhibited.
         (b) If the thyroid hormone blood level or cellular metabolism decreases, TRH secretion will be stimulated.
      (2) Long-term exposure to severe cold is an environmental factor that stimulates the release of TRH. This enhances thyroid hormone production to increase the metabolic rate, which warms the body.

3. **Adrenocorticotropic hormone (ACTH)** is also known as corticotropin.
   a. **Physiologic effects.** ACTH stimulates secretion of adrenocortical hormones by the adrenal cortex, particularly of the glucocorticoids.
   b. **Control of secretion.** ACTH is regulated by corticotropin-releasing hormone (CRH) from the hypothalamus. The feedback mechanisms for the stimulation or inhibition of CRH, ACTH, and adrenal cortical hormones function in the same manner as those for TRH, TSH, and thyroid hormones.

4. **Hormones related to ACTH.** ACTH, **endorphins**, and **melanocyte-stimulating hormone (MSH)** all are derived from **proopiomelanocortin (POMC)**, a large precursor molecule produced by the anterior and intermediate lobes of the pituitary.
   a. **Endorphins** are called endogenous opiates because they originate in the body and mimic the effect of heroin and morphine. They are linked to natural pain relief (analgesia) and may function in response to stress or exercise.
   b. **MSH** stimulates pigment formation and dispersal in the pigment-producing cells (melanocytes) of the epidermis.

5. **Gonadotropins.** Follicle-stimulating hormone (**FSH**) and **luteinizing hormone (LH)** are called gonadotropic hormones because they regulate the function of the gonads. (Also see Chapter 18.)
   a. **Physiologic effects of FSH**
      (1) In females, FSH stimulates growth of **ovarian follicles** and helps stimulate the ovarian production of **estrogens**.
      (2) In males, FSH stimulates growth and development of **spermatozoa** in the seminiferous tubules of the testes.
   b. **Physiologic effects of LH**

(1) In females, LH, in conjunction with FSH, stimulates **estrogen** production. LH is responsible for **ovulation** and the secretion of **progesterone** from the ruptured follicle.

(2) In males, LH stimulates the **interstitial cells** of the seminiferous tubules of the testes to produce **androgen** (testosterone).

c. **Control of FSH and LH secretion**

(1) Pituitary gonadotropins are regulated by gonadotropin-releasing hormone (GnRH) from the hypothalamus.

(2) GnRH causes release of both FSH and LH, which, in turn, cause secretion of gonadal hormones (estrogens, progesterone, and testosterone).

(3) Both negative and positive feedback mechanisms are involved in the secretion of GnRH, pituitary gonadotropins, and gonadal hormones.

6. **Prolactin (PRL)** is secreted during **pregnancy** and during **breastfeeding** after delivery.

   a. **Physiologic effects.** Prolactin initiates and sustains **milk secretion** from the mammary glands when they have been previously readied for lactation by the action of other hormones.

   b. **Control of secretion** of PRL involves two hypothalamic hormones.

   (1) Release is **inhibited** by **prolactin-inhibiting hormone (PIH)**, which is identical to the neurotransmitter, **dopamine**.

   (2) Release is believed to be **stimulated** by a prolactin-releasing factor (PRF), but the identification or chemical synthesis of PRF in humans has not been established.

B. **Posterior lobe hormones: ADH and oxytocin.** Two hormones are synthesized by nerve cells in the hypothalamus, transported along their axons (axoplasmic transport), and stored in the neurohypophysis to be released at the axonal endings. Each is secreted by a separate group of neurons.

1. **ADH**, also called **vasopressin**, is synthesized in the neurons of the **supraoptic nucleus** of the hypothalamus. (See Chapter 16).

   a. **Physiologic effects**

   (1) Antidiuretic hormone (ADH) promotes **water retention**. It decreases the amount of water lost in the urine (antidiuresis) by increasing water reabsorption from the distal convoluted tubules and collecting ducts in the kidneys.

   (2) ADH helps to **elevate blood pressure** by causing a constriction of peripheral blood vessels.

   b. **Control of secretion.** The release of ADH is regulated by changes in blood osmolarity (concentration of electrolytes) and changes in blood volume and pressure.

   (1) An **increased concentration of the body fluids** or a decrease in blood volume causes the **secretion of ADH**, which acts on the kidneys to conserve body water.

   (2) A **decreased concentration of the body fluids** or an increase in blood volume (for example, through ingestion of water) causes the **inhibition of ADH** to allow greater water loss through the kidneys.

   (3) ADH release is inhibited (causing water loss) by alcohol and caffeine.

   (4) ADH release is stimulated (causing water retention) by pain, anxiety, and trauma, and by such drugs as nicotine, morphine, and barbiturates.

   c. **Abnormal secretion of ADH**

   (1) **Hyposecretion** results in **diabetes insipidus**, which is characterized by excessive thirst and excessive urine production. It may be due to damage to the hypothalamus or posterior lobe or by failure of the renal response to ADH. The condition is treated by administration of small amounts of ADH.

   (2) **Hypersecretion** rarely may follow hypothalamic injury or be caused by a tumor. It results in water retention, dilution of the body fluids, and increase in blood volume.

2. **Oxytocin** is synthesized in the cell bodies of neurons in the **paraventricular nucleus** of the hypothalamus.

a. **Physiologic effects of oxytocin** in females. (Oxytocin has no known function in males, although it is released during sexual stimulation.)
  (1) Oxytocin stimulates the **contraction of smooth muscle cells** of the uterus during sexual intercourse and during labor and delivery in a pregnant woman.
  (2) Oxytocin causes **milk ejection** from the mammary glands in nursing mothers by stimulating the myoepithelial (contractile) cells surrounding the alveoli of the mammary glands.
b. **Control of secretion** of oxytocin
  (1) **Suckling at the breast, the sight or sound of an infant,** or **stimulation of the nipple or areola** in a nursing mother results in nerve stimuli to the hypothalamus, oxytocin secretion, and milk release. This is known as the **milk let-down reflex.**
  (2) Oxytocin and subsequent milk release is inhibited by emotional stress.

III. THYROID GLAND

A. **Morphology**
  1. The thyroid gland consists of **two lateral lobes** connected by a narrow **isthmus**. It is located over the anterior surface of the thyroid cartilage of the trachea, just below the larynx.
  2. The **follicles** are the functional units of the thyroid gland. Each follicle is lined by a single layer of epithelial **follicular cells**, which enclose a central cavity. The follicular epithelium is columnar when stimulated by TSH and cuboidal when the gland is inactive.
  3. The cavity of the follicle contains colloid, which is composed mainly of the globular protein **thyroglobulin**.
    a. Thyroglobulin is the storage form of thyroid hormone.
    b. It also functions in the synthesis of thyroid hormone.
  4. Smaller numbers of **parafollicular cells** (C cells), which secrete **calcitonin**, are present in the interfollicular spaces and between the follicle cells. Calcitonin lowers the calcium concentration in the blood.

B. **Formation, storage, and release of thyroid hormones**
  1. Two hormones are secreted by the thyroid gland.
    a. **Thyroxine**, or **tetraiodothyronine** (T_4), comprises 90% of thyroid gland secretion.
    b. **Triiodothyronine** (T_3) is secreted in small amounts.
  2. When TSH binds to follicle cell receptors, it results in the synthesis and secretion of **thyroglobulin**, which contains the amino acid **tyrosine**, into the follicle lumen.
  3. **Iodine** ingested in the diet is carried in the bloodstream in ionic form as iodide (I^-) to the thyroid gland. The follicular cells remove the iodide from the blood and convert it to molecular (elemental) iodine.
  4. The molecular iodine reacts with tyrosine in the thyroglobulin to yield monoiodotyrosine and diiodotyrosine molecules.
    a. Two diiodotyrosine molecules form T_4 (thyroxine).
    b. One monoiodotyrosine and one diiodotyrosine molecule form T_3 or triiodothyronine.
  5. Large amounts of T_4 and T_3 are stored in thyroglobulin form for many weeks. When thyroid hormones are to be released under the influence of TSH, proteolytic enzyme action separates the hormones from thyroglobulin. The hormones diffuse from the follicle lumen through the follicular cells to enter the blood circulation.
  6. Most of the circulating thyroid hormones combine with plasma proteins (mainly **thyroxine-binding globulin** produced by the liver) for transport.

C. **Physiologic effects of thyroid hormones**
1. Thyroid hormones **increase the metabolic rate** of nearly all cells of the body. They stimulate oxygen consumption and raise energy output, particularly in the form of heat.
2. **Normal growth and maturation** of bone, teeth, connective tissue, and nervous tissue are dependent on thyroid hormones.

D. **Control of secretion**
1. Thyroid function is regulated by pituitary thyroid-stimulating hormone (TSH) under the control of hypothalamic thyrotropin-releasing hormone (TRH) through a pituitary-hypothalamus feedback system. (See I B 2 above.)
2. The primary factors affecting the rate of TRH and TSH secretion are the **level of circulating thyroid hormones** and the **metabolic rate** of the body.

E. **Abnormalities of secretion** may result from iodine deficiency, or hypothalamic, pituitary, or thyroid gland malfunction.
1. **Hypothyroidism** is a diminished production of thyroid hormones. It results in depressed metabolic activity, constipation, lethargy, sluggish mental reactions, and increased fat deposition.
   a. In adults, the condition causes **myxedema**, which is characterized by an accumulation of water and mucin under the skin resulting in an edematous appearance.
   b. In a young child, hypothyroidism results in the physical and mental retardation known as **cretinism**.
2. **Hyperthyroidism** is an overproduction of thyroid hormones. It results in increased metabolic activity, weight loss, nervousness, tremor, diarrhea, increased heart rate, and, in severe hyperthyroidism, symptoms of hormone toxicity.
   a. Severe hyperthyroidism may cause **exophthalmic goiter (Graves' disease)**. A symptom may be swelling of the tissues in the eye sockets, producing eye bulging.
   b. Treatment of hyperthyroidism is by surgical removal of the thyroid gland or with radioactive iodine, which localizes in the gland and destroys the tissue.
3. **Goiter** is a two- to threefold enlargement of the thyroid gland. It may occur in association with hypothyroidism or hyperthyroidism.
   a. Simple (endemic) goiter associated with hypothyroidism has occurred in areas of the world deficient in iodine.
   b. Decreased dietary iodine results in the accumulation of thyroglobulin (colloid) in the follicles but a decreased thyroid hormone production.
   c. The supplementation of salt with iodine has reduced the incidence of endemic goiter.

IV. **PARATHYROID GLANDS**

A. **Morphology**
1. The parathyroid glands are four small bodies, each approximately the size of an apple seed, located on the posterior surface of the thyroid gland and separated from the thyroid gland by their connective tissue capsules.
2. Histologically, there are two types of cells in the parathyroid glands: the **chief cells**, which secrete **parathyroid hormone** (PTH), and the **oxyphil cells**, which may be a developmental stage of the chief cells.

B. **Physiologic effects of parathyroid hormone**
1. PTH controls the balance of **calcium** and **phosphate** in the body by increasing the blood level of calcium and decreasing the blood level of phosphate.

a. **Calcium ions** are vital to proper **bone and teeth formation**, blood coagulation, muscle contraction, cell membrane permeability, and neuromuscular excitability.
b. **Phosphate ions** are vital to **cellular metabolism**, acid-base buffering systems of the body, and as a component of **nucleotides** and **cell membranes**.

2. **PTH increases the blood level of calcium by three mechanisms.**
   a. PTH stimulates **osteoclast (bone-destroying cell)** activity, thereby causing withdrawal of calcium from the bones to the extracellular fluid.
   b. PTH indirectly enhances the **intestinal absorption of calcium** and reduces calcium loss in the feces. It primarily acts by **activating vitamin D**, which is required for the absorption of calcium from food.
   c. PTH stimulates the **reabsorption of calcium from the kidney tubules** in exchange for phosphorus, thus decreasing the loss of calcium ions in the urine and increasing the calcium level in the blood.

C. **Control of secretion** is via a feedback control system with calcium ion concentration in the blood.
   1. Decreased blood levels of calcium cause increased levels of PTH secretion. As the blood calcium levels increase, the PTH level decreases.
   2. **Calcitonin** (thyrocalcitonin), produced by the parafollicular cells of the thyroid gland, is a direct antagonist to PTH and **decreases blood calcium**.
      a. Calcitonin is released from the thyroid gland when blood levels of calcium are greatly increased.
      b. Calcitonin inhibits the effect of PTH on calcium resorption from bone and stimulates osteoblastic activity to result in calcium uptake into bone.

D. **Abnormalities of secretion**
   1. **Hypersecretion (hyperparathyroidism)** is rare and may be the result of a parathyroid tumor. It results in increased osteoclast activity, bone resorption, and decalcification and weakening of the bones.
   2. **Hyposecretion (hypoparathyroidism)** results in a lowered blood calcium level and an increased irritability of the neuromuscular system. If the hyposecretion is severe, it results in tetany (spasms of skeletal muscle), which is fatal unless it is treated.

V. ADRENAL GLANDS

A. **Morphology**
   1. The adrenal glands (suprarenal glands) are two flattened, triangular, yellow masses embedded in adipose tissue. They lie on top of the upper poles of the kidneys.
   2. Each adrenal gland is composed of an **outer cortex** and an **inner medulla** (Figure 10–5).
      a. The **cortex** secretes steroid hormones. It is divided into three layers from the outside in: **zona glomerulosa, zona fasciculata,** and **zona reticularis**.
      b. The **medulla** originates embryologically from the same kind of neuroectoderm (neural crest cells) that give rise to sympathetic neurons. (See Chapter 9, Part II, I A.) The cells of the medulla are actually modified postganglionic sympathetic neurons.

B. **Hormones of the adrenal gland**
   1. **Medullary** hormones are secreted by the chromaffin cells of the adrenal medulla in response to preganglionic sympathetic stimuli. They consist of the catecholamines, **epinephrine (80%)**, and **norepinephrine (20%)**.

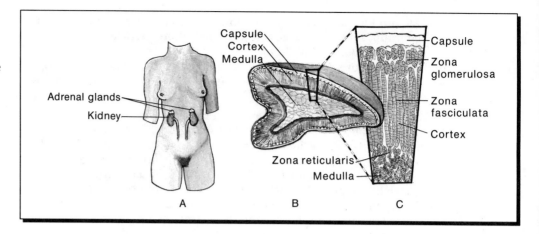

Figure 10–5. *Adrenal glands. A,* Location of the adrenal glands. *B,* Section through the gland illustrating the cortex, the medulla, and the three zones of the cortex.

      a. Epinephrine and norepinephrine have differences in physiological effects due to two types of receptors, alpha (α) and beta (β), which are located on the membranes of the target cells.
      b. Overall, the function of these hormones is to prepare the body for physical activity in response to stress, excitement, injury, exercise, and decreased blood sugar levels.
        (1) **Effects of epinephrine**
          (a) Heart rate, metabolism, and oxygen consumption are increased.
          (b) Blood glucose levels are increased through stimulation of glycogenolysis from liver and muscle glycogen stores.
          (c) Blood vessels in the skin and visceral organs are constricted while blood vessels in skeletal and cardiac muscle are dilated.
        (2) **Effects** of norepinephrine are to increase blood pressure and to stimulate cardiac muscle.
   2. **Adrenal cortical hormones,** in contrast to the medullary hormones, are essential to life.
      a. **Mineralocorticoids** are synthesized in the **zona glomerulosa**.
        (1) **Aldosterone,** the most important mineralocorticoid, regulates electrolyte and water balance by control of sodium and potassium levels in the blood.
        (2) **Control of secretion.** Aldosterone secretion is regulated by the level of blood sodium but mostly by the **renin-angiotensin mechanism,** which is explained in Chapter 17.
      b. **Glucocorticoids** are synthesized in the **zona fasciculata**. They include **corticosterone, cortisol,** and **cortisone**. The most important is cortisol.
        (1) **Physiologic effects**
          (a) Glucocorticoids **affect glucose, protein, and fat metabolism** to create a reservoir of building block molecules that can readily be metabolized.
          (b) They promote **glucose synthesis** from noncarbohydrate sources (gluconeogenesis), **glycogen storage** in the liver (glycogenesis), and an **increase in blood glucose level.**
          (c) They increase the **breakdown of fats** and proteins and **inhibit amino acid uptake and protein synthesis.**
          (d) They **stabilize lysosomal membranes** to prevent further tissue damage.
        (2) **Control** of glucocorticoid secretion is by **ACTH** operating through a typical negative feedback mechanism. The primary stimulus to ACTH secretion is any type of **physical or emotional stress.**
          (a) Stress (for example, trauma, infection, or tissue damage) initiates nerve impulses to the hypothalamus.
          (b) The hypothalamus secretes **corticotropin-releasing hormone (CRH),** which passes via the hypothalamic-hypophyseal portal system to the anterior pituitary gland, which releases **ACTH.**

(c) ACTH circulates in the blood to the adrenal gland and elicits secretion of **glucocorticoids**.
(d) Glucocorticoids result in an increased availability of amino acids, fats, and glucose in the blood to help repair the damage caused by stress and in stabilization of lysosomal membranes to prevent further tissue damage.

c. **Gonadocorticoids** (sex steroids) are synthesized by the **zona reticularis** in relatively small amounts. They function primarily as **precursors** for conversion to testosterone and estrogen by other tissues.

3. **Abnormalities of adrenocortical secretion**
   a. **Hyposecretion** is due to destruction of cortical tissue through disease or atrophy and is known as **Addison's disease**. It results in sodium-potassium imbalances, skin darkening (caused by an increase in ACTH, which is similar to MSH), and a reduced ability to respond to physiological stresses.
   b. **Hypersecretion** can result from an adrenal tumor or from an increased production of ACTH. The effects of hypersecretion depend on which cells of the adrenal cortex are secreting excessive quantities of hormone.
      (1) **Primary aldosteronism** is excessive secretion of **aldosterone** by the zona glomerulosa. It results in an increase in body sodium, extracellular fluid volume, cardiac output, and blood pressure.
      (2) **Cushing's disease** is the result of excessive production of **glucocorticoids** by the zona fasciculata. It causes an increased mobilization of protein and fat, which leads to weak muscles and an accumulation of fat in the neck, face, and trunk. The increased gluconeogenesis results in very high blood sugar levels (**adrenal diabetes**).
      (3) **Adrenogenital syndrome** (adrenal virilism) is the result of excessive **androgen production** by the zona reticularis.
         (a) The condition results in **precocious puberty** if it occurs in a prepubertal child.
         (b) In adult females, **masculinization** in the form of facial hair, voice deepening, and increased muscle development may occur.
         (c) Masculinization of a female fetus may occur if the pregnant woman has an adrenal tumor or is given androgen-like hormones (progestins) during pregnancy.
      (4) Greater than physiologic levels of glucocorticoids may be injected therapeutically to lessen inflammatory and allergic responses.
         (a) The positive effects of glucocorticoid injection include the stabilization of the lysosomal membrane and the decrease in capillary permeability, which inhibits inflammation.
         (b) The negative effects include the inhibition of the white blood cell response to infection and the decrease of antibody production, thus decreasing wound healing.

VI. THE ENDOCRINE PANCREAS

A. **Morphology**
   1. The pancreas is a flattened organ located behind and slightly below the stomach in the abdomen. It has both an exocrine and endocrine function.
   2. The **exocrine portion** of the pancreas is a function of the pancreatic acinar cells, which produce pancreatic juices secreted via the pancreatic duct into the small intestine.
   3. The **endocrine cells** are found in the **islets of Langerhans**, small clusters of cells scattered throughout the organ. Four hormone-producing cell types have been identified in the islets.
      a. The **alpha cells** secrete **glucagon**, which raises the blood sugar level.
      b. The **beta cells** secrete **insulin**, which lowers the blood sugar level.

c. The **delta cells** secrete **somatostatin**, or growth-hormone-inhibiting hormone, which inhibits the secretion of both glucagon and insulin.
d. The **F cells** secrete **pancreatic polypeptide**, a digestive hormone of uncertain function that is released after a meal.

B. **Physiologic effects of insulin**
1. Insulin **makes glucose available to most body cells**, especially muscle and adipose, by promoting the passage of glucose through the cell membrane via a carrier mechanism. (It does not facilitate the passage of glucose into brain tissue, kidney tubules, intestinal mucosa, or red blood cells.)
2. Insulin **increases the body's stores of proteins and fats.**
   a. It promotes the transport of amino acids and fatty acids from the blood into the cells.
   b. It increases protein and fat synthesis and decreases protein and fat catabolism.
3. Insulin **promotes the utilization of carbohydrates** for energy.
   a. It facilitates the storage of glucose as glycogen in skeletal muscle and the liver.
   b. It enhances storage of excess glucose as fat in adipose tissue.

C. **Physiologic effects of glucagon**
1. Glucagon **enhances the breakdown of liver glycogen** to glucose (glycogenesis), thereby increasing the blood glucose level.
2. Glucagon **increases the synthesis of glucose** from noncarbohydrate sources (gluconeogenesis) in the liver.

D. **Control of insulin secretion**
1. **Effect of blood level of glucose**
   a. Increased blood glucose levels, such as following a meal, stimulate the beta cells to produce insulin. The insulin causes the glucose to diffuse into cells where it is used for energy, converted to glycogen in the liver, or converted into fat in the adipose tissue.
   b. When the blood glucose level falls, the rate of insulin secretion is decreased.
2. **Effect of glucagon**
   a. Glucagon influences the secretion of insulin by elevating the blood glucose concentration. The effects of glucagon and insulin are antagonistic in order to maintain normal blood glucose levels during fasting or eating.
   b. Glucagon secretion is controlled by the blood glucose level.
      (1) Low blood glucose stimulates the alpha cells to produce glucagon.
      (2) Glucagon causes the release of glucose from the liver, thereby increasing blood glucose.
      (3) The increased blood glucose level inhibits the release of glucagon through negative feedback.
3. **Other hormones indirectly affect insulin secretion.**
   a. **Growth hormone, ACTH, and gastrointestinal hormones**, such as gastrin, secretin, and cholecystokinin, all stimulate insulin secretion.
   b. **Somatostatin**, produced by the pancreatic delta cells and the hypothalamus, inhibits the secretion of insulin and glucagon and blocks the intestinal absorption of glucose.

E. **Abnormalities of secretion**
1. **Diabetes mellitus** is a relative deficiency of insulin.
   a. Types of diabetes mellitus
      (1) In **Type I diabetes**, also referred to as insulin-dependent diabetes mellitus (IDDM), the pancreas fails to secrete insulin, either from degeneration or inactivation of the beta cells.
      (2) In **Type II diabetes**, also called non-insulin-dependent diabetes mellitus (NIDDM), insulin is produced by the beta cells in normal or near normal amounts, but the body cells are unable to use it due to a deficiency or disorder of insulin receptors.

b. **Causes** of diabetes mellitus are not completely known, but **genetic factors**, **obesity**, **autoimmune disease**, and **viruses**, as well as **environmental, economic,** and **cultural factors** all have been implicated.
c. **Symptoms** of diabetes
   (1) Diabetes mellitus is characterized by **hyperglycemia** (elevation of blood glucose) and impaired carbohydrate metabolism, which results in the following effects:
      (a) **Glycosuria** (loss of glucose in the urine) occurs because the renal threshold for glucose reabsorption is exceeded.
      (b) **Polyuria** (excessive loss of sodium and water in the urine) occurs because the osmotic pressure created by the excess glucose in the kidney tubules diminishes water reabsorption.
      (c) **Polydipsia** (excessive thirst and water consumption) occurs because the reduced blood volume activates hypothalamic thirst centers.
      (d) **Polyphagia** (excessive appetite and voracious eating) occurs because of the lack of carbohydrate in the body cells.
      (e) **Ketonemia** and **ketonuria**, or accumulation of fatty acids and ketones in the blood and urine, occur as a result of the abnormal catabolism of fats as an energy source. This may result in acidosis and coma.
d. Diabetic individuals have a statistically higher rate of coronary heart disease, blindness, circulatory disorders, infection, slow healing, gangrene, and kidney disorders.
2. **Hyperinsulinism** is much rarer than hypoinsulinism. The decreased blood sugar level (**hypoglycemia**) produces symptoms of weakness, anxiety, sweating, and mental disorientation.

VII. PINEAL GLAND

A. **Morphology**
   1. The **pineal gland** (epiphysis cerebri) is derived from neural tissue and is located in the roof of the third ventricle of the brain.
   2. It consists of **pinealocytes** and supporting neuroglial cells.
   3. With age, the gland accumulates calcium deposits known as "brain sand."
B. The hormone secreted by the pineal gland is **melatonin**, which has several postulated effects.
   1. In laboratory animals, melatonin affects endocrine functions of the thyroid, adrenal cortex, and gonads and influences mating behavior.
   2. In humans, melatonin appears to have an inhibiting effect on the release of gonadotropins and inhibits melanin production by melanocytes in the skin.
C. **Control of melatonin production**
   1. The intensity and duration of **environmental light**, which reaches the gland by collaterals of the visual pathway, affects the release of melatonin. Melatonin production is lowest during daylight and greatest during the night.
   2. The cyclical nature of melatonin production may be involved in circadian (daily) rhythmicity of some physiological processes.

VIII. THYMUS GLAND

A. **Morphology.** The thymus is located in the thorax posterior to the sternum and overlying the upper part of the heart. It is large in childhood and diminishes in size with age.

B. **Hormones**, or factors, produced by the thymus gland include six peptides, which collectively are called **thymosins.**

C. **Function of thymosins**
1. Thymosins control the development of the thymic-dependent immune system by stimulating the differentiation and proliferation of **T-cell lymphocytes.**
2. Thymosins may play a role in **congenital immunodeficiency diseases** such as agammaglobulinemia, a total inability to produce antibodies.

Study Questions

Directions: Each question below contains four suggested answers. Choose the **one best** response to each question.

1. All of the following statements are true about hormones EXCEPT
   (A) Hormones affect cells that are remotely located from the endocrine gland secreting the hormone.
   (B) Hormones are involved in the control and integration of many body functions.
   (C) Hormones must be present in high concentration in the blood to exert their physiologic effect.
   (D) Most hormones are steroids or amino acid derivatives.

2. Which of the following hormones has no known function in males?
   (A) follicle-stimulating hormone
   (B) oxytocin
   (C) antidiuretic hormone
   (D) gonadotropin-releasing hormone

3. Hypothalamic hormones reach the anterior pituitary by way of
   (A) the systemic circulation
   (B) a portal system
   (C) nerve tracts
   (D) various pathways, depending on the hormone

4. An individual voids very frequently with production of large quantities of very dilute urine. There is no evidence of bacteria in the urine, and blood tests for abnormal levels of any of the substances ordinarily carried in the blood are normal. Such an individual would be likely to benefit most from administration of hormone extracted from the
   (A) adrenal gland
   (B) pancreas
   (C) anterior pituitary gland
   (D) posterior pituitary gland

5. All of the following statements regarding adrenal medulla secretion are true EXCEPT
   (A) It increases heart rate and force of contraction.
   (B) It is necessary for survival.
   (C) It enhances liver glycogenolysis.
   (D) It dilates blood vessels in skeletal muscles.

6. Parathyroid hormone acts to
   (A) decrease blood and bone calcium
   (B) increase blood calcium and decrease bone calcium
   (C) increase blood and bone calcium
   (D) decrease blood calcium and increase bone calcium

Questions 7–9 refer to the following situation:

A 55-year-old farm worker says that his shoe size has increased and that he can no longer purchase any leather gloves that will fit him. He is coarse-featured in appearance, has a prominent jaw and a prominent forehead. he states that in the last 20 years he has had a gradual weight gain of 60 pounds.

7. The above symptoms are suggestive of
   (A) myxedema
   (B) acromegaly
   (C) cretinism
   (D) inactivity and excessive appetite

8. Hypersecretion of which of the following is likely to be causing the problem?
   (A) thyroid hormone
   (B) adrenocorticotropic hormone
   (C) parathyroid hormone
   (D) growth hormone

9. A physical examination, x-rays and other imaging techniques, and laboratory tests would most likely show which one of the following?
   (A) goiter
   (B) pituitary gland tumor and an abnormally enlarged sella turcica
   (C) very high blood glucose level
   (D) anemia (low blood cell count)

10. The thyroid gland has the ability to remove which of the following from the blood?
    (A) iodine
    (B) glucose
    (C) iron
    (D) thyroglobulin

Questions 11–14. Match each of the following hormones with the phrase that describes its action.
   (A) aldosterone
   (B) melatonin
   (C) growth hormone
   (D) luteinizing hormone

11. Secretion is influenced by environmental light intensity

12. Regulates plasma sodium and potassium levels

13. Promotes expulsion of the ovum (ovulation) from the ovaries

14. Stimulates protein synthesis

Answers and Explanations

1. **The answer is C.** (I A 1–4) A very small amount of hormone is able to exert a significant effect because the hormone binds to specific hormone receptors on cell membranes or in the interior of the cell. Once bound to the receptor, the effects of the hormone is biologically amplified within the cell. Hormones are transported in the bloodstream and affect target cells or tissues located at a distance from the secreting gland. Hormones interact with the nervous system and form the second mechanism of control and regulation of body activities. Chemically, hormones are peptides, polypeptides, proteins, or steroids.

2. **The answer is B.** (II B 2 a) Oxytocin, although secreted, has no recognized function in males. FSH in males is responsible for growth and development of spermatozoa in the testis. ADH from the posterior pituitary gland and GnRH from the hypothalamus do not differ in effect between males and females.

3. **The answer is B.** (II A 4 a, b) The hypothalamic-hypophyseal portal system consists of venules, which carry blood containing hypothalamic hormones between the capillaries of the hypothalamus and those of the anterior pituitary. Hormones produced by neurons in the supraoptic and paraventricular nuclei of the hypothalamus are carried by nerve tracts to be stored in the posterior pituitary until release. Hormones from other endocrine glands are transported in the systemic general circulation.

4. **The answer is D.** (II B 1 a–c) A hyposecretion of ADH results in diabetes insipidus, characterized by the production of large quantities of very dilute urine. Diabetes mellitus, treated with insulin from the pancreas, also may have symptoms of excessive thirst and urine production, but high levels of glucose would be present in the blood. Hypersecretion of glucocorticoids by the adrenal glands may lead to adrenal diabetes, but there would be greatly increased blood levels of glucose.

5. **The answer is B.** (V B 1, 2) The hormones secreted by the adrenal medulla are the catecholamines epinephrine and norepinephrine. These hormones prolong the effects of the sympathetic portion of the ANS. Although they are important in body responses to short-term stressors or to an emergency, they are nonessential to survival of the individual.

6. **The answer is B.** (IV B 2) Parathyroid hormone increases the blood level of calcium and decreases bone calcium by stimulating osteoclastic activity in the bones. The calcium in bone matrix is thereby released to the extracellular fluid and the blood circulation. Calcitonin, a hormone produced by the parafollicular cells of the thyroid gland, acts in an opposite fashion. Calcitonin stimulates osteoblastic activity in the bones, thereby causing bone to be withdrawn from the blood to be deposited in bone matrix.

7–9. **The answers are 7–A, 8–D, and 9–B.** (II B 1 c) Hypersecretion of growth hormone in adulthood results in acromegaly. The symptoms include growth in thickness of the flat bones of the face, an increase in hand and foot size, and a general increase in body mass. The most likely cause of acromegaly is a pituitary tumor, which also would produce enlargement of the sella turcica. A goiter is an abnormal enlargement of the thyroid gland as a result of hyper- or hyposecretion of thyroxine. A very high blood glucose level is a symptom of insulin deficiency. A low blood cell count is not associated with any of these endocrine disorders.

10. **The answer is A.** (III B) The thyroid gland removes ionic iodine (iodide) from the blood and converts it to iodine to be combined with tyrosine to form the thyroid hormones. Insulin has the effect of removing glucose from the blood by facilitating the entry of glucose into body cells. Thyroglobulin is stored in the lumen of the thyroid follicles and functions both in the synthesis and release of thyroid hormones.

11–14. **The answers are 11–B, 12–A, 13–D, and 14–C.** (VII C; V B 2; II B 1; II B 5 b) Blood levels of melatonin, produced by the pineal gland, are affected by the duration and intensity of light. Aldosterone, a mineralocorticoid produced by the adrenal glands, regulates electrolyte (sodium and potassium) concentrations in the body fluids. Luteinizing hormone, a gonadotropin produced by the anterior pituitary gland, causes ovulation in the ovaries. Growth hormone, produced by the anterior pituitary gland, results in growth by enhancing amino acid uptake into the cells to result in protein synthesis.

Circulatory System 11

I. INTRODUCTION

A. The **circulatory system** is the link between the external environment and the internal fluid environment of the body. It carries nutrients and gases to all the cells, tissues, organs, and organ systems, and it carries metabolic end products away from them.

B. **Components**
   1. The **cardiovascular system** is a subset of the circulatory system. It consists of the **heart**, the **blood vessels** (arteries, capillaries, and veins), and the **blood** that flows within them.
      a. The heart is a muscular pump to propel the blood.
      b. The blood vessels are a series of tubes in which the blood is conducted.
      c. The blood is the fluid transported in the vessels. No cell of the body is more than a fraction of a millimeter away from a source of supply.
   2. The **lymphatic system** is also part of the circulatory system. It consists of lymph vessels and lymph nodes positioned within the larger lymph vessels.
   3. Blood-forming and blood-storing organs, such as the spleen, liver, bone marrow, thymus gland, and lymph tissue, are associated with the circulatory system.

C. **Functions**
   1. **Transport.** Food, gases, hormones, minerals, enzymes and other vital substances are carried in the blood to all the cells of the body. All waste materials are carried in the blood from cells to the lungs, kidneys, or skin for elimination from the body.
   2. **Maintenance of body temperature.** The blood vessels constrict to retain body heat and dilate to dissipate heat at the skin surface.
   3. **Protection.** The blood and the lymphatic system protect the body against injury and foreign invasion through the immune system. The blood clotting mechanism protects against blood loss.
   4. **Buffering.** The blood proteins provide an acid-base buffer system to maintain the optimum pH of the blood.

II. BLOOD

A. Characteristics

1. Blood is a type of connective tissue in which cells (formed elements) are suspended and carried in a fluid matrix (plasma).
2. Blood is heavier than water and more viscous. It has a characteristic taste and odor, and has a pH of 7.4 (7.35 to 7.45).
3. The **color** of blood varies from bright red to dark bluish-red, depending on the amount of oxygen carried by the red blood cells.
4. **Total blood volume** is approximately 5 liters (10 pints) in a male adult of average size and less in female adults. It varies directly with body size and inversely with the amount of adipose tissue present. It also varies with changes in blood fluid and electrolyte concentration.

B. Components

1. **Blood plasma** is a clear, straw-colored liquid with essentially the same constituents as cytoplasm. It is 92% water and contains a complex mixture of organic and inorganic substances.
   a. **Plasma proteins** comprise 7% of plasma. They are the only plasma constituents that cannot pass through the capillary membrane to reach the cells. There are three main types of plasma proteins: albumins, globulins, and fibrinogen.
      (1) **Albumins** are the most numerous of the plasma proteins, comprising 55 to 60%, but are the smallest in size. Albumins are synthesized in the liver and are responsible for the **colloid osmotic pressure** of the blood.
         (a) **Colloids** are substances between 1 nm and 100 nm in diameter, whereas crystalloids are substances less than 1 nm in diameter. Plasma contains both colloids and crystalloids.
         (b) Colloid osmotic pressure (also called oncotic pressure) is determined by the number of colloid particles in solution. It is a measure of the "pulling power" of plasma for the diffusion of water from the extracellular fluid across the capillary membranes. (See Chapter 3, III B 2.)
      (2) **Globulins** make up approximately 30% of the plasma proteins.
         (a) **Alpha and beta globulins** are synthesized in the liver and function primarily as transport molecules for lipids, some hormones, various substrates, and other important body substances.
         (b) **Gamma globulins** (immunoglobulins) are antibodies. There are five types of immunoglobulins which are produced by lymphoid tissues and function in immunity.
      (3) **Fibrinogen** makes up 4% of plasma proteins. It is synthesized by the liver and is an essential component of the blood clotting mechanism.
   b. Plasma also contains nutrients, blood gases, electrolytes, minerals, hormones, vitamins, and waste products.
      (1) **Nutrients** include amino acids, sugars, and lipids that have been absorbed from the digestive tract.
      (2) **Blood gases** include oxygen, carbon dioxide, and nitrogen.
      (3) **Plasma electrolytes** include sodium, potassium, calcium, magnesium, chloride, bicarbonate, phosphate, and sulfate ions.
2. **Formed elements of the blood** include the **red blood cells (erythrocytes)**, the **white blood cells (leukocytes)**, and the **platelets**.

C. Hematopoiesis (production) of the formed elements

1. Site of production
   a. **During embryonic development,** hematopoiesis occurs first in the yolk sac and continues in the liver, spleen, lymph nodes, and all bone marrow of the developing fetus.

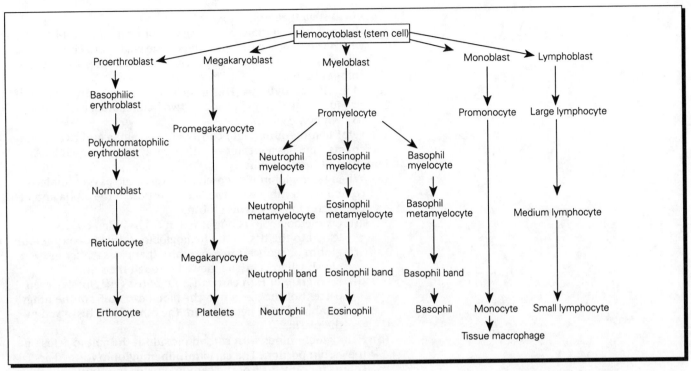

Figure 11-1. *Diagram of blood cell development.*

    b. After birth and during childhood, blood cells are formed in the marrow of all bones.

    c. In adults, blood cells are formed only in the **red bone marrow,** which is found in membranous bones such as the sternum, ribs, vertebrae, and the ilia of the pelvic girdles. Mature blood cells enter the general circulation from the bone marrow via skeletal veins.

2. Differentiation of blood cells. All blood cells derive from **hemocytoblasts** (primitive stem cells) found in the marrow, which divide and differentiate into **five types** of cells: proerythroblasts, myeloblasts, lymphoblasts, monoblasts, and megakaryoblasts (Figure 11–1).

    a. Proerythroblasts pass through a number of stages (basophilic erythroblast, polychromatophilic erythroblast, normoblast, and reticulocyte) and mature to become **erythrocytes.**

        (1) During development, erythrocytes synthesize **hemoglobin,** an oxygen-carrying pigment, and lose their organelles. The nucleus becomes smaller and eventually is extruded from the cell.

        (2) After loss of the nucleus, the erythrocyte remains in the marrow for several days until fully mature. It is then released into the circulation.

    b. Myeloblasts give rise to promyelocytes, which diverge in development to ultimately give rise to three types of white blood cells called **granulocytes: neutrophils, eosinophils,** and **basophils.**

    c. Lymphoblasts give rise to **lymphocytes. Monoblasts** give rise to **monocytes.** Lymphocytes and monocytes are called **agranulocytes.**

    d. Megakaryoblasts form megakaryocytes, which give rise to **platelets.**

D. Erythrocytes or red blood cells (RBCs)

  1. Characteristics

    a. Erythrocytes are biconcave discs, rounded with a central depression, that measure approximately 7.65 μm in diameter.

    b. Erythrocytes are enclosed by a highly permeable cell membrane. The membrane is also elastic and flexible, which allows the passage of erythrocytes through capillaries (the smallest blood vessels).

    c. Each erythrocyte contains nearly 300 million molecules of **hemoglobin,** a respiratory pigment that binds to oxygen. Hemoglobin makes up one third of the cell by volume.

(1) **Chemical structure of hemoglobin**
    (a) Hemoglobin is a molecule consisting of a protein, **globin**, which is composed of four polypeptide chains attached to four iron-containing **heme** groups. The heme is responsible for the color of blood.
    (b) In **adult hemoglobin (HgA)**, the polypeptide chains consist of two identical **alpha chains** and two **beta chains**, each having its own heme group.
    (c) **Fetal hemoglobin (HgF)** consists of two alpha and two gamma chains. HgF has a greater affinity for oxygen than does HgA.
(2) **Function of hemoglobin**
    (a) When hemoglobin is exposed to oxygen, the oxygen molecules combine reversibly with the heme portion of the alpha and beta chains to form **oxyhemoglobin**.
        (i) Oxyhemoglobin is bright red in color. When oxygen is released to the tissue, the hemoglobin is called **deoxyhemoglobin** or **reduced hemoglobin**. It appears darker or even bluish when veins are viewed through the skin.
        (ii) Each gram of HgA carries 1.3 ml of oxygen. Ninety-seven percent of the oxygen in the blood carried from the lungs is combined with hemoglobin; the other 3% is dissolved in the plasma.
    (b) Hemoglobin binds with carbon dioxide at the amino acids of the globin portion. The **carbaminohemoglobin** formed accounts for only 20% of the carbon dioxide transported in the blood. The other 80% is carried in the form of bicarbonate ions.

2. **Number**
   a. The RBC count in a healthy average male is 4.2 to 5.5 million cells per cubic millimeter (mm^3). The RBC count in a healthy average female is about 3.2 to 5.2 million cells per mm^3.
   b. The hematocrit is the percentage of total blood volume that consists of erythrocytes. It is determined by centrifuging a blood sample in a special tube and measuring the packed cells on the bottom.
      (1) The hematocrit is 42% to 54% in males and 38% to 48% in females.
      (2) The hematocrit may increase or decrease, depending on the number of erythrocytes or factors that affect blood volume, such as fluid intake or water loss.
      (3) The **sedimentation rate** is the speed with which the red cells sink to the bottom of a tube without centrifugation.

3. **Functions**
   a. RBCs **transport oxygen** to all tissues through a combination of oxygen with the hemoglobin.
   b. The hemoglobin in RBCs combines with **carbon dioxide** for tranport to the lungs, but most of the carbon dioxide is carried in the plasma in the form of bicarbonate ions. An enzyme (**carbonic anhydrase**) in the erythrocytes enables the blood to react with carbon dioxide to form bicarbonate ions. The bicarbonate ions diffuse out of the RBCs to the plasma.
   c. RBCs are important in the **regulation of blood pH**, because bicarbonate ions and hemoglobin are acid-base buffers.

4. **Regulation of RBC production**
   a. Erythrocyte production is regulated by **erythropoietin**, a glycoprotein hormone primarily produced by the kidneys. The rate of production of erythropoietin is inversely related to the oxygen supply of the tissues.
   b. Any factor that causes the tissues to receive less oxygen (anoxia) will result in increased erythropoietin production, which stimulates greater RBC production. For example,
      (1) Blood loss from **hemorrhage** results in increased RBC production.
      (2) Living at a **high altitude** where there is less oxygen in the atmosphere for long periods of time results in an increased RBC production.

(3) **Heart failure**, which decreases the blood flow to the tissues, or **lung disease**, which reduces the amount of oxygen absorbed into the blood, results in increased RBC production.
  c. Other hormones, such as **cortisone, thyroid hormone**, and **growth hormone**, also influence RBC production.
5. **Dietary factors essential to RBC production**
  a. **Iron** is vital for hemoglobin synthesis by erythrocytes. It is absorbed in the daily diet and stored in many tissues, particularly the liver.
  b. **Copper** is an essential part of a protein necessary for the conversion of ferric iron to ferrous iron.
  c. Certain **vitamins**, such as **folic acid, vitamin C**, and **vitamin B_{12}**, are necessary for the normal development and maturation of red cells.
    (1) Vitamin B_{12} cannot be synthesized by the body and must be ingested in the diet. For B_{12} to be absorbed from the digestive tract, the stomach lining must produce **intrinsic factor.**
    (2) In the absence of intrinsic factor, vitamin B_{12} is not absorbed, RBCs do not mature properly, and **pernicious anemia** (deficiency of RBCs) results. Vitamin B_{12} injections are used for treatment.
6. **Survival and destruction of erythrocytes**
  a. RBCs normally circulate approximately 120 days before they become fragile and fragment. Although mature RBCs lack nuclei, mitochondria, or endoplasmic reticulum, their cytoplasmic enzymes produce ATP for this limited period.
  b. Damaged or disintegrating RBC fragments undergo **phagocytosis** by **macrophages** in the spleen, liver, bone marrow, and other tissues.
    (1) The **globin** (protein portion) of HgA is degraded into amino acids, which are recycled for cellular protein synthesis.
    (2) The **heme** (iron-containing portion) is converted to **biliverdin** (green pigment) and then to **bilirubin** (yellow pigment), which is released into the plasma. Bilirubin is taken up by the liver and secreted in the bile.
    (3) Most of the **iron** liberated from heme is salvaged to be recycled in the synthesis of more HgA.
7. **Clinical considerations**
  a. **Anemia** is a deficiency of RBCs or a lack of hemoglobin. It may result from decreased numbers of RBCs, or normal numbers of RBCs with a subnormal amount of hemoglobin. Because the oxygen-carrying capacity of the blood is reduced, the individual may appear pale or lack energy. Some types of anemia are as follows:
    (1) **Hemorrhagic anemia** results from acute blood loss. The marrow gradually produces new RBCs to restore the normal condition.
    (2) **Iron-deficiency anemia** results from decreased dietary intake, deficient absorption, or excessive loss of iron.
    (3) **Aplastic anemia** (inactive bone marrow) is characterized by greatly decreased production of RBCs. It can result from overexposure to radiation, chemical poisoning, or cancer.
    (4) **Pernicious anemia** is due to the absence of vitamin B_{12}.
    (5) **Sickle cell anemia** is an inherited disease in which the hemoglobin molecule differs from normal hemoglobin by the replacement of one amino acid in the beta polypeptide chain. As a result, the RBCs are distorted into a sickle shape under conditions of low oxygen concentration. The distorted cells block capillaries and interrupt blood flow.
  b. **Polycythemia** is an increase in the number of RBCs in circulation, which results in an increase in blood viscosity and blood volume. Blood flow through blood vessels is impeded and capillary flow may be blocked.
    (1) **Compensatory (secondary) polycythemia** may occur as a result of **hypoxia** (oxygen deficiency) due to the following:
      (a) Permanent residence at high altitudes

(b) Prolonged physical activity
(c) Heart disease or lung disease
(2) **Polycythemia vera** is a disorder of the bone marrow.

E. **Leukocytes or white blood cells (WBCs)**
1. **Characteristics**
   a. **Number**
      (1) The normal WBC count in the blood is 7,000 to 9,000 per mm^3.
      (2) Infection or tissue damage results in an increase in the total number of leukocytes.
   b. **Function**
      (1) Leukocytes function to protect the body against foreign invaders, including bacteria and viruses.
      (2) Most of the leukocytic activity takes place in the tissues and not in the bloodstream.
   c. **Diapedesis.** Leukocytes have the property of diapedesis, which is the ability to squeeze through pores in the capillary membrane and enter the tissues.
   d. **Ameboid movement.** Leukocytes move under their own power with ameboid movement (locomotion in the manner of an ameba organism). Some are able to move as rapidly as three times their own length in a minute.
   e. **Chemotaxic ability.** Chemical substances released by damaged tissues cause leukocytes to move toward (positive chemotaxis) or away from (negative chemotaxis) the source of the substance.
   f. **Phagocytosis.** All leukocytes are phagocytic, but the ability is most highly developed in neutrophils and monocytes.
   g. **Life span.** After production in the bone marrow, leukocytes spend less than one day in circulation before they migrate to the tissues. They may remain there for days, weeks, or months, depending on the type of leukocyte.
2. **Classification of leukocytes.** There are five types of leukocytes in circulating blood, which are distinguished by size, shape of the nucleus, and the presence or absence of cytoplasmic granules. Those with granular cytoplasm are **granulocytes**; those lacking granules are **agranulocytes**.
   a. The **granulocytes** are subdivided into the **neutrophils, eosinophils,** and **basophils,** based upon the stainability of their cytoplasmic granules when treated with Wright's blood stain.
      (1) **Neutrophils** constitute 60% of WBCs.
         (a) **Structure.** Neutrophils have small pinkish granules in their cytoplasm. The nucleus has three to five lobes connected by thin strands of chromatin. The cell is 9μm to 12 μm in diameter.
         (b) **Function.** Neutrophils are highly phagocytic and very mobile. They arrive at infected tissue to attack and destroy bacteria, viruses, or other injurious agents.
      (2) **Eosinophils** constitute 1 to 3% of WBCs.
         (a) **Structure.** Eosinophils have coarse, large cytoplasmic granules, which stain reddish-orange. They have a two-lobed nucleus. The cell is 12 μm to 15 μm in diameter.
         (b) **Function**
            (i) Eosinophils are weakly phagocytic. Their numbers markedly increase in allergies and in parasitic diseases but decrease during prolonged stress.
            (ii) They may function in detoxification of the histamine produced by mast cells and injured tissue during inflammation.
            (iii) Eosinophils contain peroxidases and phosphatases, which are enzymes able to break down proteins. Their enzymes may be involved in detoxifying bacteria and removing antigen-antibody complexes, but their exact function is unknown.
      (3) **Basophils** comprise less than 1% of leukocytes.

- (a) **Structure.** Basophils have numerous large, irregular cytoplasmic granules that stain purplish to black and obscure the S-shaped nucleus. The cell is 12 μm to 15 μm in diameter.
- (b) **Function.** Basophils resemble mast cells. They contain **histamine**, possibly to increase the blood flow to injured tissues, and also the anticoagulant **heparin**, perhaps to help prevent intravascular blood clots. Their actual function is unknown.
  b. The **agranulocytes** are leukocytes that lack cytoplasmic granules. They include **lymphocytes** and **monocytes**.
  (1) **Lymphocytes** account for 30% of the total number of leukocytes in the blood. The majority of lymphocytes in the body is found in lymphatic tissues. Their life span may reach several years.
  - (a) **Structure.** Lymphocytes contain a round, deeply staining blue nucleus surrounded by a thin rim of cytoplasm. They vary in size; the smallest lymphocytes are 5 μm to 8 μm; the largest are 15 μm.
  - (b) **Origin and function.** Lymphocytes originate from stem cells in the red bone marrow, but continue their differentiation and proliferation in other organs. They function in **immunologic reactions.** (See Chapter 12.)
  (2) **Monocytes** comprise 3% to 8% of the total number of leukocytes.
  - (a) **Structure.** Monocytes are the largest of the blood cells, averaging 12 μm to 18 μm in diameter. The nucleus is large, ovoid or kidney-shaped, and is surrounded by cytoplasm that stains pale grayish-blue.
  - (b) **Function.** Monocytes are intensely **phagocytic** and highly mobile. They readily migrate through blood vessel walls. Once monocytes leave the bloodstream, they become tissue **histiocytes** (fixed macrophages).
 3. **Clinical considerations**
    a. **Leukemia** is a type of cancer characterized by uncontrolled proliferation of WBCs. The type of leukemia is classified by the dominant cell type, such as **myelocytic (granulocytic), lymphocytic,** or **monocytic** leukemia, and by the duration of the disease from its onset, such as **chronic** or **acute**.
    b. **Infectious mononucleosis,** caused by the Epstein-Barr virus, is characterized by a moderate increase in lymphocytes with a disproportionate number of immature and abnormal cells.
    c. **Acquired immune deficiency syndrome (AIDS),** caused by human immunodeficiency virus (HIV), damages the immune system by its attack on a subset of lymphocytes known as T cells.
F. **Platelets** (thrombocytes) number 250,000 to 400,000 per mm^3. They are nonnucleated cell fragments that originate from giant multinucleated megakaryocytes in bone marrow.
 1. **Structure.** Each platelet is approximately half the size of a red blood cell. The cytoplasm is surrounded by a plasma membrane and contains different types of granules associated with the process of blood coagulation.
 2. **Function.** Platelets function in **hemostasis** (arrest of bleeding) and the repair of ruptured blood vessels.
G. **Hemostasis** and **blood clotting** mechanisms involve a rapid sequence of processes.
 1. **Vasoconstriction.** When a blood vessel is cut, platelets at the site of damage release **serotonin** and **thromboxane A$_2$** (prostaglandin), which causes the smooth muscle in blood vessel walls to constrict. This initially reduces the blood loss.
 2. **Platelet plug**
    a. The platelets swell, become sticky, and adhere to the collagenous fibers of the wall of the damaged blood vessel, forming a **platelet plug**.
    b. The platelets release ADP, which activates other platelets and results in **platelet aggregation** to strengthen the plug.

(1) If the blood vessel damage is small, the platelet plug is able to stop the bleeding.
(2) If the blood vessel damage is extensive, the platelet plug reduces blood loss until the clot can be formed.

3. **Formation of the blood clot**
    a. The **extrinsic mechanism** for blood clotting is initiated by factors external to the blood vessels themselves.
        (1) **Thromboplastin** (membrane lipoprotein) released by damaged tissue cells acts on **prothrombin** (plasma protein) in the presence of **calcium ions** to produce **thrombin**.
        (2) Thrombin converts the soluble **fibrinogen** (plasma protein) to insoluble **fibrin**. The fibrin strands form a **fibrin clot**, or network, which traps RBCs and platelets and blocks further blood flow through the damaged vessel.
    b. The **intrinsic mechanism** for blood clotting expands on the simplified scheme described above. It involves 13 **clotting factors** found only in the blood plasma. Each of the protein factors (designated by Roman numerals) exists in an inactive state; as one is activated, its enzymatic activity activates the next one in the series, thus achieving a **cascade of reactions** to form a clot (Table 11–1).

4. **Dissolution of the blood clot**
    a. Soon after its formation, the clot **retracts** (shrinks) as a result of a contractile protein contained in platelets. The fibrin network is contracted to pull the cut surfaces closer together and provide a framework for tissue repair.
    b. Simultaneously with clot retraction, a fluid called **serum** is squeezed from the clot. Serum is blood plasma minus fibrinogen and other factors involved in the clotting mechanism.

5. **Source of clotting factors**
    a. The **liver** synthesizes most of the clotting factors and thus plays a major role in blood clotting. Liver diseases that interfere with this synthesis can cause clotting problems.
    b. **Vitamin K** is necessary for the synthesis of prothrombin and other clotting factors by the liver. Its absorption from the intestine depends on bile salts produced by the liver. If the bile ducts are obstructed (by gallstones, for example), the ability to form clots may be reduced.

6. **Prevention of clots in uninjured blood vessels**
    a. The anticoagulants **antithrombin** and **heparin**, which are found in circulating blood, prohibit clotting. Heparin, secreted by basophils and mast cells, activates antithrombin. Antithrombin blocks thrombin action on fibrinogen.
    b. The **smooth endothelial lining of blood vessels** repels platelets and the coagulation factors.
    c. **Prostacyclin (PGI_2)** is a prostaglandin that inhibits platelet aggregation. It is antagonistic to thromboxane, a prostaglandin that activates platelet aggregation. Both prostaglandins help to regulate blood clotting.

7. **Clotting abnormalities**
    a. An abnormal clot is called a **thrombus**. A thrombus that breaks loose to travel in the bloodstream is an **embolus**. Either may occlude blood flow.
        (1) Conditions that favor thrombus formation
            (a) Vessels with roughened areas due to **cholesterol plaques** (atherosclerosis) may trap platelets to initiate a clot.
            (b) Sluggish blood flow allows the accumulation of thromboplastin. Because blood flow decreases with immobility, bed patients must move frequently or be moved.
        (2) **Treatment** of individuals prone to form thrombi
            (a) Anticoagulants such as **coumarin compounds** impair vitamin K activity, thus preventing prothrombin synthesis.

Table 11–1. Clotting factors.

Factor Number	Name	Origin and Function
I	Fibrinogen	Plasma protein synthesized in liver; converted to fibrin
II	Prothrombin	Plasma protein synthesized in liver; converted to thrombin
III	Thromboplastin	Lipoprotein released from damaged tissues; activates factor VII for thrombin formation
IV	Calcium ions	Inorganic ion in plasma from dietary sources and bone; required in all stages of clotting process
V	Proaccelerin (labile factor)	Plasma protein synthesized in liver; required in both extrinsic and intrinsic mechanisms
VI	(Number no longer used)	Believed to be the same as factor V
VII	Proconvertin (serum prothrombin conversion accelerator)	Plasma protein (globulin) synthesized in liver; required in intrinsic mechanism
VIII	Antihemophilic factor	Plasma protein (enzyme) synthesized in liver; requires vitamin K); functions in extrinsic mechanism
IX	Plasma thromboplastin (Christmas factor)	Plasma protein (enzyme) synthesized in liver (requires vitamin K); functions in intrinsic mechanism
X	Stuart-Prower factor	Plasma protein synthesized in liver (requires vitamin K); functions in extrinsic and intrinsic mechanisms
XI	Plasma thromboplastin antecedent	Plasma protein synthesized in liver; functions in intrinsic mechanism
XII	Hageman factor	Plasma protein synthesized in liver; functions in intrinsic mechanism
XIII	Fibrin stabilizing factor	Protein found in plasma and platelets; cross-links fibrin filaments

Platelet factors	
Platelet accelerator	Platelets; same as plasma factor V
Thrombin accelerator	Platelets; accelerates thrombin and fibrin production
Platelet thromboplastin factor	Platelets; phospholipids necessary for intrinsic mechanism
Platelet factor 4	Binds heparin (anticoagulant) so clotting can occur

(b) **Aspirin** blocks platelet aggregation and interferes with the synthesis of prostacyclin.
   b. **Thrombocytopenia** is a condition in which there is an abnormally small number of platelets in the circulating blood (below 100,000 per mm^3). It results in an increased coagulation time and an increased tendency to bleed from small vessels throughout the body. Thrombocytopenia may be caused by an adverse reaction to drugs, bone marrow malignancy, or ionizing radiation damage to bone marrow.
   c. **Hemophilia** is a hereditary sex-linked disorder that results from the absence of several clotting factors. Transfusions are necessary to replace missing factors when excessive bleeding follows minor injury.
H. **Blood groups and blood typing**
   1. Prior to birth, genetically determined protein molecules called **antigens** appear on the surface membranes of RBCs. These antigens, type A and type B, react with their corresponding antibodies, which begin to appear at about 2 to 8 months after birth.
      a. Because the antigen-antibody reactions cause **agglutination** (clumping) of red cells, the antigens are called **agglutinogens** and the corresponding antibodies are called **agglutinins**.
      b. An individual may inherit neither type A nor type B agglutinogens, only one of them, or both simultaneously.
   2. The **ABO blood group classification** is based on the presence or absence of **agglutinogens (antigens type A and type B)** found on the erythrocyte surface and the agglutinins (antibodies), **anti-A and anti-B**, found in the blood plasma.
      a. **Type A blood** contains **type A** agglutinogen and **anti-B** agglutinin.
      b. **Type B blood** contains **type B** agglutinogen and **anti-A** agglutinin.
      c. **Type AB blood** contains **both type A and type B** agglutinogens and **neither anti-A nor anti-B** agglutinin.
      d. **Type O blood** contains **no** agglutinogens but **both anti-A and anti-B** agglutinins.
   3. **Blood typing** is necessary before a blood transfusion because the mixing of incompatible blood groups causes the agglutination and destruction of RBCs.
      a. In the usual slide technique of ABO blood typing, two separate drops of blood from the person to be typed are placed on a microscope slide.
      b. A drop of serum containing anti-A agglutinin (obtained from type B blood) is placed on one of the drops of blood, while a drop of serum containing anti-B agglutinin (obtained from type A blood) is placed on the other drop of blood.
         (1) If anti-A serum causes agglutination of the drop of blood, the individual has type A agglutinogens (blood type A).
         (2) If anti-B serum causes agglutination, the individual has type B agglutinogens (blood type B).
         (3) If both anti-A serum and anti-B serum cause agglutination, the individual has both type A and type B agglutinogens (blood type AB).
         (4) If neither anti-A nor anti-B sera cause agglutination, the individual has no agglutinogens (blood type O).
      c. **Blood transfusions**
         (1) When a blood transfusion is given, the donor plasma is diluted by the recipient's plasma, so the **agglutinins of the donor cannot cause agglutination**.
         (2) The **agglutinogens** on the **donor cells**, however, are significant in transfusions. If the blood group of the donor is of a different type than that of the recipient, the **agglutinin** of the **recipient's plasma** will agglutinate the donor's foreign RBCs.
         (3) **Transfusion reactions** are caused by the agglutination of donor RBCs.
            (a) Blood flow in small vessels is **blocked** by the clumped cells.

- (b) **Hemolysis** (rupture) of RBCs leads to release of hemoglobin into the bloodstream.
- (c) Hemoglobin carried to the kidney tubules precipitates, blocks the tubules, and results in **renal shutdown.**
- (4) **Cross-matching** of the blood types of recipient and donor is performed prior to a transfusion to ensure compatibility of the blood.
- (5) Concept of **universal donor** and **universal recipient**
  - (a) **Universal donor.** Type O blood has no agglutinogens to be agglutinated and can be given to any recipient, as long as the volume of the transfusion is small. Type O blood is called a universal donor.
  - (b) **Universal recipient.** Type AB individuals have no agglutinins in their plasma and can receive any donor erythrocytes. Type AB blood is called a universal recipient.

4. The **Rh system** is another group of inherited antigens found in humans. It was discovered and named for the Rhesus monkey. The RhD antigen is the most significant in immune reactions.
   a. If the RhD factor is present, the individual is said to be **Rh positive**; if it is absent, the individual is **Rh negative**. There are more Rh positive than Rh negative individuals.
   b. The system differs from the ABO groups in that **Rh negative** individuals have **no anti-Rh agglutinins** in their plasma.
   c. If an Rh-negative individual is given Rh-positive blood, the anti-Rh agglutinins are produced. Although an initial transfusion usually produces no harm, subsequent introduction of Rh-positive blood will result in agglutination of the **donor's** RBCs.
   d. **Erythroblastosis fetalis**, or hemolytic disease of the newborn, may occur after the first pregnancy of an Rh-negative mother with an Rh-positive fetus.
      (1) At the time of birth (or spontaneous or induced abortion), the mother is exposed to some of the fetal Rh-positive antigens and may produce antibodies against them.
      (2) If antibodies against the Rh factor are produced by the woman, in subsequent pregnancies they cross the placenta to the fetal bloodstream and cause hemolysis of the fetal RBCs. An affected infant is born anemic.
      (3) **Prevention.** If an Rh-negative woman is injected with antibodies against the Rh-positive factor within 72 hours after birth, miscarriage, or abortion of each Rh-positive fetus, the antigens are inactivated. The woman will not produce antibodies against them.

III. THE HEART

A. **Gross anatomy** (See Color Plate 8.)
   1. **Size and shape**
      a. The heart is a four-chambered, hollow organ that lies between the lungs in the middle of the thoracic cavity. Two thirds of the heart is to the left of the midsternal line. It is protected by the mediastinum.
      b. The heart is about the size of the owner's fist. It is shaped like a blunt cone. The broad upper end (**base**) is directed toward the right shoulder; the pointed lower end (**apex**) is directed toward the left hip.
   2. **Coverings**
      a. The **pericardium** is the loose-fitting double-walled sac that encloses the heart and the great blood vessels. It is attached to the diaphragm, the sternum, and the pleurae enclosing the lungs.
         (1) The **outer fibrous layer** of the pericardium is composed of collagenous fibers that form a dense, connective tissue layer to protect the heart.

(2) The **inner serous layer** is composed of two layers.
    (a) The **visceral** membrane (epicardium) covers the surface of the heart.
    (b) The **parietal** membrane lines the inner surface of the fibrous pericardium.
  b. The **pericardial cavity** is the potential space between the visceral and the parietal membranes. It contains pericardial fluid secreted by the serous layer to lubricate the membranes and reduce friction.
3. The **wall of the heart** consists of three layers.
  a. The outer **epicardium** (described above) consists of a layer of mesothelial cells resting on connective tissue.
  b. The middle **myocardium** consists of cardiac muscle tissue, which contracts to pump the blood.
    (1) The myocardium varies in thickness from heart chamber to chamber.
    (2) The muscle fibers are arranged in **spiral bundles** that wrap around the heart chambers. Contraction of the myocardium "wrings" the blood out of the chambers in the direction of the great arteries.
  c. The inner **endocardium** consists of an endothelial lining resting on connective tissue. It lines the heart, covers the valves, and is continuous with the endothelial lining of the blood vessels entering and leaving the heart.

B. **Heart chambers**
1. There are four chambers, the upper **right and left atria**, separated by the interatrial septum; and the lower **right and left ventricles**, separated by the **interventricular septum**.
2. The **atria** are relatively thin walled. They receive blood from veins, which carry blood back to the heart.
  a. The **right atrium**, in the right superior portion of the heart, receives blood from all tissues except the lungs.
    (1) The **superior vena cava** and the **inferior vena cava** bring **deoxygenated** blood back from the body to the heart.
    (2) The **coronary sinus** brings blood back from the heart wall itself.
    (3) The **left atrium**, in the left superior portion of the heart, is smaller than the right atrium, but it has a thicker wall. It receives **four pulmonary veins** that return **oxygenated blood** from the lungs.
  b. The **ventricles** are thick walled. They force blood out of the heart into the **arteries**, which carry blood away from the heart.
    (1) The **right ventricle** is located in the right inferior portion of the apex of the heart. Blood leaves the right ventricle via the **pulmonary trunk** and flows only a short distance to the lungs.
    (2) The **left ventricle** is located in the left inferior portion of the apex of the heart. Its walls are three times as thick as that of the right ventricle. Blood leaves the left ventricle through the **aorta** and flows to all parts of the body except the lungs.
  c. The **trabeculae carneae** are round or irregular muscle ridges that project from the inner surfaces of both ventricles into the ventricular cavity.
    (1) The **papillary muscles** are elevations of the trabeculae carneae to which the collagenous cords of the heart valves (**chordae tendineae**) are attached.
    (2) The **moderator band** (septomarginal trabecula) is a curved band of muscle in the right ventricle that extends transversely from the interventricular septum to the anterior papillary muscle. It assists in transmission of impulse conduction for heart contraction.

C. **Heart valves** (Figure 11–2)
1. The **tricuspid valve** lies between the **right atrium** and the **right ventricle**. It consists of three irregular flaps (cusps) of fibrous connective tissue covered by endocardium.

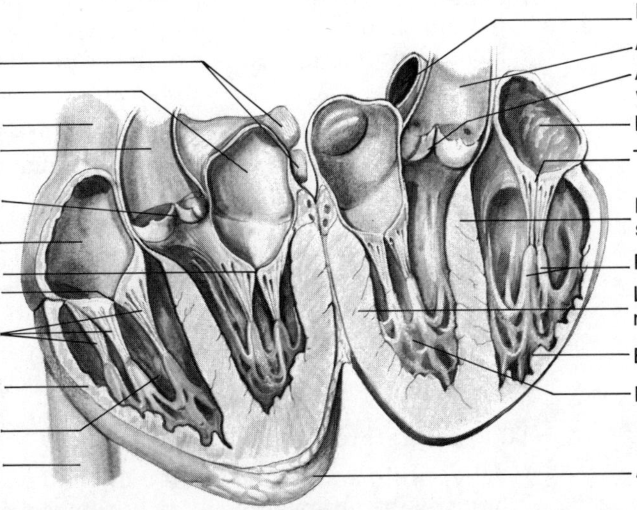

Figure 11-2. *Internal anatomy of the heart, which has been frontally bisected.* The anterior portion is flipped over to the left.

    a. The pointed ends of the flaps are attached to fibrous connective tissue cords, the **chordae tendineae** ("heart strings"), which are attached to the **papillary muscles**. The chordae tendineae prevent eversion of the valve flaps backward into the atria.

    b. When the blood pressure in the right atrium is greater than that in the right ventricle, the flaps of the tricuspid valve open and the blood flows from the right atrium to the right ventricle.

    c. When the blood pressure in the right ventricle is greater than the blood pressure in the right atrium, the flaps of the valve snap shut and prevent backflow of blood into the right atrium.

2. The **bicuspid (mitral) valve** lies between the left atrium and the left ventricle. It is attached to chordae tendineae and papillary muscles and functions in the same manner as the tricuspid valve.

3. The **aortic and pulmonary semilunar valves** are present at **ventricular exits** of the heart into the aorta and the pulmonary trunk. The semilunar valves are comprised of three half-moon-shaped cusps, which are attached by their convex borders to the insides of the blood vessels. Their free borders extend into the lumens of the vessels.

    a. The **pulmonic semilunar valves** are located between the right ventricle and the pulmonary trunk.

    b. The **aortic semilunar valves** are located between the left ventricle and the aorta.

    c. Pressure changes in the ventricles and in the aorta and pulmonary vessels permit blood flow only into the vessels and prevent backflow into the ventricles.

D. **Surface markings**

1. The **coronary (atrioventricular) sulcus** encircles the heart between the atria and the ventricles.

2. The **anterior and posterior interventricular sulci** mark the position of the interventricular septum that separates the right and left ventricles.

E. The **fibrous skeleton of the heart** consists of nodules of fibrocartilage in the upper part of the interventricular septum and rings of dense connective tissue around the bases of the pulmonary and aortic trunks. The framework serves as attachments for cardiac muscle and the valves.

F. **Passage of blood through the heart** (Color Plate 8). The pathway to and from the lungs is the **pulmonary circuit**; the pathway to and from the rest of the body is the **systemic circuit**.

1. **Pulmonary circuit.** The right side of the heart receives deoxygenated blood from the body and sends it to the lungs to be oxygenated. The oxygenated blood returns to the left side of the heart. The circulation through the heart is as follows:

   Right atrium → tricuspid valve → right ventricle → semilunar valve → pulmonary trunk → right and left pulmonary arteries → lung capillaries → pulmonary veins → left atrium

2. **Systemic circuit.** The left side of the heart receives oxygenated blood from the lungs and sends it to the body. The circulation through the heart is as follows:

   Left atrium → bicuspid valve → left ventricle → semilunar valve → aortic trunk → body regions and organs (muscles, kidneys, brain, etc.)

G. **Fetal bypasses through the heart**

1. Before birth, most of the blood is diverted from the fetal nonfunctioning lungs by the **foramen ovale**, an opening in the interatrial septum between the right atrium and the left atrium. Oxygenated blood from the umbilical vein enters the right atrium and flows to the left atrium, thus bypassing the pulmonary circuit.

2. At birth, the lungs become functional and the foramen ovale closes. The site is marked by a depression in the interatrial septum called the **fossa ovalis**. A foramen ovale that does not close is an **interatrial septal defect**.

3. Fetal oxygenated blood that has not been shunted through the foramen ovale passes into the right ventricle. It is then diverted from the pulmonic trunk into the aorta by a channel called the **ductus arteriosus**. The ductus arteriosus closes after birth leaving a fibrous remnant, the **ligamentum arteriosum**. If the shunt does not close, the heart defect is a **patent (open) ductus arteriosus**.

H. The **coronary circulation** forms the blood supply to the heart wall.

1. The **right and left coronary arteries** branch off from the aorta just beyond the aortic semilunar valve. They lie in the coronary sulci.
   a. The major branches of the **left coronary artery** are the following:
      (1) The **anterior interventricular (descending) artery**, which supplies the anterior portions of the left and right ventricles and gives off a branch, the **left marginal artery**, which supplies the left ventricle.
      (2) The **circumflex artery**, which supplies the left atrium and the left ventricle. Posteriorly, the circumflex artery anastomoses (joins together) with the right coronary artery.
   b. The major branches of the **right coronary artery** are the following:
      (1) The **posterior interventricular (descending) artery**, which supplies the walls of both ventricles.
      (2) The **right marginal artery**, which supplies the right atrium and the right ventricle.

2. **Cardiac veins** (great, middle, and oblique) drain the blood from the myocardium into the **coronary sinus**, which empties into the right atrium.

3. Blood flows through the coronary arteries primarily during heart muscle relaxation because the coronary vessels are compressed during contraction.

4. The anatomy of the coronary circulation varies in the population. The majority of individuals have a balanced coronary circulation, but a certain proportion have "right coronary dominance" or "left coronary dominance."

IV. CARDIAC PHYSIOLOGY

A. **Conducting system of the heart**
1. **Purkinje fibers.** Specialized cardiac muscle fibers, the Purkinje fibers, are able to conduct impulses at a speed five times that of cardiac muscle fibers. Rapid conduction along the Purkinje system permits simultaneous atrial contraction, followed by simultaneous ventricular contraction, for coordinated pumping action.
2. The **sinoatrial node (S-A node)**
    a. **Location.** The S-A node is a mass of specialized cardiac muscle tissue located on the posterior wall of the right atrium just below the opening of the superior vena cava.
    b. The S-A node spontaneously discharges impulses at approximately 72 times per minute, a faster rate of rhythm than that in the atria (40 to 60 times per minute), and the ventricles (20 times per minute). It is influenced by the parasympathetic and sympathetic divisions of the ANS, which slow and accelerate its rhythmicity.
    c. The S-A node sets the pace of the rhythmic contraction and is called, therefore, the **pacemaker of the heart.**
3. **Atrioventricular node (A-V node)**
    a. **Location.** The impulse spreads along bands of Purkinje fibers in the atria to the **A-V node** located lower in the posterior wall of the right atrium.
    b. The A-V node delays the impulse a few hundredths of a second to allow for complete ejection of atrial blood prior to ventricular contraction.
4. **A-V bundle (of His)**
    a. **Location.** The A-V bundle is a large group of Purkinje fibers that arises from the A-V node and carries the impulse along the interventricular septum to the ventricles. It divides into the right and left bundle branches.
    b. The **right bundle branch** continues along the inner side of the right ventricle. The fibers branch into smaller Purkinje fibers that merge into cardiac muscle fibers to continue the impulse.
    c. The **left bundle branch** continues along the inner side of the left ventricle and branches into those cardiac muscle fibers.

B. **Conduction abnormalities**
1. **Abnormal heart rhythms** (arrhythmias) may be caused by S-A node or A-V node irregularities or by disturbances of the conduction system.
2. **Heart block** is an **interruption in conduction** so that some or all of the impulses fail to reach the ventricles, which may then **beat independently** at their own rhythm.
    a. In a **partial (incomplete) heart block**, the atria beat normally, but the rate of conduction through the A-V node is slowed. The ventricles may contract only after every second, third, or fourth atrial contraction.
    b. In **complete heart block**, the conduction from the A-V node or bundle is severely hampered. The atria beat normally, but the ventricles beat independently at 20 to 40 beats per minute.
    c. Heart block of varying degrees is treated by implantation under the skin of an **artificial pacemaker**, a battery-operated stimulator with electrodes connected to the ventricles.
3. **Ectopic foci.** In some conduction defects, a site other than the pacemaker may become excitable and initiate a beat or beats on its own between the normal beat.
    a. The extra beat is called a **premature ventricular contraction (PVC)**, or **extrasystole.**
    b. Ectopic foci may be precipitated by certain medications, stimulants (such as caffeine), lack of sleep, or anxiety.
    c. **Flutter and fibrillation.** Uncoordinated rapid contractions of the atria or the ventricles are called flutter at frequencies of 200 to 300 beats per minute and fibrillation at higher frequencies.

Figure 11-3. *A typical electrocardiogram showing the various waves and segments.* The ECG is correlated with a phonogram tracing of the heart sounds.

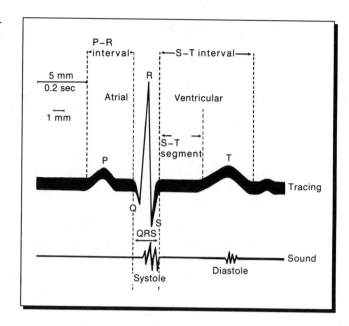

(1) **Atrial flutter or fibrillation** is abnormal, but it is not a threat to life. Although the effectiveness of the ventricles is reduced, they can still pump blood to the lungs and the body.

(2) **Ventricular fibrillation** does not allow blood to be pumped and rapidly is fatal unless the heart is electrically defibrillated (shocked) into normal rhythm.

C. Electrocardiography

1. The **electrocardiogram (ECG)** is a graphic record of the electrical activity accompanying the contraction of the atria and the ventricles of the heart (Figure 11-3).
2. The depolarization and repolarization of the cardiac muscle produce potentials at the body surface that can be recorded by a polygraph or oscilloscope after appropriate placement of surface electrodes.
    a. The position of the electrodes relative to each other and to the heart are called **leads**.
    b. Twelve conventional leads are used for recording the ECG.
        (1) The three standard **limb leads** are right arm to left arm, right arm to left leg, and left arm to left leg. These are bipolar because they detect electric variations as two points and display the difference.
        (2) The three **augmented limb leads** are augmented by an electric connection that results in a deflection of increased amplitude. These are unipolar because they register the changes in voltage at one point (right arm, left arm, or left leg) with respect to another point that does not change significantly in electrical activity during heart contraction.
        (3) The **precordial leads**, recorded in six chest positions called V_1 through V_6, are unipolar.
3. The individual **deflections** of the cardiac action potential are designated by the letters P, Q, R, S, and T.
    a. The **P wave** represents the electrical activity associated with **depolarization of the atria** after the initial depolarization of the S-A node.
    b. The **QRS complex** (of approximately 0.12 sec duration) represents the **spread of depolarization through the ventricles.**
        (1) A small amount of atrial repolarization also occurs at the same time, but it is masked by the QRS deflection.
        (2) The shape, amplitude, and direction of the QRS depends on variables such as the position of the heart and the mass of the ventricles.

c. The **P-R interval** (approximately 0.1 sec) is the time from the beginning of the P wave to the beginning of the QRS complex. It represents the interval between activation of the S-A node and the beginning of ventricular depolarization.

d. The **T wave** represents the repolarization of the ventricles. It is of longer duration and of lower amplitude than the depolarization wave (QRS complex), which indicates that ventricular repolarization is less synchronized and slower than depolarization.

D. **The cardiac cycle**

1. **Definition.** The cardiac cycle includes the period from the end of one heart contraction (**systole**) and relaxation (**diastole**) to the end of the next systole and diastole.

   a. Contraction of the heart generates **blood pressure and volume changes** in the heart and major vessels that regulate the opening and closing of the heart valves and the flow of blood through the chambers and into the arteries.

   b. Although the right and left sides of the heart have different atrial and ventricular pressures, they contract and relax at the same time and eject equal volumes of blood simultaneously.

2. **Mechanical events of the cardiac cycle** (Figure 11–4)

   a. During **diastole** (relaxation), the pressure is **low** in both atria and ventricles, but it is greater in the atria than in the ventricles.

      (1) The atria passively receive blood continuously from the veins (inferior and superior venae cavae, pulmonary veins).

      (2) The blood flows from the atria through the open A-V valves into the ventricles.

      (3) The ventricular pressure begins to rise as the ventricles expand to receive the incoming blood.

      (4) The aortic and pulmonary semilunar valves are closed because the pressure in those vessels is greater than the pressure in the ventricles.

      (5) About 70% of ventricular filling takes place prior to atrial systole.

   b. **Late in ventricular diastole**, the S-A node discharges, the atria contract, and the increased pressure in the atria forces the additional 30% of blood into the ventricles.

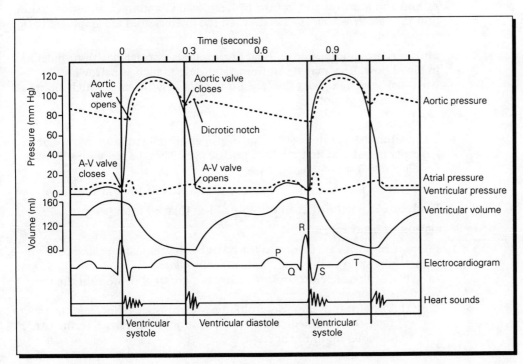

Figure 11–4. The cardiac cycle showing the relationship between pressure changes in the left atrium, left ventricle, and aorta, and the resultant opening and closing of the different valves. Note that the ventricular volume decreases as the ventricular pressure increases.

(1) The pressure in the left atrium rises to 7 to 8 mm Hg; the pressure in the right atrium rises to 4 to 6 mm Hg.
(2) The **end diastolic volume (EDV)** is the volume of blood in each of the ventricles at the end of diastole. It is normally about 120 ml.
c. **Ventricular systole.** The electrical activity spreads to the ventricles, which begin to contract. The pressure in the ventricles rises rapidly and pushes against the A-V valves, which snap shut.
  (1) The ventricles are now closed cavities and the volume of blood cannot change. This is the period of **isovolumetric contraction**.
  (2) The sound of the valves closing is the **first heart sound**.
  (3) As ventricular contraction continues, the pressure rapidly increases to 80 mm Hg in the left ventricle and 8 mm Hg in the right ventricle, forcing open the **aortic and pulmonary semilunar valves**.
d. **Ventricular ejection of blood** into the arteries
  (1) Not all of the ventricular blood is emptied on contraction. The **end systolic volume (ESV)** of blood remaining at the end of systole is about 50 ml.
  (2) The **stroke volume** (70 ml) is the difference between the EDV (120 ml) and the ESV (50 ml).
e. **Ventricular diastole**
  (1) The ventricles repolarize and stop contracting. The pressure in the ventricles abruptly falls below that in the aorta and the pulmonary trunk and the semilunar valves snap shut (the second heart sound).
  (2) The dicrotic notch indicated in Figure 11–4 indicates the brief rise in aortic pressure caused by the closing of the aortic semilunar valve.
  (3) The ventricles are again closed cavities in the period of **isovolumetric relaxation** as both inlet and outlet valves are closed. When the pressure in the ventricles drops from approximately 100 mm Hg to nearly zero, well below the pressure in the atria, the A-V valves open and the cycle begins again.

E. **Heart sounds**
  1. Heart sounds traditionally are refered to as **lup-dup** and can be heard with a stethoscope. The "lup" corresponds to the closing of the A-V valves and the "dup" to the closing of the semilunar valves.
  2. A third and fourth sound, caused by heart wall vibration when blood rushes into the ventricles, may be heard when the heart sounds are amplified by a microphone.
  3. **Murmers** are abnormal or unusual heart sounds associated with turbulence in blood flow. They may result from valve defects such as narrowing (stenosis), which impedes the forward flow, or insufficiency, which allows a backward flow of blood.

F. **Heart rate**
  1. The **normal heart rate** is 60 to 100 beats per minute, with an average of 75 beats per minute. At that rate, the cardiac cycle lasts 0.8 sec, with 0.5 sec in systole and 0.3 sec in diastole.
  2. **Tachycardia** is an increase in heart rate that exceeds 100 beats per min.
  3. **Bradycardia** is defined as a heart rate of less than 60 beats per min.

G. **Regulation of heart rate**
  1. **Efferent impulses** are carried to the heart by the sympathetic and parasympathetic divisions of the autonomic nervous system.
    a. The **cardioaccelerator reflex center** is a group of neurons in the medulla oblongata.
      (1) The effects of impulses from these neurons are to increase the heart rate. They are mediated by **sympathetic fibers** carried in the **cardiac nerves** to the heart.

(2) The nerve fiber endings secrete **norepinephrine** (noradrenalin), which increases the rate of discharge of the S-A node, decreases conduction time through the A-V node and the Purkinje system, and **increases excitability** throughout the heart.
  b. The **cardioinhibitory reflex center** is also in the medulla oblongata.
   (1) The effects of impulses from these neurons are to decrease the heart rate. They are mediated by **parasympathetic fibers** carried in the vagus nerves. The right vagus nerve ends on the S-A node; the left vagus on the A-V node.
   (2) The nerve fiber endings secrete **acetylcholine**, which decreases the rate of discharge of the S-A node and increases conduction time through the A-V node.
  c. The heart rate at any given time is determined by the **balance of acceleratory and inhibitory impulses** from the sympathetic and parasympathetic nerves.
 2. **Afferent** (sensory) impulses to the cardiac control centers originate from **receptors**, which are located in various parts of the cardiovascular system.
  a. **Pressoreceptors (baroreceptors)** in the **carotid artery (carotid sinus)** and aorta (**aortic sinus** and **aortic arch**) are sensitive to **changes in blood pressure**.
   (1) An **increase** in blood pressure will result in a reflex **slowing** of the heart rate mediated through the medullary centers. The cardioacceleratory center is inhibited and the cardioinhibitory center is stimulated.
   (2) A **decrease** in blood pressure will result in a reflex stimulation of heart rate mediated through the medullary centers in a similar way.
  b. Pressoreceptors in the **venae cavae** are sensitive to a **decrease** in blood pressure. If blood pressure falls, a reflex increase of heart rate to restore pressure will occur.
 3. **Additional influences on heart rate**
  a. The heart rate is affected by stimulation of almost any cutaneous nerve, such as receptors for pain, heat, cold, and touch, or by emotional inputs from the CNS.
  b. Normal heart function also depends on the balance of electrolytes such as calcium, potassium, and sodium, which affect the heart rate if their levels are increased or decreased.
H. **Cardiac output** is the amount of blood ejected by **either ventricle** per minute. It is sometimes called the **minute volume** of the heart. It is approximately **5 L per min** for an average man at rest and 20% less in a woman.
 1. The calculation of cardiac output is as follows:

   Cardiac output = heart rate × stroke volume (EDV minus ESV)

 2. **Major factors that affect cardiac output**
  a. **Strenuous activity** increases cardiac output to 25 L per min; in a trained athlete, to 35 L per min. **Cardiac reserve** is the ability of the heart to increase cardiac output.
  b. **Venous return to the heart.** The heart is able to adjust its output in accord with the input for the following reasons:
   (1) An increase in venous return increases the EDV.
   (2) An increased EDV stretches the myocardial fibers of the ventricles.
   (3) The more that cardiac muscle fibers are stretched at the beginning of contraction (within physiological limits), the more the ventricles are filled, and the greater is the force of contraction. This is known as the **Frank-Starling law of the heart**.
  c. Factors that **assist venous return and increase cardiac output** include the following:
   (1) **Skeletal muscle pumps.** Muscular veins contain valves, which allow the flow of blood only toward the heart and prevent backflow. The contraction of the limb muscles helps push the blood up against the force of gravity.

(2) **Respiration.** During inspiration, the increase in negative pressure inside the thoracic cavity sucks air into the lungs and also venous blood into the atria.
(3) **Venous reservoirs.** Under sympathetic stimulation, blood that is stored in the spleen, the liver, and the large veins is returned to the heart when cardiac output falls.
(4) The force of **gravity** from areas above the heart assists in venous return.

d. Factors that **decrease venous return** and affect cardiac output include the following:
(1) **A change in body position** from the supine to the erect position shifts blood from the pulmonary circulation to veins in the legs. Reflex increases in heart rate and blood pressure compensate for the decrease in venous return.
(2) **Abnormally low venous pressure** (for example, as a result of hemorrhage and low blood volume) results in decreased venous return and cardiac output.
(3) **High blood pressure.** Increased aortic and pulmonary blood pressure forces the ventricles to work harder to push out the blood against resistance. The greater the resistance against which the ventricles contract, the lower the cardiac output.

e. **Additional influences** on cardiac output include the following:
(1) **Adrenal medullary hormones.** Epinephrine (adrenaline) and norepinephrine increase heart rate and force of contraction, which increases cardiac output.
(2) **Ions.** The concentrations of potassium, sodium, and calcium in the blood and interstitial fluids affect heart rate and cardiac output.
(3) The **age or size** of the individual can influence cardiac output.
(4) **Cardiovascular disease.** Some examples of heart abnormalities, which make the heart pump less effectively and decrease cardiac output, include the following:
   (a) **Atherosclerosis,** the development of plaques within the coronary blood vessel walls, results in eventual impedance of blood flow.
   (b) **Ischemic heart disease,** an insufficiency of blood to the myocardium, usually results from atherosclerosis in the coronary arteries and leads to heart failure.
   (c) **Myocardial infarction** (heart attack), usually is the result of a sudden decrease in blood supply to the myocardium.
   (d) **Disease of the valves** of the heart will diminish cardiac output, especially during exercise.

V. HEMODYNAMICS: BLOOD FLOW AND BLOOD PRESSURE

A. **Structure and function of blood vessels.** The blood vessels are a series of branching closed tubes that carry blood from the heart to the tissues and back to the heart. The three major types of vessels are **arteries, capillaries,** and **veins.**

1. **Arteries** function to carry blood away from the heart.
   a. **Structure of artery walls.** All arteries have three layers in their walls: an outer **adventitia** of fibrous connective tissue; a middle **media** of smooth muscle and/or elastic fibers; and an inner **intima,** a thin tube of endothelial cells.
   b. **Types of arteries**
      (1) **Elastic arteries.** The largest arteries from the heart contain primarily elastic tissue in their walls. Their distention during systole and recoil during diastole are responsible for the continuity of blood flow despite the pulsatile nature of the heart beat.

- (2) **Muscular arteries.** The elastic arteries branch into medium-sized muscular arteries with smooth muscle fibers in their walls that respond to nerve stimuli. These are **distributing (conducting) arteries**; the size of their lumens is regulated by the nervous system so that the amount of blood sent to different parts of the body to suit particular needs can be controlled.
- (3) **Small arteries** contain varying amounts of muscle and elastic fibers, the proportion depending on their size and position. They cushion the pulsatile flow of blood into a smooth flow.
- (4) **Arterioles,** which are small arteries with narrow lumens and thick, muscular walls, deliver blood to the capillary beds in the tissues. These are the **resistance arteries**; under sympathetic stimulation, they provide the major site of resistance for the increase of blood pressure.

2. **Capillaries** are microscopic channels for the exchange of nutrients and wastes between the blood and the tissues. Capillaries connect arterioles and venules. All tissues except cartilage, hair, nails, and the cornea of the eye have capillaries.
   a. At the site of capillary origin from an arteriole, a **precapillary sphincter** of smooth muscle controls the flow of blood into a capillary bed. The sphincters contract and relax intermittently (**vasomotion**) and are open more in active tissue.
   b. The velocity of blood flow in the capillary bed is very slow to allow for the exchange of nutrients, wastes, and gases.
   c. The total capillary area is very large, with an estimated surface area of 7,000 square meters in the adult body.
   d. **Arteriovenous anastomoses** (AV shunts) are alternative channels that allow blood to pass directly from the arterial to the venous circulation without having to go through the capillaries.

3. **Veins** carry blood back to the atria of the heart.
   a. **Structure of vein walls.** Veins have the same layers in the walls as the arteries, but they have less smooth muscle and elastic tissue and more fibrous connective tissue.
      (1) The venous system is thin-walled and distensible. The veins hold 75% of the total blood volume and return blood to the heart under very low pressure.
      (2) Veins have valves, which arise as flaps from their innermost layer, to prevent backflow.
   b. **Types of veins.** The venous system begins at the venous end of the capillary bed with **postcapillary venules**, which merge into **venules**, and then into **small, medium,** and **large veins**.

B. **Blood pressure**
   1. **Definition.** Blood pressure is the force exerted by the blood in all directions against all enclosed surfaces; that is, on the inner walls of the heart and the blood vessels.
   2. **Origin** of the blood pressure. The pumping action of the heart provides the pressure that drives the blood through the vessels. Blood flows through the system of closed vessels because of the pressure differences, or **pressure gradient**, between the left ventricle and the right atrium.
      a. The **left ventricular pressure** changes from a high of 120 mm Hg at **systole** to a low of 0 mm Hg at **diastole**.
      b. The **aortic pressure** changes from a high of 120 mm Hg at **systole** to a low of 80 mm Hg at **diastole**. The diastolic pressure is maintained in the arteries because of the rebound effect from the aorta's elastic walls. The mean of the aortic pressure is 100 mm Hg.

c. **Systemic circulation pressure changes.** Blood flows from the aorta (with pressure of 100 mm Hg) to the arteries (with a pressure change from 100 to 40 mm Hg) to the arterioles (with a pressure change from 40 to 25 mm Hg) to the capillaries (with a pressure of 25 mm Hg at the arterial end to 10 mm Hg at the venous end) to the veins (with a pressure change from 10 to 5 mm Hg) to the superior and inferior venae cavae (with pressure of 2 mm Hg) to the right atrium (with pressure of 0 mm Hg).

3. **Factors that affect blood pressure**
    a. **Cardiac output.** Blood pressure is **directly proportional** to the cardiac output (determined by stroke volume and heart rate).
    b. **Peripheral resistance to blood flow.** Blood pressure is inversely proportional to the resistance in the vessels. The peripheral resistance has several determinants:
        (1) **Blood viscosity.** The more proteins and blood cells in the plasma, the greater is the resistance to blood flow. An increase in the hematocrit causes an increase in viscosity; in anemia, the hematocrit and viscosity decrease.
        (2) **Vessel length.** The longer the vessel, the greater the resistance to blood flow.
        (3) **Vessel radius.** The peripheral resistance is inversely proportional to the vessel radius to the fourth power.
            (a) If the radius of the vessel is doubled, as would occur in **vasodilation**, the flow would increase sixteen-fold. The blood pressure would decrease.
            (b) If the radius of the vessel is halved, as would occur in **vasoconstriction**, the **impedance** to flow would increase 16 times and the blood pressure would increase.
        (4) Because vessel length and blood viscosity are normally constants, changes in blood pressure are produced by changes in the radius of blood vessels.

C. **Regulation of blood pressure**
    1. **Neural regulation.** The **vasomotor center** in the medulla of the brain regulates blood pressure. The **cardioacceleratory** and **cardioinhibitory** centers regulate cardiac output.
        a. The **vasomotor center**
            (1) **Vasomotor tone** is the continuous low-level stimulation of smooth muscle fibers in vessel walls. It maintains blood pressure through vasoconstriction of the vessels.
            (2) The maintenance of vasomotor tone is mediated by impulses from **vasomotor nerve fibers**, which are the efferent fibers of the sympathetic division of the ANS.
            (3) **Vasodilation** generally occurs because of **reduced vasoconstrictor impulses**. The exception is in the blood vessels of the heart and the brain.
                (a) Cardiac and brain vessels have adrenergic **beta receptors** that respond to circulating epinephrine released by the adrenal medulla.
                (b) This mechanism ensures an adequate blood supply to vital organs during stressful situations that induce sympathetic stimulation and vasoconstriction elsewhere in the body.
                (c) Parasympathetic stimulation causes vasodilation of vessels only in a few areas; for example, the erectile tissue of the genitalia and in certain salivary glands.
        b. The **cardiac accelleratory** and **inhibitory centers** and the **aortic and carotid baroreceptors** described previously regulate blood pressure through the ANS.
    2. **Chemical and hormonal regulation.** A number of chemical substances directly or indirectly affect blood pressure. These include the following:

a. **Adrenal medulla hormones.** Norepinephrine is a vasoconstrictor. As described above, **epinephrine** may act as a vasoconstrictor or vasodilator, depending on the kind of smooth muscle receptors in the blood vessels of the organ.
b. **Antidiuretic hormone** (vasopressin) and **oxytocin** secreted from the posterior pituitary gland are vasoconstrictors.
c. **Angiotensin** is a blood peptide that is a powerful vasoconstrictor in active form. (See Chapter 16.).
d. Various amines and peptides such as **histamine, glucagon, cholecystokinin** (CCK), **secretin,** and **bradykinins,** which are produced by a number of tissues, are vasoactive chemicals.
e. **Prostaglandins** are locally produced, hormone-like agents that can act as vasodilators or vasoconstrictors.

D. **Measurement of arterial systolic and diastolic blood pressure**
1. Blood pressure is measured indirectly by the auscultatory method by means of a **sphygmomanometer.**
   a. The apparatus consists of an arm cuff to cut off the flow of blood through the brachial artery, a mercury manometer to read the pressure, a bulb for inflation of the cuff to close off the blood flow in the brachial artery, and a valve to release the air from the cuff.
   b. A stethoscope is used to detect the beginning and the ending of **Korotkoff's sounds,** which is the noise of the blood spurting through a partly closed vessel. The sounds and the simultaneous readings of the mercury column are the criteria for establishing systolic and diastolic pressure.
2. The average blood pressure in a young male adult is 120 mm Hg systolic and 80 mm Hg diastolic, usually expressed as 120/80. The average blood pressure in a young female adult is usually about 10 mm Hg less, both systolic and diastolic.

E. **Pulse.** The arterial pulse is a **pressure wave** that travels 6 to 9 m per/sec, about 15 times faster than the blood.
1. The pulse can be felt at any point where an artery lies near the surface and over a firm background. It is commonly felt at the radial artery at the wrist.
2. Two heart sounds equal one arterial pulse beat.
3. The pulse rate reveals information about heart action, blood vessels, and circulation.

VI. **CIRCULATORY PATHWAYS** (Color Plates 6 and 7)

A. **Vessels of the pulmonary circulation**
1. The **pulmonary trunk** leaves the right atrium. It is a short, thin-walled vessel, 5 cm in length and 3 cm in diameter.
   a. In the concavity of the aortic arch, the pulmonary trunk divides into the **right pulmonary artery** and the **left pulmonary artery.**
   b. The right pulmonary artery divides into **three lobar branches** in the right lung; the left pulmonary artery divides into **two lobar branches** in the left lung.
   c. Further arterial divisions end in capillary networks that surround the air sacs of the lungs. After gas exchange, the blood is collected into venules that lead into veins.
2. Two **right pulmonary veins** (from the right lung) and two **left pulmonary veins** (from the left lung) return oxygenated blood into the left atrium.

B. **Major arteries of the systemic circulation**
1. The **aorta,** which emerges from the left ventricle, is the largest vessel in diameter in the body.

2. The **ascending aorta** is the first portion of the aorta. It is 5 cm in length and begins at the level of the junction of the sternum with the second rib.
    a. The aortic body, which is a chemoreceptor to changes in carbon dioxide and oxygen, lies between the ascending aorta and the pulmonary trunk.
    b. The **right and left coronary arteries** are the only branches off the ascending aorta. They arise just beyond the cusps of the semilunar valves. The coronary circulation has been described.
3. The **arch of the aorta** begins at the sternal angle. It has three important branches: the **brachiocephalic artery**, the **left common carotid artery**, and the **left subclavian artery**.
    a. The brachiocephalic artery branches into the **right common carotid artery** and the **right subclavian artery**.
    b. Each **common carotid artery** (the right from the brachiocephalic artery and the left from the arch of the aorta) gives rise to the **external and internal carotid arteries**.
        (1) The **external carotid artery** supplies the head and neck outside of the cranial cavity. Branches are the superior thyroid artery, pharyngeal artery, lingual artery, facial artery, occipital artery, auricular artery, superficial temporal artery, and maxillary artery.
        (2) The **internal carotid artery** enters the cranial cavity through the carotid canal in the temporal bone. It branches into the ophthalmic artery, anterior cerebral artery, and middle cerebral artery. The cerebral arteries form part of the **circle of Willis** that supplies the base of the brain.
        (3) The **carotid sinus** (pressoreceptor) and the **carotid body** (chemoreceptor) are located at the fork of the internal and external carotid arteries.
    c. The **right subclavian artery** (from the brachiocephalic artery) and the **left subclavian artery** (from the arch of the aorta) give rise to the following branches:
        (1) The **vertebral artery** branches further into the **basilar artery**, which gives rise to the **right and left posterior cerebral arteries** as well as the **spinal** and the **cerebellar arteries**.
        (2) The **thyrocervical trunk** supplies the thyroid, cervical, and scapular regions.
        (3) The **costocervical trunk** supplies the upper intercostal muscles and the muscles in the back of the neck.
        (4) The **internal thoracic artery** (internal mammary artery) supplies the thoracic and intercostal muscles, the mediastinum, and the diaphragm.
    d. The right and left subclavian arteries continue into the upper limb on each side as the **axillary artery**, which gives off branches as the superior thoracic, thoraco-acromial, lateral thoracic, subscapular, anterior and posterior circumflex humeral arteries to structures in the axilla.
    e. The axillary artery continues as the **brachial artery**, which runs down the arm and divides into the **radial artery** and the **ulnar artery**.
    f. The radial and ulnar arteries join in the hand via **superficial and deep palmar arches** and give off **digital arteries** to the fingers.
4. The **thoracic aorta** gives off visceral and parietal branches to the organs and muscles of the thoracic region.
    a. The **pericardial arteries** supply the pericardium of the heart.
    b. The **bronchial arteries** supply the lungs.
    c. The **esophageal arteries** supply the esophagus as it passes through the mediastium.
    d. The **intercostal arteries** supply the intercostal muscles and thoracic wall.
    e. The **phrenic arteries** supply the diaphragm.
5. The **abdominal aorta** begins at the level of the diaphragm and ends at the level of the fourth lumber vertebra 1 cm below and to the left of the umbilicus. It divides into two common iliac arteries. Branches of the abdominal aorta are as follows:

a. The **celiac artery** (celiac trunk) arises just below the diaphragm and gives off three branches.
   (1) The **left gastric artery** joins the right gastric artery and supplies the stomach;
   (2) The **splenic artery** is the largest of the three branches off the celiac artery. It gives off smaller branches to the spleen, pancreas, and stomach.
   (3) The **common hepatic artery** branches into the **right gastric artery**; the **gastroduodenal artery**, which supplies the stomach, duodenum, and part of the pancreas and bile duct; the **hepatic artery** proper to the liver; and the **cystic artery**, which supplies the gall bladder.
b. The **superior mesenteric artery** arises just below the celiac artery. It supplies the whole of the small intestine (except the superior part of the duodenum) as well as the cecum, ascending and transverse colon.
c. The paired **suprarenal (adrenolumbar) arteries** supply the adrenal glands.
d. The paired **renal arteries** supply the kidneys.
e. The paired **testicular (internal spermatic) arteries** or **ovarian arteries** supply the gonads.
f. The **inferior mesenteric artery** arises 3 cm to 4 cm above the division of the abdominal aorta into the common iliac arteries. It supplies the left one third of the transverse colon, the whole descending colon, the sigmoid colon, and the rectum.
g. The paired **lumbar and sacral arteries** branch off the abdominal aorta and supply the musculature and spinal cord of the lumbosacral region.

6. The **right and left common iliac arteries** pass downward for 5 cm. Each divides into the external and internal iliac arteries.
   a. The **internal iliac artery** supplies the gluteal region and pelvic organs (urinary bladder and internal reproductive organs).
      (1) The **internal pudendal artery**, which serves the pelvic muscles and external genitalia, is a branch important in reproduction.
      (2) The erection and engorgement of the sex organs during sexual excitement are vascular events controlled by the ANS. They involve branches of the internal pudendal artery, which are the deep and dorsal arteries of the penis and clitoris.
   b. The **external iliac artery** becomes the **femoral artery** in the thigh and gives off branches to supply the thigh region.
   c. The **popliteal artery** is a continuation of the femoral artery as it passes across the posterior knee.
   d. The **anterior and posterior tibial arteries** are divisions of the popliteal artery and supply the knee joint, leg, and ankle.
      (1) The anterior tibial artery becomes the **dorsalis pedis artery** and **arcuate artery** of the ankle and foot.
      (2) The posterior tibial artery gives rise to the **peroneal artery** and divides into the **lateral and medial plantar arteries** at the foot. The dorsalis pedis artery and the lateral medial plantar artery join to form the **plantar arch**, which sends digital arteries to the toes.

C. Major veins of the systemic circulation
   1. Overview. The venous system begins at the venous end of the capillary beds with venules that converge to form progressively larger veins. All systemic veins return blood to the right atrium by three routes: from the heart wall into the **coronary sinus**, from the upper parts of the body into the **superior vena cava**, and from the lower parts of the body into the **inferior vena cava**.
      a. **Deep veins** are those that drain deeper tissues and organs. They accompany arteries and they generally have the same name as the artery. Exceptions are certain deep veins of the head and spinal column.
      b. **Superficial veins** are located in the hypodermis of the skin and drain into deep veins. Their names do not correspond to the names of arteries.
      c. **Venous sinuses** are blood-collecting spaces found in certain organs. They are lined with endothelium, which is continuous with the endothelium of the capillaries and veins.

2. **Major veins from the head, brain, and neck**
   a. The **external jugular veins** drain blood from the superficial regions of the head and neck.
   b. The **internal jugular veins** are larger and deeper than the external jugular veins. They drain blood collected from the brain by way of venous sinuses.
   c. Each internal jugular vein joins with a **subclavian vein** to form a **brachiocephalic vein** on either side.
   d. The two brachiocephalic veins join to form the **superior vena cava**, which enters the right atrium.
3. **Major veins from the upper limbs**
   a. The deep veins of the arm accompany the arteries and are given the same names as the arteries: **axillary**, **brachial**, **radial**, and **ulnar** veins. They drain into the subclavian veins.
   b. The superficial veins of the arm begin from venous anastomoses in the hand and wrist and drain into the deep veins.
      (1) The **cephalic vein** runs upward on the lateral side of the arm and empties into the axillary vein at the shoulder.
      (2) The **basilic vein** courses upward on the medial-posterior side of the arm, crosses forward to the front of the arm just below the elbow, and joins with the brachial vein.
      (3) The **median cubital vein** connects the basilic vein and the cephalic vein in front of the elbow. It is a favored site for blood sampling through venipuncture.
4. **Veins of the thorax.** Venous blood from the thorax empties into the superior vena cava via the **brachiocephalic veins** and the **azygos** group of veins.
   a. The **brachiocephalic vein** drains the upper thorax and anterior thoracic wall.
   b. The **azygos vein** drains thoracic muscles and organs. It arises as a continuation of the right ascending lumbar vein and empties into the superior vena cava.
   c. The **hemiazygos vein** drains thoracic muscles and organs on the left side of the vertebral column. It arises as a continuation of the left ascending lumbar vein and joins the azygos vein. The **accessory hemiazygos vein** is a superior continuation of the hemiazygos vein.
5. **Veins of the abdomen and pelvis**
   a. The **inferior vena cava** returns blood to the heart. It is formed by the union of the **right and left common iliac veins** at the level of the fifth lumbar vertebra.
   b. The inferior vena cava receives tributaries from abdominal and pelvic veins. These correspond to most of the arteries that arise from the abdominal aorta, with the exception of blood returning from the digestive tract, pancreas, or spleen.
   c. The **hepatic portal system** is a circulatory system modification by which blood absorbed from the digestive tract is carried directly to the liver before it is returned to the heart. It ensures that nutrients and potentially harmful substances can be removed from the blood to be stored, metabolized, or detoxified. The major vessels of the hepatic portal system include the following components:
      (1) The **splenic vein** drains the spleen, pancreas, part of the stomach, and, via the tributary **inferior mesenteric vein**, much of the large intestine.
      (2) The **superior mesenteric vein** receives blood from the small intestine and parts of the colon and stomach.
      (3) The **hepatic portal vein** is a short vein formed by the junction of the splenic and superior mesenteric veins.
         (a) It passes to the underside of the liver and divides into right and left branches, which accompany the right and left branches of the hepatic artery into the liver.

- **(b)** In the liver, the hepatic portal branches break up into liver **sinusoids**, which drain into venules, **central veins**, and **hepatic veins**.
- **(4)** The hepatic veins exit the liver to join the inferior vena cava.

6. **Veins of the lower limbs**
   a. Deep veins accompany the arteries and have corresponding names: external iliac, femoral, popliteal, anterior and posterior tibial, and peroneal veins.
   b. Superficial veins arise from a venous anastomosis, the **dorsal venous arch**.
      (1) The **small saphenous vein** ascends the posterior leg, divides into two veins, and empties into the popliteal and the deep femoral veins. It contains 7 to 13 valves to prevent backflow.
      (2) The **great saphenous vein** is the longest vein in the body. It ascends along the medial side of the foot, leg, and thigh to join the femoral vein below the inguinal ligament. There are 10 to 20 valves in the great saphenous vein.

D. **Fetal circulation**
   1. In the fetus, the digestive, urinary, and respiratory systems are nonfunctional.
   2. The **umbilical vein** from the placenta carries nutrients and oxygen to the fetus through the umbilical cord and enters the fetus at the umbilicus.
      a. The umbilical vein joins the fetal hepatic portal system. Some of the blood from the umbilical vein supplies the liver, but most enters the **ductus venosus**, which bypasses the liver to enter the inferior vena cava and mix with unoxygenated blood returned from the fetal body.
         (1) The ductus venosus is obliterated after birth and remains as the **ligamentum venosum** in the liver.
         (2) The umbilical vein becomes the **ligementum teres** (round ligament) of the liver.
      b. The foramen ovale and the ductus arteriosus pathways through the fetal heart and their subsequent postnatal fate have been described previously. (See III G.)
   3. Unoxygenated blood containing wastes returns to the placenta by way of two **umbilical arteries**, which are continuations of the iliac arteries of the fetus. The proximal parts of the umbilical arteries remain as part of the arterial supply to the urinary bladder; the distal parts become fibrous umbilical ligaments.

VII. CAPILLARY EXCHANGE AND THE LYMPHATIC SYSTEM

A. **Mechanisms of fluid exchange at the capillaries**
   1. All transfer of gases, nutrients, and metabolic waste products between the blood and the tissue cells is carried out across the capillary membranes by means of the physical processes of **diffusion, osmosis,** and **filtration**. (See Chapter 3, III, A,B.)
      a. The two-way transfer can occur only at the level of the capillaries, which have walls thin enough for the passage of water and particles.
      b. Substances move through spaces or capillary "pores" between adjacent endothelial cells, and through molecular pores in the cell membrane.
   2. The capillary walls hold back the formed elements of the blood and, under normal circumstances, the large protein macromolecules in the plasma.
   3. The exchange of water and dissolved substances is dependent on several opposing forces or pressures.
      a. **Blood hydrostatic pressure** (filtration pressure) in the capillary tends to force fluid and dissolved substances out of the capillaries.

b. **Blood colloid osmotic (oncotic) pressure** is created by the plasma proteins. This pressure tends to pull the interstitial fluid surrounding cells into the capillaries.
c. **Tissue (interstitial) fluid colloid osmotic pressure** is generated by a small amount of proteins that have leaked out of the capillaries. It tends to pull fluid out of the capillaries into the interstitial spaces.
d. **Tissue (interstitial) fluid pressure** is the pressure of fluid in the spaces between the cells. It opposes blood hydrostatic pressure.

4. Whether fluids leave or enter the capillaries thus depends on the balance of opposing pressures along the length of the capillary from the arterial end to the venous end.
   a. Blood hydrostatic pressure and tissue fluid colloid pressure, which move fluid out of the capillaries, is opposed by blood colloid osmotic pressure and tissue fluid pressure, which move fluid back into the capillaries.
   b. **Effective filtration pressure** is the algebraic sum of the opposing pressures. It is the net force moving fluid out of the blood into the tissues.

5. **Starling's law of the capillaries** is the hypothesis for the mechanism of fluid exchange based on an arterial/venous pressure gradient.
   a. The effective filtration pressure at the **arterial** end of the capillary favors net movement of water and dissolved substances **outward** to the tissues. Tissue fluid moves out at the arterial end.
   b. As the blood pressure drops through the capillary bed, it no longer can overcome the opposing pressure for **inward** absorption. Tissue fluid, therefore, moves into the capillary at the **venous** end.

6. The Starling mechanism assumed idealized capillaries. It currently is theorized that rhythmic vasomotion (intermittent constriction and dilation of precapillary sphincters) in the capillary bed alternately increases and decreases the pressure in an individual capillary. Therefore, inward and outward fluid movement may occur throughout the entire length of the capillary.

7. A slight imbalance in the capillary exchange mechanism results in nonabsorption of approximately one-tenth of the tissue fluid at the venous end of the capillary bed. The small excess is drained by the **lymphatic system**.

B. **Lymphatic system** (Color Plate 10)
   1. **Definition.** The lymphatic system is an accessory component of the circulatory system. It consists of organs that produce and store lymphocytes; a circulating fluid (lymph), which is derived from tissue fluid; and lymphatic vessels that return the lymph to the circulation.

   2. **Functions**
      a. The lymphatic system **returns the excess tissue fluid** that has leaked out from the capillaries. If the fluid were not removed, it would collect in the spaces between the cells and result in **edema**.
      b. The lymphatic system also **returns plasma proteins** to the circulation. Any **plasma proteins** that have leaked out of the capillaries to pass into the tissue spaces are absorbed into the lymphatic vessels. If the proteins were allowed to accumulate, the osmotic pressure of the tissue fluid would increase and upset capillary dynamics.
      c. Specialized lymphatic vessels **transport absorbed nutrients**, especially fats, from the digestive system to the blood.
      d. The lymphatic system **removes toxic substances and cellular debris** from tissues after infection or tissue damage.
      e. The lymphatic system controls the quality of the tissue fluid overflow by **filtering it through the lymph nodes** before returning it to the circulation.

   3. **Anatomy**
      a. Lymphatic vessels originate as microscopic blind-ended sacs called **lymphatic capillaries.** (Those originating in the villi of the small intestine are called **lacteals.**)

(1) Lymphatic capillaries are larger and more irregular than blood capillaries, but they have the same basic structure.
(2) **Lymph** is tissue fluid that is absorbed into a lymphatic capillary.
  b. **Circulation of lymph.** Lymph flows from the lymphatic capillaries into collecting lymphatics to progressively larger vessels that join to form the **main lymphatic trunks**.
    (1) The **thoracic duct** is the principal lymphatic trunk that drains the entire body except for the upper right quadrant. It enters the left subclavian vein at its junction with the internal jugular vein.
      (a) The thoracic duct originates as the dilated, sac-like **cisterna chyli** in the lumbar region of the abdominal cavity. The cisterna chyli is the collecting duct for all lymphatics leading from the liver, intestines, pelvis, and lower limbs.
      (b) As it ascends through the thorax, the thoracic duct collects lymph from the posterior chest wall, the left side of the head and neck, and the left arm.
    (2) The **right lymphatic duct** is a smaller lymphatic trunk. It empties into the junction of the right internal jugular and subclavian veins. It receives lymph drainage from the right side of the head and neck and the right arm.
    (3) The **right bronchomediastinal trunk** collects lymph from the mediastinal structures and lungs and joins the right lymphatic duct.

4. **Mechanisms of lymph flow**
  a. Movements of skeletal muscle adjacent to lymph vessels move the lymph forward toward the lymphatic trunks.
  b. Periodic contraction of the lymphatic vessels act like a lymph pump.
  c. Negative intrathoracic pressure, which occurs during inspiration, provides a suction effect on lymph in the thoracic duct.

5. **Lymph nodes**
  a. **Structure.** Lymph nodes are interspersed in the channels of a number of lymph vessels. They filter lymph before it is returned to the venous circulation.
    (1) Lymph nodes or "glands" are oval, or bean-shaped, stuctures that range from 1 mm to 20 mm in size. The cortex is the outer part of a node; the medulla is the inner part.
    (2) Lymph enters a node through a fibrous capsule via several **afferent lymph vessels** and exits through one **efferent lymph vessel.**
    (3) One-way **valves** in the afferent and efferent vessels keep the lymph flowing in one direction.
  b. **Function**
    (1) In moving from the afferent vessels to the efferent vessels, the lymph in a node percolates through a meshwork of reticular fibers that forms irregular spaces called sinuses, which are partially lined by reticular cells and macrophages. The **macrophages phagocytize foreign particles** such as bacteria.
    (2) Densely packed areas of lymphocytes called **primary (cortical) nodules** are found in the cortex of the node. Activated lymphocytes in the nodules provide an **immune response** for destroying foreign invaders.
    (3) When many bacteria are filtered out of the lymph, the node may swell to many times its normal size because of proliferating lymphocytes and other cells.
    (4) Lymphatic organs whose functions are related to those of the lymph nodes include the thymus gland, tonsils, and spleen.
  c. **Locations of some lymph nodes of clinical significance**
    (1) The **submaxillary nodes** are located in the floor of the mouth.
    (2) The **cervical nodes** are located in the neck along the sternocleidomastoid muscle.
    (3) The **supratrochlear nodes** are located just above the bend of the elbow.

(4) The **axillary nodes** are clustered deep within the underarm and upper chest regions.
(5) The **inguinal nodes** are located in the groin.

6. **Disorders of capillary dynamics and lymphatic flow**
   a. **Edema** is the accumulation of abnormal amounts of interstitial fluid in the spaces surrounding cells. It may be caused by any factor that increases fluid passage from the capillaries into the tissues or decreases its return to the capillaries.
   b. Factors that can disrupt the normal production and absorption of interstitial fluid and result in edema include the following:
      (1) **Increased hydrostatic pressure (filtration pressure) in the capillaries** occurs as a result of heart failure.
         (a) The hypoeffective heart pumps less blood out, which causes damming up of blood in the venous system and venous obstruction. A failing heart also results in less blood flow to the kidneys, which causes kidney retention of fluid. This results in an increase in blood volume, which the failing heart cannot pump out.
         (b) The result is an increase of pressure in the capillaries, which forces more fluid out into the tissues.
      (2) **Decreased plasma colloid osmotic pressure**, which disturbs capillary dynamics, can be caused by a loss of plasma proteins through kidney disease or by the lack of protein in the diet (malnutrition).
      (3) **Lymphatic obstruction** prevents the normal return of interstitial fluid or proteins to the circulation. It can be caused by surgical procedures or infection that results in an interruption of the lymph channels.
      (4) **Increased capillary membrane permeability** as a result of an inflammation process causes fluid and protein leakage into interstitial spaces. Histamine and related substances released by damaged tissues cause increased capillary permeability.

Study Questions

Directions: Each question below contains four suggested answers. Choose the **one best** response.

1. A blood test performed on an individual shows a hematocrit reading of 45. This means the individual
   - (A) is anemic
   - (B) has more formed elements than plasma in the blood
   - (C) has 45% RBCs and other solids in every 100 mL of whole blood
   - (D) must be female

2. All of the following statements describing erythrocytes are true EXCEPT
   - (A) Erythrocytes are shaped like biconcave discs.
   - (B) The heparin in erythrocytes is the substance of major importance in their function.
   - (C) There are no nuclei in mature erythrocytes.
   - (D) Erythrocyte production by the bone marrow is regulated by the oxygen requirements of the tissues.

3. Which of the following cells formed during hematopoiesis is the most mature?
   - (A) reticulocyte
   - (B) hemocytoblast
   - (C) basophilic erythroblast
   - (D) neutrophilic metamyelocyte

4. Which of the following is most likely to occur if total body irradiation or adverse drug reactions damage the bone marrow?
   - (A) a decrease in the susceptibility to infections
   - (B) an increase in blood viscosity
   - (C) anemia
   - (D) an increase in thrombus (internal clot) formation

5. The final event in the formation of a blood clot is the conversion of
   - (A) fibrinogen to fibrin
   - (B) vitamin K to prothrombin
   - (C) thrombin to fibrinogen
   - (D) molecular calcium to calcium ions

6. An individual who has only anti-A agglutinins in the plasma has what blood type?
   - (A) type A
   - (B) type B
   - (C) type AB
   - (D) type O

7. If the chordae tendineae of the heart were damaged, which of the following effects would be most likely?
   - (A) The blood in the venae cavae would not empty into the heart at a normal rate.
   - (B) The rhythm of the cardiac cycle would be interrupted.
   - (C) The efficency of the heart valves would be reduced.
   - (D) The closing of the foramen ovale would be prevented.

8. Mitral stenosis is a thickening of the cusps of the mitral valve that increases the resistance to the flow of blood from the left atrium to the left ventricle. Which of the following would be a consequence of mitral stenosis?
   - (A) The blood pressure in the entire systemic cirulation would increase.
   - (B) The blood pressure in the pulmonary veins would decrease.
   - (C) The blood volume in the left ventricle would increase.
   - (D) The blood pressure in the left atrium would increase.

9. All other factors remaining constant, which of the following would **not** increase arterial pressure?
   - (A) arterial constriction
   - (B) increased heart rate
   - (C) hemorrhage
   - (D) increased venous return

10. Which statement concerning the cardiac cycle is correct?
    - (A) The second heart sound is due to the closing of the A-V valves.
    - (B) The aortic semilunar valves open when the pressure in the left ventricle becomes greater than the pressure in the aorta.
    - (C) During each cardiac cycle, the muscles attached to the A-V valves contract to open and close the valves.
    - (D) At no point in the cardiac cycle are all four heart valves closed.

11. All of the following are normal components of lymph EXCEPT
    - (A) protein molecules
    - (B) lymphocytes
    - (C) tissue fluid
    - (D) RBCs

12. All of the following may be a cause of tissue edema (accumulation of fluid in the interstitial spaces) EXCEPT

    (A) heart failure
    (B) severe protein malnutrition
    (C) increased permeability of capillary membranes
    (D) increased heart rate as a result of exercise

Questions 13–15 refer to the diagram below.

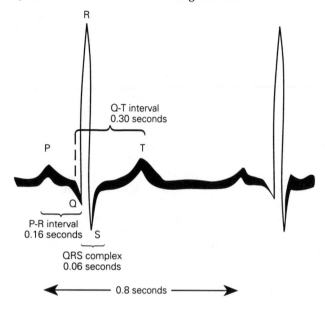

13. The length of time that elapses between the beginning of the contraction of the atria and the beginning of the contraction of the ventricles is

    (A) 0.8 sec
    (B) 0.46 sec
    (C) 0.16 sec
    (D) 0.06 sec

14. Which time period would be lengthened if the transmission of the action potential were slowed between the atria and the ventricles?

    (A) the P-R interval
    (B) the QRS complex
    (C) the period between one P-wave and the next P-wave
    (D) the period between the P-wave and the beginning of the T-wave

15. What is the heart rate (total number of beats per minute) in an individual with the ECG depicted?

    (A) 48 per min
    (B) 60 per min
    (C) 72 per min
    (D) 75 per min

Questions 16–19. For each of the larger blood vessels listed below, match the smaller blood vessel branch to which it provides blood.

    (A) celiac trunk
    (B) subclavian artery
    (C) brachial artery
    (D) femoral artery

16. vertebral artery

17. popliteal artery

18. radial artery

19. hepatic artery

Questions 20–24. For each of the following pairs of statements, choose the answer

    (A) if the item on the left is usually or always greater
    (B) if the item on the right is usually or always greater
    (C) if both items are approximately of the same magnitude

20. amount of blood leaving the right ventricle | amount of blood leaving the left ventricle

21. oxygen concentration in the pulmonary artery | oxygen concentration in the pulmonary vein

22. movement of tissue fluid out at the arterial end of the capillary bed | movement of tissue fluid out at the venous end of the capillary bed

23. systolic blood pressure | diastolic blood pressure

24. carbon dioxide content in arteries | carbon dioxide content in veins

Answers and Explanations

1. **The answer is C.** (II D 2 b) The hematocrit is the volume percentage of RBCs in whole blood. Normally, the hematocrit for an average healthy man is 42 to 54 and about 38 to 48 for an average healthy woman. Thus, a hematocrit of 45 would indicate neither anemia (decreased number of RBCs) nor the gender of the individual.

2. **The answer is B.** (II D 1–5) Erythrocytes contain hemoglobin, which combines loosely with oxygen for transport to the body tissues. Mature RBCs lack nuclei, although they contain cytoplasmic enzymes and ATP to ensure survival for approximately four months. Erythrocytes are shaped like biconcave discs, which gives them a large surface area relative to volume. The production of erythrocytes is stimulated by tissue hypoxia (oxygen deficiency), which acts to increase the secretion of erythropoietin, a hormone produced by the kidneys. Erythropoietin stimulates the bone marrow to produce RBCs.

3. **The answer is A.** (II C 2, Figure 11–1) Reticulocytes are newly formed erythrocytes, which have a stainable network, or reticulum, formed by remnants of ribosomal nucleoprotein. They remain in the marrow for several days until they are released as fully mature RBCs. A few reticulocytes (0.5% to 1.5% of the RBCs) are normally found in the circulation. Hemocytoblasts are the primitive stem cells located in the bone marrow from which the blood cells in the circulation form. Basophilic erythroblasts are a stage in the development of erythrocytes in which the cytoplasm stains deeply with a basophilic (blue) dye. Neutrophilic metamyelocytes are a stage in the development of neutrophils, which follows the myelocyte stage and precedes the band stage.

4. **The answer is C.** (II C–E) Damage to bone marrow would result in damage to blood cell development and the most likely consequence is anemia, or a deficiency of RBCs. Because of a decrease in production of leukocytes, the susceptibility to infections would increase. Fewer cells would result in a decrease in blood viscosity. Fewer megakaryocytes would result in fewer platelets and there would be no increase in thrombus formation.

5. **The answer is A.** (II G 1–7) The conversion of the soluble plasma protein fibrinogen to insoluble fibrin strands is the completion of the coagulation mechanism. Coagulation factors IV, V, VII, and X act with tissue thromboplastin to convert prothrombin to the enzyme thrombin, which acts to convert fibrinogen to fibrin. Vitamin K is a requirement for liver synthesis of several clotting factors. Calcium ions are required throughout the entire clotting sequence.

6. **The answer is B.** (II H 1–3) Type B blood contains type B agglutinogen on the red cells and anti-A agglutinin in the plasma. This individual could not be blood type A because the anti-A agglutinins present in the plasma would react with his or her own agglutinogens. A person with blood type O has both anti-A and anti-B agglutinins in the plasma, and one with blood type AB has no agglutinins in the plasma.

7. **The answer is C.** (III C–1 a) The chordae tendineae are the fibrous connective tissue cords that attach the flaps of the A-V valves to the papillary muscles of the ventricles and function to prevent the valves from everting backward into the atria. Their damage would reduce the efficiency of the valves. There are no valves or attachments at venous entries into the atria. The foramen ovale is a fetal opening between the atria. It closes at birth to become the fossa ovalis.

8. **The answer is D.** (III B C; IV D 1–4) An obstruction to blood flow from the left atrium to the left ventricle would result in a damming up of blood in the left atrium. The result would be an increase of blood pressure in the left atrium, an increased back pressure in the pulmonary veins entering the left atrium, and a decrease in left ventricular blood volume. An obstruction to blood flow in the right side of the heart would increase blood pressure in the systemic circulation.

9. **The answer is C.** (V B 3) Factors that increase blood pressure include an increase in cardiac output, which is affected by increased heart rate and increased venous return, and an increase in peripheral arterial resistance, which is affected by blood viscosity, blood volume, and arterial vessel radius (constriction). Hemorrhage would result in a decrease in blood volume, which would decrease cardiac output and reduce arterial pressure.

10. **The answer is B.** (IV D) During ventricular contraction, ventricular pressure increases until it is greater than the pressure in the aorta or pulmonary arteries, forcing open the aortic and pulmonic semilunar valves. The second heart sound is the result of the closing of the semilunar valves when ventricular pressure drops below that in the aorta and the pulmonic trunk. All four heart valves are closed at the beginning of ventricular systole during the period of isovolumetric contraction.

11. **The answer is D.** (VII B 2 a–e) Lymph contains plasma protein and fat molecules, lymphocytes, and tissue fluid as well as toxic substances, bacteria, and cellular debris to be cleared from the tissues. It does not contain RBCs.

12. **The answer is D.** (VII B 6) An abnormal accumulation of fluid in the interstitial spaces may result from any factor that disturbs capillary dynamics according to the arterial-venous pressure gradient. Such factors would include failure of the heart to pump all the blood returned to it; a decrease in the concentration of plasma protein, which decreases plasma oncotic pressure; blockage of lymph flow; an increase in the capillary permeability to plasma protein; increased capillary hydrostatic pressure; increased venous blood pressure; local release of histamine following tissue damage; increased retention of body water, and so on.

13–15. **The answers are 13–C, 14–A, and 15–D.** (IV C 1–3) The P-R interval, 0.16 sec on the ECG depicted, is the time from the beginning of the depolarization (beginning of contraction) of the atria to the beginning of ventricular contraction. If the passage of an action potential from the atria to the ventricles were delayed, the P-R interval would be lengthened. The interval from one P-wave to the next P-wave is the entire cardiac cycle, which begins with the onset of contraction and ends with the onset of the next contraction. That interval depicted is 0.8 seconds, which means that the heart rate is 75 beats per minute (60 sec divided by 0.8 sec).

16–19. **The answers are 16–B, 17–D, 18–C, and 19–A.** (VI B 1–6) The right and left subclavian arteries give rise to the vertebral arteries. The femoral artery in the thigh continues as the popliteal artery across the knee joint. The brachial artery divides to form the radial and ulnar arteries. The celiac trunk branches into the common hepatic artery, which gives rise to the hepatic artery to the liver.

20–24. **The answers are 20–C, 21–B, 22–A, 23–A, and 24–B.** (IV D 1; VII A; V B 2; III F 1–2) Both the right and left sides of the heart eject equal volumes of blood although they have different atrial and ventricular pressures. The pulmonary artery carries deoxygenated blood from the right ventricle to the lungs; the pulmonary vein carries oxygenated blood from the lungs to the left ventricle. Because the effective filtration pressure is greater at the arterial end of the capillary bed, the tissue fluid moves out at the arterial end. Blood pressure is the force exerted by blood against the blood vessel walls and is measured in terms of mm Hg. Systolic blood pressure (average 120 mm Hg) is greater than diastolic pressure (average 80 mm Hg). Arteries, with a few exceptions, carry oxygenated blood away from the left heart to the tissues; veins return deoxygenated blood with a high carbon dioxide content back to the right heart.

Nonspecific Defenses and the Immune System   12

I. INTRODUCTION

A. **Overview.** The nonspecific body defenses and the immune system protect the body against environmental agents that are foreign to the body. Such foreign agents in the external environment may be **pathogens** (viruses, bacteria, fungi, protozoa, or their products), **plant and animal products** (certain foods, pollen, or animal hair or dander), or **chemicals** (drugs or pollutants).

1. **Nonspecific defenses** provide general protection against many kinds of agents. They are classified as nonimmune defenses by some immunologists. Others refer to them as **innate** immune defenses or **natural immunity**.
   a. The nonspecific defenses consist of all the physical, mechanical, and chemical barriers against foreign matter that are present from birth.
   b. They include skin, mucous membranes, phagocytic cells, and substances released by leukocytes.
2. **Acquired immunity** is a **specific defense**, which is induced (acquired) by exposure to specific infectious agents. The lymphatic tissues and organs of the body constitute the **immune system.**
   a. The **components of the immune system** include the primary lymphoid organs (bone marrow and thymus glands), the secondary lymphoid tissue (lymph nodes, spleen, adenoids, tonsils, Peyer's Patches of the small intestine, and the appendix), as well as several other cells and cell products.
   b. **The two types of immune responses** are humoral immunity and cellular (cell-mediated) immunity.
      (1) **Humoral immunity** is mediated by antibodies, which are produced by bone-marrow derived lymphocytes (B cells) and are found in blood plasma.
      (2) **Cellular immunity** is mediated by thymus-derived lymphocytes (T cells).

II. NONSPECIFIC DEFENSES

A. **Physical, chemical, and mechanical barriers** to infectious agents
   1. **Intact skin** provides one of the first lines of defense since it is impermeable to infection by most organisms.
      a. Although some microorganisms can enter the body through sebaceous glands and hair follicles, the antimicrobial effect of sweat and sebaceous secretions (due to lactic acid and fatty acids) minimizes the significance of this route.

b. When there is skin loss, as from burns, or when the skin is broken, infections can occur. Minor wounds, however, rarely cause severe infection because they provoke a skin immune response.
 2. **Mucous membranes** that line the inner surfaces of the body secrete mucus, which traps microbial and other foreign particles and blocks their access to epithelial cells.
   a. For example, large particles that enter the nasal chambers are filtered by hairs in the nose and become trapped in mucus. Those particles that enter the upper respiratory tract are expelled by sneezing and coughing.
   b. Small particles and microorganisms, which may elude the mucus barrier, enter the respiratory passages but are removed by the cilia of the epithelial lining. The particles are swept upward away from the lungs to be expelled or swallowed with mucus into the digestive tract.
 3. Some **body fluids** contain antimicrobial agents. For example, microorganisms can be destroyed by the enzyme lysozyme in saliva, nasal secretions, and tears; by enzymes and acid in the digestive fluids; by proteolytic enzymes and bile in the small intestine; and by acidity in the vagina. These protective chemicals create an unfavorable environment for some, but not all, organisms.
 4. **Mechanical factors** such as the washing action of tears, saliva, and urine contribute to protection.

B. **Phagocytosis** is the second line of defense against infectious agents. It is the engulfment and digestion of microorganisms and toxins once they have penetrated the body. (See Chapter 3, section III C 2.)
 1. The major phagocytes of the body are the blood **neutrophils** and tissue **macrophages**, which are derived from blood monocytes.
 2. Neutrophils and macrophages move throughout the tissues by **chemotaxis**, the chemically directed cell movement of these leukocytes. The chemotaxins that attract them are produced by microorganisms, other leukocytes, or other blood components.
   a. Connective tissue macrophages (histiocytes) are fixed or wandering (mobile), which refer to different states of the same cell.
   b. Macrophages and their precursors (monocytes) may fuse to form **foreign body giant cells**, which are large multinucleated cells that act as a barrier between large masses of foreign material and the body tissues. Such cells are common in tuberculosis, for example.
   c. Macrophages also play an important role in **facilitating immune responses**.
 3. The **mononuclear phagocytic system (MPS)**, formerly known as the **reticuloendothelial system (RES)**, includes the combination of phagocytic monocytes, mobile macrophages, and fixed tissue macrophages. Fixed macrophages are known by special names in various tissues. They include the following:
   a. **Alveolar macrophages** in the lungs
   b. **Kupffer** cells in the liver
   c. **Langerhans cells** in the epidermis
   d. **Microglia** in the CNS
   e. **Mesangial cells** in the kidneys
   f. **Reticular cells** in the spleen, lymph nodes, bone marrow, and thymus

C. **Inflammation** is the tissue response to injury by infection, punctures, abrasions, burns, foreign objects, or toxins (bacterial products that damage host cells or host tissues). It includes a complex sequence of events and may be **acute** (short term) or **chronic**.
 1. The local **signs** of the inflammatory response include **redness, heat, swelling**, and **pain**. Occasionally a fifth symptom is loss of function, depending on the extent of injury.

2. The **events** of inflammation are as follows:
   a. The first stage is the production of **vasoactive chemical factors** by damaged cells at the site of injury. They include **histamine** (from mast cells), **serotonin** (from platelets), arachidonic acid derivatives (**leukotrienes, prostaglandins,** and **thromboxane**), and **kinins** (activated plasma proteins). The factors cause effects that include the following:
      (1) **Vasodilation,** or an increase of the diameter of the blood vessels in the area of damage, increases blood flow and is responsible for the redness (erythema), throbbing pain, and heat.
      (2) **Increased capillary permeability** results in fluid loss from the vessels into the intercellular spaces. The accumulation of fluid in the tissues causes swelling, or edema.
      (3) **Walling-off of the site of injury** occurs as a result of fibrinogen leaking into the tissues from the plasma. Fibrinogen is converted into fibrin to form a clot, which isolates the area from the undamaged tissue.
   b. The second stage is **chemotaxis** (movement of phagocytes to the site of injury), which generally occurs within an hour after the beginning of the inflammatory process.
      (1) **Margination** is the adherence of phagocytes (neutrophils and monocytes) to the endothelial walls of the capillaries at the damaged site.
      (2) **Diapedesis** is the migration of the phagocytes through the capillary walls to the injured area. Neutrophils arrive first; the monocytes follow into the tissue and become macrophages.
   c. **Phagocytosis** of harmful agents occurs at the site of injury.
      (1) The neutrophils and macrophages break down enzymatically and die after engulfing large numbers of microorganisms.
      (2) The dead leukocytes, dead tissue cells, and various body fluids form **pus,** which continues to be formed until the infection subsides. Pus moves to a body surface for dispersal or to an internal cavity, eventually to be destroyed and absorbed by the body.
      (3) An abscess or granuloma may form if the inflammatory response is unable to overcome the injury or invasion.
         (a) An **abscess** is a circumscribed sac of pus surrounded by inflamed tissue. It generally does not subside spontaneously and must be drained.
         (b) A **granuloma** generally is the result of a chronic inflammatory process in response to a persistent irritation. It is an accumulation of phagocytic cells and microorganisms surrounded by a fibrous capsule.
   d. **Repair** through tissue regeneration or scar formation is the final stage of the inflammatory response.
      (1) In **tissue regeneration,** the healthy cells within the affected tissue divide mitotically to proliferate and restore tissue mass.
      (2) **Scar formation** by fibroblasts is an alternative response to tissue regeneration. It provides a substitute for the tissue originally damaged.
      (3) The nature of the damaged tissue and the extent of the injury determines whether regeneration or scar formation takes place. Skin has a high capacity for complete regeneration unless the injury was deep or wide.
3. **Systemic effects of inflammation** include fever and leukocytosis.
   a. **Fever,** or an abnormally high body temperature, may occur in association with inflammation.
      (1) **Exogenous pyrogens** (fever inducers), which are released by bacteria, and **endogenous pyrogens,** which are released by a variety of leukocytes, act on the hypothalamus to reset the normal thermoregulatory controls to a higher temperature.

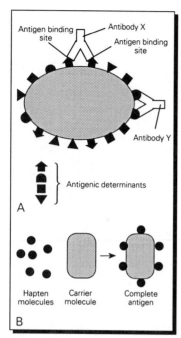

Figure 12-1. *A,* Antigenic determinants (epitopes). Most antigens have more than one antigenic determinant; four are show on this drawing. Each antibody molecule has at least two binding sites that can attach to its specific complementary determinant site on an antigen. *B,* Haptens and carrier. When a hapten molecule, which is too small to be antigenic by itself, combines with a larger carrier molecule, both hapten and carrier function together as an antigen and can stimulate an immune response.

(2) The body adjustments to the increased setting include **vasoconstriction** to reduce heat loss, **chills** and **shivering** to increase body heat, and an **increased metabolic rate**. The result is increased body temperature.

(3) The fever subsides when the infection subsides, pyrogen levels decline, and normal thermoregulatory control is regained.

b. Leukocytosis (increased number of leukocytes in the blood) results from the increased demand for additional numbers of WBCs and their increased production by the bone marrow.

D. **Nonspecific antiviral and antibacterial substances** are produced by the body to protect itself from infection. Their action does not require antigen-antibody interaction for initiation.

1. **Interferons** (IFNs) are antiviral proteins that can be synthesized by almost any type of host cell in response to virus infection, immune stimulation, or a variety of chemical stimulants.
   a. Types of IFNs
      (1) **Alpha interferon** (α-IFN) is produced by virus-infected leukocytese.
      (2) **Beta interferon** (β-IFN) is produced by virus-infected fibroblasts.
      (3) **Gamma interferon** (γ-IFN) is produced by two types of immune lymphocytes.
   b. Functions of IFNs. Interferons block viral multiplication and also play a role in modulating immunological activities.

2. **The complement system** is a group of plasma proteins that circulate in the blood in inactive forms. Complement components are named for their ability to enhance, or complement, the body's defense systems.
   a. **Function.** The overall function of the complement system is to attack and destroy invading microorganisms.
   b. **Mechanisms of activation.** The pathway for complement action is a **cascade**. Each protein in the group that is activated affects the next protein in the pathway. The cascade allows amplification and activation of large amounts of complement from a small initiation signal.
      (1) The **classical pathway** for complement activation requires antigen-antibody interactions for initiation.
      (2) The **alternate pathway**, a form of nonspecific resistance, is triggered by bacteria or their products without antigen-antibody interaction.
      (3) **Properdin** is a serum protein that enhances the complement activation by stabilizing certain of the complement proteins.

III. THE IMMUNE SYSTEM: SPECIFIC DEFENSES

A. **Definition.** The immune system is a complex system that provides immune responses (humoral and cellular) directed against specific foreign agents such as bacteria, viruses, toxins, or any substances that are recognized by the body as "nonself."

B. **Characteristics**

1. **Specificity.** The immune system discriminates among different foreign substances and responds specifically when needed.

2. **Memory and amplification.** The immune response has the ability to recall previous contact with a particular agent so that subsequent exposure leads to a more rapid and greater response.

3. **Recognition of self and nonself.** The immune system can distinguish between foreign agents and the body's own cells and proteins. An immune response to "self" is possible, however, and may create a state of **autoimmunity**. Autoimmunity may have pathological effects on the body.

Nonspecific Defenses and the Immune System

C. **Components** of immune responses
1. An **antigen** is a substance that elicits a specific immune response. Antigens usually are of high molecular weight and are chemically complex substances, such as proteins and polysaccharides.
   a. An **antigenic determinant (epitope)** is the smallest chemical group of an antigen that can invoke an immune response. An antigen can possess two or more antigenic determinants, any one of which under appropriate conditions can stimulate a distinct response (Figure 12–1).
   b. **Haptens** are small compounds that cannot by themselves induce the immune response, but they become immunogenic when they are coupled to **carriers** of high molecular weight, such as serum proteins.
   c. Haptens may include drugs, antibiotics, food additives, or cosmetics. Many types of low molecular weight compounds that are conjugated to carriers in the body may achieve immunogenicity. For example, in some people penicillin is not antigenic until it combines with serum proteins and is able to initiate an immune response.
2. An **antibody** is a soluble protein produced by the immune system in response to the presence of an antigen, and that will react specifically with that antigen.
   a. **Structure** (Figure 12–2)
      (1) An antibody molecule is comprised of four polypeptide chains: **two identical heavy (H) chains** and **two identical light (L) chains.** The terms light and heavy refer to the relative molecular weights.
      (2) The chains are joined by disulfide (–S–S–) and other bonds to form a Y-shaped molecule, which has a flexible hinge area. This allows a change in shape to react with a maximum number of antigens.
      (3) The **variable regions** on the H and the L chains are located on the ends of the Y arms. They form the two **antigen-binding sites**. Each antibody has at least two binding sites and is termed **bivalent**.
         (a) The variable regions in different antibodies are dissimilar in their amino acid sequences.
         (b) The specificity of an antibody for a particular antigen depends on the structure of the variable regions.
      (4) The **constant** region, consisting of the rest of the Y arms and the stem, is always identical in all antibodies of the same class.
   b. **Antibody classes.** Antibodies are a group of plasma proteins called **immunoglobulins** (Ig). There are five classes (isotypes) of immunoglobulins: IgA, IgD, IgE, IgG, and IgM.
      (1) IgA molecules constitute about 15% of the antibodies in blood serum and are found in body secretions such as perspiration, saliva, tears, respiratory, genitourinary, and intestinal secretions, and breast milk. Their main function is to **fight microorganisms** at potential points of entry into the body.

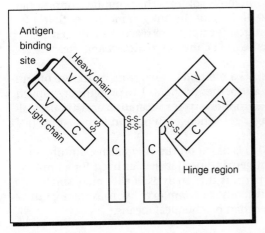

Figure 12–2. *General structure of the simplest type of antibody molecule.* The Y-shaped molecule is composed of two heavy and two light chains linked by disulfide bridges. The variable regions (V) differ in their amino acid sequence from molecule to molecule and form a receptor for a specific antigen. The constant regions (C) stabilize the antigen binding site and are the same in all antibodies of a given class. The hinge region allows the two arms of the Y to open and close. This can accommodate binding to two antigenic determinants separated by a fixed distance, as may be found on a bacterium's surface.

(2) **IgD** molecules are present in small numbers in blood serum and lymph but are found on a large proportion of B lymphocytes. Little is known about their function; they may help to **initiate the immune response**.

(3) **IgE** molecules usually are found in extremely low blood concentration. Their levels increase during allergic reactions and certain parasitic diseases. They bind to receptors on mast cells and basophils and cause the **release of histamine** and other chemical mediators.

(4) **IgG** molecules account for 80% to 85% of circulating antibodies and are the only antibodies to **cross the placenta** and provide the newborn with immunity. They are produced abundantly at the second and subsequent exposure to a specific antigen and they protect against **circulating microorganisms and toxins, activate the complement system**, and **enhance phagocytic cell effectiveness.**

(5) **IgM** molecules are the first antibodies to arrive at an infection site at the initial exposure to an antigen. A second exposure results in increased production of IgG. IgM antibodies **activate complement** and **enhance phagocytosis**, but they are relatively short lived. Because of their size, they generally remain in the blood vessels and do not enter surrounding tissues.

3. **Antibody-antigen interaction.** The antigen-binding sites on the variable regions of the antibody bind with their corresponding antigenic-determinant sites on the antigen to form an antigen-antibody (or immune) complex. This allows the inactivation of the antigen by complement fixation, neutralization, agglutination, or precipitation.

   a. **Complement fixation** occurs when portions of the antibody molecule bind complement. The bound complement molecules are activated through the "classical pathway," which initiates a cascade of effects that prevent damage by the invading organism or toxin. The more important effects include the following:

      (1) **Opsonization.** The antigen particle is coated by antibody or complement components, which facilitates the phagocytosis of the particle. In addition, a cleaved protein product of the complement cascade, C3b, interacts with special receptors on neutrophils and macrophages and enhances phagocytosis.

      (2) **Cytolysis.** A combination of multiple complement factors ruptures the plasma membranes of bacteria or other invaders and causes the cellular contents to leak out.

      (3) **Inflammation.** Complement products contribute to acute inflammation by activating mast cells, basophils, and blood platelets.

   b. **Neutralization** occurs when the antibodies cover the toxic sites of antigen and render them harmless.

   c. **Agglutination** (clumping) occurs if the antigen is particulate matter, such as bacteria or red cells.

   d. **Precipitation** occurs if the antigen is soluble. The immune complex becomes so large through cross-linking of the antigen molecules that it becomes insoluble and precipitates. Precipitation reactions between antigens and antibodies can be used clinically to detect and measure either component.

      (1) **Immunoelectrophoresis** is a method of analyzing a mixture of antigens (proteins) and the antibodies to them. The proteins are moved in an electrical field (electrophoresis) for separation and are then allowed to diffuse in an agar gel, where each protein forms a precipitin line with its antibodies.

      (2) **Radioimmunoassay** (RIA) is based on the competitive binding of radioactively labeled antigen and nonlabeled antigen for a limited amount of antibody. It permits the analysis of extremely small quantities of antigen, antibody, or complexes by measurement of their radioactivity rather than by chemical means.

D. **Types of immunity**
   1. **Active immunity** is acquired from direct contact with microorganisms or toxins and the body produces its own antibodies.
      a. **Naturally acquired active immunity** occurs when a person has been exposed to a disease and the immune system produces antibodies and specialized lymphocytes. The immunity may be lifelong (measles, chicken pox) or short lived (pneumococcal pneumonia, gonorrhea).
      b. **Artificially acquired (induced) active immunity** results from vaccination. A **vaccine** is made from weakened or dead pathogens or altered toxins. It can stimulate the immune response, but it does not cause the disease.
   2. **Passive immunity** results when antibodies are transferred from one individual to another.
      a. **Natural passive immunity** occurs in the fetus when IgG antibodies are passed across the placenta from the mother. It provides short-term (weeks to months) protection for the immature immune system.
      b. **Artificial passive immunity** is that provided by the injection of antibodies produced by an already immunized person or animal who had been exposed to the antigen. For example, antibodies from a horse that was immunized to a particular snake venom may be injected after an individual is bitten by that snake.

E. **Cells involved in the immune response.** Three types of cells play major roles in immunity: **B cells (B lymphocytes), T cells (T lymphocytes),** and **macrophages.**
   1. **Functions of the cells**
      a. **B cells** are antigen-specific and they proliferate in response to a particular antigen. B cells differentiate into nonproliferating **plasma cells,** which synthesize and secrete antibodies.
      b. **T cells** also exhibit antigen specificity and proliferate in the presence of antigen, but they do not produce antibodies.
         (1) T cells recognize and interact with antigens through **T-cell receptors,** which are membrane-bound cell surface proteins that are analogous to antibodies.
         (2) T cells produce immunologically active substances called **lymphokines.** Subtypes of T lymphocytes function in **helping** B lymphocytes respond to antigen, **killing** specific foreign cells, and **regulating** the immune response.
      c. **Macrophages** phagocytically engulf foreign substances and, by enzymatic action, break down trapped materials for excretion or reutilization.
         (1) Macrophages **process** phagocytosed antigens by denaturing or partially digesting them to yield fragments that contain the antigenic determinants.
         (2) They **present** these fragments of antigens on their cell surfaces for exposure to specific T lymphocytes. This is a necessary step in T-cell activation.
   2. **The B cell response** (Figure 12–3)
      a. B cells are named for the bursa of Fabricius, which is lymphoid tissue found in chickens. The equivalent tissue in mammals is believed to be bone marrow, intestinal lymph tissue, and spleen.
      b. After differentiation from the precursor stem cells, mature B cells migrate to peripheral lymph organs such as the spleen, lymph nodes, Peyer's patches in the digestive tract, and tonsils.
      c. Mature B cells carry surface immunoglobulin (S-Ig) molecules bound to their cell membranes. When activated by specific antigen and with appropriate help from T lymphocytes, B cells differentiate in two ways.
         (1) **Plasma cells** are the fully differentiated B cells. They are capable of synthesizing and secreting antibody to destroy the specific antigen.

Figure 12–3. *Humoral (antibody-mediated) immunity.* A bacteria (pathogen) invades the body and is brought and presented to a competent B lymphocyte in a lymph node. A helper T cell facilitates the activation of the B lymphocyte, which multiplies by mitosis to produce a clone of competent B cells. Many differentiate to become plasma cells, which secrete antibodies to be transported to the site of infection; other become memory cells, which remain and secrete small amounts of antibody long after the infection is over. At the site of invasion, antigen-antibody complexes directly inactivate the bacteria and also activate the complement system.

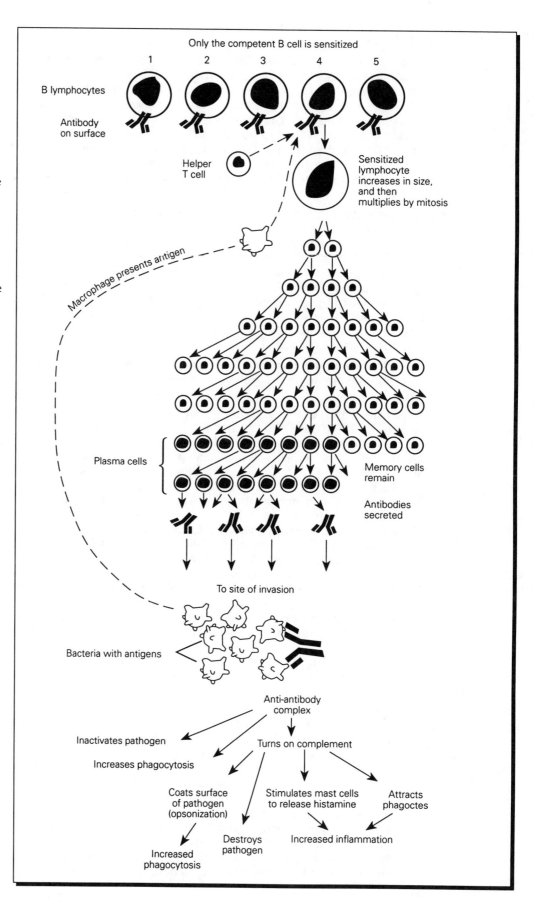

(2) **Memory B cells** are nondividing cells that arise from a fraction of the original antigen-activated B lymphocytes. Memory cells remain in the lymphoid tissue ready to respond to the sensitizing antigen the next time it appears with a faster and greater **secondary immune response**.

d. **The clonal selection theory** of antibody formation, proposed and developed by Jerne, Burnet, Talmadge, and Lederberg, is the working hypothesis that explains the complexities of immune system function. The main points of the theory are as follows:
  (1) B cells are genetically programmed to respond to a specific antigen **prior** to contact with that antigen. Before differentiating to antibody-secreting plasma cells, they carry their antibodies as membrane-bound surface receptors. These are Ig molecules of the same specificity as the antibodies that will be produced once the B cells are activated and differentiated.
  (2) Each individual has millions of B cells. Each B cell carries a different membrane-bound antibody capable of reacting with a single antigenic determinant.
  (3) When an antigen or antigenic determinant meets its appropriate antibody receptor on an immature B cell, it binds to the receptor and triggers the proliferation and maturation of that particular B cell into plasma cells and memory cells.
  (4) The result is a clone, or group of genetically identical cells descended from a single B cell. The antibody produced reacts specifically with the antigen that provoked the response.
  (5) Each plasma cell produces a single kind of antibody for as long as the antigen is present.
  (6) Any lymphocytes that carry antibodies against self-antigens are destroyed during fetal life. All B cells in an immunocompetent individual are thus tolerant of self and usually produce no subsequent immune response against self-antigens.
  (7) The clonal selection theory applies equally to T cells. Antigen binding to a T-cell receptor triggers the proliferation of a clone of mature cells descended from the single immature T cell.
  (8) The clonal selection theory explains immunological memory.
    (a) The **primary immune response** is slow because, initially, only a small number of cells has surface antibody molecules or T-cell receptors to respond to the antigen.
    (b) The **secondary response** to the next encounter with that antigen is rapid and more intense because expanded clones of long-lived memory B and T cells can respond to it.

3. **The T cell response** (Figure 12–4)
  a. T cells, like B cells, derive from precursor stem cells in the bone marrow. Late in fetal development or shortly after birth, the precursor cells migrate to the thymus gland, where they proliferate, differentiate, and acquire the ability to recognize self.
    (1) Each individual has a unique set of cell surface protein markers (antigens), which are coded by genes known as the major histocompatibility complex (MHC). Class I and class II MHC-encoded proteins are important in T cell activation.
      (a) **Class I MHC-encoded antigens** are produced on the surface of **all nucleated cells** of the body.
      (b) **Class II MHC-encoded antigens** are found only on the surfaces of **B cells** and **macrophages**.
    (2) During early life, T cells become imprinted with their own MHC-encoded self-antigens in the thymus gland. Thereafter, T cells recognize any other MCH-encoded antigens as foreign. This is the basis for the immune rejection of grafts and transplanted organs.
  b. After differentiation and maturation, T cells migrate to lymphoid organs such as the spleen or lymph nodes. They are specialized to operate against cells bearing intracellular organisms.

Figure 12–4. *Cell-mediated immunity.* A macrophage digests (processes) an antigen and presents it on its surface along with major histocompatibility complex (MCH) antigen for recognition by receptors on a specific T cell. The sensitized T cell increases in size and divides to form a large clone of cells, which differentiate into mature T cell types.

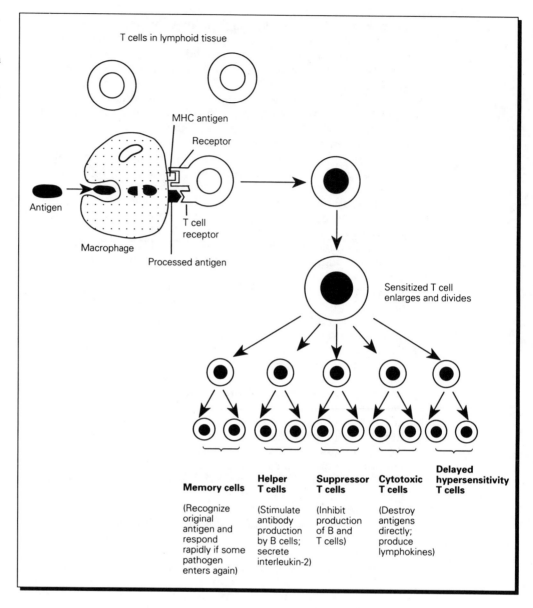

(1) Each T cell has one kind of cell surface receptor molecule (T cell receptor) that is antigen specific. There are millions of variants of the T cell receptor, but each can recognize a specific foreign antigen only when it is associated with MHC-encoded antigen. This informs the T cell that it has made contact with another cell.
(2) Upon recognition of the foreign antigens, T cells differentiate into **memory cells**, which persist after antigen inactivation, and three types of effector T cells.
c. **Effector T cells**
  (1) **Cytotoxic T cells** (killer T cells) recognize and destroy cells that display foreign antigens on their surfaces, such as cancer cells, transplanted tissue cells, and viruses and some bacteria that reproduce within host cells.
    (a) Cytotoxic T cells leave the lymphoid tissue and migrate to the site of their target cells. There they bind to the target cells and destroy them.
    (b) Because the T cell receptor of the cytotoxic T cell recognizes the **foreign** antigen of a target cell only when it also recognizes its **normal** cell surface MHC-encoded antigens (class I self-antigens), cytotoxic T cell function is said to be **MHC-restricted**.

(2) **Helper T cells** do not participate directly in killing cells. They recognize class II MHC antigens, which are present on B cells and macrophages, and must "see" those antigens to become activated. After activation by antigen-presenting macrophages, helper T cells have several functions.
  (a) They are required for normal **antibody synthesis**.
    (i) The activated helper T cell interacts with a B cell whose antibody recognizes the same antigen that stimulated the helper T cell.
    (ii) The B cell is triggered to divide and differentiate into a clone of antibody-producing plasma cells.
  (b) Upon recognition of a foreign antigen, T cells and helper T cells release **interleukin-2**, which induces proliferation of cytotoxic T cells. (The secretion of interleukin-2 is stimulated by interleukin-1, which is released by macrophages).
  (c) Some helper T cells assist other T cells to respond to antigens.
  (d) The **delayed hypersensitivity T cell** is a type of helper T cell that produces substances (lymphokines) important in certain allergic (hypersensitivity) reactions and rejection of transplants.
(3) **Suppressor T cells**, after being activated by helper T cells, suppress B and T cell responses. In a self-regulatory feedback circuit, helper T cells themselves are inhibited by suppressor T cells.
(4) **Lymphokines** are a variety of substances produced by T cells that function as immune response modifiers. Their release mobilizes cell-mediated immunity against foreign material. Their effects are summarized in Table 12–1.

4. **Natural killer (NK) cells** are a distinct population of non-B, non-T lymphocytes with cytotoxic properties.
   a. NK cells need not interact with antigens or lymphocytes to destroy certain cells. They can spontaneously lyse a target cell in the absence of MHC restriction, antibodies, complement, or lymphokines.
   b. NK cells are believed to be a critical component of the body's **natural surveillance system** against cancer cells that arise at primary or metastatic sites. They also participate in protection against certain viruses, fungi, and parasites.

Table 12–1. Some of the T cell lymphokines and their effects.

Lymphokine	Function
Macrophage activation factor	Activates macrophages to improve phagocytosis
Macrophage chemotactic factor	Attracts macrophages to site of infection
Macrophage migration inhibition factor	Prevents macrophages from leaving site of infection
Interleukin-1	Promotes activation and proliferation of B cells and T cells; assists in triggering fever response
Interleukin-2	Activates proliferation and differentiation of T cells
Interferon	Inhibits replication of virus
Lymphotoxin	Function uncertain in the body; destroys target cells in vitro
Transfer factor	Enhances the action of sensitized T cells

F. Damaging effects of the immune response
 1. **Hypersensitivity, or allergy,** is the immune response that occurs in some people against certain substances that, although foreign, are harmless to the body. The individual whose immune system acts excessively or inappropriately to produce pathological changes is said to be **hypersensitive**.
    a. Antigens that provoke a hypersensitivity response are known as **allergens**. Exposure to an allergen immunizes or **sensitizes** the individual so that a subsequent exposure results in an **allergic reaction**.
    b. **Immediate hypersensitivity** is an allergic reaction that appears within minutes or hours after reexposure to an antigen. There are three subdivisions of immediate hypersensitivity reactions.
       (1) **Type I (anaphylaxis) reactions** occur within a few minutes of reexposure in a sensitized person and results from the binding of host IgE to mast cells and basophils.
          (a) The allergen binds to the IgE-coated cells, which triggers their explosive release of vasoactive substances (anaphylaxis mediators) such as histamine, serotonin, and leukotrienes.
          (b) The mediators collectively cause an increase in capillary permeability, smooth muscle contraction, and mucus secretion.
          (c) Because mast cells and basophils are located in so many areas of the body, anaphylactic reactions may include **localized reactions** such as hives, eczema, red eyes, nasal congestion, itching, respiratory difficulties, GI tract distress, or severe cramping.
          (d) **Acute anaphylaxis (anaphylactic shock)** is a life-threatening reaction that includes the inability to breathe due to bronchiolar constriction and cardiovascular failure. It must be treated by immediate injection of epinephrine or antihistamines.
       (2) **Type II (cytotoxic) reactions** usually are mediated by complement. They involve the combination of antibodies (IgG or IgM) with antigens on blood cells or tissue cells. Examples of type II reactions are transfusion reactions or Rh incompatibility (erythroblastosis fetalis).
       (3) **Type III (immune complex) reactions** are mediated by aggregates (complexes) of antibody and antigen that accumulate and activate complement, platelets, and phagocytes to result in tissue damage. Examples of type III reactions include rheumatoid arthritis, systemic lupus erythematosus, and serum sickness.
    c. **Delayed hypersensitivity reactions (type IV reactions)** take 24 hours or longer to develop and are mediated by T cells and macrophages rather than by B cells and antibodies. Type IV examples include reaction to the tuberculin skin test, tissue transplant rejection, and allergic contact dermatitis.
 2. **Autoimmune diseases** result from the failure of immunological, self-tolerance, which causes an immune system response against the body's own cells. Examples of some diseases believed to be caused by autoimmune mechanisms include Addison's disease of the adrenal gland, thyroiditis, rheumatoid arthritis, multiple sclerosis, myasthenia gravis, noninsulin dependent diabetes, pernicious anemia, and systemic lupus erthyematosus.
 3. **Immunodeficiencies** are conditions that decrease the effectiveness of the immune system or in which there is inability to respond to antigens.
    a. **Congenital immune deficiency.** In rare cases, an individual is born with neither B nor T cells. Such a person has no protection against infection and must live in a sterile environment.
    b. **Acquired immune deficiency syndrome (AIDS)** is a viral disease caused by the human immunodeficiency virus (HIV). In HIV-infected individuals, the number of helper T cells is decreased and the immune system is weakened. The affected person thus is vulnerable to microorganisms that normally do not cause problems in healthy individuals (opportunistic infections) and to the development of cancers such as Kaposi's sarcoma.

Study Questions

Directions: Each question below has four suggested answers. Choose the **one best** response to each question.

1. All of the following are part of the nonspecific defense against infection EXCEPT
   (A) There is an increase in blood flow and capillary permeability to proteins.
   (B) Leukotrienes, kinins, histamine, and other chemicals are released by damaged tissue.
   (C) B cells are released from the bone marrow.
   (D) Macrophages engulf microorganisms.

2. T cells and B cells are similar in all of the following ways EXCEPT
   (A) Both B cells and T cells secrete antibodies.
   (B) Both B cells and T cells originate from stem cells in the bone marrow.
   (C) Both types of cells are lymphocytes.
   (D) Both are integral to the immune system.

3. The memory cells of the immune system
   (A) remain in circulation for a short time before they are destroyed
   (B) are responsible for a more rapid and greater response upon second exposure to an antigen
   (C) are nonspecific and respond to any antigenic determinant
   (D) function solely in cell-mediated responses

4. Which of the following is a true statement concerning antibodies?
   (A) Antibodies may be circulating proteins, polysaccharides, or large lipids.
   (B) Each antibody is made up of two polypeptide chains joined by a disulfide bond.
   (C) The constant regions of an antibody are the most important in antigen specificity.
   (D) The variable regions of an antibody form the antigen-binding sites.

5. IgG molecules
   (A) can pass from the mother to the fetus across the placenta to provide a newborn with immunity
   (B) are the least prevalent of the circulating antibodies
   (C) induce the formation of WBCs
   (D) are found in secretions such as saliva and tears

6. The immune system's ability to distinguish between "self" and "not self"
   (A) is not subject to failure throughout the lifetime of the individual
   (B) is acquired at puberty as a result of hormonal changes
   (C) largely is due to the presence of MHC antigens on T cells
   (D) increases with the age of the individual

Questions 7–10. Match each description with the appropriate term.
   (A) anaphylactic shock
   (B) autoimmunity
   (C) delayed hypersensitivity reaction
   (D) passive immunity

7. Gamma globulins obtained from an individual who has been exposed to a disease such as mumps are transferred to an individual who has not had the disease.

8. An acute and life-threatening allergic reaction mediated by IgE antibodies.

9. An immune response to one's own tissues or components that has consequences leading to disease.

10. A reaction to antigen, which is mediated by T cells and macrophages rather than by antibodies and which is exemplified by a tissue graft rejection.

Questions 11–14. Match the description with the appropriate term.
   (A) complement
   (B) pyrogens
   (C) lymphokines
   (D) Interferons

11. A group of proteins produced by cells in response to a viral stimulus, which are able to enhance and modify the immune response.

12. Soluble substances secreted by lymphocytes, which can act on the immune system in a variety of ways.

13. A series of circulating plasma proteins involved in the mediation of immune responses, which is activated in a cascade of reactions.

14. Chemical agents that cause a rise in body temperature.

Answers and Explanations

1. **The answer is C.** (II C; III B, E) B-cell proliferation is a specific response to a particular antigen. B cells are produced in the bone marrow and are transported by the circulation to secondary lymph organs (spleen, lymph nodes, and tonsils) where they encounter and respond to antigens. An increased blood flow and capillary permeability and the release of vasoactive chemicals by damaged tissue is a stage in the inflammatory response. Phagocytosis is a part of both nonspecific and specific responses.

2. **The answer is A.** (III E 1–4) Only B cells secrete antibodies to inactivate antigens. Both B and T cells stem from hematocytoblasts in the bone marrow, both are lymphocytes, and both play roles in the specific immune system.

3. **The answer is B.** (III E 2, 3) Memory cells are resting cells that persist and are capable of being activated for a subsequent accelerated response to a previously seen specific antigen. Memory cells are formed by both mature B cells and T cells.

4. **The answer is D.** (III B 2) The variable regions on the tips of the heavy and light chains of an antibody have different sequences of amino acids (variability) and are where the antigen-binding sites are localized. They are, therefore, the most important in the antigen specificity of the antibody. Each antibody is a protein composed of four polypeptide chains: two identical heavy chains and two identical light chains. The constant region of an antibody does not have varying sequences of amino acids; it is identical for all antibodies within a particular class of immunoglobulins.

5. **The answer is A.** (III B 2 b) IgG molecules are the most prevalent antibodies in the circulation and are the only class of immunoglobulin that passes through the placenta to provide maternal immunity to the fetus. IgA molecules are found in tears, saliva, and other body secretions. None of the classes of immunoglobulins induces the formation of leukocytes.

6. **The answer is C.** (III E 3, F 2) The ability to recognize "self" from "not self" takes place early in life when the T cells of a person become imprinted with the MHC-encoded antigens specific to that individual in the thymus gland. Thereafter, the T cells recognize their own MHC-encoded antigens on body cells and do not attack them unless a foreign antigen is displayed on their surface in association with their own MHC antigen.

7-10. **The answers are 7–D, 8–A, 9-B, and 10–C.** (III D 2; III F 1 b; III F 2; III F 1 c) Passive immunity is acquired when antibodies are transferred from one individual to another. It occurs in the fetus when maternal IgG antibodies cross the placenta to confer immunity on the newborn and can also be acquired when antibodies produced by an individual or animal immunized to a particular antigen are injected into another individual. Anaphylactic shock is a life-threatening acute allergic reaction, which involves respiratory and cardiovascular failure and must be treated immediately. Autoimmunity is the result of a breakdown of the self–not-self recognition and is a condition in which antibodies are produced against an individual's own tissues. A delayed hypersensitivity reaction is mediated by T cells, takes up to 48 hours to develop fully, and involves macrophages and the release of lymphokines.

11-14. **The answers are 11–D, 12–C, 13–A, and 14–B.** (II D 1; III E 3; II D 2, III B 3; II C 3) Interferons are produced by almost any type of cell in response to viral infection, immune stimulation, or a variety of chemical stimulants. They function to block viral replication and to modulate immune responses. Lymphokines are released by T cells upon their activation and have a variety of effects on other T cells and on the immune response. Complement refers to a group of blood-borne proteins, which are activated in a cascade of reactions and which enhance the inflammatory and immune response. Pyrogens are chemicals usually secreted by injured tissue cells and leukocytes in response to bacterial infection, but they can also be any agents that are able to increase body temperature. Pyrogens adjust the body's temperature control center in the hypothalamus to result in fever.

Respiratory System 13

I. INTRODUCTION

A. The respiratory system functions to supply **oxygen** (O_2) from the atmosphere to the body cells and to transport **carbon dioxide** (CO_2), produced by the body cells, to the atmosphere. The respiratory organs also function in the **production of speech** and play a role in **acid-base balance, body defenses** against foreign material, and the **hormonal regulation of blood pressure.**

B. Respiration involves the following processes:
   1. **Pulmonary ventilation** (breathing) is the passage of air into and out of the respiratory passages and the lungs.
   2. **External respiration** is the diffusion of O_2 and CO_2 between the air in the lungs and the pulmonary capillaries.
   3. **Internal respiration** is the diffusion of O_2 and CO_2 between the blood and the tissue cells.
   4. **Cellular respiration** is the utilization of O_2 by the body cells in the production of energy, and the release of the products of oxidation (CO_2 and water) by the body cells.

C. The respiratory tract consists of the passageways from the environment to the lungs.

II. FUNCTIONAL ANATOMY OF THE RESPIRATORY TRACT (Color Plate 13)

A. Nose and nasal cavity
   1. The **external nose** is pyramidal, with a root and a base. It consists of a framework of bone, hyaline cartilage, and fibroareolar tissue.
      a. The **nasal septum** divides the nose into the right and left sides of the nasal cavity. The anterior portion of the septum is cartilage.
      b. The **external nares** (nostrils) are bordered by the **nasal cartilages.**
         (1) The **lateral** nasal cartilages are located below the bridge of the nose.
         (2) The **greater alar** and the **lesser alar** nasal cartilages surround the nostrils.
      c. Bones of the nose
         (1) The **nasal bones** form the bridge and superior portion of both sides of the nose.
         (2) The **vomer** and the **perpendicular plate** of the **ethmoid** form the posterior part of the nasal septum.

(3) The floor of the nasal cavity is the **hard palate**, which is formed by the **maxilla** and the **palatine** bone.
(4) The roof of the nasal cavity is formed medially by the **cribriform plate of the ethmoid bone**, anteriorly by the **frontal** and **nasal** bones, and posteriorly by the **sphenoid** bone.
(5) The **superior, middle, and inferior nasal conchae (turbinates)** project medially from the lateral wall of the nasal cavity. Each concha is covered with mucous membrane (pseudostratified ciliated columnar epithelium), which contains mucus-producing glands and is richly supplied with blood vessels.
(6) The **superior, medial, and inferior meatuses** are the air passages of the nasal cavity, which are located beneath the conchae.
d. The four pairs of **paranasal sinuses** (frontal, ethmoid, maxillary, and sphenoid) are blind-ending sacs in the frontal ethmoid, maxillary, and sphenoid bones. They are lined with mucous membrane.
(1) The sinuses serve to **lighten the cranial bones, provide additional surface area** to the nasal passages for warming and humidifying the air, **produce mucus**, and **provide resonance** in the production of speech.
(2) The paranasal sinuses drain into the meatuses of the nasal cavity through small ducts, which lie higher in the body than the level of the sinus floors. In the erect position, the drainage of mucus into the nasal cavity may be impeded, especially in cases of sinus infection.
(3) The nasolacrimal duct from the tear gland opens into the inferior meatus.

2. **Nasal mucous membrane**
    a. **Structure**
        (1) The skin on the external surface of the nose, which contains hair follicles, sweat, and sebaceous glands, continues into the **vestibule** located just inside the nostrils. It contains hairs (**vibrissae**), which serve to filter particles out of the inhaled air.
        (2) Further back in the nasal cavity, the **respiratory epithelium** forms the mucosal lining of the rest of the nasal chamber. It consists of ciliated epithelium with goblet cells lying on a vascularized connective tissue layer and continues to line the respiratory tract down to the bronchi.
    b. **Functions**
        (1) **Filtering of small particles.** The cilia in the respiratory epithelium wave back and forth within a sheet of mucus. The movement and the mucus form a trap for particles and sweep them up to be swallowed, expectorated, or sneezed out.
        (2) **Warming and humidifying of incoming air.** Dry air is moisturized by evaporation of the serous and mucous secretions and warmed by heat radiation from the underlying blood vessels.
        (3) **Odor reception.** The olfactory epithelium, which is located in the upper part of the nasal cavity below the cribriform plate, contains specialized olfactory cells for the sense of smell.

B. The **pharynx** is a 5-in. muscular tube that extends from the base of the skull to the esophagus. It is divided into the nasopharynx, oropharynx, and laryngopharynx.

1. The **nasopharynx** is posterior to the nasal cavity, which opens into it through two **internal nares** (choanae).
    a. Two **eustachian (auditory)** tubes connect the nasopharynx with the middle ear. They serve to equalize air pressure on both sides of the tympanic membrane.
    b. The **pharyngeal tonsils (adenoids)** are accumulations of lymphatic tissue that lie close to the internal nares. Enlargement of the adenoids can obstruct the passage of air.

2. The **oropharynx** is separated from the nasopharynx by the muscular soft palate, an extension of the bony hard palate.
    a. The **uvula** ("little grape") is a small, conical process that hangs down from the middle of the lower border of the soft palate.
    b. The **palatine tonsils** are located on either side of the posterior oropharynx.
3. The **laryngopharynx** surrounds the openings of the esophagus and of the **larynx**, which is the gateway to the rest of the respiratory system.

C. The **larynx** (voice box) connects the pharynx with the trachea. It is a short tube shaped like a triangular box and is supported by nine cartilages; three paired and three unpaired (Figure 13–1).
   1. Unpaired cartilages
       a. The **thyroid cartilage** (Adam's apple) is proximal to the thyroid gland. It is usually larger and more prominent in males as a result of male hormones secreted during puberty.
       b. The **cricoid cartilage** is a smaller and thicker anterior ring that lies below the thyroid cartilage.
       c. The **epiglottis** is an elastic cartilage flap that is attached to the inner anterior border of the thyroid cartilage. During swallowing, the epiglottis reflexively covers the opening to the larynx to prevent the entrance of food and fluid.
   2. Paired cartilages
       a. The **arytenoid** cartilages are located above and on either side of the cricoid cartilage. They are attached to the true vocal cords, which are paired folds of stratified squamous epithelium.
       b. The **corniculate** cartilages are attached to the tips of the arytenoid cartilages.
       c. The **cuneiform** cartilages are tiny rods that help support the soft tissues.
   3. Two pairs of lateral folds divide the laryngeal cavity.
       a. The upper pair are the **ventricular folds** (false vocal cords), which do not function in sound production.
       b. The lower pair are the **true vocal cords**, which are attached to the thyroid cartilage above and to the arytenoid and cricoid cartilages below. The opening between them is the **glottis**.
           (1) During breathing, the vocal cords are **abducted** (drawn apart) by laryngeal muscles and the glottis is triangular-shaped.
           (2) During swallowing, the vocal cords are **adducted** (drawn together) and the glottis is a narrow slit.
           (3) Thus, the contraction of skeletal muscles regulates the size of the glottis aperture and the degree of tension of the vocal cords necessary for sound production.

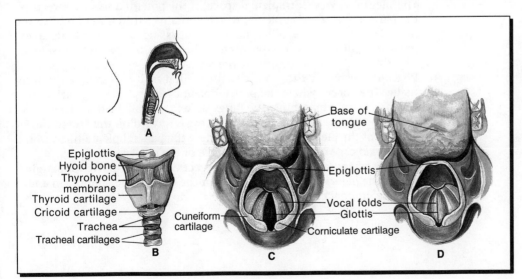

Figure 13–1. *Larynx. A*, Position in relation to the head and neck. *B*, Anterior view. *C*, Superior view of larynx with vocal cords open (glottis abducted). *D*, Superior view of larynx with vocal cords closed (glottis adducted).

D. The **trachea** (windpipe) is a tube that is 10 cm to 12 cm long and 2.5 cm in diameter and lies against the anterior surface of the esophagus. It extends from the larynx at the level of the sixth cervical vertebra to the level of the fifth thoracic vertebra where it divides into two primary bronchi.
   1. The **trachea** is kept permanently open by 16 to 20 C-shaped cartilagenous rings. The open posterior ends of the rings are bridged by connective tissue and muscle, which permit expansion of the esophagus.
   2. The trachea is lined with respiratory (pseudostratified ciliated columnar) epithelium that contains numerous goblet cells.

E. **Bronchial tree**
   1. The **right primary (main) bronchus** is shorter, thicker, and straighter than the **left primary bronchus** because the arch of the aorta displaces the lower trachea to the right. Foreign objects falling into the trachea are more likely to lodge in the right bronchus.
   2. Each primary bronchus branches 9 to 12 times to form **secondary and tertiary bronchi** and become progressively smaller in diameter. As the tubes become narrower, the cartilage rings are replaced by bars or plates of cartilage.
   3. Bronchi are termed **extrapulmonary** until they enter the lungs, after which they are called **intrapulmonary**.
   4. The essential structures of the two lungs are the rest of the bronchial tree: bronchi, bronchioles, terminal bronchioles, respiratory bronchioles, alveolar ducts, and alveoli. Cartilage disappears in the bronchioles; cilia remain until the smallest respiratory bronchioles.

F. **Lungs**
   1. The lungs are pyramid-shaped, spongy, air-filled organs, which are molded to the thoracic cavity that contains them.
      a. The right lung has three lobes; the left lung has two lobes.
      b. Each lung has an **apex** reaching above the first rib, a **diaphragmatic surface** (base) resting on the diaphragm, a **mediastinal** (medial) surface separated from the other lung by the mediastinum, and a **costal** surface lying against the rib cage.
      c. The mediastinal surface bears the **hilus** (root), through which the bronchi and the pulmonary and bronchial blood vessels enter and leave the lung.
   2. The **pleura** are the membrane coverings that enclose each lung.
      a. The **parietal pleura** lines the thoracic cavity (i.e., rib cage, diaphragm, and mediastinum).
      b. The **visceral pleura** covers the lung and is continuous with the parietal pleura at the root of the lung.
      c. The **pleural cavity** (intrapleural space) is the potential space between the parietal and visceral pleura, which contains a thin film of lubricating fluid. The fluid is secreted by the pleural cells and allows the lungs to expand without friction. The pressure of the fluid (intrapleural pressure) is slightly negative to atmospheric pressure.
      d. **Pleural recesses** are those areas of the pleural cavity unoccupied by lung tissue. They occur where the parietal pleura crosses from one surface to another. During breathing, the lungs slide in and out of these recesses.
         (1) The **costomediastinal pleural recess** is located on the anterior margin of the pleura on each side, where the parietal pleura turns from the rib cage onto the lateral surface of the mediastinum.
         (2) The **costodiaphragmatic pleural recess** is on the posterior margin of the pleura on each side between the edge of the diaphragm and the internal costal surfaces of the thorax.

III. MECHANICS OF BREATHING (PULMONARY VENTILATION)

A. Basic principles

1. The thorax is an airtight, closed cavity surrounding the lungs, which are open to the atmosphere through the passages of the respiratory system.
2. Breathing is the inspiration (inhalation) of air into the lungs and the expiration (exhalation) of air from the lungs to the outside.
3. Before inspiration begins, the air pressure in the atmosphere (approximately 760 mm Hg) is equal to the air pressure inside the alveoli, which is known as the intra-alveolar (intrapulmonary) pressure.
4. The intrapleural pressure in the pleural cavity (space between the pleura) is subatmospheric, or less than the intra-alveolar pressure.
5. An increase or decrease in the volume of the thoracic cavity **changes the intrapleural and intra-alveolar pressures**, which mechanically causes the lungs to inflate or deflate.
6. The muscles of inspiration enlarge the thoracic cavity and increase its volume. The muscles of expiration decrease the volume of the thoracic cavity.
   a. **Inspiration** requires muscle contraction and energy.
      (1) The **diaphragm**, which is a dome-shaped muscle when relaxed, flattens upon contraction and **enlarges the thoracic cavity inferiorly**.
      (2) The **external intercostal muscles** lift the ribs outward and upward upon contraction, which **enlarges the thoracic cavity anteriorly and superiorly**.
      (3) In active or deep breathing, the **sternocleidomastoids, pectoralis major, anterior serratus,** and **scalene** muscles further enlarge the thoracic cavity.
   b. **Expiration** in quiet breathing is effected by relaxation of muscles and is a passive process. In deep expiration, the internal intercostal muscles pull the rib cage downward and the abdominal muscles contract, which pushes the abdominal contents up against the diaphragm.

B. Factors in the inflation and deflation of the lungs

1. The negative **intrapleural pressure** in the pleural cavity holds the lungs in tight contact with the thoracic wall, because it produces a vacuum (suction) between the parietal pleura, which are attached to the thoracic wall, and the visceral pleura, which cover the surface of the lungs.
2. The **elastic tissue in the lungs** is responsible for their tendency to recoil away from the thoracic wall and collapse. They do not collapse in the body because the suction holding them against the thoracic wall is greater than the elastic forces in the lungs.
3. During inspiration and expansion of the thorax, the negative intrapleural pressure decreases still further (becomes more negative). The increased suction, assisted by the cohesion of the pleural fluid, pulls the lung surface outward toward the thoracic walls and assists in lung expansion.
4. As the lungs expand, the pressure of the air inside the lungs (intra-alveolar pressure) **drops below** the atmospheric pressure outside the body. The outside air is drawn through the respiratory passages into the lungs until the intra-alveolar pressure again equals the atmospheric pressure.
5. When the inspiratory muscles relax, the thoracic cavity decreases in size, the elasticity of the lungs pulls them inward, the intra-alveolar pressure **rises above** atmospheric pressure, and the air is expelled from the lungs.
6. **Surfactant** is a lipoprotein secreted by epithelial cells in the alveoli of the mature lung. The surfactant layer lies between the moist lining and the air in the alveolus. Surfactant reduces the surface tension of the fluid, which decreases the collapsing tendency of the alveoli and allows them to inflate under less pressure.
   a. Surfactant diminishes the surface tension more in smaller alveoli than in larger ones.

b. Because surfactant is not produced until late in fetal life, premature infants may be born with insufficient surfactant, collapsed alveoli, and breathing difficulties.
c. The condition, called **respiratory distress syndrome** (hyaline membrane disease), is treated by mechanical ventilating machines until the infant matures enough to produce sufficient surfactant.

7. **Compliance** refers to the distensibility of the lungs, or the ease with which they can be inflated (stretched). It is defined as a measure of the increase in lung volume produced by each unit of change in the intra-alveolar pressure. It is expressed as liters (volume of air) per centimeter of water (pressure).
    a. Decreased lung compliance requires a greater than normal difference in pressure created during inspiration to inflate the lungs. Any condition that impedes lung expansion and contraction decreases compliance and requires more work to inflate the lungs.
    b. Compliance is decreased by pulmonary diseases that cause changes in lung elasticity, pulmonary congestion or edema in the lungs, alteration of surface tension in the alveoli, or obstruction of the airway passages. It is also affected by deformity of the thoracic cage.

8. **Pneumothorax and atelectasis.** Normally, no air enters the pleural cavity. If air is allowed to enter the intrapleural space (as a result of a stab wound or broken rib) the condition is called **pneumothorax** ("air in the chest"). The consequent elimination of the negative pressure in the intrapleural space results in collapse of the lungs, which is known as **atelectasis**.

C. **Lung volumes and capacities.** The volume of air in the lungs and the rate at which it is exchanged during inspiration and expiration is measured with a **spirometer**. Lung volume values assume standard body temperature and ambient pressure and are measured in milliliters of air (Figure 13–2).

1. **Volumes**
    a. **Tidal volume** (TV) is the volume of air moved into and out of the lungs during effortless normal ventilation. In healthy young adults, TV is approximately 500 ml in males and 380 ml in females.
    b. **Inspiratory reserve volume** (IRV) is the extra volume of air moved into the lungs with maximum inspiration beyond a tidal inspiration. IRV amounts to approximately 3,100 ml in males and 1,900 ml in females.
    c. **Expiratory reserve volume** (ERV) is the extra volume of air that can be forcefully expired at the end of normal tidal expiration. ERV is usually approximately 1,200 ml in males and 800 ml in females.

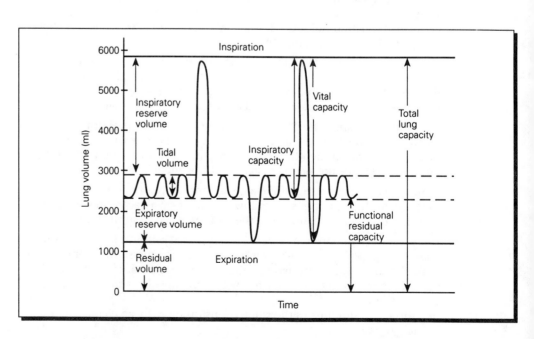

Figure 13–2. *Spirogram of lung volumes and capacities.*

d. **Residual volume** (RV) is the volume of air that remains in the lungs after the most forceful expiration. The residual volume is necessary to continue to aerate the blood between breaths. It averages 1,200 ml in males and 1,000 ml in females.

2. **Capacities**
   a. **Functional residual capacity** (FRC) is the sum of the residual volume and the expiratory reserve volume (**FRC = ERV + RV**). It is the amount of air remaining in the respiratory system after normal expiration. The average value is 2,200 ml.
   b. **Inspiratory capacity** (IC) is the sum of the tidal volume and the inspiratory reserve volume (**IC = TV + IRV**). The average value is 3,500 ml.
   c. **Vital capacity** (VC) is the sum of tidal volume, inspiratory reserve volume, and expiratory reserve volume (**VC = TV + IRV + ERV**). As measured by spirometry, it is the maximal amount of air that can be forcefully exhaled after a maximum inspiration. It is affected by such factors as **posture, size of the thoracic cavity**, and **lung compliance**, but an average value is 4,500 ml.
   d. **Total lung capacity** (TLC) is the total amount of air that the lungs can hold and equals the vital capacity plus the residual volume (**TLC = VC + RV**). The average value is 5,700 ml.

3. The **forced expiratory volume in one second** (FEV_1) is the volume of air that can be expired from maximally inflated lungs during the **first second** of forcible exhalation. Normally, FEV_1 is about 80% of VC.

4. The **minute respiratory volume** is the tidal volume times the number of breaths per minute.

D. **Alveolar ventilation rate** is the volume of new air that enters the alveolae per minute, and equals the TV times the respiration rate.

1. The air filling the conducting airway passages (nose, pharynx, trachea, bronchi, and bronchioles) does not participate in gas exchange and is referred to as **dead space air** and amounts to approximately 150 ml Upon inspiration of an average 500 ml TV, the first air entering the alveoli is the 150 ml of old dead space air, leaving only 350 ml of new air per breath.
   a. The **anatomical dead space** refers to conducting passages filled with dead space air.
   b. The **physiological dead space** includes all areas of non- or partially functioning alveoli as well as the anatomical dead space. In normal individuals, the physiological dead space equals the anatomical dead space.

2. The functional residual capacity of the lungs is about 2,400 ml, and only about 350 ml of new air is brought in with each breath; thus, only about **one seventh of the old alveolar air is exchanged with each breath.**

IV. GAS EXCHANGE

A. **Composition of atmospheric air.** Atmospheric air, at 760 mm Hg pressure and on a warm day, is composed of oxygen (21%), nitrogen (79%), carbon dioxide (0.04%), and various inert gases.

B. **Gas properties and the concept of partial pressures**
   1. In a mixture of gases, each of the gases exerts its own pressure in proportion to its percentage in the mixture, regardless of the presence of the other gases (**Dalton's law**).
   2. This pressure is the **partial pressure** (tension) of a gas in a mixture. It is designated by the symbol P in front of the chemical symbol for the gas. It is expressed in millimeters of mercury (mm Hg).

   Partial pressure of oxygen (PO_2) in the atmosphere:

   $$21/100 \times 760 \text{ mm Hg} = 160 \text{ mm Hg } PO_2$$

Partial pressure of carbon dioxide (PCO_2) in the atmosphere:

$$0.04/100 \times 760 \text{ mm Hg} = 0.3 \text{ mm Hg } PCO_2$$

3. The **solubility of a gas in water** varies with its pressure and temperature. Solubility increases with an increase in its partial pressure and decreases with an elevated temperature (**Henry's law**).
4. The **volume of a gas** varies inversely with the pressure upon it (**Boyle's law**). As the pressure is increased, the gas molecules are compressed and volume is reduced.

C. Pulmonary gas exchange
1. The **respiratory membrane**, through which gaseous exchange occurs, is composed of **surfactant layer, simple squamous epithelium of the alveolar wall, basement membrane of the alveolar wall**, the **interstitial space** containing connective tissue fibers and tissue fluid, the **capillary basement membrane**, and the **capillary endothelium**. Gas molecules must pass by diffusion through these six layers.
2. O_2 **and** CO_2 **move down their own partial pressure gradients** across the respiratory membrane.
   a. Gas molecules diffuse from an area of higher partial pressure to an area of lower pressure regardless of the concentration of other gases in solution; thus, the rate at which a gas diffuses across a membrane is determined by its partial pressure.
   b. The PO_2 in alveolar air is 100 mm Hg, while the PO_2 in the deoxygenated blood in the pulmonary capillaries surrounding the alveoli is 40 mm Hg. Thus, O_2 diffuses from the alveolar air across the respiratory membrane into the lung capillaries.
   c. The PCO_2 in alveolar air is 40 mm Hg and the PCO_2 in the surrounding capillaries is 45 mm Hg. Thus, CO_2 diffuses from the capillaries into the alveoli.
3. **Factors that affect gas diffusion** in addition to their pressure gradients include the following:
   a. **Thickness of the respiratory membrane.** Anything that increases the thickness, such as edema in the interstitial space or fibrous infiltration of the lungs as a result of pulmonary disease, significantly decreases diffusion.
   b. **Surface area of the respiratory membrane.** In diseases such as emphysema, large portions of the surface available for gas exchange are decreased and the gaseous exchange is severely impaired.
   c. **Solubility of the gases in the respiratory membrane.** CO_2 is 20 times more soluble than O_2. Thus, CO_2 diffuses through the membrane 20 times as rapidly as O_2.

V. TRANSPORT OF GASES BY THE BLOOD

A. **Oxygen transport.** Ninety-seven percent of the oxygen in the blood is carried in erythrocytes combined with hemoglobin (Hb). The other 3% is dissolved in the plasma.
1. Each of the four iron molecules in hemoglobin combines with one molecule of oxygen to form scarlet-red **oxyhemoglobin (HbO_2)**. The combination is loose and reversible. **Reduced hemoglobin (HHb)** is bluish-red in color.
2. The **oxygen capacity** is the maximum amount of oxygen that can combine with a given amount of hemoglobin in the blood.
   a. Each red blood cell contains 280 million molecules of hemoglobin. Each gram of hemoglobin can combine with 1.34 ml of oxygen.
   b. One hundred milliliters of blood contains an average 15 gm of hemoglobin for a maximum of 20 ml O_2 per 100 ml blood (15 × 1.34). This hemoglobin concentration is usually expressed as **volume percent** and is an amount sufficient for the body's needs.

3. The **oxygen saturation** of the blood is the **ratio** between the actual amount of oxygen bound to hemoglobin and the oxygen capacity:

$$\text{Oxygen saturation} = \frac{\text{oxygen content}}{\text{oxygen capacity}} \times 100$$

   a. Oxygen saturation is limited by the amount of hemoglobin or PO_2.

4. The **oxygen-hemoglobin dissociation curve** (Figure 13–3). The graph shows the percent of hemoglobin saturation on the vertical axis and the partial pressures of oxygen on the horizontal axis.
   a. The curve is S-shaped (sigmoidal) because hemoglobin's oxygen loading capacity (affinity for binding to oxygen) increases as its saturation increases. Similarly, its oxygen unloading (release of bound oxygen) increases as its saturation increases. Hemoglobin is 97% saturated at a PO_2 of 100 mm Hg, as occurs in alveolar air.
   b. The dissociation curve slopes steeply between 10 mm Hg and 50 mm Hg and flattens out between 70 mm Hg and 100 mm Hg. Thus, at higher PO_2 levels, hemoglobin saturation is affected very little by large changes, such as a drop in PO_2 to 50 mm Hg.
   c. As PO_2 decreases below 50 mm Hg, as in the body tissues, even slight changes in PO_2 greatly change hemoglobin saturation and the amount of oxygen released.
   d. Arterial blood normally carries oxygen at 97% of its capacity to do so.
      (1) Therefore, deep breathing or inhaling pure oxygen cannot increase significantly the **hemoglobin saturation with oxygen**.
      (2) Inhaling pure oxygen can increase **oxygen delivery to the tissues** because the amount of dissolved oxygen in the plasma would increase.
   e. In venous blood, the PO_2 is 40 mm Hg and the hemoglobin is still 75% saturated, indicating that blood gives up only about one fourth of its load of oxygen as it passes through the tissues. This provides a large margin of safety should breathing be interrupted or the tissue requirements for oxygen be increased.

5. The **affinity of hemoglobin for oxygen and the oxygen-hemoglobin dissociation curve** is affected by pH, temperature, and the concentration of 2,3-diphosphoglycerate (2,3-DPG).

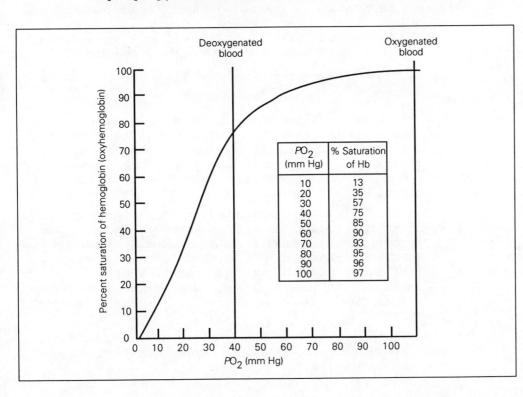

Figure 13–3. *Oxygen-hemoglobin dissociation curve at standard body conditions.* The pO_2 curve would shift to the right with increasing pCO_2, decreasing pH, increasing temperature, or increasing diphosphoglycerate concentration.

a. **Hemoglobin and pH.** An increase in blood PCO_2 or an increase in blood acidity (decrease in blood pH and increase in hydrogen ion concentration) weakens the bond between hemoglobin and oxygen and **shifts the curve to the right.** For any given PO_2, increased acidity of the blood causes hemoglobin to give up more oxygen to the tissues.
   (1) Actively metabolizing cells, as in exercise, release more CO_2 and hydrogen ions.
   (2) The effect of increased CO_2 and decreased pH is called the **Bohr effect.** The Bohr effect is greater at lower PO_2 values, as occurs in the tissues, and enhances oxygen release from hemoglobin for their use.
b. **Hemoglobin and temperature. Temperature elevation**, which occurs in the vicinity of actively metabolizing cells, also shifts the curve to the right and increases the delivery of oxygen to exercising muscles.
c. **Hemoglobin and DPG.** An increased concentration of **2,3-DPG**, a metabolite of glycolysis found in RBCs, decreases hemoglobin's affinity for oxygen and shifts the oxygen-hemoglobin dissociation curve to the right.
   (1) 2,3-DPG concentration is gradually increased when oxygen levels are chronically decreased, as in anemia or cardiac insufficiency. It reacts with hemoglobin and reduces the hemoglobin affinity for oxygen, thus making more oxygen available to the tissues.
   (2) 2,3-DPG concentration also is important in oxygen transfer from maternal to fetal blood. Fetal hemoglobin (hemoglobin F) has a greater affinity for oxygen than does adult hemoglobin (hemoglobin A), an alteration produced by 2,3-DPG in hemoglobin F.
6. P_{50} is a convenient index of shifts of the oxygen-hemoglobin dissociation curve. It is the PO_2 at which the hemoglobin is 50% saturated with oxygen. **The higher the P_{50}, the lower the affinity of hemoglobin for oxygen.**

B. **Carbon dioxide transport.** Carbon dioxide that diffuses into the blood from the tissues is carried to the lungs in the following ways:
1. A small amount of carbon dioxide (7% to 8%) remains **dissolved** in the plasma.
2. The remainder of the carbon dioxide moves into the RBC where approximately 25% combines in a loose, reversible form with the amino groups on the globin part of hemoglobin to form **carbaminohemoglobin.**
3. Most of the carbon dioxide is carried in the form of **bicarbonate**, mainly in the plasma.
   a. The carbon dioxide in the RBC combines with water to form carbonic acid in a reversible reaction catalyzed by **carbonic anhydrase.**

   $$CO_2 + H_2O \rightleftharpoons H_2CO_3 \rightleftharpoons H^+ + HCO_3^-$$
   Carbonic anhydrase

   b. The above equation proceeds in either direction, depending on the concentration of the compounds. When the concentration of CO_2 is high, as in the tissues, the reaction goes to the right and more hydrogen and bicarbonate ions are formed. In the lung, where the CO_2 concentration is lower, the reaction goes to the left and releases carbon dioxide to be expired.
4. **Chloride shift.** The negatively charged bicarbonate ions formed in the RBC diffuse out into the plasma, leaving excess positively charged ions inside.
   a. To maintain electrochemical neutrality, other negatively charged ions, which are mostly chloride ions, move into the red cells to restore the ionic equilibrium. This is known as the **chloride shift.**
   b. The chloride content of the RBCs in venous blood, which has a higher concentration of carbon dioxide, is necessarily greater than in arterial blood.
5. The positively charged hydrogen ions, which are released as a result of carbonic acid dissociation, combine with the hemoglobin within the RBC to minimize changes in pH.

VI. CONTROL OF RESPIRATION

A. **Nervous control.** Respiration is controlled by two separate neural mechanisms: a **voluntary system**, which originates in the **cerebral cortex** and controls breathing during activities such as speaking and eating, and an **involuntary system**, which is located in the medulla and pons and regulates respiration according to the metabolic needs of the body.

1. The **medullary respiratory centers** contain inspiratory and expiratory neurons that are located as a loose aggregation in the reticular formation of the medulla. They discharge to produce automatic respiration.
   a. The **inspiratory neurons** are located in the dorsal medulla. They send impulses to motor neurons that end on the muscles of inspiration. When the inspiratory neurons stop their activity, the inspiratory muscles relax and expiration occurs.
   b. The **expiratory neurons** are located in the ventral medulla (ventral medullary group). During active or forceful breathing, they discharge impulses to motor neurons that end on internal intercostal and abdominal muscles and facilitate in expiration.
   c. The exact mechanism to explain respiratory rhythm is unknown, but the basic pattern (two seconds for inspiration and three seconds for expiration) is believed to be generated by the inspiratory neurons with a reciprocal inhibition of impulses that occurs between the inspiratory and the expiratory neurons.

2. Pons respiratory centers
   a. The **pneumotaxic center** in the upper pons **limits the duration** of inspiration but **increases the rate** of respiration, resulting in rapid, shallow breathing.
   b. The **apneustic center** in the lower pons has a facilitating effect on inspiration.

3. Respiratory reflexes
   a. The **inflation reflex** (Hering-Breuer reflex, vagal reflex). Stretch receptors in the smooth muscle of the lungs are stimulated when the lungs inflate. They send inhibitory impulses along afferent vagus fibers to the medullary inspiratory neurons.
      (1) The reflex protects against overinflation of the lungs that might occur during vigorous exercise. It is not believed to be important in quiet breathing.
      (2) The reflex acts like the pneumotaxic center by decreasing the depth of respiration and increasing the rate.
   b. **Spinal reflexes.** Muscle spindles in the muscles of respiration monitor the length of the muscle fibers. Decreased shortening of the fibers is sensed and relayed to the spinal cord, which results in motor impulses for increased contraction.
   c. Air passage **irritation** from smoke, fumes, or particles in inhaled air leads to coughs and sneezes to dislodge the irritant.
   d. **Proprioceptor input** to the CNS from joints and tendons helps regulate respiration during exercise.

B. **Chemical control.** Chemoreceptors detect changes in the levels of oxygen, carbon dioxide, and hydrogen ions in arterial blood and cerebrospinal fluid, and cause appropriate adjustments in respiration rate and depth.

1. The **central chemoreceptors** are neurons located on the lateral ventral surface of the medulla.
   a. An increased CO_2 level in arterial blood and cerebrospinal fluid stimulates an increased rate and depth of respiration.
      (1) The primary stimulus to the central chemoreceptors is an increased hydrogen ion concentration (decreased pH) inside the neurons. The cell membranes of the neurons, however, are not very permeable to the inward diffusion of hydrogen ions.

(2) Carbon dioxide diffuses rapidly into the neurons, reacts with water, and forms carbonic acid, which dissociates into bicarbonate and **hydrogen ions**, which stimulate the central chemoreceptors.
   b. Decreased levels of oxygen have little effect on the central chemoreceptors.
 2. The **peripheral chemoreceptors** are located in the aortic and carotid bodies in the arterial system. They respond to changes in oxygen, carbon dioxide, and hydrogen ion concentrations.
    a. The peripheral receptors are particularly sensitive to decreases in oxygen. The aortic bodies respond to changes in oxygen bound to hemoglobin; the carotid bodies respond to changes in oxygen dissolved in the plasma.
    b. An increased hydrogen ion concentration (decreased pH) stimulates the peripheral chemoreceptors directly. Increased carbon dioxide also stimulates them, but the major effect of carbon dioxide is on the central chemoreceptors.

VII. RESPIRATORY PROBLEMS

A. **Hypoxia** (anoxia) is oxygen deficiency, which is a condition of less than physiologically normal oxygen in tissues and organs.
   1. Hypoxia may be the result of insufficient oxygen in the atmosphere; anemia (insufficient RBCs); impairment of blood circulation; lung diseases, which impair pulmonary ventilation; or the presence of toxic substances, such as carbon monoxide or cyanide, in the body.
   2. Carbon monoxide (CO) is toxic because it combines with hemoglobin at the same site as oxygen. Its tenacity for hemoglobin is about 230 times greater than that of hemoglobin for oxygen and it is liberated very slowly. Therefore, small amounts of carbon monoxide in air can be lethal.
B. **Hypercapnia** is increased CO_2 in the body fluids and it often accompanies hypoxia. Excessive CO_2 increases respiration and hydrogen ion concentration, which leads to acidosis (excessive amounts of acid).
C. **Hypocapnia** is decreased CO_2 in the blood, which is usually the result of hyperventilation (rapid breathing) and the blowing off of CO_2. Reduced CO_2 leads to alkalosis (excessive amounts of bicarbonate) in the body fluids.
D. **Asphyxia** is suffocation, a condition of hypoxia and hypercapnia, which results from inadequate pulmonary ventilation.
E. **Chronic obstructive pulmonary diseases (COPD)** are a group of diseases that include asthma, chronic bronchitis, and emphysema, as well as a group of industrial diseases including asbestosis, silicosis, and black lung. Prolonged exposure to cigarette smoking and/or environmental and industrial pollutants can cause COPD.
F. **Lung cancer (pulmonary carcinoma)** is often associated with smoking, but it also occurs in nonsmokers.
G. **Tuberculosis** is a disease caused by a bacterium that can affect any tissue of the body but most commonly lodges in the lungs.
H. **Pneumonia** is an acute infectious inflammatory process that results in fluid filling the alveoli. It may be caused by bacteria, fungi, protozoa, viruses, or chemicals.

Study Questions

Directions: Each question below contains four suggested answers. Choose the **one best** response.

1. The passageway common to both the respiratory and the digestive systems is the

   (A) pharynx
   (B) trachea
   (C) nasal cavity
   (D) esophagus

2. The C-shaped cartilage rings of the trachea function to prevent the trachea from collapsing and also to

   (A) provide a surface for gaseous exchange
   (B) allow for expansion of the esophagus during swallowing
   (C) remove foreign materials from the airway passages
   (D) minimize the formation of mucus

3. Which of the following has the smallest diameter?

   (A) right primary bronchus
   (B) left primary bronchus
   (C) respiratory bronchiole
   (D) bronchiole

4. The intrapleural cavity is

   (A) the mediastinum
   (B) normally filled with blood
   (C) a fluid-filled potential space
   (D) the first site of oxygen diffusion

5. When the thoracic cavity enlarges in both a superior-inferior and an anterior-posterior direction

   (A) the diaphragm is contracted
   (B) the glottis is closed
   (C) the air pressure in the alveoli increases
   (D) the intrapleural pressure increases

6. If an average man with normal tidal volume has a respiratory rate of 13 breaths per minute, what is his minute respiratory volume?

   (A) 3,800 ml/min
   (B) 6,500 ml/min
   (C) 7,800 ml/min
   (D) 1,300 ml/min

7. In major urban areas, tons of carbon monoxide are emitted into the atmosphere daily as a result of normal traffic. The following reasons why such levels of air pollution constitute a danger to public health all are true EXCEPT

   (A) Carbon monoxide and oxygen compete for the same binding site on the hemoglobin molecule.
   (B) Carbon monoxide reduces the amount of oxygen that is dissolved in plasma.
   (C) Carried per 100 ml of blood.
   (D) Carbon monoxide can be lethal because it interferes with the capacity of the blood to deliver oxygen to the tissues.

8. The chloride shift occurs when

   (A) oxygen is released from hemoglobin
   (B) oxygen combines with hemoglobin
   (C) bicarbonate ions leave the red blood cells
   (D) carbonic acid combines with water to form carbonic acid

9. What is most likely to happen when atmospheric air enters the left intrapleural space as a result of a bullet or stab wound?

   (A) Intrapleural pressure would decrease immediately.
   (B) The left lung would collapse.
   (C) Both lungs would collapse.
   (D) No adverse consequences would occur because the air slowly would diffuse out.

Questions 10–13 refer to the following diagram:

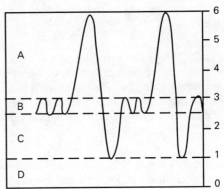

10. Tidal volume is represented by letter

    (A) A   (C) C
    (B) B   (D) D

11. Vital capacity is represented by letters

    (A) A plus B plus C plus D   (C) C plus D
    (B) B plus C   (D) A plus B plus C

12. Functional residual capacity is represented by letter(s)

    (A) D   (C) B
    (B) D plus C   (D) A plus B plus C

13. For the above spirogram, the inspiratory reserve volume equals

    (A) 0.5 l   (C) 3.0 l
    (B) 1.0 l   (D) 6.0 l

Answers and Explanations

1. **The answer is A.** (II B, see also Color Plate 13). The pharynx serves both the respiratory and the digestive systems. Because the trachea lies anterior to the esophagus, food must pass from the oral cavity posteriorly across the opening to the trachea to gain entrance to the esophagus and the stream of incoming air from the nasal chambers must pass anteriorly to gain access to the trachea.

2. **The answer is B.** (II D 1) The open posterior ends of the rings of tracheal cartilages are completed by bands of smooth muscle and connective tissue, which allow the esophagus to expand to receive food from the pharynx without being compressed by the trachea. Gaseous exchange occurs in the lungs. The cilia and the mucus in the upper respiratory tract facilitate the removal of foreign bodies from the respiratory tract.

3. **The answer is C.** (II E 1–4) All bronchi within the lung are intrapulmonary and they become progressively smaller in diameter as they branch. The respiratory bronchioles are the smallest branches of the bronchiole tree and give rise to the alveolar ducts of the alveoli. The right and left primary bronchi are extrapulmonary.

4. **The answer is C.** (II F 2) The intrapleural cavity between the parietal and visceral pleurae is a potential space that contains a thin film of pleural fluid formed by the pleural membranes. The fluid lubricates the pleural surfaces as they move against each other during breathing.

5. **The answer is A.** (III A 1–6, B 1–5) The thoracic cavity enlarges superiorinferiorly and antero-posteriorly when the diaphragm flattens on contraction and the external intercostal muscles elevate the rib cage. The increase in thoracic volume causes the lungs to inflate. The glottis is open, the intra-alveolar pressure decreases below atmospheric pressure, the intrapleural pressure decreases further, and the lungs expand in inspiration.

6. **The answer is B.** (III C 4) The tidal volume in males averages 500 ml. The minute respiratory volume is the TV times the number of breaths per minute.

7. **The answer is B.** (VII A 2) Carbon monoxide combines with hemoglobin at the same site on the hemoglobin molecule as does oxygen but about 230 times more tenaciously than does oxygen. Thus, it reduces the amount of oxygen carried per 100 ml of blood and interferes with the capacity of the blood to deliver oxygen to the tissues. Carbon monoxide has no effect on the amount of oxygen dissolved in plasma.

8. **The answer is C.** (V B 4) Almost all of the carbon dioxide released into venous blood enters the RBCs, where most of it combines with water to form carbonic acid in the presence of carbonic anhydrase. The carbonic acid dissociates to form hydrogen and bicarbonate ions. As the negatively charged bicarbonate ions diffuse out of the RBCs, they are replaced by negatively charged chloride ions, a move known as the chloride shift.

9. **The answer is A.** (III B 8) The elasticity of the lungs accounts for a part of their tendency to recoil and pull away from the chest wall. The pressure produced by the fluid in the intrapleural space is normally about -4 mm Hg, which is the amount of negative pressure required to keep the lungs expanded to their normal size and to prevent collapse. The immediate effect of opening the left intrapleural space to atmospheric pressure is collapse of the left lung due to its elastic nature.

10–13. **The answers are 10–B, 11–D, 12–B, and 13–C.** (III C 1,2) Tidal volume is the volume of air moved into and out of the lungs during normal pulmonary ventilation. Vital capacity is the sum of the tidal volume, the inspiratory reserve volume, and the expiratory reserve volume, or the maximal amount of air that can forcefully be exhaled after a maximum inspiration. The functional residual capacity of the lung is the sum of the residual volume and the expiratory reserve volume, and is the amount of air remaining in the respiratory system after a normal expiration. The inspiratory reserve volume is the extra amount of air moved into the lungs beyond a normal inspiration and is 3.0 L on the spirogram depicted.

Digestive System   14

I. **INTRODUCTION.** The digestive system consists of the **digestive (alimentary) tract**, which is a long muscular tube extending from the mouth to the anus, and the **accessory organs**, which include the teeth, tongue, salivary glands, liver, gallbladder, and pancreas. Below the level of the diaphragm, the digestive tract may be referred to as the **gastrointestinal (GI) tract** (Color Plate 14).

   A. **Functions of the digestive system.** The primary function is to provide the body with food, water, and electrolytes by digesting nutrients to prepare them for absorption. Digestion is both mechanical and chemical, and includes the following processes:

   1. **Ingestion** is the taking of food into the mouth.
   2. **Mechanical mincing or grinding** of food takes place by action of the teeth. The food is then mixed with saliva to prepare it for swallowing (deglutition).
   3. **Peristalsis** is the involuntary waves of smooth muscle contraction that moves the ingested materials through the digestive tract.
   4. **Digestion** is the chemical hydrolysis (breakdown) of large molecules into small molecules so that absorption can occur.
   5. **Absorption** is the movement of the end products of digestion from the lumen of the digestive tract into the blood and lymphatic circulations so they can be utilized by body cells.
   6. **Egestion** (defecation) is the elimination of undigested wastes, along with bacteria, as feces from the digestive tract.

   B. **General plan of the digestive tract**
   1. The **wall of the tract** consists of four basic tissue layers from the lumen (central cavity) outward. The components of the layers in each region vary with the function of the region.
      a. The **mucosa** (mucous membrane) is composed of three layers.
         (1) The **epithelium** lining functions in protection, secretion, and absorption. At the oral and anal ends of the tract, the lining is nonkeratinized stratified squamous epithelium for protection. It consists of simple columnar epithelium with goblet cells in those areas specialized for secretion and absorption.
         (2) The **lamina propria** is areolar connective tissue that supports the epithelium. It contains blood vessels, lymphatics, lymph nodules, and several types of glands.
         (3) The **muscularis mucosae** consists of a thin inner circular layer and an outer longitudinal layer of smooth muscle.

b. The **submucosa** consists of areolar connective tissue containing blood vessels, lymphatics, some submucosal glands, and a plexus of nerve fibers and ganglion cells called **Meissner's plexus** (submucosal plexus). The submucosa binds the mucosa to the muscularis externa.
c. The **muscularis externa** is comprised of two layers of muscle, an **inner circular layer** and an **outer longitudinal layer.** Contraction of the circular layer constricts the lumen of the tract and contraction of the longitudinal layer shortens and widens it. These contractions result in waves of **peristalsis** that move the contents of the tract ahead of them.
    (1) The muscularis externa consists of skeletal muscle in the mouth, pharynx, and upper esophagus, and of smooth muscle in the rest of the tract.
    (2) **Auerbach's plexus** (myenteric plexus), which consists of nerve fibers and parasympathetic ganglion cells, is located between the inner circular and outer longitudinal muscle layers.
d. The **serosa (adventitia)**, the fourth and outermost layer, is also called the **visceral peritoneum**. It consists of a serous membrane of loose connective tissue covered by simple squamous epithelium. Below the level of the diaphragm and in locations where the squamous epithelium is missing and the connective tissue merges with the surrounding connective tissue, it is known as **adventitia**.

2. The **abdominopelvic peritoneum, mesenteries, and omenta** are the most extensive serous membranes in the body.
    a. The **parietal peritoneum** lines the abdominopelvic cavity.
    b. The **visceral peritoneum** covers the organs and is connected by various folds to the parietal peritoneum.
    c. The **peritoneal cavity** is the potential space between the visceral and parietal peritoneums.
    d. **Mesenteries** and **omenta** are double-layered folds of peritoneal tissue that reflect back from the visceral peritoneum. They serve to bind the abdominal organs to each other and anchor them to the back abdominal wall. Blood vessels, lymphatics, and nerves are carried in the peritoneal folds.
        (1) The **greater omentum** is a large, double fold attached to the duodenum, stomach, and large intestine. It hangs down like an apron over the intestine.
        (2) The **lesser omentum** suspends the stomach and duodenum from the liver.
        (3) The **mesocolon** attaches the colon to the back abdominal wall.
        (4) The **falciform ligament** attaches the liver to the front abdominal wall and the diaphragm.
    e. Organs that are not enclosed by peritoneum but are only covered by it are **retroperitoneal** (behind the peritoneum). They include the pancreas, duodenum, kidneys, rectum, urinary bladder, and some of the female reproductive organs.

C. **Nervous control of the digestive tract.** The ANS innervates the entire digestive tract, except for the upper and lower ends, which are under voluntary control.
    1. **Parasympathetic impulses,** which are carried in the vagus nerve (CN X), exert a constant stimulating effect on smooth muscle tone and are responsible for **increasing overall activity**. This includes motility and the secretion of digestive juices.
    2. **Sympathetic impulses,** which are carried from the spinal cord in the splanchnic nerves, inhibit contractions of smooth muscle of the tract, decrease motility, and inhibit the secretion of digestive juices.
    3. **Meissner's and Auerbach's plexuses** are the sites of synapses for parasympathetic preganglionic fibers. They also function in local regulation of the contractile and secretory activity of the tract.

II. ORAL CAVITY, PHARYNX, AND ESOPHAGUS

A. The **oral cavity** is the entry to the digestive system and contains accessory organs that function in the initiation of digestion. The **vestibule** (buccal) cavity lies between the lips and cheeks on the outside and the teeth. The main oral cavity is bounded by the teeth and gums in front, the hard and soft palate above, the tongue below, and the oropharynx at the rear.

1. The **lips** are comprised of skeletal muscle (orbicularis oris) and connective tissue. They function in the acquisition of food and in the production of speech.
    a. The **outer surface** of the lips is covered by skin with hair follicles, and sweat and sebaceous glands.
    b. The **transitional area** has a transparent epidermis. It appears red because of the numerous capillaries that are visible through it.
    c. The **inner surface** is mucous membrane. The **labial frenulum** attaches the mucous membrane to the gums in the midline.

2. The **cheeks** contain the buccinator muscles of mastication. The epithelial lining of the cheeks is subject to abrasion and the cells are constantly sloughed off to be replaced by rapidly dividing new cells.

3. The **tongue** is attached to the floor of the mouth by the **lingual frenulum**. It functions in manipulating food for chewing and swallowing, for tasting, and in speech.
    a. The **extrinsic muscles** of the tongue originate on bone and tissue outside the tongue and function in movements of the whole tongue.
    b. The **intrinsic muscles** of the tongue have fibers that run in many directions at right angles to each other. This provides great mobility to the tongue.
    c. The **papillae** are elevations of mucosal and connective tissue on the dorsal surface of the tongue. They give it a rough texture.
        (1) The **fungiform** and **circumvallate papillae** bear **taste buds**.
        (2) A watery secretion from Von Ebner's glands, located in the tongue muscle, mixes with food on the surface and assists in taste reception.
    d. The **lingual tonsils** are aggregations of lymphoid tissue on the rear third of the tongue.

4. The **salivary glands** secrete saliva into the oral cavity. Saliva is composed of a watery enzyme-containing fluid and a thicker mucus-containing fluid.
    a. There are three pairs of salivary glands.
        (1) The **parotid glands** are the largest salivary glands. They lie below and in front of the ears and open via the **parotid (Stensen's) duct** into a little elevation (papilla), which is located opposite the second molar tooth on either side.
        (2) The **submaxillary (submandibular) glands** are approximately the size of walnuts. They lie against the inner surface of the mandibles and open by **Wharton's ducts** into the floor of the mouth on either side of the lingual frenulum.
        (3) The **sublingual glands** lie in the floor of the mouth and open by the **lesser sublingual ducts** into the floor of the mouth.
    b. Composition of saliva. Saliva is primarily a **serous secretion**, which is 98% water and contains the enzyme amylase and various ions (sodium, chloride, bicarbonate, and potassium), and a lesser, more viscous **mucous secretion**, which contains glycoprotein (mucin), ions, and water.
    c. Functions of saliva
        (1) Saliva **dissolves food chemicals** for taste perception.
        (2) Saliva **moistens and lubricates food** so it can be swallowed, and it provides moisture for the lips and tongue to prevent them from drying.
        (3) Salivary amylase **breaks starch down to polysaccharides and maltose**, a disaccharide.
        (4) Wastes, such as uric acid and urea, and many other substances, such as drugs, viruses, and metals, are **excreted** into the saliva.

(5) **Antibacterial and antibody substances** in saliva have a cleansing action in the oral cavity and help to maintain oral health and prevent tooth decay.
   d. **Neural control** of salivary secretion
      (1) The flow of saliva may be initiated by **psychic** (the thought of food), **mechanical** (the presence of food), or **chemical** (the kind of food) stimuli.
      (2) The stimuli are carried by afferent fibers in cranial nerves V, VII, IX, and X to **inferior** and **superior salivatory nuclei** in the medulla. All salivary glands are supplied with both sympathetic and parasympathetic fibers.
      (3) The amount and composition of the saliva vary with the kind of stimulus and with whether the sympathetic or parasympathetic system is innervated.
         (a) **Parasympathetic stimulation** results in vasodilation of blood vessels and copious amounts of **watery (serous) secretion**.
         (b) **Sympathetic stimulation** results in vasoconstriction of blood vessels and thicker, more scant, **mucous secretion**. Drugs that contain cholinergic (parasympathetic neurotransmitter) blockers produce a dry mouth sensation.
         (c) Approximately 1 ml per minute of saliva is secreted in a normal individual. This may reach 1 L to 1.5 L in 24 hours.
5. **Teeth** are set into bony sockets (alveoli) in the mandible and maxilla.
   a. **Anatomy of teeth**
      (1) Each of the curved rows of teeth in the jaws makes up a **dental arch**. The upper arch is larger than the lower so the upper teeth normally overlap the lower teeth.
      (2) Humans develop two sets of teeth: the primary teeth (deciduous, milk teeth) and the secondary teeth (permanent).
         (a) The **primary teeth** in one half of the dental arch (beginning from the space between the two front teeth) consist of two incisors, one canine, and two molars, for a total of 20 teeth.
         (b) The **secondary teeth** begin to erupt in the fifth to sixth year. One half of the dental arch consists of two incisors, one canine, two premolars (bicuspids), and three molars (tricuspids), for a total of 32 teeth. The third molar is the "wisdom tooth."
      (3) **Components of a tooth** (Figure 14–1)
         (a) The **crown** is the visible portion of the tooth. One to three buried **roots** comprise the portion of a tooth embedded into the **alveolar process** (socket) of the jawbone.

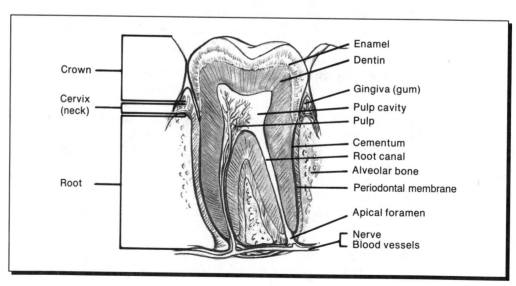

Figure 14–1. *Section through a molar tooth.*

- (b) The crown and root meet at the **neck**, which is surrounded by the **gingiva** (gum).
- (c) The **peridontal membrane** is connective tissue that lines the alveolar socket and attaches to the **cementum** on the root(s). It anchors the tooth in the jaw.
- (d) The **pulp chamber** within the crown extends into **root canals**, which are filled with dental pulp containing blood vessels and nerves. The root canal opens into the bone through the **apical foramen.**
- (e) **Dentin** surrounds the pulp chamber and forms the bulk of the tooth. It is covered in the crown portion by **enamel** and by **cementum** in the root portion. Enamel is 97% inorganic (mostly calcium phosphate) and is the hardest substance in the body. It is protective, but can be eroded by enzymes and acids produced by mouth bacteria to result in **dental caries**. Fluoride in drinking water or applied to the teeth strengthens enamel.
  - b. **Function of teeth.** Teeth function in **mastication** (chewing). The food taken into the mouth is minced into smaller pieces and mixed with the saliva to form a **bolus** of food that can be swallowed.
- B. **Swallowing (deglutition) moves the food from the pharynx into the esophagus.** The act of swallowing involves three phases.
  1. **Voluntary phase.** The tongue pushes up against the hard palate with the jaws closed and directs the bolus of food to the **oropharynx.**
  2. **Pharyngeal phase.** The bolus of food in the pharynx stimulates receptors in the oropharynx that send impulses to the **swallowing center** in the medulla and lower pons. Reflexes are initiated that seal off all other openings and leave the esophagus open to receive the food.
     - a. The **tongue presses against the hard palate** and blocks the food from returning to the mouth.
     - b. The **muscles of the soft palate and uvula raise the soft palate** to cover the openings to the nasal passages so the food cannot enter the nasal cavity.
     - c. The **larynx is elevated**, the **glottis is closed**, and the **epiglottis tips backward** over the opening to the larynx, which deflects food from entering the respiratory tract.
     - d. The **upper esophageal sphincter** at the entrance to the esophagus normally narrows to prevent air from entering the esophagus, and relaxes reflexly when the pharyngeal muscles contract and the larynx is elevated.
     - e. A **peristaltic wave of contraction** that begins in the pharyngeal muscles moves the bolus into the esophagus.
  3. **Esophageal phase.** The **lower esophageal sphincter**, a narrowed area of smooth muscle at the lower end of the esophagus in constant tonic contraction, relaxes ahead of the peristaltic wave and allows the food to be propelled into the stomach. The sphincter then constricts to prevent regurgitation (reflux) of stomach contents back into the esophagus.
- C. **Esophagus**
  1. **Anatomy.** The esophagus is a muscular tube, 9 in. to 10 in. (25 cm) long and one in. (2.54 cm) in diameter. It originates at the level of the laryngopharynx, passes through the diaphragm and through the **esophageal hiatus** (opening) at the level of the tenth thoracic vertebra, and opens into the stomach.
  2. **Function.** The esophagus transports food from the pharynx to the stomach by peristalsis. The esophageal mucosa produces large amounts of mucus to lubricate and protect the esophagus. No digestive enzymes are produced by the esophagus.

III. STOMACH

A. **Anatomy** (Figure 14–2)
1. The stomach is a J-shaped organ located in the superior left part of the abdominal cavity under the diaphragm. All but a small portion lies left of the median line. Its size and shape varies with the individual. The regions of the stomach are the cardiac, fundus, body, and pyloric portions.
   a. The **cardiac portion** of the stomach is the area surrounding the junction of the esophagus and the stomach (gastroesophageal junction).
   b. The **fundus** is the portion that bulges upward to the left above the entrance of the esophagus.
   c. The **body** is the dilated portion below the fundus, which forms two thirds of the stomach. The concave medial border of the body is the **lesser curvature**; the convex lateral border of the body is the **greater curvature**.
   d. The **pyloric portion** of the stomach narrows at the lower end of the stomach and opens into the duodenum. The **pyloric antrum** leads into the **pyloric orifice**, which is encircled by the thick, muscular **pyloric sphincter**.
2. **Histology of the stomach wall.** The three basic tissue layers (mucosa, submucosa, and muscularis) are present with modifications.
   a. The **muscularis externa** in the fundic and body portions contains an additional oblique muscle layer. The extra muscle layer assists effective churning and mixing of the stomach contents.
   b. The **mucosa** is thrown up into prominent longitudinal folds (**rugae**), which permit stretching of the stomach wall. Rugae appear when the stomach is empty and are smoothed out when the stomach is distended with food.
   c. Located between the rugae are approximately 3 million **gastric pits** into which empty about 15 million **gastric glands**. Gastric glands, which are named for their location, produce daily 2 L to 3 L of **gastric juice**. Gastric juice contains digestive enzymes, hydrochloric acid, mucus, salts, and water.

B. **Functions of the stomach**
1. **Food storage.** The capacity of the stomach normally allows a substantial time interval between meals and the ability to store large quantities of food until it can be accommodated in the lower parts of the tract. The stomach is nonessential to life and can be removed, provided that meals are small and frequent.

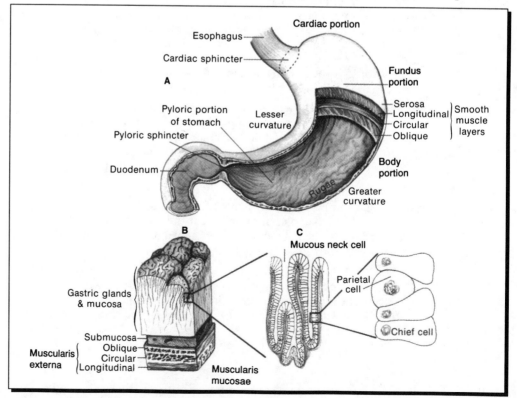

Figure 14–2. *Stomach. A,* Frontal section of the stomach and duodenum showing internal and external anatomy. *B,* Gastric mucosa. *C,* Gastric glands from the fundus. *D,* Detail showing chief and parietal cells.

2. **Chyme production.** Stomach activity results in the formation of chyme (the semifluid, acidic, homogeneous mass that was the former bolus) and propels it into the duodenum.
3. **Protein digestion.** The stomach initiates protein digestion through secretion of pepsin and hydrochloric acid.
4. **Mucus production.** The mucus produced by the glands forms a barrier over 1 mm thick that protects the stomach from the digestive action of its own secretions.
5. **Intrinsic factor production**
   a. Intrinsic factor is a glycoprotein secreted by the parietal cells.
   b. Vitamin B_{12}, liberated from ingested food in the stomach, is bound to intrinsic factor. The intrinsic factor vitamin B_{12} complex is carried to the ileum of the small intestine, where the vitamin B_{12} is absorbed.
6. **Absorption.** Little absorption of nutrients takes place in the stomach. Some lipid-soluble drugs (aspirin) and alcohol are absorbed through the stomach wall. Water-soluble substances are absorbed in insignificant amounts.

C. **Stomach secretion**
   1. **Types of gastric glands**
      a. **Cardiac glands** are found in the region of the cardiac orifice. They secrete only mucus.
      b. **Fundic (gastric) glands** contain three types of cells
         (1) **Chief (zymogenic) cells** secrete **pepsinogen**, the precursor of the enzyme pepsin. They also secrete **gastric lipase** and **rennin**, which are of lesser importance.
         (2) **Parietal cells** secrete **hydrochloric acid** (HCl) and **intrinsic factor.**
            (a) In the production of HCl, CO_2 moves into the cell where it combines with water to form carbonic acid (H_2CO_3) in a reaction catalyzed by carbonic anhydrase.
            (b) H_2CO_3 is ionized to form H^+ and HCO_3^-. The bicarbonate ion leaves the cell in exchange for chloride (CL^-) and enters the systemic circulation.
            (c) The hydrogen ions, accompanied by the chloride ions, are actively pumped into the stomach.
         (3) **Mucous neck cells** are found in the necks of all stomach glands. They secrete a barrier of **mucus**, which is over 1 mm thick and protects the stomach lining from damage by HCl or autodigestion.
      c. **Pyloric glands** are located in the pyloric antrum region. They secrete **mucus** and **gastrin**, a peptide hormone that is the major influence on stomach secretion.
   2. The **three stages of stomach secretion** are named for the regions in which the stimuli act. Both neuronal and hormonal factors are involved.
      a. The **cephalic stage** occurs before food reaches the stomach. It is elicited by food in the mouth or the sight, smell, or thought of food, and it stimulates gastric secretion.
      b. The **gastric stage** occurs when food reaches the stomach and lasts as long as food is present.
         (1) **Distention of the stomach wall** stimulates nerve receptors in the gastric mucosa and initiates **gastric reflexes**. Afferent fibers are carried to the medulla by the vagus nerve. Parasympathetic efferent fibers travel in the vagus to the gastric glands and stimulate production of HCl, digestive enzymes, and **gastrin**.
         (2) **Amino acids and proteins** in the partially digested food and chemicals (alcohol and caffeine) also promote gastric secretion through **local reflexes.**
         (3) **Functions of gastrin** include the following:
            (a) Gastrin **stimulates gastric secretion.**
            (b) It **increases gastric and intestinal motility.**
            (c) It **constricts the lower esophageal sphincter** and **relaxes the pyloric sphincter.**

(d) Additional effects, such as stimulation of pancreatic secretion and the enhancement of intestinal motility, also have been attributed to gastrin.

(4) Regulation of gastrin release in the stomach is via feedback inhibition based on the pH of the stomach contents.

(a) When no food is in the stomach between meals, the pH is low and gastric secretion is limited.

(b) Food entering the stomach has a buffering effect, which produces an increase in pH and an increase in gastric secretion.

c. The **intestinal stage** occurs after chyme has left the stomach and has entered the small intestine, where it initiates neural and hormonal factors.

(1) Gastric secretion is stimulated to continue for several hours by the secretion of **duodenal gastrin**, which is produced by the upper part (duodenum) of the small intestine and is carried in the circulation to the stomach.

(2) Gastric secretion is inhibited by polypeptide hormones produced by the duodenum. The hormones, which are carried in the circulation to the stomach, are secreted in response to stomach acidity below pH 2 and in the presence of fatty foods. They include **gastric inhibitory polypeptide (GIP), secretin, cholecystokinin (CCK)**, and the putative hormone **enterogastrone**.

D. **Digestion in the stomach.** The gastric juices **initiate** the digestion of proteins and fats.

1. **Digestion of proteins. Pepsinogen** (secreted by the chief cells) is converted to pepsin by **hydrochloric acid** (secreted by the parietal cells). Pepsin is a proteolytic enzyme, which can only act below pH 5. It hydrolyzes proteins to polypeptides. Infant stomachs produce **rennin**, an enzyme that coagulates milk protein, breaking it down to form curds.

2. **Fats.** Gastric lipase (secreted by chief cells) hydrolyzes milk fats to fatty acids and glycerol, but it has limited activity at low pH levels.

3. **Carbohydrates.** The salivary amylase that initiated the hydrolysis of starch in the mouth functions at a neutral pH. It is carried down with the bolus and continues to act in the stomach until the stomach acidity penetrates into the bolus. **No enzymes to digest carbohydrates are secreted by the stomach.**

E. **Control of stomach emptying**

1. Emptying is stimulated reflexly in response to **distention, gastrin release, fluidity of the chyme,** and **type of food.** Carbohydrates pass through rapidly, proteins less rapidly, and fats remain in the stomach for 3 to 6 hours.

2. **Gastric emptying is inhibited** by the same duodenal hormones that inhibit gastric secretion and by the **enterogastric feedback reflex** from the duodenum. These hormonal and nervous factors protect against intestinal overloading and provide more time for digestion in the small intestine.

3. The feedback signals allow chyme to enter the small intestine only at the rate it can be processed.

IV. SMALL INTESTINE (Figure 14–3)

A. **General features.** The entire small intestine is a coiled tube that extends from the pyloric sphincter to the ileocecal valve, where it joins the large intestine. It measures 2.5 cm in diameter and approximately 3 m to 5 m in length in life. The length is 7 m in a cadaver when the muscularis externa coat is relaxed.

B. **Divisions**

1. The **duodenum** is the shortest portion (25 cm to 30 cm). Both the common bile duct and the pancreatic duct open together into the posterior wall of the duodenum a few centimeters below the pyloric opening.

2. The **jejunum** is the next portion. It extends for about 1 m to 1.5 m.
3. The **ileum** (2 m to 2.5 m) continues to the junction with the large intestine.

C. **Motility.** Movements of the small intestine mix the contents with enzymes for digestion, allow digestive end products to come into contact with absorptive cells, and move the residue further down into the large intestine. They are initiated by distention and reflexly controlled by the ANS.

1. **Rhythmic segmentation** is the main mixing movement. Segmentation mixes chyme with digestive juices and exposes it to absorptive surfaces. It refers to alternating rings of constriction and relaxation of the muscle of the intestinal wall, which divides the contents into segments and pushes the chyme back and forth from one relaxed segment to the other.
2. **Peristalsis** is the rhythmic contraction of the longitudinal and circular smooth muscle. It is the main propelling force that moves chyme down the length of the tract.

D. **Microscopic anatomy of the intestinal wall**

1. Three structural specializations increase the absorptive surface of the small intestine by approximately 600 times.
    a. The **plicae circulares** are large, permanent, circular folds of the mucous membrane. They extend almost entirely around the lumen.
    b. **Villi** are millions of finger-like projections (0.2 mm to 1.0 mm tall) that extend into the lumen from the mucosal surface. Villi are found exclusively in the small intestine; each villus contains a capillary network and a lymph vessel called a **lacteal**.
    c. **Microvilli** are microscopic projecting folds of the cell membrane that occur on the exposed borders of the epithelial cells.

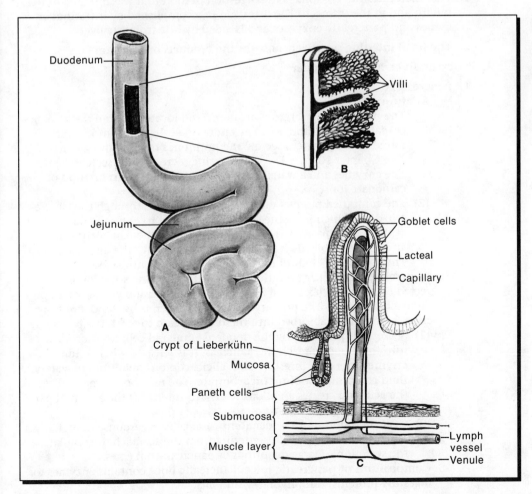

Figure 14–3. *Portion of the small intestine. A,* Part of the duodenum and jejunum. *B,* Villi. *C,* Section through a villus illustrating the capillary network, lacteal, and the relationship of the intestinal glands.

2. **Glands**
   a. **Intestinal glands (crypts of Lieberkühn)** dip into the mucosa and open between the bases of the villi. They secrete enzymes and hormones.
      (1) The **enzymes** formed by intestinal epithelial cells are required for the completion of digestion. They will be described later.
      (2) The **hormones** influence digestive tract secretion and motility. They include the following:
         (a) **Secretin, CCK, and GIP** are involved in the inhibition of gastric gland secretion.
         (b) **Vasoactive intestinal peptide** has vasodilator and smooth muscle relaxation effects.
         (c) **Substance P** affects smooth muscle motor activity.
         (d) **Somatostatin** inhibits gastrin and hydrochloric acid secretion as well as the hypothalamic release of growth hormone releasing factor.
   b. **Mucus-producing glands**
      (1) **Goblet cells** are located in the epithelium along the entire length of the small intestine. They produce protective mucus.
      (2) **Brunner's glands** are located in the submucosa of the duodenum. They produce mucus to protect the duodenal mucosa from the acidic chyme and gastric juice that enters the pylorus from the stomach.
   c. **Enteroendocrine glands** produce gastrointestinal hormones.
3. **Lymphatic tissue.** Leukocytes and isolated lymph nodules occur throughout the small intestine to protect the intestinal wall against foreign invasion. Aggregations of lymph nodules called **Peyer's patches** occur in the ileum.

E. **Functions of the small intestine**
   1. The small intestine **completes** the digestion of food that was initiated in the mouth and in the stomach. The digestion is accomplished by **intestinal enzymes** and **pancreatic enzymes** and is aided by **bile** from the liver.
   2. The small intestine selectively **absorbs** the products of digestion.

F. **Pancreas, liver, and gallbladder**
   1. **Pancreas**
      a. **Anatomy**
         (1) The **pancreas** is a large, elongated gland located behind the greater curvature of the stomach. The endocrine cells (**islets of Langerhans**) of the pancreas secrete the hormones **insulin** and **glucagon** (See Chapter 10 VI A–D). The exocrine (**acinar**) cells secrete digestive enzymes and a watery solution with a high concentration of bicarbonate ions.
         (2) The combined product of the acinar cells flows through the pancreatic duct, which joins the common bile duct to enter the duodenum at the **hepatopancreatic ampulla**, although the pancreatic duct and common bile duct may open separately into the duodenum. The **sphincter of Oddi** normally keeps the opening closed.
      b. **Control of pancreatic secretion.** Pancreatic exocrine secretion is affected by neural reflex activity during the cephalic and gastric stages of stomach secretion. The primary control, however, is by **duodenal hormones**, which are absorbed into the bloodstream to reach the pancreas.
         (1) **Secretin** is produced by the duodenal mucosal cells and absorbed into the blood to reach the pancreas. It is released when **acidic chyme** enters the intestine and elicits copious quantities of watery fluid containing **sodium bicarbonate**. The bicarbonate neutralizes the acid and creates an **alkaline environment** for the action of pancreatic and intestinal enzymes.
         (2) **CCK** is produced by duodenal mucosal cells in response to partially digested fats and proteins entering from the stomach. It stimulates the secretion of large quantities of pancreatic enzymes.
      c. **Composition of pancreatic juice.** Pancreatic juice contains enzymes for digesting proteins, carbohydrates, and fats.

(1) **Pancreatic proteolytic enzymes (proteases)**
    (a) **Trypsinogen** secreted by the pancreas is activated to **trypsin** by **enterokinase** produced by the small intestine. Trypsin digests proteins and large polypeptides to form smaller polypeptides and peptides.
    (b) **Chymotrypsin** is activated from chymotrypsinogen by trypsin. Chymotrypsin has the same action as trypsin on proteins.
    (c) **Carboxypeptidase, aminopeptidases,** and **dipeptidases** are enzymes that continue protein digestion to result in free amino acids.
(2) **Pancreatic lipase** hydrolyzes fats to **fatty acids** and **glycerol** after the fats have been emulsified by bile salts.
(3) **Pancreatic amylase** hydrolyzes starch not digested by salivary amylase to **disaccharides** (maltose, sucrose, and lactose).
(4) **Ribonuclease** and **deoxyribonuclease** hydrolyze RNA and DNA to their nucleotide building blocks.

2. **Liver and the secretion of bile**
   a. **Anatomy of the liver.** The liver is the largest visceral organ and is located under the rib cage. It weighs 1,500 gm (3 lbs) and in living condition is dark red because of its rich blood supply. It receives oxygenated blood from the hepatic artery and unoxygenated blood rich in nutrients from the hepatic portal vein. It is divided into **right and left lobes.**
      (1) The right lobe of the liver is larger than its left lobe and has three main parts: the right lobe proper, the caudate lobe, and the quadrate lobe.
      (2) The **falciform ligament** separates the right from the left lobe. Between the lobes is the **porta hepatis**, the entrance and exit for blood vessels, nerves, and ducts.
      (3) Within the lobes, plates of liver cells branch and anastomose to form a three-dimensional network. Sinusoidal blood spaces lie between the plates of cells. The **portal canals**, each containing a branch of the portal vein, hepatic artery, and bile duct, delineate a **portal lobule**.
   b. **Major functions of the liver**
      (1) **Secretion.** The liver produces **bile**, which acts in the emulsification and absorption of fats.
      (2) **Metabolism.** The liver **metabolizes digested protein, fat, and carbohydrate.**
         (a) The liver plays the major role in the homeostatic **maintenance of blood sugar.** It stores glucose in the form of **glycogen** and converts it back to glucose when it is needed by the body.
         (b) The liver **breaks down protein** from worn-out tissue cells and RBCs. It **forms urea from excess amino acids and nitrogenous wastes.**
         (c) The liver **synthesizes fats** from carbohydrate and protein, and is involved in the **storage and utilization of fats.**
         (d) The liver synthesizes the constituents of cell membranes (**lipoproteins, cholesterol,** and **phospholipids**).
         (e) The liver **synthesizes** plasma proteins and blood clotting factors. It synthesizes **bilirubin** from products of hemoglobin breakdown and secretes it into the bile.
      (3) **Storage.** The liver **stores minerals,** such as iron and copper, and **fat-soluble vitamins** (A, D, E, and K), and it stores certain toxins (e.g., pesticides) and drugs that cannot be broken down and excreted.
      (4) **Detoxification.** It **inactivates** hormones and **detoxifies** toxins and drugs. The liver **phagocytizes** disintegrating erythrocytes and foreign substances in the blood.
      (5) **Heat production.** The liver's many chemical activities make it a major **source of heat** for the body, especially during sleep.
      (6) **Blood storage.** The liver is a **reservoir** for about 30% of cardiac output and, along with the spleen, it regulates blood volume to body needs.

c. **Bile**
  (1) **Anatomy of bile secretion**
    (a) Bile produced by liver cells enters **bile canaliculi**, which converge into **right and left hepatic ducts.**
    (b) The hepatic ducts combine to form the **common hepatic duct**, which is joined by the **cystic duct** from the gallbladder to emerge from the liver as the **common bile duct.**
    (c) The common bile duct, along with the pancreatic duct, empties the duodenum or is diverted for storage to the gallbladder.
  (2) **Composition of bile.** Bile is a greenish-yellow solution consisting of 97% water, bile pigments, and bile salts.
    (a) **Bile pigments** are biliverdin (green) and bilirubin (yellow). They are breakdown products of the hemoglobin released from disintegrating RBCs.
      (i) The predominant pigment is bilirubin, which provides the yellow color to urine and feces.
      (ii) **Jaundice**, or yellowish color in tissues, is the result of an elevated blood bilirubin level. It is an indication of **liver dysfunction** and may be caused by liver cell damage (hepatitis), increased destruction of RBCs, or obstruction of the bile ducts by gallstones.
    (b) **Bile salts** are formed from bile acids combined with cholesterol and amino acids. After being secreted into the intestine, they are reabsorbed from the lower ileum back to the liver and recycled. This is known as the **enterohepatic circulation** of the bile salts.
  (3) **Functions of bile salts** in the small intestine
    (a) **Emulsification of fats.** Bile salts emulsify large fat globules in the small intestine, which results in smaller fat globules and a greater surface area for enzymatic action.
    (b) **Absorption of fats.** Bile salts aid in the absorption of fat-soluble substances by facilitating their passage through cell membranes.
    (c) **Cholesterol removal from the body.** Bile salts combine with cholesterol and lecithin to form small aggregates called micelles to be eliminated in the feces.
  (4) **Control of bile secretion and flow.** The secretion of bile is regulated by the same neural (parasympathetic impulses) and hormonal (secretin and CCK) factors that control the secretion of pancreatic juice. When fatty acids and amino acids reach the small intestine, CCK is released to contract the muscles of the gallbladder and relax the sphincter of Oddi. Bile is then forced into the duodenum.

3. **Gallbladder**
  a. **Anatomy.** The gallbladder is a muscular, pear-shaped, green sac approximately 10 cm long. It is located in a depression under the right lobe of the liver. The total capacity of the gallbladder is about 30 ml to 60 ml.
  b. **Functions**
    (1) The gallbladder **stores the bile**, which is continually secreted by the liver cells, until it is needed in the duodenum. Between meals, the sphincter of Oddi is closed and bile flows into the relaxed gallbladder. Its release is stimulated by CCK.
    (2) The gallbladder **concentrates bile** by reabsorbing water and electrolytes. It is thus able to hold as much as 12 hours of liver biliary secretion.

G. **Absorption in the small intestine**
  1. **Digestion by intestinal enzymes.** Intestinal enzymes complete the digestion of chyme so that the products can be easily and thoroughly absorbed. The enzymes and their actions include the following (Table 14-1):
    a. **Enterokinase** activates pancreatic trypsinogen to **trypsin**, which breaks proteins and peptides into smaller peptides.
    b. **Aminopeptidases, tetrapeptidases, tripeptidases,** and **dipeptidases** break peptides down to **free amino acids.**

c. **Intestinal amylase** hydrolyzes starch to **disaccharides** (maltose, sucrose, and lactose).
   d. **Maltase, isomaltase, lactase, and sucrase** act on the disaccharides maltose, lactose, and sucrose, respectively, to produce **monosaccharides** (simple sugars).
   e. **Intestinal lipase** breaks monoglycerides into **fatty acids** and **glycerol**.
2. **Absorptive pathway.** The products of digestion (monosaccharides, amino acids, fatty acids, and glycerol) as well as water, electrolytes, vitamins, and digestive juices are absorbed across the epithelial cell membranes of the duodenum and jejunum. Little absorption takes place in the ileum except for bile salts and vitamin B_{12}.
3. The **transport mechanisms** for absorption include diffusion, facilitated diffusion, active transport, and pinocytosis. The major mechanism is active transport. Substances transported from the lumen of the intestine to the blood or lymph must cross through the following cells and intercellular fluids:
   a. The plasma membrane of the epithelial columnar cell of the villus, its cytoplasm, and its basement membrane
   b. The connective tissue between the epithelial cell and the capillary or lacteal within the villus
   c. The wall of the capillary or lacteal located in the core of the villus
4. **Carbohydrate absorption.** Each simple sugar is believed to have its own transport mechanism. Sugars move from the intestine to the villus capillary network and are carried to the liver by the hepatic portal vein.

TABLE 14–1. Digestion of carbohydrates, proteins, and fats.

Enzyme	Source of Secretion	Action
Carbohydrates		
Salivary amylase (ptyalin)	Salivary glands	Starch→Maltose
Pancreatic amylase	Pancreas	Starch→Disaccharides and maltose
Maltase	Small intestine	Maltose→Glucose
Sucrase	Small intestine	Sucrose→Glucose and fructose
Lactase	Small intestine	Lactose→Glucose and galactose
Proteins		
Pepsin	Stomach (Pepsinogen activated by gastric HCl)	Protein→Polypeptides
Trypsin	Pancreas (Trypsinogen activated by enterokinase)	Protein and peptides →Smaller peptides
Chymotrypsin	Pancreas (Chymotripsinogen activated by trypsin)	Protein and peptides →Smaller peptides
Peptidases	Small intestine	Dipeptides →Amino acids
Fats		
Pancreatic lipase	Pancreas (with bile salts)	Triglycerides →Monoglycerides and fatty acids
Intestinal lipase	Small intestine (with bile salts)	Monoglycerides →Fatty acids and glycerol

a. The absorption of **glucose** is coupled to the active transport of sodium ions (co-transport).
b. **Fructose** is transported by a carrier-mediated facilitated diffusion.
c. Other monosaccharides may be absorbed by simple diffusion.
5. **Protein absorption.** Amino acid active transport into the intestinal cells is also coupled to the active transport of sodium, with separate carrier systems for different amino acids. From the villus capillaries the amino acids are carried to the liver.
6. **Fat absorption.** Lipid-soluble fatty acids and glycerol are absorbed in the form of **micelles**, which are spherical globules of bile salts surrounding the fatty portion. The micelles ferry the fatty acids and monoglycerides to the epithelial cell, where they are released and absorbed by passive **diffusion** into the intestinal cell membrane.
    a. Short carbon-chain fatty acids (less than 10 to 12 carbon atoms) are small molecules and move into the villus capillaries along with the amino acids and monosaccharides.
    b. Long carbon-chain fatty acids (over 90% of fatty acids) and glycerol molecules move to the endoplasmic reticulum, are resynthesized back to triglycerides, packaged with lipoproteins, phospholipids, and cholesterol, and escape as **chylomicrons** from the intestinal cell lateral borders.
    c. Chylomicrons pass to the central lacteal of the villus to the lymphatic system and the systemic circulation, initially bypassing the liver.
7. **Water, electrolyte, and vitamin absorption**
    a. Only 0.5 L of the 5 L to 10 L of fluid that enters the small intestine reaches the large intestine. **Water** is passively absorbed by the laws of **osmosis** as it follows the absorption of electrolytes and digested foods.
    b. **Ions and trace elements** are absorbed by **diffusion** or **active transport**.
        (1) The absorption of **calcium** varies with dietary intake, plasma levels, and body needs and is regulated by parathyroid hormone and vitamin D ingestion.
        (2) **Iron** absorption is determined by metabolic requirements. It is bound to a globulin (**transferrin**) in the blood and stored in the body as **ferritin** to be released when required.
        (3) **Water-soluble vitamins (C and B)** are absorbed by diffusion. **Fat-soluble vitamins (A, D, E, and K)** are absorbed with fats. B_{12} absorption depends on gastric intrinsic factor and occurs in the ileum.

V. **LARGE INTESTINE.** By the time the material within the digestive tract gets to the large intestine, most of the nutrients have been digested and absorbed, leaving only undigestible substances. A typical meal takes 2 to 5 days to travel from one end of the digestive tract to the other: 2 to 6 hours in the stomach, 6 to 8 hours in the small intestine, and the rest of the time in the large intestine.

A. **General features**
    1. The large intestine has no villi, has no plicae circulares, and is wider in diameter, shorter in length, and more distensible than the small intestine.
    2. The longitudinal muscle fibers in the muscularis externa form three bands, the **taeniae coli**, that draw the colon into bulging sacs called **haustra**.
    3. The **ileocecal valve** is the sphincter opening between the small and large intestines. It is normally closed and opens in response to a peristaltic wave, which allows the passage of about 15 ml of chyme at a time, for a total of 500 ml per day.

B. **Parts of the large intestine** (Figure 14–4)
    1. The **cecum** is the pouch that hangs down below the level of the ileocecal valve. The **vermiform appendix**, a narrow blind-ending tube filled with lymphoid tissue, projects from the end of the cecum.

2. The **colon** is the portion of the large intestine from the cecum to the rectum. It has three divisions.
    a. The **ascending colon** extends from the cecum to the lower border of the liver on the right and turns horizontally at the **hepatic flexure**.
    b. The **transverse colon** extends across the abdomen below the liver and the stomach to the lateral border of the left kidney, where it turns downward at the **splenic flexure**.
    c. The **descending colon** extends down on the left side of the abdomen and becomes the **S-shaped sigmoid colon**, which empties into the rectum.
3. **The rectum** is the last 12 cm to 13 cm of the digestive tract. It terminates in the **anal canal** and opens to the exterior in the **anus**.
    a. The mucosa of the anal canal is arranged in rectal (anal) columns, which are vertical folds each containing an artery and vein.
    b. The **internal anal sphincter** of smooth muscle (involuntary) and the **external anal sphincter** of skeletal muscle (voluntary) surround the anus.

C. **Functions of the large intestine**
1. The large intestine **absorbs 80% to 90% of the water and electrolytes** from the remaining chyme and reduces the chyme from fluid to a semisolid mass.
2. The large intestine produces only mucus. Its secretions contain no digestive enzymes or hormones.
3. Numerous bacteria present in the colon are capable of digesting small amounts of cellulose and produce a few calories of nutrients to the body each day. The bacteria also produce vitamins (K, riboflavin, and thiamin) and various gases.
4. The large intestine excretes waste as feces.
    a. Feces are 75% to 80% water. The solid material is about one third bacteria and the rest consists of 2% to 3% nitrogen, and organic and inorganic residues from digestive secretions, which includes mucus and fat.
    b. Feces also contain variable amounts of roughage, or indigestible fibers and cellulose. The brown color is from bile pigments; the odor from bacterial action.

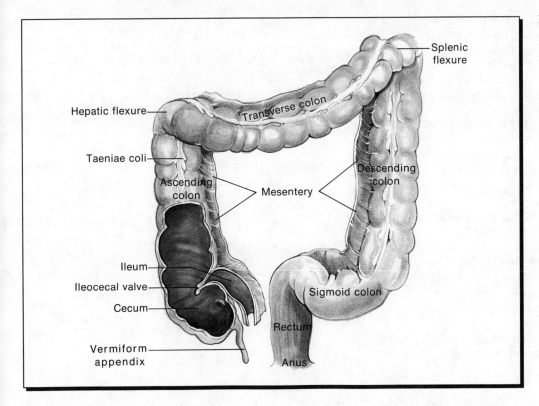

Figure 14–4. *Large intestine.*

Study Questions

Directions: Each question below contains four suggested answers. Choose the **one best** response.

1. If an incision were to be made in the stomach to remove a gastric ulcer, which of the following tissues would be cut first?

    (A) muscularis mucosa
    (B) visceral peritoneum
    (C) submucosa
    (D) lamina propria

2. Injury to which of the following nerves would lead to a decrease in digestive tract motility and secretions?

    (A) phrenic nerve
    (B) splanchnic nerve
    (C) intercostal nerve
    (D) vagus nerve

3. The composition of saliva is primarily

    (A) water
    (B) mucus
    (C) amylase
    (D) enzymes

4. All of the following statements concerning the esophagus are true EXCEPT

    (A) The esophagus is a muscular tube, 25 cm in length, that originates at the laryngopharynx and ends at the stomach.
    (B) The esophagus transports food from the oral cavity to the stomach by peristalsis.
    (C) The esophagus produces mucus and digestive enzymes to initiate the breakdown of proteins.
    (D) The lower esophageal sphincter muscle contracts to prevent regurgitation of gastric contents.

5. Which state best describes the relationship between the gastric mucosa and vitamin B_{12}?

    (A) The gastric mucosa secretes vitamin B_{12}.
    (B) The gastric mucosa stores vitamin B_{12}.
    (C) The gastric mucosa converts vitamin B_{12} into other B vitamins.
    (D) The gastric mucosa secretes a substance that promotes the absorption of vitamin B_{12}.

6. Cardiac glands of the stomach

    (A) are restricted to the initial portion of the stomach near the esophageal entry
    (B) are found throughout the stomach but are most numerous near the pyloric sphincter
    (C) contain chief and parietal cells
    (D) produce hydrochloric acid and enzymes

7. Stomach secretion is increased by all of the following EXCEPT

    (A) gastrin
    (B) protein
    (C) secretin
    (D) distention

8. The rapidity with which the stomach empties its contents into the duodenum is increased by all of the following EXCEPT

    (A) the volume of food in the stomach
    (B) the fat content of a meal
    (C) the secretion of gastrin
    (D) the stimulation of peristaltic action

9. The effective surface area for absorption in the small intestine is increased by all of the following EXCEPT

    (A) villi
    (B) plicae circulares
    (C) crypts of Lieberkühn
    (D) microvilli

10. The pancreas

    (A) secretes enzymes in response to stimulation by stomach gastrin from the duodenum
    (B) produces a secretion that neutralizes acidic chyme after it enters the small intestine
    (C) produces bile for the emulsification of fats
    (D) empties its secretion into the pyloric region of the stomach

Questions 11–14. For each of the questions, refer to the following reactions:

Reaction 1: protein —pepsin→ peptides

Reaction 2: starch —amylase→ maltose

Reaction 3: maltose —maltose→ glucose

Reaction 4: fat —lipase→ fatty acids and glycerol

11. In order for Reaction 4 to proceed

    (A) The pH must be 2 or lower.
    (B) Bile must be present.
    (C) Saliva must be secreted.
    (D) The gastric phase of stomach secretion must occur.

12. Which reaction takes place in a highly acid environment?

    (A) Reaction 1
    (B) Reaction 2
    (C) Reaction 3
    (D) Reaction 4

13. Which reaction occurs in both the mouth and the small intestine?

    (A) Reactions 1 and 3
    (B) Reaction 1
    (C) Reaction 2
    (D) Reactions 3 and 4

14. Reaction 1 occurs in the

    (A) oral cavity
    (B) stomach
    (C) small intestine
    (D) stomach and small intestine

Question 15–18. For each of the following organs, choose the one with which the activity is most likely associated.

   (A) stomach
   (B) duodenum
   (C) liver
   (D) colon

15. an organ from which the least absorption occurs

16. the site of digestion of all three foodstuffs

17. an area in which absorption is limited to water and electrolytes

18. the source of bile salts

Answers and Explanations

1. **The answer is B.** (I B 1a–d) The wall of the stomach consists of four layers; the fourth and outermost layer is the visceral peritoneum, which covers all the organs of the digestive tract.

2. **The answer is D.** (I C 1–3) The vagus nerve carries nerve fibers of the parasympathetic portion of the autonomic nervous system to the digestive tract. These impulses stimulate motility and secretion of digestive juices. The splanchnic nerves carry sympathetic fibers, which inhibit motility and secretion. The phrenic nerve innervates the diaphragm. The intercostal nerve(s) innervate the intercostal muscles.

3. **The answer is A.** (II A 4 b) The secretions of the salivary glands are about 98% water and 2% electrolytes and protein. The proteins are mucin, which produces more viscous mucus, and salivary amylase, which initiates the digestion of starch.

4. **The answer is C.** (II B 2; C 1) No digestive enzymes are produced by the esophagus.

5. **The answer is D.** (III B 5) The parietal cells of the gastric mucosal glands secrete intrinsic factor, a glycoprotein that binds to vitamin B_{12} in food and promotes its absorption in the small intestine.

6. **The answer is A.** (III C 1) Cardiac glands of the stomach are located only near the cardiac orifice and produce only mucus. Fundic glands contain chief and parietal cells. The produce HCl and pepsinogen, which is converted to pepsin by HCl.

7. **The answer is C.** (III C 2 a–c) Gastric secretion is inhibited by duodenal hormones, which include secretin, gastric inhibitory polypeptide, and cholecystokinin. It is stimulated by stomach distention, stomach and duodenal gastrin, amino acids and proteins in the partially digested food, and chemicals such as alcohol and caffeine.

8. **The answer is B.** (III E 1–3) Gastric emptying is inhibited by the fat and acid content of chyme. This effect is mediated by both neural and hormonal responses in the duodenum. These include the enterogastric reflex and secretion of secretin, cholecystokinin, enterogastrone, and gastric inhibitory polypeptide.

9. **The answer is C.** (IV D 1) The plicae circulares, villi, and microvilli are specializations to increase the effective surface area for absorption in the small intestine. The crypts of Lieberkühn are the intestinal glands. They dip down into the mucosa between the villi.

10. **The answer is B.** (IV F 1 a–c) Duodenal secretin, released when acidic chyme enters the small intestine, is carried in the circulation to the pancreas, where it stimulates the release of large quantities of fluid containing sodium bicarbonate to neutralize the acid. The pancreas secretes enzymes upon stimulation by cholecystokinin in response to partially digest fats and proteins in chyme. The pancreatic duct joins the common bile duct to enter the duodenum at the hepatopancreatic ampulla. The liver produces bile for the emulsification and absorption of fats.

11-14. **The answers are 11–B, 12–A, 13–C, and 14–B.** (Table 14–1; II A 4 c; III D 1; IV E,F) The emulsification of fats by bile is necessary for the enzymatic breakdown of fats to monoglycerides, fatty acids, and glycerol. Pepsin acts only at a pH below 2; pepsin must be activated from pepsinogen by hydrochloric acid. Polysaccharides are digested to disaccharides by salivary amylase in the oral cavity and by pancreatic amylase in the small intestine. The activity of pepsin occurs in the stomach.

15-18. **The answers are 15–A, 16–B, 17–D, and 18–C.** (III B 6; IV G 1; V C 1) Only water, alcohol, and some drugs are absorbed from the stomach. The digestion of proteins, carbohydrates, and fats are completed in the small intestine, primarily in the duodenum. The large intestine (colon) absorbs water and electrolytes from the remaining chyme. The liver is the source of bile and bile salts.

Metabolism, Nutrition, and Body Temperature Regulation 15

I. INTRODUCTION

A. **Energy** is required for the physiological processes that take place in the body cells. These processes include muscular contraction, nerve impulse generation and conduction, glandular secretion, temperature maintenance by the production of heat, active transport mechanisms, and synthesis and degradation reactions of all kinds.

1. **Sunlight** is the ultimate source of energy. During photosynthesis, the chlorophyll (green pigment) in plants converts sunlight to energy. The energy is stored in the chemical bonds of carbohydrate molecules.
2. When food from plants and animals is eaten and digested, the stored energy in the food must be released so it can be utilized by the cells.
3. **Cellular respiration** is the stepwise enzymatic process within the body cells by which the energy from the absorbed monosaccharides, fatty acids, glycerol, and amino acids is extracted. The energy is temporarily stored in high energy phosphate compounds such as ATP to be used directly for cellular activities or to be transformed to another kind of energy, such as electrical energy for nerve impulse conduction or mechanical energy for movement, depending on the type of cell and the body's needs.

B. **Metabolism** is the sum of all the physical and chemical reactions and energy transformations in the body that support and maintain life.

1. **Anabolism** includes chemical reactions by which complex molecules required for the growth and maintenance of the individual are **synthesized** from simpler substances with the use of energy.
2. **Catabolism** includes the chemical reactions that break down complex molecules into smaller molecules with the release of energy.
3. Anabolic and catabolic reactions take place in cells simultaneously and continually.

C. A **metabolic pathway** is a specific sequence of chemical reactions involving anabolism and catabolism.

1. The reactions in the cell are mainly **oxidation-reduction reactions**, which involve the transfer of one or more electrons from one reactant to another (see Chapter 2 II A 4).
    a. **Oxidation** is a chemical reaction in which an atom or molecule **loses electrons** and is said to be oxidized.

b. **Reduction** is a chemical reaction in which an atom or molecule gains **electrons** and is said to be reduced.
c. Because electrons released during an oxidation reaction cannot exist in a free state in living cells, every oxidation reaction is accompanied by a reduction reaction in which the electrons are accepted by another atom or molecule.
d. In living cells, oxidation generally involves the removal of an entire **hydrogen atom** (rather than an electron) from a compound and reduction generally means a gain in a hydrogen atom.

2. **Enzymes** catalyze (speed up) each of the steps in a metabolic pathway. Most enzymatic reactions need the presence of a **coenzyme**, which is an organic compound that combines with the enzyme for it to be effective. Coenzymes may act to carry electrons from one reaction to the next and may be alternately oxidized and reduced in the process.
    a. **Nicotinamide adenine dinucleotide (NAD)** is a coenzyme that accepts electrons and becomes reduced. It functions with the enzyme lactic dehydrogenase in the metabolism of glucose and ATP formation. Nicotinamide is formed from niacin, a B vitamin.
    b. **Flavin adenine dinucleotide (FAD)** is another coenzyme electron acceptor. It functions with succinic dehydrogenase and is formed from riboflavin, or vitamin B_2.

3. **ATP** is the most widely distributed high-energy phosphate compound that stores energy for the body. ATP is formed from the nucleotide adenosine plus two additional phosphate groups attached by high-energy bonds (see Chapter 3 IV B 1,2).
    a. The hydrolysis of ATP removes one phosphate, leaves adenosine diphosphate (ADP), and liberates energy. Loss of another phosphate to form adenosine monophosphate (AMP) liberates more energy.
    b. The energy released by catabolism of foods is used by ADP to recombine and form ATP for the storage of energy. The energy released from ATP is used for cellular work and to power anabolic reactions.
    c. The **ATP-ADP system** is the major means of energy transfer in cells.

D. **Hormonal regulation of metabolism.** Metabolism in the body is regulated and coordinated by hormones from the pancreas, adrenal gland, anterior pituitary gland, hypothalamus, and thyroid gland (Chapter 10). The hormones determine the metabolic pathways taken by absorbed nutrients and ensure efficient use of energy sources based on the body's needs.

II. METABOLISM OF ABSORBED CARBOHYDRATES

A. **Role of the liver**

1. The glucose, fructose, and galactose absorbed from the small intestine are transported to the liver by the hepatic portal vein. The liver cells convert fructose and galactose to glucose, which is then either stored in the liver as glycogen or released into the blood to be transported to other body cells.
2. Glucose can be converted to fat by the liver and by adipose tissue when there is an excess of glucose. The liver may also convert glucose to amino acids.
3. The glucose molecule enters a body cell from the blood by diffusion. **Insulin** facilitates glucose transport into cells by increasing the affinity of the membrane carrier molecule for glucose.

B. **Glucose catabolism.** The extraction of energy from glucose can be divided into three sequential parts: **glycolysis**, which is anaerobic and occurs in the cytoplasm of cells; the **citric acid cycle** (Krebs cycle, tricarboxylic acid cycle), which is aerobic and occurs in the mitochondria; and **electron transport**, which also takes place in the mitochondria and accounts for most of the ATP yield.

C. **Metabolic pathway of glycolysis.** Glycolysis splits the six-carbon glucose molecule into two three-carbon molecules (**pyruvate**) with the release of a small amount of energy in the form of **two ATP molecules** (Figure 15-1).
1. **Glucose to glucose-6-phosphate.** Once in the cell, glucose is almost immediately **phosphorylated** to **glucose-6-phosphate** by the addition of a phosphate group (P) removed from one molecule of ATP to the sixth carbon of glucose. Glucose-6-phosphate is the key compound that prepares glucose for catabolic reactions for the release of energy or anabolic reactions for synthesis to glycogen.
    a. Although the phosphorylation reaction to glucose-6-phosphate is irreversible in most body cells, the liver cells, intestinal epithelial cells, and kidney tubule cells have the necessary enzyme to remove the phosphate from glucose-6-phosphate and re-form glucose.
    b. Only the liver, however, plays a major role in regulating glucose levels by releasing free glucose into the blood.
2. **Glucose-6-phosphate to fructose-6-phosphate.** In this reaction, the hydrogen and oxygen atoms are rearranged to form the isomer of glucose-6-phosphate.
3. **Fructose-6-phosphate to fructose 1,6-diphosphate.** The addition of another phosphate group to glucose-6-phosphate from the second molecule of ATP results in a new six-carbon compound with a phosphate at either end.
4. **Fructose 1,6-diphosphate to PGAL.** Fructose 1,6-diphosphate is split between the third and fourth carbons forming two different three-carbon sugar molecules: **glyceraldehyde 3-phosphate (PGAL)**, and **dihydroxyacetone phosphate**, which is an isomer and can be converted to PGAL. Therefore, both molecules can be considered to be two PGAL molecules.
5. **PGAL to two diphosphoglycerate molecules.** Energy is harvested from the two PGAL molecules in the next series of steps, which involve a coupled oxidation-reduction reaction.
    a. In the reaction, two hydrogen electrons are removed from each molecule of PGAL, and an inorganic phosphate group (P_i) is added to each molecule of PGAL.
    b. The hydrogen electrons from PGAL, which has been oxidized by their loss, are picked up by a hydrogen acceptor, **nicotinamide adenine dinucleotide (NAD)**, which thereby becomes reduced as NADH.

$$2 \text{ PGAL} + 2 \text{ NAD} + 2 P_i \xrightarrow{\text{Enzyme}} (2) \text{ 1,3-diphosphoglycerate} + 2 \text{ NADH}$$

6. **Two diphosphoglycerate molecules to two 3-phosphoglycerate molecules.** The new phosphate bond in 1,3-diphosphoglycerate is energy rich. When the phosphate in each of the two molecules is transferred over to ADP, two molecules of 3-phosphoglycerate and two molecules of ATP are formed. The transfer of energy from a compound with an energy-rich phosphate is called **substrate-level phosphorylation.**

$$(2) \text{ 1,3-diphosphoglycerate} + 2 \text{ ADP} \xrightarrow{\text{Enzyme}} (2) \text{ 3-phosphoglycerate} + 2 \text{ ATP}$$

7. In the next few steps, the two 3-phosphoglycerate molecules undergo an internal rearrangement, which relocates the phosphate bonds and changes them into high-energy bonds. Also, a water molecule is removed from each of the two molecules.

$$(2) \text{ 3-phosphoglycerate} \xrightarrow{\text{Enzyme}} (2) \text{ 2-phosphoglycerate}$$

$$(2) \text{ 2-phosphoglycerate} \xrightarrow{\text{Enzyme}} (2) \text{ 2-phosphoenolpyruvate (PEP)}$$

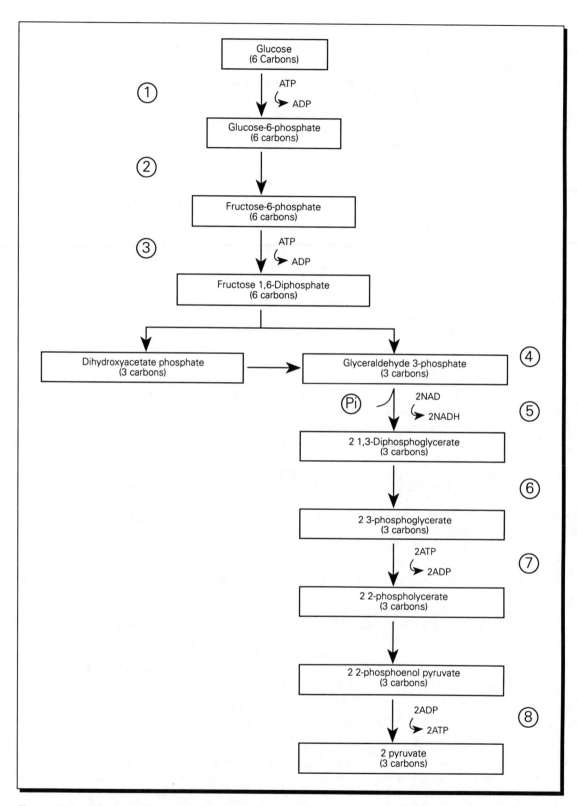

Figure 15–1. *Glycolysis.* The steps in the glycolytic reactions are numbered 1 through 8.

8. **PEP to pyruvate.** Finally, by substrate-level phosphorylation, the high-energy phosphate group of each PEP molecule is transferred to two ADPs, forming two ATPs. The remaining products are two molecules of pyruvic acid.

$$2 \text{ PEP} + 2 \text{ ADP} \xrightarrow{\text{Enzyme}} 2 \text{ pyruvate} + 2 \text{ ATP}$$

9. **Energy yield of glycolysis**
   a. Glycolysis generates **four** molecules of ATP through substrate-level phosphorylation. Because two are consumed to initiate glucose breakdown, there is a **net gain of two molecules of ATP**. This represents only 5% of the total energy in a glucose molecule.
   b. Two molecules of reduced coenzyme **NADH**, the hydrogen acceptor and energy carrier, are produced. The hydrogen electrons they carry will be available for the synthesis of more ATP.

10. **Pathways after glycolysis**
    a. **Oxidative respiration.** In the presence of oxygen, the two pyruvic acid molecules are completely oxidized to carbon dioxide and water during the reactions of the citric acid cycle.
    b. **Fermentation.** When sufficient oxygen is unavailable, as may occur in the body during strenuous exercise, pyruvate is reduced to lactate (lactic acid) in a **fermentation pathway**. The lactate molecules may be converted back to pyruvate molecules in the cell when oxygen is available, or may be transported to the liver for conversion back to pyruvate or all the way back to glycogen for storage.

D. **Oxidative respiration.** The **citric acid cycle** is the second part of glucose catabolism. It takes place in the cell **mitochondria** and is a series of decarboxylation (removal of carbon dioxide) and oxidation-reduction reactions (Figure 15–2).

1. **Oxidation of pyruvic acid to acetyl coenzyme A.** The pyruvic acid molecules enter a mitochondrion and are oxidized in its inner compartment, or matrix. A preparatory step reduces the 3-carbon pyruvic acid to a 2-carbon acetyl group known as acetyl co-enzyme A or **acetyl CoA**.
   a. A carboxyl group is removed from three-carbon pyruvic acid as carbon dioxide, which diffuses out of the cell.
   b. The remaining two-carbon fragment is oxidized. The removed hydrogen atoms are accepted by NAD^+.
   c. The oxidized fragment, an acetyl group, is combined with CoA to form a molecule of acetyl CoA.

   $$\text{pyruvate} + NAD^+ + \text{CoA} \longrightarrow \text{acetyl CoA} + NADH + CO_2$$

2. **Formation of citric acid (citrate).** CoA carries the two-carbon acetyl group into the citric acid cycle where it reacts with four-carbon **oxaloacetic acid** to form six-carbon **citric acid**. The CoA is freed to combine with another acetyl group and one molecule of water is used in the citric acid synthesis.

3. **Isocitric acid (isocitrate).** The six-carbon citric acid is rearranged, forming isocitric acid. Isocitrate is oxidized and loses two hydrogen electrons to become a short-lived molecule, **oxalosuccinic acid**. The hydrogens are accepted by NAD forming NADH plus H^+ ($NADH_2$).

4. **Alpha-ketoglutaric acid (alpha-ketoglutarate).** The oxalosuccinic acid loses a carbon, which enters into the production of the first of two CO_2 molecules released in the cycle as a waste product. The remaining five-carbon molecule, **alpha ketoglutaric acid**, is oxidized. Another $NADH_2$ is formed as two hydrogen electrons are accepted by NAD and another CO_2 is formed.

5. **Succinyl coenzyme A.** In the next step, which is catalyzed by a multienzyme complex, alpha-ketoglutaric acid undergoes oxidative decarboxylation. A second CO_2 is released, NAD is reduced to $NADH_2$, and the remaining four-carbon compound is linked to CoA as succinyl CoA. The attachment is an energy-rich unstable bond; that is, it has enough energy to phosphorylate ADP.

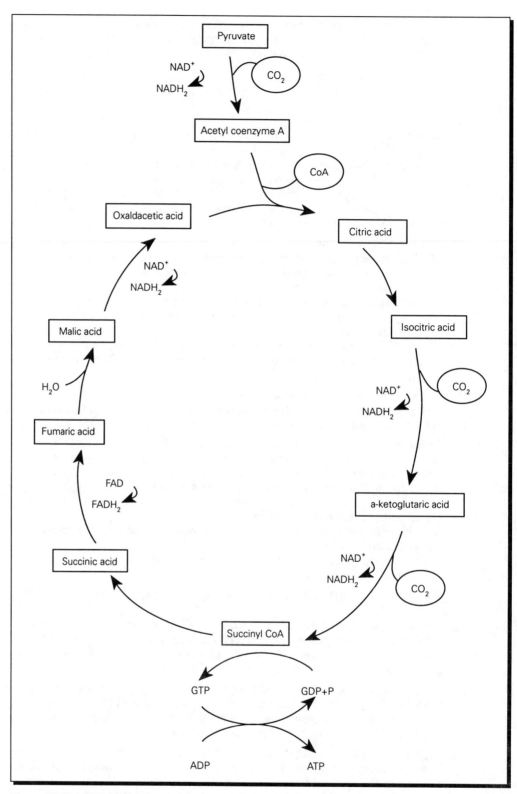

Figure 15-2. *Citric acid cycle.*

6. **Succinic acid (succinate).** The energy in the succinyl-CoA bond is transferred to an energy-rich phosphate bond in guanosine triphosphate (GTP) by guanosine diphosphate (GDP). From GTP, the high-energy phosphate group is transferred to ADP, forming ATP by substrate-level phosphorylation.
7. **Fumaric acid (fumarate).** The succinic acid is oxidized to fumaric acid, but instead of the hydrogens being passed to NAD, they are picked up by another coenzyme, **FAD (flavin adenine dinucleotide)**, which contains the vitamin riboflavin.
8. **Malic acid (malate).** With the addition of water, fumaric acid is converted to malic acid.
9. **Oxaloacetic acid (oxaloacetate).** Malic acid loses hydrogens and is converted to oxaloacetic acid. The two hydrogens are transferred to NAD, and oxaloacetic acid can combine with another molecule of CoA to start the cycle again.
10. **Summary of energy yield of the citric acid cycle.** For each glucose molecule (two acetyl CoAs) that enter the citric acid cycle, **two additional ATP** molecules are formed. The total of anaerobic glycolysis and the aerobic citric acid cycle is four ATP molecules.

E. **Electron (hydrogen transport) system and oxidative phosphorylation.** The electron transport system consists of a chain of electron acceptors located in the inner membrane of the mitochondrion. The transport of electrons is coupled to the formation (phosphorylation) of ATP from ADP, which is referred to as oxidative phosphorylation.
   1. **The source of the electrons** is molecules of NADH + H$^+$ (NADH$_2$) and FADH$_2$. As the hydrogens are transferred along the electron acceptor molecules, the hydrogen protons become separated from their electrons and are released into the surrounding medium. The electrons are transferred along the electron acceptors in a series of oxidation-reduction reactions.
   2. **The electron acceptors in the chain** are **flavin mononucleotide (FMN), ubiquinone (coenzyme Q)**, and a series of **cytochromes (cytochromes c, b, a, and a$_3$)**. Cytochromes are proteins characterized by a heme (iron) group. The iron combines with the electrons from the hydrogen atoms.
      a. The electrons are passed from cytochrome to cytochrome, losing energy on the way.
      b. The last cytochrome, cytochrome a$_3$, passes two electrons to molecular oxygen. The electrons combine with protons to re-form hydrogen and the union of hydrogen and oxygen produces water.
   3. The **chemiosmotic model** for electron transport and ATP synthesis was postulated by Mitchell, who received the Nobel Prize in 1978. The hypothesis proposes that as the electrons are transferred along the chain, the separated protons are pumped from the inner surface of the mitochondrial membrane (matrix) to the outer surface (cytoplasmic side). This results in an increased H$^+$ concentration on the cytoplasmic side and sets up an electrochemical (proton) gradient across the mitochondrial membrane. The force generated, which is chemical and osmotic (chemiosmotic), drives the synthesis of ATP.

F. **Summary of energy released by glucose catabolism.** The ATP yield from the complete oxidation of one molecule of glucose depends on the cell and the pathways involved, but the usual figure is taken as **36 ATP molecules**. This represents approximately 38% of the total energy available in a molecule of glucose. The remainder is released as heat, which is either dissipated or used for the maintenance of body temperature.
   1. Glycolysis results in the production of 2 ATP.
   2. The citric acid cycle results in the production of 2 ATP.
   3. Oxidative phosphorylation via electron transport chain results in the production of 32 ATP.

G. **Gluconeogenesis** is the formation of glucose from noncarbohydrate sources such as lactic acid, some amino acids (which are termed glucogenic amino acids), glycerol, and some fatty acids.
   1. **Sites of gluconeogenesis.** Almost all gluconeogenesis occurs in the liver, but the kidney will make glucose during starvation. It also occurs to a very limited extent in the epithelial cells of the small intestine.
   2. **Functions.** Gluconeogenesis maintains an adequate blood sugar level during starvation, periods of limited dietary intake of carbohydrates, or after strenuous exercise, when the lactic acid formed in skeletal muscles is converted back to glucose in the liver.
   3. **Regulation.** Gluconeogenesis is stimulated by a low cellular concentration of carbohydrate and a decreased blood sugar. It is also stimulated hormonally by **glucagon**, the peptide hormone secreted by the alpha cells of the pancreatic islets, by **epinephrine** from the adrenal medulla, and by **glucocorticoids** from the adrenal cortex.
H. **Glycogenesis** is the anabolic process of **glycogen formation** for storage of glucose when the blood glucose level is high, such as after a meal. Glycogenesis occurs especially in **liver** and **skeletal muscle** cells, but does not occur in brain cells, which must depend on a constant supply of blood glucose for energy.
I. **Glycogenolysis** is the **breakdown of glycogen to glucose** for release into the bloodstream by the liver when the body needs energy. The breakdown is accelerated by glucagon and epinephrine.

III. METABOLISM OF FATS

A. **Transport of fats (lipids) in the bloodstream.** Fats are transported as **chylomicrons, free fatty acids,** and **lipoproteins.**
   1. **Chylomicrons** are formed in the intestinal mucosa from fatty acids and glycerol, absorbed into the lacteals, and enter the blood circulation. They are composed of 90% **triglycerides**, plus **cholesterol, phospholipids,** and a thin coating of **protein**. Within four hours after a meal (postabsorptive state), most of the chylomicrons are removed from the blood by adipose tissue and the liver.
      a. The enzyme **lipoprotein lipase**, which is found in the liver and adipose tissue capillaries, breaks down the triglycerides in the chylomicrons to free the fatty acids and glycerol. The fatty acids and glycerol are recombined into triglyceride (neutral fat) for storage by adipose tissue. The cholesterol-rich chylomicron remnants are metabolized by the liver.
      b. Stored fat is withdrawn from adipose tissue when needed for energy. The enzyme **hormone-sensitive lipase** breaks the triglycerides back down to fatty acids and glycerol.
      c. The amount of stored fat depends on the total food intake. Both adipose tissue and the liver can synthesize fat from excess dietary fat, carbohydrate, or protein.
   2. **Free fatty acids (FFA)** are fatty acids that are bound to albumin, one of the plasma proteins. They are the form in which fatty acids are transported from the cells of adipose tissue to be used by other tissues for energy.
   3. **Lipoproteins** are smaller particles similar in composition to chylomicrons. They are synthesized primarily in the liver. The lipoproteins are used to transport fat between the tissues and they circulate in the blood in the postabsorptive state after the chylomicrons have been removed. They are divided into three classes according to their density.
      a. **VLDL (very low density lipoproteins)** contain approximately 60% triglycerides and 15% cholesterol and have the smallest mass. They transport triglycerides and cholesterol away from the liver to the tissues for storage or utilization.

b. **LDL (low density lipoproteins)** contain almost 50% cholesterol and carry 60% to 75% of the plasma cholesterol that they deposit in adipose tissue and in smooth muscle. Their concentration depends on many factors, but the dietary intake of cholesterol and saturated fat is significant. A high level of blood LDL is associated with a high incidence of coronary heart disease.

c. **HDL (high density lipoproteins)** contain 20% cholesterol, less than 5% triglycerides, and 50% protein by weight. They are important in clearing triglycerides and cholesterol from the plasma because they bring cholesterol back to the liver for metabolism instead of depositing it in other tissues. A high concentration of blood HDL is associated with a low incidence of coronary heart disease.

B. **Fat catabolism**

1. **Glycerol** enters the cell and is changed by enzymes to glyceraldehyde-3-phosphate, which enters the glycolysis pathway. It may continue through the citric acid cycle or be used to resynthesize glucose.

2. **Fatty acids** enter the cell and are transported to the mitochondria by a carrier protein. In the mitochondrial matrix, the fatty acids are converted by the **beta oxidation** process into acetyl CoA, which is then metabolized via the citric acid cycle.
   a. The fatty acids are oxidized in a cyclic sequence of steps. The process is termed beta-oxidation because an oxygen atom is added at the beta-carbon of the chain, which is the second carbon atom from the end carboxyl group.
   b. The energy yield from the breakdown of a fat is very high, with a net gain of 135 to 145 ATP molecules from a typical long-chain fatty acid molecule.

3. **Ketone bodies.** Acetyl molecules may condense to form **acetoacetic acid**, which is converted to **beta-hydroxybutyric acid** and **acetone**. These are called ketone bodies.
   a. Ketone bodies are normal products of fatty acid oxidation. The blood level of ketone bodies is low because most tissues, except the liver, can metabolize them back to acetyl CoA as rapidly as they are formed.
   b. When the rate of fat catabolism is high and many acetyl CoAs are formed, the liver produces and releases more ketones than the tissues can handle. The excess ketones accumulate in the bloodstream (**ketosis**). In severe conditions of ketosis, the acidosis and lowered pH that develops may lead to coma and death.
   c. There are three primary reasons for a diminished glucose supply and an excessive rate of fatty acid oxidation and ketone production.
      (1) **Starvation** results in excessive beta-oxidation of fatty acids because of the lack of glucose for energy.
      (2) A **high fat, low carbohydrate diet** increases the ketone level in the blood because there is no biochemical pathway for conversion of fats to carbohydrates and fatty acids become the primary source of energy.
      (3) In **uncontrolled diabetes mellitus**, the lack of insulin, which stimulates glucose entry and storage in cells, results in excessive fatty acid oxidation in lieu of glycolysis.

C. **Fat anabolism**

1. **Essential fatty acids.** Although many tissue cells can synthesize most of the fatty acids from acetyl CoA and the liver can convert one fatty acid to another, three unsaturated fatty acids (**linolenic, linoleic,** and **arachidonic acids**) cannot be synthesized or converted. They must be obtained from the diet and are termed **essential fatty acids**.

2. When there are more carbohydrates in the diet than can be stored as glycogen or utilized for energy, or more proteins in the diet than the tissues need, triglycerides are synthesized from excess glucose and amino acids (**lipogenesis**). Thus, most of the fat in the body does not come from fat in the diet.

D. **Regulation of fat metabolism**
   1. **Hormones** control the balance between fat breakdown and fat deposition.
      a. **Insulin** is the most important controlling factor.
         (1) Insulin promotes glucose passage into cells so glucose is preferentially used for energy.
         (2) Insulin also prevents fat breakdown in adipose cells by inhibiting hormone-sensitive lipase, which catalyzes the hydrolysis of triglycerides to fatty acids and glycerol.
         (3) Insulin and glucagon secretion are regulated by blood glucose levels. Thus, glucose is also a regulator of fat metabolism.
      b. **Epinephrine, glucagon, growth hormone, ACTH, and thyroxine** stimulate the breakdown and release of fatty acids from triglycerides stored in adipose tissue.
   2. **Neural regulation** of fat metabolism occurs through parasympathetic stimulation, which increases fat deposition, and sympathetic stimulation, which accelerates the breakdown of fatty acids from stored fat.

IV. **PROTEIN METABOLISM**

   A. **Transport and fate of absorbed amino acids.** The amino acids from dietary protein are absorbed from the intestine via active transport and carried to the liver, where they may be synthesized into protein molecules or released into the circulation to be transported to other cells.
      1. After entry into the body cells, amino acids are combined by peptide linkages to form the cellular proteins used for tissue growth and regeneration.
      2. Little storage of amino acids occurs in body cells, except by the liver cells. The body's own intracellular proteins continuously are hydrolyzed to amino acids and resynthesized back to protein. Both dietary amino acids and those from intracellular protein breakdown form a common **amino acid pool** that supplies the body's needs.
   B. **Protein catabolism** (breakdown of amino acids for energy) occurs in the liver. When the cells have enough protein for their needs, any additional amino acids are used for energy or stored as fat.
      1. **Deamination** of the amino acid, which is the first step, involves the removal of a hydrogen and an amino group, which results in the formation of ammonia (NH_3).
      2. **Urea formation by the liver.** The ammonia is converted to **urea** through the urea cycle (orthinine cycle) by the liver. The urea is excreted by the kidneys in the urine.
      3. **Oxidation of deaminated amino acids.** The remaining non-nitrogen part of the amino acid is a keto acid product, which may be oxidized for energy through the citric acid cycle. Some keto acids may be converted to glucose (gluconeogenesis) or fat (lipogenesis).
      4. Carbohydrates and fats are "protein-sparers" and are utilized by the body in preference to protein for energy. During starvation, the body uses up carbohydrates and fats first and then starts catabolizing protein.
   C. **Protein anabolism**
      1. **Protein synthesis** from amino acids occurs in most of the body cells. The amino acids are linked by peptide bonds in a particular sequence, which is determined by gene regulation.
      2. **Transamination**, which occurs in the liver, is the synthesis of nonessential amino acids by the conversion of one amino acid to another. It involves the transfer of an amino group (NH_2) from one amino acid to a keto acid, thus forming a new amino acid and a new keto acid.

3. **Essential and nonessential amino acids.** Nine amino acids (phenylalanine, valine, tryptophan, threonine, lysine, leucine, isoleucine, methionine, and histadine) are essential amino acids. These cannot be synthesized by cells and must be obtained in the diet. The other 11 amino acids can be synthesized and are termed nonessential amino acids.
    a. Animal protein is more likely to contain all essential amino acids and is termed a **complete protein.**
    b. Plant protein is likely to be lacking some essential amino acids and is termed **incomplete protein.** Plant proteins can be combined in the diet to obtain all essential amino acids.
D. **Nitrogen balance** exists when the amount of nitrogen lost through excretion equals the nitrogen content of protein ingested. The minimum amount of protein required by an individual to maintain this equilibrium is approximately 0.4 g per lb of body weight.
    1. A **positive** nitrogen balance (the amount of protein ingested is greater than the amount eliminated) normally occurs in a growing child, during repair of injured portions of the body, and during pregnancy and lactation.
    2. A **negative** nitrogen balance (the amount of tissue breakdown and excretion of protein far exceeds the amount ingested) occurs during periods of starvation, high fevers, or wasting disease.
E. **The regulation of protein metabolism**, like carbohydrate and fat metabolism, is primarily by hormones.
    1. **Growth hormone** stimulates active transport of amino acids into cells, especially muscle cells, and stimulates protein synthesis.
    2. **Testosterone**, the male sex hormone, stimulates protein synthesis and an increased deposition of protein in the tissues. **Estrogen**, the female sex hormone, also stimulates protein synthesis to a lesser degree.
    3. **Thyroid hormone** increases the rate of metabolism of all cells and is necessary for protein synthesis and growth.
    4. **Glucocorticoids** stimulate protein catabolism in cells other than the liver and enhance the use of amino acids by the liver in gluconeogenesis.
    5. **Insulin** enhances entry of amino acids into cells and stimulates protein synthesis.

V. **ABSORPTIVE (FEASTING) AND POSTABSORPTIVE (FASTING) STATES**

A. The **absorptive state** is the period following a meal and during the absorption of nutrients. All nutrients are in high concentration in the blood and the main energy source is glucose.
    1. Any glucose that is not oxidized for energy undergoes glycogenesis or lipogenesis to replenish glycogen and fat stores.
    2. Amino acids are used for protein synthesis or converted to fatty acids to be synthesized into fats and stored in adipose tissue.
    3. Most of the fat is stored in adipose tissue.
B. The **postabsorptive state** is the period between meals. The body must maintain the blood glucose at a normal level to satisfy energy needs. Glucose is obtained from glycogen, tissue protein, and fats.
    1. The liver cells release glucose from stored glycogen (glycogenolysis) and provide a maintenance of blood sugar levels for about four hours.
    2. Gluconeogenesis provides glucose from other sources, such as glycerol, pyruvic acid, lactic acid, and keto acids. Fatty acids cannot be converted to glucose, but the body can shift to fatty acid catabolism for energy and spare the glucose for the nervous system.
    3. When fasting is prolonged and after glycogen and fat stores are depleted, tissue proteins become a source of glucose.

VI. METABOLIC RATE AND NUTRITION

A. The **metabolic rate** refers to the amount of energy released in the body by catabolism per unit time. It may be measured in kilocalories or expressed as a percentage above or below normal.

1. A **kilocalorie**, also termed a large **Calorie (C)** or kcal, is the amount of heat required to raise one kilogram of water 1° Celsius. It is equal to 1,000 small calories (c) and is used to express the caloric content of food and the energy expenditure by the individual.

2. Kilocalories, which are liberated by the oxidation of foods in the body, can be measured directly by **bomb calorimetry** (placing the body in a calorimeter and measuring the amount of heat liberated in the water surrounding the body) or, indirectly, by measuring the oxygen consumption of the body with a respirometer. The body produces 4.8 C of heat for every liter of oxygen used.
   a. In the body, 1 g of carbohydrate liberates 4.1 C, 1 g of fat liberates 9.5 C, and 1 g of protein liberates 4.1 C.
   b. Different foods have different caloric contents depending on their composition. Most foods are a combination of carbohydrates, fats, and proteins.

3. The **specific dynamic action (SDA)** of food is the elevation of the metabolic rate by 10% to 20% over the basal rate by the energy used in digestion of the food. The SDA of protein is greater than that of fats or carbohydrates so a protein-rich meal generally increases the metabolic rate slightly.

4. **Muscular activity** produces the greatest increase in metabolic rate. Strenuous exercise increases metabolism by 15 times.

B. The **basal metabolic rate (BMR)** is the energy expenditure per unit time under basal conditions, which are standardly accepted: awake and resting, in the postabsorptive state (12 hours after a meal), and in a comfortable and warm environment.

1. The BMR represents the minimal energy required for the work of respiration, circulation, digestion, excretion, metabolism of food, nervous and muscular activities, and maintenance of body temperature when the individual is awake.

2. Factors influencing the BMR
   a. **Hormonal.** By increasing cellular metabolism, thyroid hormone is the major factor that affects the BMR.
   b. **Body size and surface area.** The BMR increases with an increase in body weight and height, which increases the surface area.
   c. **Age.** The rate is highest in childhood and decreases thereafter with age.
   d. **Gender.** Men have a slightly higher BMR than women of the same age, which probably is related to size.
   e. **Body temperature.** The rate rises about 14% for each Celsius degree increase in fever (7% per Fahrenheit degree).
   f. Other factors that increase the BMR include anxiety, certain drugs, and a lowered environmental temperature. Depression decreases the BMR, as does prolonged starvation.

C. Energy balance and its relation to body weight

1. When the caloric value of the food ingested (energy intake) equals the energy expended as heat and work (the total metabolic rate), body weight remains the same.

2. If the caloric intake is less than the energy output, body stores of glycogen, then fat, and, finally, body protein are catabolized and weight reduction occurs.

3. When the caloric intake exceeds the energy output, the excess energy is stored as fat.
   a. Approximately 3,500 C are used to synthesize one pound of adipose tissue. Thus, eating 500 C daily above energy requirements for one week will result in a gain of a pound.
   b. To reduce weight, caloric intake should decrease and energy output should increase.

D. **Essential dietary components.** An optimal, balanced diet satisfies the nutritional requirements. It includes carbohydrate, fat, protein, vitamins, minerals, and water (see Chapter 2 III C 2–4).
  1. **Dietary carbohydrate sources**
     a. Except for lactose in milk and glycogen (animal starch) in meat, all carbohydrates come from plants.
     b. Monosaccharides (glucose, fructose) are found in vegetables, fruits, honey, and syrups. Disaccharides, such as sucrose (table sugar), are derived from sugar cane and sugar beets.
     c. Polysaccharides (complex carbohydrates) include starch, glycogen, and cellulose. Starch is found in vegetables, fruits, and grains; cellulose (fiber) is a component of the cell walls of plants.
  2. **Dietary fat sources**
     a. **Saturated fats** are found in meats, dairy products, egg yolk, nuts, and some vegetable fats, such as cocoa butter, coconut oil, and palm oils.
     b. **Monounsaturated fats** are found in olive and peanut oils. **Polyunsaturated fats** are found in fish, safflower, sunflower, canola, and corn oils.
  3. **Dietary protein sources**
     a. Complete proteins, which contain all essential amino acids, are found in meat, fish, poultry, cheese, and eggs.
     b. Incomplete proteins are found in vegetables, grains, and legumes.
  4. **Vitamins** are organic compounds that are necessary in small quantities for normal metabolic processes. They must be components of the diet because most of them cannot be synthesized within body cells (Table 15–1).
  5. **Minerals** required in the diet at levels of 100 mg daily or more include calcium, phosphorus, sodium, potassium, chlorine, magnesium, and sulfur (Table 15–2). Trace minerals are elements that are vital to metabolic processes in smaller quantities (Table 15–3).
  6. Dietary guidelines issued by the U.S. government in 1990 specify the following recommendations:
     a. Eat a variety of foods.
     b. Maintain a healthy weight.
     c. Choose a diet low in fat, saturated fat, and cholesterol.
     d. Choose a diet with adequate starch and fiber.
     e. Use sugars only in moderation.
     f. Use salt and sodium only in moderation.
     g. Drink alcoholic beverages only in moderation.

VII. TEMPERATURE REGULATION

A. A **constant body temperature** is necessary for normal enzymatic activities. Enzymes function within a narrow range of normal temperature of 36.1° to 37.8° Celsius (97° to 100° Fahrenheit).

B. **Factors affecting body temperature** include the individual's diurnal rhythm, gender, and age.

C. The **determinant of body temperature** is the balance between heat production and heat loss. It is maintained by homeostatic mechanisms.
  1. **Heat production** occurs by catabolism of foods and muscular activity. Under basal conditions, the liver produces 20% of body heat; the brain, 15%; the heart, 12%; and the muscles, the rest.
  2. **Heat loss** occurs to the air and to nearby objects through the physical processes of radiation, conduction, convection, and evaporation. Eighty percent of heat loss is transferred through the skin. The remainder of the loss occurs through the mucous membranes of the digestive, respiratory, and urinary tracts.

Table 15–1. Vitamins.

Name	Rich Sources	Function	Recommended Dietary Allowances (RDA)	Deficiency	Potential Toxicity When Large Amounts Consumed
Water-soluble vitamins					
Thiamine (B_1)	Pork; whole grains, enriched cereal grains; legumes (lost when cereals are milled and refined)	Component of coenzyme; energy release	1.2 to 1.5 mg	Beriberi; impairment of cardiovascular, nervous, and gastrointestinal systems	None; excreted in urine
Niacin (Nicotinic acid)	Lean meats, liver; peanuts; yeast; cereal bran and germ	Component of two coenzyme systems; energy release	15 to 20 mg	Pellagra; "4 Ds:" dermatitis, diarrhea, depression, death	None; may cause harmless symptoms of flushing of skin, dizziness, and nausea
Riboflavin (B_2)	Milk; eggs; liver; kidney; heart; green leafy vegetables (lost in dehydrated vegetables)	Component of various enzymes involved in energy release	1.1 to 1.8 mg	Dermatitis; light sensitivity of eyes; sores at corners of mouth	None
Pantothenic acid (B_3)	Liver; kidney; egg yolk; wheat bran; fresh vegetables (especially broccoli and sweet potatoes); molasses	Component of coenzyme A; energy release	Unknown, probably 5 to 10 mg	Fatigue; gastrointestinal distress; personality changes; numbness and tingling of hands and feet; muscle cramps	None
Biotin	Milk; liver; kidney; egg yolk; yeast; also synthesized by bacteria in intestine	Coenzyme carrier of carbon dioxide	0.15 to 0.3 mg	Scaly skin; seborrheic dermatitis in infants	None
Folic acid (folacin; pteroylglutamic acid)	Green leafy vegetables; liver; kidney; lima beans; asparagus; whole grains; nuts; legumes; yeast	Blood-cell formation	0.05 mg	Anemia	Generally none; reports of folate hypersensitivity, possible neurotoxicity at 15 mg daily
Cobalamin (B_{12}; cyanocobalamin)	Only foods of animal origin; beef, liver, kidney; milk; eggs; oysters; shrimp; pork; chicken	Essential for function of all cells	0.003 mg	Anemia; nerve fiber degeneration	None
Pyridoxine (B_6)	Yeast; wheat and corn; egg yolk; liver; kidney; muscle meats (20% lost in processing grain)	Components of coenzymes for protein synthesis; central nervous system metabolism	2 mg	Anemia	None until doses of 600 mg; neurotoxicity, depression at megadoses
Ascorbic acid (C)	Citrus fruits; tomatoes; green vegetables	Collagen formation; capillary integrity; synthesis of adrenal cortical hormones; aids in iron absorption from intestine	45 to 80 mg	Scurvy; bleeding gums; easy bruising; swollen joints; impaired wound healing	Inconclusive; reports of kidney stones, diarrhea, nausea at megadoses over 2 to 5 g; interference with absorption of B_{12} and trace minerals

Table 15–1. Vitamins (con't.).

Name	Rich Sources	Function	Recommended Dietary Allowances (RDA)	Deficiency	Potential Toxicity When Large Amounts Consumed
Fat-soluble vitamins					
A (A alcohol = retinol; A aldehyde = retinal; A acid = retinoic acid)	Whole milk; liver; kidney; cream, butter; egg yolk; yellow and green vegetables; fruits	Night vision; growth; reproduction; health of epithelial cells; cell membrane maintenance	1,000 retinol equivalents (5,000 I.U.)	Night blindness; skin lesions	Very toxic in high doses of 20 to 30 × requirement; effects reverse on discontinuation
D (ergocalciferol = D_2; cholecalciferol = D_3)	Fatty fish; eggs; liver; butter, fortified milk; cod liver oil; also through exposure to ultraviolet light	Normal bone formation; promotes absorption and retention of calcium and phosphorus	300 to 400 I.U.	Rickets in children; osteomalacia in adults	Highly toxic in large doses; possibly fatal in children
E (alpha-tocopherol)	Wheat germ oils; other vegetable oils; beef liver; milk; eggs; butter; leafy vegetables	May function as antioxidant in tissues; possible role in prevention of cell degeneration	12 to 15 mg	None known in humans; sterility; muscular dystrophy; red cell fragility in animals	Reportedly nontoxic up to 1 g/day; effects of long-term megadoses unknown
K (naphthoquinones)	Synthesized by intestinal bacteria; also lettuce, spinach, kale, cauliflower	Essential for blood clotting	Unknown	Increased clotting time	None

Reproduced by Permission. *Biology of Women*, 3E, by Ethel Sloane, Delmar Publishers Inc, Albany, NY. Copyright ©1993

Table 15–2. Essential minerals needed in amounts of 100 mg/day or more.

Element	RDA (adults)	Rich Sources	Functions in Body
Calcium	800 mg	Milk, cheese; leafy vegetables, legumes, nuts; whole grain cereals; bones from sardines and other canned fish	Normal bone and teeth structure, muscular contraction, blood coagulation, nerve membrane stability
Phosphorus	800 mg	Protein-rich foods	Normal bone and teeth structure, production and transfer of high-energy phosphates, absorption and transportation of other nutrients, regulation of acid-base balance
Sodium	5 g (five times more than actual physiological need)	All food, table salt	Osmotic pressure of body fluids, muscle function, permeability of all cells
Potassium	4 g	All foods	Muscular activity, especially heart, and proper nerve function
Chlorine	Same as sodium	All foods, table salt	Osmotic pressure regulation, water balance, acid balance
Magnesium	350 mg	Most foods, especially vegetables; milk, meat, cocoa, nuts, soybeans	Enzyme activity, energy release, nerve and muscle function
Sulfur	0.6 to 1.6 g	All proteins, particularly those rich in cystine and methionine	Component of vitamins, hormones, enzyme systems, important in detoxification mechanisms

Reproduced by Permission. *Biology of Women*, 3E, by Ethel Sloane, Delmar Publishers Inc., Albany, NY. Copyright © 1993.

Table 15–3. Essential trace elements.

Element	Recommended Dietary Allowances (RDA)	Rich Sources	Functions in Body
Iron	10 mg (males) 18 mg (females)	Organ meats (liver, heart, kidney, spleen), egg yolk, fish, oysters, whole wheat, beans, figs, dates, molasses, green vegetables	Oxygen transport, cellular respiration
Copper	2.5 mg	Liver, kidney, shellfish, nuts, raisins, dried legumes	Enzyme component, hemoglobin formation
Cobalt	Unknown	Animal protein sources	Part of vitamin B_{12}
Zinc	15 mg	Meat, especially liver, eggs, seafoods, milk, grain	Enzyme component, part of insulin molecule
Manganese	300 to 350 mg (?)	Bananas, whole grains, leafy vegetables	Normal bone structure, normal function of reproductive and nervous systems
Iodine	100 to 140 µg	Fish, iodized table salt	Necessary for normal thyroid function
Molybdenum	Unknown	Beef kidney, legumes, some cereals	Enzyme component
Selenium	Unknown, probably 50 to 100 µg is adequate	Seafood, meat, grains raised in selenium-rich soil	Enzyme component, similar to vitamin E in function
Chromium	Unknown, probably 20 to 50 µg adequate	Meat, corn oil	Normal glucose metabolism
Fluorine	1 to 2 mg	Fluoridated water, milk	Resistance to dental caries

Reproduced by Permission. *Biology of Women*, 3E, by Ethel Sloane, Delmar Publishers Inc., Albany, NY. Copyright © 1993.

    a. **Radiation** is the transfer of heat in the form of infrared rays between objects that are not in contact. Normally, more than half of the heat loss from the body is by radiation.

    b. **Conduction** is the transfer of heat between objects that are in contact. Unless the heat gradients are large (for example, a cold compress placed on skin), conduction results in relatively little heat loss.

    c. **Convection** is the transfer of heat to a moving medium such as air or water when the temperature of the air or water is lower than that of the body.

    d. **Evaporation** is the transfer and loss of heat by diffusion of water molecules from the body surface to the air. Water is lost from the body surface through **insensible perspiration**, which occurs continually through diffusion from underlying tissues and evaporates without being perceived on the skin, and **sweating**, which is controlled by thermoregulation.

  D. **Regulation of body temperature**

    1. The **hypothalamic thermoregulatory centers** are a group of neurons in the preoptic area and posterior hypothalamus that function as a thermostat. The hypothalamic thermostat has a setpoint, which is adjusted to maintain body temperature. When body temperature falls below or rises above this value, the center initiates impulses to conserve heat or increase heat loss.

      a. **Peripheral thermoreceptors**, which are located in the skin, detect changes in skin temperature and certain mucous membranes and transmit the information to the hypothalamus.

      b. **Central thermoreceptors**, which are located within the anterior hypothalamus, spinal cord, abdominal organs, and other internal structures, detect changes in blood temperature.

2. **Mechanisms for heat conservation**
    a. **Vasoconstriction of peripheral blood vessels,** as a result of sympathetic stimulation, reduces blood flow and heat loss through the skin and retains warm blood in the body's core.
    b. **Increased muscular activity,** such as voluntary muscle contraction and involuntary shivering, increases heat production.
    c. **Hormonal mechanisms,** which include increased production of epinephrine, norepinephrine, thyroxine, and glucocorticoids, enhance metabolism and heat production.
3. **Mechanisms for heat loss**
    a. **Vasodilation of peripheral blood vessels,** as a result of sympathetic inhibition, causes an increased blood flow to the body surface to increase heat loss and a decreased muscle tone to decrease heat production.
    b. **Increased secretion of sweat glands** results in an increased heat loss through evaporation.
4. **Fever** is an elevation of body temperature above normal as a result of physiological stress such as allergic reaction, tissue trauma, dehydration, CNS lesions, or bacterial or viral infections. It occurs as a nonspecific defense against infection (see Chapter 12, II C 3).

Study Questions

Directions: Each question below contains four suggested answers. Choose the **one best** response to each question.

1. Which of the following is an example of a catabolic reaction?
   - (A) protein synthesis
   - (B) the formation of glycogen in the liver
   - (C) the hydrolysis of sucrose to glucose and galactose
   - (D) the combination of oxaloacetic acid and coenzyme A to form citric acid

2. Which of the following statements is true about glycolysis?
   - (A) It takes place in the mitochondria of the cell.
   - (B) It requires the presence of molecular oxygen.
   - (C) It yields a net of four ATP molecules.
   - (D) It consumes energy in its initial stages.

3. Which of the following is true of oxidation of an organic molecule?
   - (A) Oxidation of a molecule involves a gain of electrons.
   - (B) An oxidation reaction is generally accompanied by a reduction reaction.
   - (C) Oxidation is an anabolic process.
   - (D) Oxygen atoms are lost in an oxidation reaction.

4. All of the following are true about the citric acid (Krebs) cycle EXCEPT
   - (A) NAD is reduced.
   - (B) Carbon dioxide is produced.
   - (C) Pyruvic acid is formed.
   - (D) Hydrogen is released.

5. Cyanide is a compound that acts by blocking the transfer of electrons from cytochrome a_3 to oxygen. Which of the following effects on cellular respiration would occur as a result of cyanide ingestion?
   - (A) Glycolysis would be inhibited.
   - (B) The production of water as an end product would increase.
   - (C) Electron transport and oxidative phosphorylation could not continue.
   - (D) The reactions of the citric acid cycle would be speeded up in compensation.

6. The presence of an increased level of ketone bodies in the blood and in the urine indicates an increased metabolism of
   - (A) amino acids
   - (B) fatty acids
   - (c) lactic acid
   - (d) glucose

7. All of the following would be true during a prolonged fast or a starvation diet EXCEPT
   - (A) The blood glucose level would be elevated above normal.
   - (B) There would be an increased release of fatty acids from adipose tissue.
   - (C) The nitrogen balance would be negative.
   - (D) Glycogen synthesis in the liver would be decreased.

8. When energy expended is greater than the energy intake over a period of time, an individual is likely to
   - (A) gain weight
   - (B) lose weight
   - (C) maintain weight
   - (D) increase the basal metabolic rate

9. Compared with meats and dairy products, plant foods
   - (A) contain a better distribution of essential amino acids
   - (B) are not as necessary in a balanced diet
   - (C) contain more cholesterol
   - (D) are a better source of complex carbohydrates and fiber

10. When an individual cools off on a hot day by swimming in a lake or pool, body heat has been lost by
    - (A) convection
    - (B) conduction
    - (C) radiation
    - (D) evaporation

Answers and Explanations

1. **The answer is C.** (I B 1–2) Catabolism is the breakdown of more complex molecules to simpler molecules. Protein synthesis, the formation of glycogen, and the formation of six-carbon citric acid from four-carbon oxaloacetic acid and two-carbon acetyl coenzyme A are all anabolic reactions.

2. **The answer is D.** (II C 1–10) Although four molecules of ATP are produced in glycolysis, two molecules of ATP are used in the initial stages: one ATP in the formation of glucose to glucose-6-phosphate and another ATP in the formation of fructose 1, 6-diphosphate from fructose-6-phosphate. Glycolysis takes place in the cytoplasm and is anaerobic.

3. **The answer is B.** (I C 1) In living cells, every oxidation reaction is accompanied by a reduction reaction in which the electrons are accepted. Oxidation is a loss of electrons or a removal of entire hydrogen atoms. It is a catabolic reaction.

4. **The answer is C.** (II D 1–10, Figure 15–2) Pyruvic acid is the product of glycolysis and is oxidized to coenzyme A, which enters the citric acid cycle. In one turn of the cycle, hydrogens are released to reduce two molecules of NAD, and two molecules of carbon dioxide are produced.

5. **The answer is C.** (II E 1–3) The last of the molecules in the electron transport chain is cytochrome a_3, which is oxidized by losing two electrons to oxygen. If that reaction is blocked, aerobic respiration and oxidative phosphorylation would come to a halt. This is why cyanide is a fast-acting, lethal poison.

6. **The answer is B.** (III B 3) Ketone bodies are normal products of fatty acid oxidation and are generally metabolized within cells as rapidly as they are formed. An increased number of ketone bodies in the blood or appearing in the urine would reflect an increased rate of fat catabolism.

7. **The answer is A.** (II G 1–2; III B 3 c; IV D 2; V B 1–3) During starvation or prolonged fasting, the body will utilize fat and protein catabolism in order to keep the blood glucose levels at a normal level. The glucose concentration would not be elevated. After glycogen and fat stores are depleted, protein catabolism will result in negative nitrogen balance. The glycogen stores in the liver are depleted first to maintain an adequate blood glucose level. The liver provides glucose from other sources and fatty acids are catabolized for energy to spare the glucose level.

8. **The answer is B.** (VI C 1–3) If energy output exceeds input, weight loss will result. The metabolic rate is increased with greater body size and surface area and tends to decrease with lesser food intake over time. This is why weight loss is initially rapid during a reducing diet and then slows down.

9. **The answer is D.** (VI D 1–5) Vegetables and fruits lack essential amino acids, but they are a source of carbohydrate and contain far less fat and no cholesterol. Because of their vitamin, low fat, and fiber content, plant foods should make up the greater part of the diet. Complete protein can be obtained by combining them with the consumption of small amounts of meats and dairy products.

10. **The answer is A.** (VII C 2 a–d) Convection is the transfer of heat to a moving medium, such as water, when the temperature of the medium is lower than the body temperature. Conduction is heat transfer between objects in contact; radiation is heat transfer by infrared rays between objects not in contact. Evaporation is the conversion of water from a liquid to a gas, which takes heat energy.

16 Urinary System

I. **INTRODUCTION.** The urinary (renal) system consists of the organs that produce urine and eliminate it from the body. It is a major system in the maintenance of homeostasis (constancy of the internal environment).

   A. **Components.** The urinary system consists of two **kidneys**, which produce urine; two **ureters**, which carry urine to a single **urinary bladder** for temporary storage; and the **urethra**, which conveys the urine to the outside of the body through the **external urethral orifice** (Color Plate 15).

   B. **Kidney functions**

   1. **Elimination of organic wastes.** The kidneys excrete urea, uric acid, creatinine, and the breakdown products of hemoglobin and hormones.

   2. **Regulation of the concentrations of important ions.** The kidneys excrete sodium, potassium, calcium, magnesium, sulfate, and phosphate ions. The excretion of these ions is balanced with their intake and excretion by other routes, such as the gastrointestinal tract or skin.

   3. **Regulation of the acid-base balance of the body.** The kidneys control the excretion of hydrogen (H^+), bicarbonate (HCO_3^-), and ammonium (NH_4^+) ions and produce an acid or alkaline urine, depending on the body's requirements.

   4. **Regulation of RBC production.** The kidneys release erythropoietin, which regulates the production of RBCs in the bone marrow.

   5. **Regulation of blood pressure.** The kidneys regulate the fluid volume, which is essential to the regulation of blood pressure, and also produce the enzyme renin. Renin is an important component of the renin-angiotensin-aldosterone mechanism, which increases blood pressure and water retention.

   6. **Limited control of blood glucose and blood amino acid concentration.** By excreting excess amounts of glucose and amino acids, the kidneys have some responsibility for the concentration of those nutrients in the blood.

   7. **Elimination of toxic substances.** The kidneys eliminate pollutants, food additives, drugs, or other chemicals foreign to the body.

   C. **Gross anatomy** of the kidneys

   1. **Appearance.** The kidneys are bean-shaped, dark red organs, approximately 5 in. long and 1 in. thick (about the size of a clenched fist). Each kidney weighs 125 to 175 g in men and 115 to 155 g in women.

2. **Location**
   a. The kidneys are located high on the posterior abdominal wall adjacent to the last two pairs of ribs. They are retroperitoneal and lie between the muscles of the back and the peritoneum of the upper abdominal cavity. Each is capped by an adrenal gland.
   b. **The right kidney** is slightly lower in position than the left because of the liver on the right side.
3. **Connective tissue coverings.** Each kidney is surrounded by three layers of connective tissue.
   a. The **renal fascia** is the outermost covering. It anchors the kidney to surrounding structures and maintains the position of the organ.
   b. The **perirenal fat** is adipose tissue enclosed by the renal fascia. It cushions the kidney and helps to hold it in place.
   c. **The fibrous (renal) capsule** is the smooth, transparent membrane that directly covers the kidney and can easily be stripped from it.

D. **Internal structure of the kidney** (Figure 16–1)
   1. The **hilus** (hilum) is a concavity on the medial border of the kidney.
   2. The **renal sinus** is a cavity filled with fat that opens at the hilus. It forms the attachment for entry and exit of the ureter, the renal artery and vein, nerves, and lymphatics.
   3. The **renal pelvis** is the expanded proximal end of the ureter. It is continuous with two to three **major calyces**, which are cavities that reach up into the glandular, urine-producing, portion of the kidney. Each major calyx branches further into several (8 to 18) **minor calyces**.
   4. The **renal parenchyma** is kidney tissue that surrounds the structures of the renal sinus. It is divided into an inner **medulla** and an outer **cortex**.
      a. The **medulla** consists of triangular masses called **renal pyramids**. The narrow end of each pyramid, the **papilla**, fits into a minor calyx and is perforated by the openings of urinary collecting ducts.
      b. The **cortex** is composed of the tubules and blood vessels of **nephrons**, which are the structural and functional units of the kidneys. The cortex dips in between adjacent medullary pyramids to form **renal columns**, which are composed of collecting tubules that drain into collecting ducts.
   5. Kidney **lobes** subdivide the kidney. Each lobe consists of a renal pyramid, the adjacent renal columns, and the overlying cortical tissue.

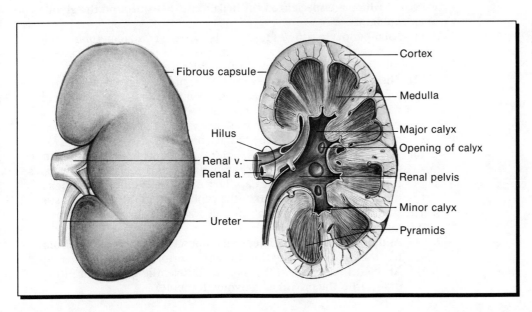

Figure 16–1. *A*, External view of the kidney. *B*, Frontal section through the left kidney.

Figure 16-2. *A*, Kidney, *B*, Enlarged section illustrating locations of the cortical and juxtaglomerular nephrons. *C*, Structure of a nephron including the blood vessels. The arrows indicate the direction of urine flow.

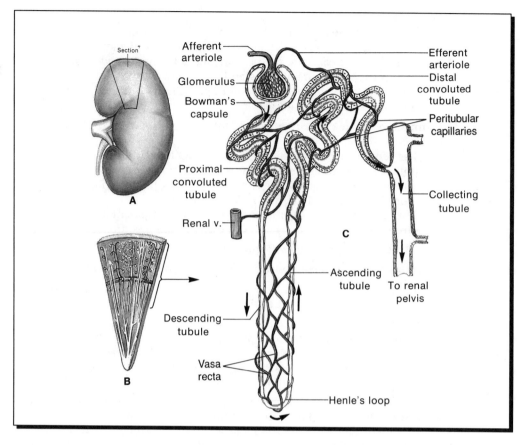

E. **Structure of a nephron.** Each kidney contains 1 to 4 million nephrons, which are urine-forming units. Each nephron has a vascular (capillary) component and a tubular component (Figure 16–2).

1. The **glomerulus** is a tuft of capillaries, which is surrounded by a double-walled, epithelial capsule called **Bowman's capsule**. Together, the glomerulus and Bowman's capsule form a **renal corpuscle**.
   a. The **visceral** layer of Bowman's capsule is the internal layer of epithelium. The visceral layer cells are modified into **podocytes** ("foot-like cells"), which are specialized epithelial cells that surround the glomerular capillaries.
      (1) Each podocyte adheres to the outer surface of a glomerular capillary by several long primary processes that bear secondary processes called foot processes or **pedicels** ("little feet").
      (2) The pedicels interdigitate with similar processes from adjacent podocytes. The narrow spaces between interdigitating pedicels are **filtration slits** (slit pores) approximately 25 nm wide. Each slit is covered by a thin membrane, which allows the passage of some molecules and restricts the passage of others.
      (3) The **glomerular filtration barrier** is the tissue barrier that separates the blood in the glomerular capillaries from the space in Bowman's capsule. It consists of the **capillary endothelium**, the **basement membrane** (basal lamina) of the capillary, and the **filtration slits**.
   b. The **parietal** layer of Bowman's capsule forms the outer limit of the renal corpuscle.
      (1) At the **vascular pole** of the renal corpuscle, the **afferent arteriole** enters the glomerulus and the **efferent arteriole** exits the glomerulus.
      (2) At the **urinary pole** of the renal corpuscle, the glomerular filtrate drains into the proximal convoluted tubule.

2. The **proximal convoluted tubule** (PCT) is approximately 15 mm in length and highly coiled. On the surface that faces the lumen of the PCT, the cuboidal epithelial cells have abundant microvilli (**brush border**), which increase the luminal surface area.
3. **Loop of Henle.** The proximal convoluted tubule leads into the **descending limb of the loop of Henle**, which dips down into the medulla, forms a sharp hairpin turn (the loop), and returns as the thick **ascending limb of the loop of Henle.**
    a. Cortical nephrons are located in the outer part of the cortex. They have short loops that extend to the upper third of the medulla.
    b. Juxtamedullary nephrons are located nearer to the medulla. They have long loops that extend down into the medullary pyramids.
4. The **distal convoluted tubule** (DCT) is also highly coiled, about 5 mm in length, and forms the last segment of the nephron.
    a. Along its path, the distal tubule establishes contact with the wall of the afferent arteriole. The portion of the tubule that contacts the arteriole contains modified cells called the **macula densa**. The macula densa functions as a chemoreceptor and is stimulated by a decrease in sodium ions.
    b. The wall of the afferent arteriole next to the macula densa contains modified smooth muscle cells called **juxtaglomerular cells**. The juxtaglomerular cells are stimulated by a decrease in blood pressure to produce renin.
    c. Together, the macula densa, the juxtaglomerular cells, and the **mesangial cells** between them form the **juxtaglomerular apparatus** (JGA), which is important in blood pressure regulation.
5. **Collecting tubules and ducts.** As each collecting tubule descends in the cortex, it drains an increasing number of distal convoluted tubules. The collecting tubules join and form larger, straight **collecting ducts**. The collecting ducts form larger tubes, which drain the urine into minor calyces, which empty into the kidney pelvis via the major calyces. From the kidney pelvis the urine is funneled into the ureter, which leads to the urinary bladder.

F. Blood supply
1. A **renal artery** branches off the abdominal aorta to supply each kidney and enters the hilus via an anterior and a posterior branch.
2. The anterior and posterior branches of the renal artery give rise to **interlobar arteries**, which pass between the renal pyramids.
3. The **arcuate arteries** arise from the interlobar arteries at the level of the junction between the cortex and the medulla.
4. **Interlobular arteries** branch off at right angles from the arcuate arteries and pass into the cortex.
5. **Afferent arterioles** arise from the interlobular arteries. An afferent arteriole gives rise to about 50 capillaries, which form the **glomerulus**.
6. An **efferent arteriole** leaves each glomerulus and gives rise to another capillary network, the **peritubular capillaries**, which surrounds the proximal and distal tubules to nourish them and carry away reabsorbed materials.
    a. Efferent arterioles from the glomeruli of cortical nephrons enter a **peritubular capillary network** that surrounds the proximal and distal convoluted tubules of that nephron.
    b. The efferent arterioles from the glomeruli of juxtaglomerular nephrons have extensions of long, straight capillary vessels called the **vasa recta**, which descend into the medullary pyramids. The loops of the vasa recta form hairpin turns that run alongside the loops of Henle. They permit an interchange of materials between the loops of Henle and the capillaries and play a role in the concentration of urine.
7. The peritubular capillaries drain into **cortical veins**, which unite to form **interlobular veins**.

8. The **arcuate veins** receive blood from the interlobular veins. They empty into **interlobar veins**, which converge to empty into **renal veins** that leave the kidney to join the **inferior vena cava**.

II. **FORMATION OF URINE.** The kidney produces urine containing metabolic wastes and regulates the composition of body fluids by three main processes: glomerular filtration, tubular reabsorption, and tubular secretion.

A. **Glomerular filtration**
   1. **Definition.** Glomerular filtration is the transfer of fluid and solutes from the glomerular capillaries along a pressure gradient into Bowman's capsule. Filtration is assisted by the following factors:
      a. The **capillary membranes of the glomerulus are more permeable** than other capillaries in the body so that filtration occurs very rapidly.
      b. The **blood pressure in glomerular capillaries is higher** than in other capillaries because the diameter of the efferent arteriole is smaller than the diameter of the afferent arteriole.
   2. **Mechanics of glomerular filtration**
      a. The glomerular **hydrostatic (blood) pressure** forces fluids and solutes out of the blood and into Bowman's capsular space.
      b. Two forces oppose glomerular hydrostatic pressure.
         (1) The **hydrostatic pressure exerted by the fluid in Bowman's capsule** tends to move fluid out of the capsule into the glomerulus.
         (2) The **colloid osmotic pressure in the glomerulus**, which is generated by the plasma proteins, is a pulling force for fluid from Bowman's capsule to enter the glomerulus.
      c. The **effective filtration pressure** (EFP) is the **net** driving force. It is the difference between the forces tending to force fluid out of the glomerulus into Bowman's capsule and the forces tending to move fluid into the glomerulus from Bowman's capsule.

      EFP = (glomerular hydrostatic pressure) − (capsular pressure)
      + (glomerular colloid osmotic pressure)

   3. The **glomerular filtration rate (GFR) is the amount of filtrate formed per minute** in all the nephrons of both kidneys. In males, the rate is about 125 ml/min or 180 L in 24 hours; in females, about 110 ml/min.
   4. **Factors affecting the GFR**
      a. **Effective filtration pressure.** The GFR is directly proportional to the EFP and change in any of the pressures will affect the GFR. The degree of constriction of the afferent and efferent arterioles determines the renal blood flow, and consequently the glomerular hydrostatic pressure.
         (1) Constriction of the **afferent** arteriole reduces blood flow and **decreases** the glomerular filtration rate.
         (2) Constriction of the **efferent** arteriole causes a backup blood pressure in the glomerulus and **increases** the GFR.
      b. **Renal autoregulation.** Intrinsic autoregulatory mechanisms in the kidney prevent changes in renal blood flow and GFR as a result of normal physiological variations in mean arterial blood pressure. Such autoregulation operates over a wide range of blood pressure (between 80 mm Hg and 180 mm Hg).
         (1) When the mean arterial pressure (normally 100 mm Hg) is increased, the afferent arteriole constricts to reduce the renal blood flow and reduce the GFR. When mean arterial pressure is decreased, vasodilation of the afferent arteriole occurs to increase GFR. Thus, major changes in GFR are prevented.
         (2) Autoregulation involves feedback mechanisms from stretch receptors in the arteriole walls and from the juxtaglomerular aparatus.

(3) Despite autoregulatory mechanisms, an increase in arterial pressure does result in a small increase in GFR. Because so much glomerular filtrate is produced daily, even a small change causes an increased urine output.
- c. **Sympathetic stimulation.** An increase in sympathetic impulses, as would occur in stress, causes constriction of the afferent arteriole, reduces renal blood flow into the glomerulus, and results in a **decrease** in GFR.
- d. **Obstruction of the urinary flow** by kidney or ureteral stones increases the hydrostatic pressure in Bowman's capsule and causes a **decrease in GFR.**
- e. **Starvation, a very low protein diet, or liver disease** would decrease the colloid osmotic pressure of the blood and thereby **increase the GFR.**
- f. **Various kidney diseases** that cause an increased permeability of glomerular capillaries would **increase the GFR.**

5. **Composition** of the glomerular filtrate
    - a. The filtrate in Bowman's capsule is identical to that of plasma with respect to **water and solutes of low molecular weight,** such as glucose, chloride, sodium, potassium, phosphate, urea, uric acid, and creatinine.
    - b. Small amounts of **plasma albumin** may be filtered, but most of it is reabsorbed and does not normally appear in the urine.
    - c. **RBCs and protein are not filtered.** Their appearance in the urine indicates an abnormality. The presence of white blood cells usually indicates a bacterial infection in the lower urinary tract.

B. **Tubular reabsorption.** Most of the filtrate (99%) is selectively reabsorbed in the kidney tubules by **passive diffusion down chemical or electrical gradients, active transport against such gradients,** or **facilitated diffusion.** Approximately 85% of the sodium chloride and water and all of the glucose and amino acids in the glomerular filtrate are absorbed in the **proximal convoluted tubule,** although reabsorption occurs from all parts of the nephron.

1. **Reabsorption of sodium ions**
    - a. Sodium ions are **passively transported** from the proximal convoluted tubule lumen into the tubule epithelial cells, which have a lower concentration of sodium ions, by **facilitated diffusion** (with a carrier).
    - b. Sodium ions then are **actively transported** out of the epithelial cells into the interstitial fluid near the peritubular capillaries via the sodium-potassium pump.

2. **Reabsorption of chlorine ions and other negative ions**
    - a. As the positive sodium ions move passively from the tubular fluid to the cells and actively from the cells to the peritubular interstitial fluid, an electrical imbalance is created that favors the **passive following of negative ions.**
    - b. Therefore, the negative chlorine and bicarbonate ions passively diffuse into the epithelial cells from the lumen and follow the sodium outward into the peritubular fluid and peritubular capillaries.

3. **Reabsorption of glucose, fructose, and amino acids**
    - a. Glucose and amino acids share carriers with sodium ions and are moved by **cotransport.**
    - b. **Transport maximum.** Carriers in the tubular cell membranes have a maximum reabsorption capacity for glucose, many amino acids, and some other reabsorbed substances. This amount is designated as the **transport maximum (Tm).**
    - c. The Tm for glucose is the maximum amount that can be transported (reabsorbed) per minute and is approximately 200 mg glucose/100 ml plasma. When the blood level of glucose exceeds the Tm, the **renal plasma threshold** for glucose is exceeded and glucose appears in the urine (glycosuria).

4. **Reabsorption of water.** Water follows the movement of sodium ions by os-

mosis. It moves from a high water concentration in the lumen of the proximal convoluted tubule to a low water concentration in the interstitial fluid and peritubular capillaries.

5. **Reabsorption of urea.** All the urea formed daily is filtered through the glomerulus. About 50% is passively reabsorbed due to a diffusion gradient created when water is reabsorbed. Therefore, about 50% of the urea filtered is excreted in the urine.

6. **Reabsorption of other inorganic ions,** such as potassium, calcium, phosphate, and sulfate, and of a number of organic ions is by active transport.

C. **Tubular secretion** mechanisms are **active processes** that transfer substances **out of the blood in peritubular capillaries** across the tubular cells **into the tubular fluid** for elimination in the urine.

1. Substances such as hydrogen, potassium, and ammonium ions, the metabolic endproducts creatinine and hippuric acid, and certain drugs (penicillin), are actively secreted into the tubules.

2. Hydrogen and ammonium ions are exchanged for sodium ions in the distal convoluted and collecting tubules. The selective tubular secretion of hydrogen and ammonium ions helps regulate plasma pH and the acid-base balance of the body fluids.

3. Tubular secretion is an important mechanism for the elimination of unwanted or foreign chemicals.

III. CONCEPT OF CLEARANCE

A. The kidney functions to clear the blood plasma of waste substances such as urea and other nonprotein nitrogenous wastes formed as a result of metabolic processes. When plasma filters through the glomerulus and passes through the nephron tubules, it becomes cleared of unreabsorbed or partially reabsorbed substances.

B. **Plasma clearance,** expressed in ml/min, is the volume of blood that is cleared of a substance per minute. It can be calculated by the following formula:

$$\text{Plasma clearance (ml/min)} = \frac{\text{urinary excretion rate (mg/min)}}{\text{plasma concentration (mg/ml)}}$$

1. **Example of plasma clearance using urea.** If the quantity of urea entering the urine per minute is 12 mg (urinary excretion rate) and the concentration of urea in the plasma is 0.2 mg/ml (plasma concentration), the plasma clearance of urea is 60 ml per minute.

2. Thus, only slightly more than half of the urea filtered through the glomeruli in each passage is excreted in the urine.

IV. URINE CONCENTRATION AND DILUTION MECHANISMS

A. **Urine volume.** The daily volume of urine produced varies from 600 ml to more than 2,500 ml.

1. When the urine volume is high, the wastes are excreted in a dilute solution, **hypotonic (hypoosmotic) to plasma.** The specific gravity of the urine is close to that of water (about 1.003).

2. When it is necessary for the body to conserve water, the urine produced is concentrated so that a small volume can still hold the same amount of waste that must be eliminated. The solute concentration is much greater, the urine is hypertonic **(hyperosmotic) to plasma,** and the specific gravity of the urine is high (over 1.030).

B. **Regulation of urine volume.** The production of a small volume of concentrated urine or a large volume of dilute urine is regulated by hormonal mechanisms and renal urine-concentrating mechanisms.
   1. **Hormonal mechanisms**
      a. **Antidiuretic hormone** (ADH) increases the permeability of the distal tubules and the collecting ducts to water and results in water reabsorption and less urine volume.
         (1) **Site of synthesis and secretion.** ADH is synthesized by nerve cell bodies in the supraoptic nucleus of the hypothalamus and stored in their nerve fibers in the posterior pituitary. It is released by impulses to the nerve fibers.
         (2) **Stimuli to ADH secretion**
            (a) **Osmotic**
               (i) The hypothalamic neurons are **osmoreceptors** and are sensitive to changes in the concentration of sodium ions and other solutes in the intercellular fluid around them.
               (ii) An increase of plasma osmolarity, as would occur during dehydration, stimulates the osmoreceptors to send impulses to the posterior pituitary gland to release ADH. Water is reabsorbed from the kidney tubules and a small volume of concentrated urine is produced.
               (iii) A decrease in plasma osmolarity results in less ADH secretion, less water reabsorption from the kidneys, and the production of a large volume of dilute urine.
            (b) **Blood volume and blood pressure.** Baroreceptors in the blood vessels (in the veins, right and left atria, pulmonary vessels, carotid sinuses, and aortic arch) monitor blood volume and blood pressure. A decrease in volume and pressure increases ADH secretion; an increase in volume and pressure decreases ADH secretion.
            (c) **Other factors.** Pain, anxiety, exercise, narcotic analgesics, and barbiturates increase ADH secretion. Alcohol decreases ADH secretion.
      b. **Aldosterone** is a steroid hormone secreted by the cortical cells of the adrenal gland. It acts on the distal tubules and the collecting ducts to promote active **absorption of sodium ions** and active **secretion of potassium ions.** The renin-angiotensin-aldosterone mechanism, which promotes salt and water retention, is described in Chapter 17.
   2. **Countercurrent systems in the loop of Henle and the vasa recta** enable the osmotic reabsorption of water from the tubules and collecting ducts into the more concentrated medullary interstitial fluid **under the influence of ADH.** The reabsorption of water allows for body conservation of water and the excretion of urine more concentrated than normal body fluids.
C. **The countercurrent multiplier system in the loop of Henle**
   1. A **countercurrent system** is one in which the **inflow** into a U-shaped tube (such as the loop of Henle) runs parallel to, in close proximity with, and in an opposite direction to the **outflow** from the tube. A countercurrent **multiplier** system is a countercurrent system assisted by active transport.
   2. As the iso-osmotic glomerular filtrate enters and passes through the loop of Henle, it becomes progressively more concentrated (hyperosmotic) at the bottom of the loop (Figure 16-3):
      a. The **descending** limb of the loop of Henle is highly permeable to water and relatively impermeable to solutes such as NaCl. It does not actively transport any substance.
      b. The **ascending** limb is impermeable to water but relatively permeable to NaCl. Chlorine ions are actively pumped out of the filtrate in the ascending limb into the peritubular interstitial fluid and the sodium ions follow because of their electrical attraction to the negative chlorine ions. This increases the osmotic concentration of NaCl in the interstitial fluid.

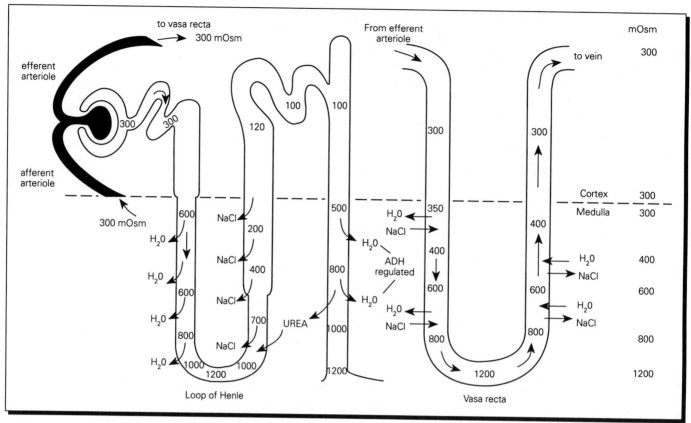

Figure 16–3. *The countercurrent system in the loop of Henle and the countercurrent exchanger system in the vasa recta. These mechanisms serve to concentrate NaCl in the interstitial fluid and allow the excretion of a concentrated urine under the influence of antidiuretic hormone.*

    c. Due to the increased osmolarity of the interstitial fluid, water moves out of the descending limb and the loop into the peritubular interstitial fluid by osmosis. This produces a greater concentration of solutes in the tubular fluid as it rounds the hairpin turn of the loop of Henle. Its osmolarity increases to a maximum concentration of 1,200 milliosmols/l, which is four times the normal concentration of body fluids.

    d. As the filtrate moves along the ascending limb, it continues to lose sodium chloride. NaCl diffuses passively out of the loop at the beginning of the ascending limb and is actively transported out as the filtrate passes up the ascending limb. Because the ascending limb is impermeable to water, the water cannot follow, and the tubular fluid becomes progressively more dilute (hypo-osmotic) as it passes upward toward the cortex.

    e. Some of the NaCl leaving the ascending limb moves by diffusion into the descending limb from the interstitial fluid, thereby increasing the solute concentration in the descending limb. Also, new NaCl in the glomerular filtrate continuously enters the tubule inflow to be transported out of the ascending limb into the peritubular interstitial fluid. Thus, the recycling mechanism multiplies the concentration of NaCl.

    f. The result is that the interstitial fluid surrounding the loop of Henle contains a high concentration of salt, as does the filtrate in the loop of Henle. A **vertical concentration gradient** from the cortex (iso-osmotic) to the medulla (hyperosmotic) is maintained.

3. The **medullary recycling of urea** helps maintain the vertical concentration gradient in the interstitial fluid and the loop of Henle.

    a. Urea diffuses passively out of the collecting ducts into the medullary interstitial fluid. Some urea diffuses from the medullary fluid into the descending limb.

    b. Thus, urea is recirculated between the collecting tubules and the descending limb. The high concentration of urea in the medullary interstitial fluid contributes to its osmolarity. This increases the osmotic movement of water out of the descending limb and increases the filtrate's concentration of NaCl in the descending limb.

D. The **countercurrent exchanger mechanism** in the blood vessels and the renal tubules aids the countercurrent multiplier mechanism. If the blood circulation removed the solutes from the medullary extracellular fluid, the concentration gradient could not be maintained. The vertical gradient of hyperosmolarity is not disturbed by the blood circulation for the following reasons:
1. The capillaries of the **vasa recta** function as countercurrent exchangers, because the direction of blood flow around the loop of Henle is opposite to the direction of filtrate flow around the loop.
2. The walls of the vasa recta are permeable to NaCl and water. As blood flows down the descending vessel of the vasa recta, which is parallel to the ascending limb of the tubule, it becomes hyperosmotic as it picks up sodium and chlorine ions and loses some water. At the bottom of the capillary loop, the plasma osmolarity is identical to that of the surrounding interstitial fluid.
3. As the blood flows back up the ascending vessel of the vasa recta, which is parallel to the descending limb of the tubule, salt diffuses back out of the capillary and the water reenters the vessel. The osmolarity of the blood decreases as it flows toward the cortex.
4. Because of the passive exchange of salt and water between the vasa recta and the medullary interstitial fluid and the fact that the blood flow in the vasa recta is relatively slow, the blood leaves the medulla only slightly hyperosmotic to arterial blood. The concentration gradient in the medullary extracellular fluid is maintained.

E. **Excretion of a concentrated urine**
1. The hypo-osmotic (dilute) filtrate in the ascending limb of Henle's loop enters the distal convoluted tubule and travels down the collecting duct toward the ureter. As a result of the countercurrent systems, the interstitial fluid surrounding the collecting ducts is hyperosmotic and the necessary concentration gradient for osmosis of water out of the ducts is established.
2. The collecting ducts are impermeable to water in the absence of ADH. Water will leave the collecting ducts by osmosis in the presence of ADH.
3. Long loops of Henle are required for the countercurrent multiplier and exchange systems to work. Fish, with no loops of Henle, and animals with short loops (beavers) cannot produce a concentrated urine.

F. **Excretion of a dilute urine.** When ADH is not present, the collecting ducts become almost impervious to water. Additional reabsorption of solutes in the distal tubules and the collecting ducts results in a decrease of osmolarity in the tubular fluid to as little as 60 to 70 mOs/L. The dilute urine enters the kidney pelvis to be excreted.

V. CHARACTERISTICS OF URINE

A. **Composition.** Urine is 95% water and contains the following principal solutes:
1. **Nitrogenous wastes** include urea from protein deamination, uric acid from catabolism of nucleic acids, and creatinine from creatine phosphate breakdown in muscle tissue.
2. **Hippuric acid** is a by-product of fruit and vegetable ingestion.
3. **Ketone bodies**, which are produced in fat metabolism, are normal constituents in small amounts.
4. **Electrolytes** include sodium, chloride, potassium, ammonium, sulfate, phosphate, calcium, and magnesium ions.
5. **Hormones** or **hormone catabolites** normally are present in the urine.
6. **Various toxins or foreign chemicals, pigments, vitamins, or enzymes** normally may be present in small quantities.

7. **Abnormal constituents** include albumin, glucose, RBCs, large amounts of ketone bodies, casts (formed when substances harden in the tubules and are flushed out), and kidney stones or calculi.

B. **Physical properties**

1. **Color.** Urine is pale yellow when dilute, darker amber when concentrated. Fresh urine is generally clear and may become cloudy on standing.

2. **Odor.** Urine has a characteristic odor and becomes more ammonia-like after standing. The odor may also vary with the diet; for example, after eating asparagus. In uncontrolled diabetics, acetone produces a sweetish odor to the urine.

3. **Acidity or alkalinity.** The pH of urine varies between 4.8 and 7.5 and is usually about 6.0, but depends on the diet. High protein ingestion increases acidity, while a vegetarian diet increases alkalinity.

4. The **specific gravity** ranges from 1.001 to 1.035, depending on the concentration of the urine.

VI. THE URETERS, URINARY BLADDER, AND URETHRA

A. The **ureters** are muscular, paired, tubular extensions of the renal pelvis that lead to the urinary bladder.

1. Each ureter is 25 cm to 30 cm long and 4 mm to 6 mm in diameter. It is narrower along its course at three places: at the origin of the ureter in the renal pelvis, at the point where it crosses the pelvic brim, and at its junction with the bladder. A kidney stone may lodge in the ureter at these sites, which produces pain known as renal colic.

2. The wall of the ureters is composed of three layers of tissue: an outer fibrous layer, a middle muscularis of inner longitudinal and outer circular smooth muscle, and an inner mucosa of epithelium that secretes a protective mucous coating.

3. The muscle layers have intrinsic peristaltic activity. Waves of peristalsis deliver urine to the bladder in spurts.

B. The **urinary bladder** is a hollow, muscular organ that functions as a storage container for urine.

1. **Location.** In males, the bladder lies just behind the pubic symphysis and in front of the rectum. In females, it lies slightly lower and is situated below the uterus in front of the vagina. The bladder is the size of a walnut and lies deep in the pelvis when empty; it becomes pear-shaped and may reach as high as the umbilicus in the abdominopelvic cavity when distended with urine.

2. **Structure.** The urinary bladder is supported in the pelvic cavity by folds of peritoneum and condensations of fascia.
    a. The **wall** of the bladder consists of four layers.
        (1) The **serosa** is the outermost layer. It is a part of the peritoneal lining of the abdominal cavity and is present only in the upper part of the bladder.
        (2) The **detrusor muscle** is the middle layer. It is arranged in bundles of smooth muscle at right angles to each other. This ensures that during urination, the bladder will contract uniformly in all directions.
        (3) The **submucosa** is a connective tissue layer that lies under the mucosa and connects it to the muscularis.
        (4) The **mucosa** is the innermost layer. It is an epithelial lining composed of transitional epithelium. In a relaxed bladder, the mucosa contains rugae (folds), which flatten and stretch out when urine accumulates in the bladder.

b. The **trigone** is a smooth, triangular, relatively unexpandable area located internally at the base of the bladder. Its corners are formed by three openings. At the upper corners of the trigone, the two ureters open into the bladder. The urethra exits from the bladder at the apex of the trigone.

C. The **urethra** conveys urine from the urinary bladder to the exterior of the body.
1. In males, the urethra carries both seminal fluid and urine but not at the same time. The male urethra is 20 cm long and passes through the prostate gland and the penis.
    a. The **prostatic urethra** is surrounded by the prostate gland. It receives two ejaculatory ducts, each of which are formed by the union of the ductus deferens and the duct of the seminal vesicle gland, and receives numerous ducts from the prostate gland.
    b. The **membranous urethra** is the shortest (1 cm to 2 cm) portion. It is thin-walled and surrounded by the skeletal muscle of the external urethral sphincter.
    c. The **cavernous (penile, spongy) urethra** is the longest portion. It receives the ducts of the bulbourethral glands and extends to the **external urethral orifice** on the end of the penis. Just prior to the opening, the urethra enlarges to form a small dilation, the **fossa navicularis**. The cavernous urethra is surrounded by the **corpus spongiosum**, which is a network of large venous spaces.
2. In females, the urethra is short (3.75 cm). It opens to the outside by the **external urethral orifice** located in the vestibule between the clitoris and the vaginal opening. The urethral glands, which are homologous to the prostate glands in the male, open into the urethra.
3. The length of the male urethra tends to inhibit bacterial invasions of the bladder (cystitis), which are much more common in females.

D. **Micturition (urination)** is dependent on parasympathetic and sympathetic innervation as well as voluntary nerve impulses. The expulsion of urine requires the active contraction of the detrusor muscle.
1. The portion of the trigone muscle that surrounds the urethral exit acts as an **internal urethral sphincter** to keep the urethra closed. It is supplied by parasympathetic neurons.
2. The **external urethral sphincter** is formed of skeletal muscle fibers from the transverse perineal muscle and is under voluntary control. The pubococcygeus portion of the levator ani muscle contributes to the formation of the sphincter.
3. The **micturition reflex** is initiated when distention of the bladder by 300 ml to 400 ml of urine stimulates stretch receptors in the bladder wall.
    a. Impulses to the spinal cord are relayed to the brain and result in parasympathetic impulses via **pelvic splanchnic** nerves to the bladder.
    b. The micturition reflex causes contraction of the detrusor muscle and relaxation of the internal and external sphincters to result in emptying of the bladder.
    c. In males, sympathetic fibers supply the area of the exit of the urethra and constrict it to prevent reflux of semen into the bladder at the time of orgasm.
4. Inhibition of the micturition reflex through voluntary control of the external sphincter is a learned response.
    a. Voluntary inhibition is dependent on the integrity of nerves to the bladder and urethra, the projection tracts in the spinal cord to and from the brain, and the motor area of the cerebrum. Injury to any of these results in incontinence.
    b. Voluntary control of urination ("toilet training") is a learned response. It cannot be exercised in an immature CNS and generally should be deferred until at least 18 months of age.

VII. DISORDERS OF THE URINARY SYSTEM

A. **Cystitis** is inflammation of the urinary bladder. It may be due to a bacterial infection (usually ***Escherichia coli***) that has spread from the urethra or may be an allergic reaction or the result of mechanical irritation of the bladder. The symptoms are frequent and painful urination (dysuria), which is often accompanied by blood in the urine (hematuria).

B. **Glomerulonephritis** is inflammation of the nephrons, especially of the glomeruli.
   1. Acute glomerulonephritis is often the result of an immune response to toxins from certain bacteria (Group A beta streptococci).
   2. Chronic glomerulonephritis damages not only the glomeruli but also the tubules. It may result from a streptococcal infection, but it may also be secondary to other systemic diseases or follow acute glomerulonephritis.

C. **Pyelonephritis** is inflammation of the kidney and renal pelvis caused by bacterial infection. It may originate in the lower urinary tract (bladder) and spread upward through the ureters, or it may be an infection carried by the blood or lymph to the kidneys. A urinary tract obstruction due to an enlarged prostate gland, a kidney stone, or a congenital defect predisposes to pyelonephritis.

D. **Kidney stones** (urinary calculi) are formed from precipitated salts of calcium, magnesium, uric acid, or cysteine. Small stones may be passed in the urine; larger stones may become lodged in the ureter and cause intense pain (renal colic), which radiates from the kidney area to the groin.

E. **Renal failure** is the loss of kidney function. It results in salt, water, and nitrogenous waste (urea and creatinine) retention and a drop in urine volume (oliguria). With treatment of the condition causing renal failure, the prognosis is good. Untreated, renal failure may result in complete renal shutdown and death.
   1. **Acute renal failure** develops suddenly and can usually be treated successfully. It is characterized by sudden oliguria that may be followed by total absence of urine production (anuria). It is caused by a decreased blood flow to the kidneys as a result of trauma or injury, acute glomerulonephritis, hemorrhage, an incompatible blood transfusion, or severe dehydration.
   2. **Chronic renal failure** is a severe, progressive condition caused by diseases that result in the destruction of the kidney parenchyma, such as chronic glomerulonephritis or pyelonephritis, trauma, or diabetic nephropathy (kidney disease resulting from diabetes mellitus). It is treated by hemodialysis ("artificial kidney") or kidney transplant.

Study Questions

Directions: Each question below contains four suggested answers. Choose the **one best** response to each question.

1. All of the following structures are in direct association with the renal hilus EXCEPT the
   (A) renal pelvis
   (B) renal artery
   (C) renal corpuscle
   (D) renal vein

2. Which of the following statements concerning kidney function is accurate?
   (A) The only substances eliminated by the kidneys are toxic and nitrogenous wastes.
   (B) In order to be excreted, all substances first must be filtered by the kidneys.
   (C) Blood pressure regulation occurs through the formation and release of renin.
   (D) Except for sodium ions, which are excreted by the kidneys, the body regulation of other important inorganic ions, such as calcium and potassium ions, occurs only through the modification of their intake.

3. The juxtaglomerular apparatus consists of cells that are associated with the afferent arteriole and the
   (A) efferent arteriole
   (B) distal convoluted tubule
   (C) ascending portion of Henle's loop
   (D) proximal convoluted tubule

4. The blood supply to the kidney changes from arterial to venous blood
   (A) before the blood enters the glomerular capillaries
   (B) within the capillaries of the glomerulus
   (C) upon leaving the glomerular capillaries
   (D) upon leaving the peritubular capillaries

5. All of the following statements concerning the formation of urine are true EXCEPT
   (A) Urine formation involves the filtration of blood plasma from the glomerulus into Bowman's capsule.
   (B) Less than half of the water that filters from the glomerulus into Bowman's capsule is reabsorbed into the blood.
   (C) The second step in urine formation is reabsorption by the tubular cells of the water and some of the solutes that have filtered from the glomerulus into Bowman's capsule.
   (D) Urine formation involves secretion by the tubular cells into the tubular fluid of certain substances in the peritubular capillaries.

6. Constriction of the afferent arteriole would result in
   (A) an increase in blood flow into the efferent arteriole
   (B) a decrease in glomerular filtration rate
   (C) an increase in hydrostatic (blood) pressure in the glomerulus
   (D) an increase in the protein concentration of the glomerular filtrate

7. Tubular reabsorption
   (A) of urea is about half of the amount filtered through the glomerulus
   (B) of sodium occurs only in the proximal convoluted tubule
   (C) of glucose does not take place in an individual with diabetes mellitus
   (D) is accomplished solely by active transport mechanisms

8. The specific gravity of a substance is the ratio of a unit volume of its weight of an equal volume of distilled water. When there is more water than usual in the blood, the urine produced would most likely have a specific gravity of
   (A) 1.002
   (B) 1.020
   (C) 1.035
   (D) 1.040

9. In a hot environment, an individual who is strenuously exercising can lose 3 L to 4 L of sweat per hour. Which of the following physiological consequences of the loss would **least** likely?
   (A) The plasma volume decreases.
   (B) The circulating levels of antidiuretic hormone increases.
   (C) The circulating levels of aldosterone decreases.
   (D) The plasma osmolarity increases.

10. The countercurrent multiplier system
    (A) results in a decrease in the concentration of sodium from the cortex to the medulla of the kidney
    (B) enables the kidney to produce a urine more concentrated than the body fluids
    (C) occurs primarily in the proximal convoluted tubule
    (D) allows the osmolarity of the urine to remain constant despite a wide variation in urine volume

11. Micturition is a reflex that occurs when _____ impulses cause contraction of the _____.

    (A) sympathetic, trigone
    (B) parasympathetic, detrusor
    (C) parasympathetic, ureters
    (D) parasympathetic, external sphincter

12. The common passageway for metabolic wastes and reproductive material in the male is the

    (A) renal pelvis
    (B) ureter
    (C) urinary bladder
    (D) urethra

Questions 13–18. For each of the following statements concerning kidney function, choose A if the statement is true and choose B if the statement is false.

13. The protein concentration in the collecting duct is half of what it is in the plasma.

14. The glucose concentration in Bowman's capsule is normally zero.

15. The presence of albumin in the collecting tubules is an indication of abnormality.

16. The concentration of sodium in the loop of Henle is greater than it is in Bowman's capsule.

17. Glucose, amino acids, fatty acids, glycerol, and urea are present in the venous end of the peritubular capillaries and in the vasa recta.

18. In the presence of ADH, water moves out of the collecting duct and enters the loop of Henle.

Answers and Explanations

1. **The answer is C.** (I C 1–3; D 1) The structures that enter and leave the kidney at the hilus are the renal artery, renal vein, and the renal pelvis, which is the funnel-shaped proximal end of the ureter. The renal corpuscle is formed by the invagination of the glomerulus into Bowman's capsule of the nephron.

2. **The answer is C.** (I B 1–7) The kidneys eliminate organic wastes; toxic substances; hydrogen, bicarbonate, and ammonium ions; and hormones. The kidneys excrete important ions such as sodium, potassium, calcium, magnesium, phosphate, and sulfate in amounts exactly balanced by the intake and secretion of these substances by other body routes. Renin is a major component of the renin-angiotensin-aldosterone mechanism, which regulates fluid volume and pressure.

3. **The answer is B.** (I E 4 a–c) The juxtaglomerular apparatus is composed of the macula densa, which is a portion of the distal convoluted tubule, the juxtaglomerular cells of the afferent arteriole, and the mesangial cells between the macula densa and the juxtaglomerular cells.

4. **The answer is D.** (I F 1–8) The afferent arteriole enters the glomerulus; the efferent arteriole exits the glomerulus and gives rise to the peritubular capillaries that surround the proximal and distal convoluted tubules. The peritubular capillaries drain into cortical veins.

5. **The answer is B.** (II A–C) Urine production occurs through glomerular filtration, tubular reabsorption, and tubular secretion. Nearly 99% of the water in the glomerular filtrate is reabsorbed as it passes through the tubules.

6. **The answer is B.** (II A 4 a) The major determinant of the glomerular filtration rate is the hydrostatic pressure within the glomerulus. Constriction of the afferent arteriole reduces glomerular hydrostatic pressure and reduces blood flow, thereby decreasing the glomerular filtration rate.

7. **The answer is A.** (II B 1–6) Tubular reabsorption occurs by the processes of passive diffusion, active transport, and facilitated diffusion. Most of the solutes in the filtrate primarily are reabsorbed in the proximal convoluted tubule, but reabsorption takes place all along the nephron. In a patient with diabetes mellitus, glucose appears in the urine when the transport maximum for the reabsorption of glucose is exceeded.

8. **The answer is A.** (IV A 1–2) The specific gravity of urine depends on the total amount of solutes. When a dilute urine is produced, which would occur when there is more water in the blood, the specific gravity would be less. The more solutes in a unit volume of urine, the higher the specific gravity.

9. **The answer is C.** (IV B 1, 2) Loss of fluid results in a decrease in plasma volume and blood pressure and an increase in plasma osmolarity. This would result in the secretion of ADH to cause reabsorption of water from the collecting ducts and the secretion of aldosterone to promote the retention of sodium ions and water and the excretion of potassium ions.

10. **The answer is B.** (IV B 2; C 1–4; D) The countercurrent multiplier system in the loop of Henle maintains a vertical concentration gradient from the isoosmotic cortex to the hyperosmotic medulla. It sets the stage for the passive reabsorption of water from the collecting tubules and ducts in the presence of ADH and allows the production of a concentrated urine.

11. **The answer is B.** (VI D 1) Micturition (urination) is the result of efferent parasympathetic impulses, which are carried in the pelvic splanchnic nerves to initiate relaxation of the internal urethral sphincter and contraction of the detrusor muscle of the urinary bladder.

12. **The answer is D.** (VI C 1) In males, the urethra, which carries urine from the urinary bladder to the exterior, also receives seminal fluid via the ejaculatory duct of the male reproductive tract.

13-18. **The answers are 13–B, 14–B, 15–A, 16–A, 17–A, and 18–B.** (II B; IV C 1–3) Proteins normally are not filtered from the blood into the nephron, so no protein (albumin) would be found in the collecting ducts and tubules. The countercurrent mechanism establishes a vertical osmotic gradient so that the concentration of sodium in the loop of Henle and the medulla is greater than the concentration of sodium in Bowman's capsule. Nutrients are reabsorbed from the nephrons and are carried back into the plasma by the peritubular capillary and vasa recta network. The presence of ADH causes water reabsorption from the collecting tubules and ducts into the capillary network.

17 Fluids, Electrolytes, and Acid-Base Balance

I. INTRODUCTION. Living cells in the body are surrounded by interstitial fluid, which contains the nutrients, gases, and the electrolyte concentrations necessary for the maintenance of normal cell function. Continued cell existence requires the constancy of this internal environment (homeostasis). Important regulatory mechanisms control the volume, composition, and acid-base balance of the body fluids during normal metabolic fluctuations or during abnormalities, such as disease or trauma.

II. BODY FLUIDS

A. **Amount and distribution**
1. The **total body water** (TBW) depends on the age, weight, sex, and degree of obesity. It progressively decreases from birth.
   a. In **infancy**, about 80% of body weight is water. Because infants have a greater surface area in proportion to their body weight, they have a greater **insensible water loss** (the diffusion of water molecules through the cells of the skin). Their fluid requirements also are higher because of their rapid growth and increased metabolism, which results in an increased urine production.
   b. In **adults**, the TBW constitutes 60% of body weight (approximately 40 L) in young males and 50% of body weight (approximately 30 L) in young females.
      (1) The TBW is less in women due to their greater amount of subcutaneous fat. Adipose tissue contains very little (only about 10%) cellular water.
      (2) Obesity may result in a TBW content of only 25% to 30% of body weight.
   c. In **people over 65** years of age, TBW may be only 40% to 50% of body weight.
   d. Infants, the elderly, and the obese are particularly vulnerable to water losses. Water deficit (dehydration) may develop more rapidly through water loss mechanisms such as sweating, fever, diarrhea, and vomiting.
2. **Distribution.** TBW is distributed 50% in muscle, 20% in skin, 20% in other organs, and 10% in blood.

B. **Body fluid compartments**
1. The **intracellular fluid (ICF) compartment** refers to the fluid within the trillions of body cells. About two thirds of the body water is intracellular.

2. The **extracellular fluid (ECF) compartment**, which consists of all the body fluids outside of the cells, contains the remaining one third of body water. It includes several subcompartments.
    a. **Interstitial fluid** is the fluid surrounding the body cells and **lymph** is the fluid in lymphatic vessels. Together they comprise three quarters of the ECF.
    b. **Blood plasma** is the fluid portion of the blood and accounts for about one quarter of the ECF.
    c. **Transcellular fluid**, which amounts to 1% to 3% of body weight, includes all body fluids that are separated from the rest of the ECF by epithelial cell layers. This subcompartment includes sweat; cerebrospinal fluid; synovial fluids; the fluid in peritoneal, pericardial, and pleural cavities; the fluid in the chambers of the eye; and the fluids within the respiratory, digestive, and urinary systems.
3. **Composition of the fluid compartments**
    a. **ECF.** Blood plasma and interstitial fluid are similar in that both contain large quantities of sodium and chloride ions and many bicarbonate ions but little potassium, calcium, magnesium, phosphate, sulfate, and organic acid ions. They are dissimilar in that plasma contains much protein and interstitial fluid contains very little protein.
    b. **ICF.** As a result of the ATP-dependent sodium-potassium pump, the intracellular sodium and potassium ion concentrations are opposite to that in ECF. That is, the concentration of intracellular potassium ions is high and the concentration of intracellular sodium ions is low. The protein concentration inside cells is high, being approximately four times as large as it is in plasma.
4. **Movement of fluid between compartments**
    a. Between the cells and the ECFs
        (1) Water distribution inside and outside of cells is dependent on **osmotic pressure**.
        (2) Osmotic pressures are related to the **total solute concentration (osmolality)** inside and outside the cells. Water will move from a region of lower osmolality to a region of higher osmolality.
        (3) Normally, there is equal osmolality (osmotic pressure) inside and outside of cells and there is no net gain or loss of water into or out of the cells.
        (4) If solute or water is either added to or lost from the extracellular compartment, the osmotic equilibrium is temporarily disturbed. Water then will move into or out of the cells until a new equilibrium is reached.
    b. Between plasma and the interstitial fluid
        (1) Water movement across capillary cell membranes is regulated by **hydrostatic** and **osmotic pressures** according to the forces described by the Starling-Landis hypothesis (see Chapter 11 VII A, B). Excess fluid and protein in the interstitial fluid are removed by the lymphatic system.
        (2) An increase in capillary hydrostatic pressure or a decrease in plasma colloid osmotic pressure results in more fluid moving from the capillaries to the interstitial fluid. Conversely, a decrease in capillary hydrostatic pressure or an increase in plasma colloid osmotic pressure favors movement of interstitial fluid into the capillaries.

C. **Regulation of water balance**
1. The **daily intake and output of water** for an individual with moderate activity and at moderate temperatures is balanced at about 2,500 ml. In the healthy body, an adjustment of the water balance occurs by an increase in water intake through the thirst mechanism or by a decrease in water output by the kidney.
    a. The water intake in 24 hours is primarily from the diet.
        (1) **Ingested foods** contain about 700 ml. Meat contains 50% to 75% water and some fruits and vegetables are 95% water.

(2) **Ingested water and other beverages** amount to about 1,600 ml.
(3) **Metabolic water** produced through catabolism amounts to about 300 ml. The catabolism of 1 g fat yields 1.07 ml water; of 1 g carbohydrate, 0.55 ml water; and of 1 g protein, 0.41 ml water.

b. The **water output** (water loss) occurs through several routes.
(1) The **kidneys** are responsible for the largest water loss (about 1,500 ml).
(2) Water also is lost through the **skin** by sweating and insensible perspiration (about 500 ml), by evaporation from the **lungs** (300 ml), and through the **GI tract** (200 ml).

2. **Thirst**, or the conscious desire for water, is the primary regulator of water intake.
   a. **Regulation** of thirst. The thirst mechanism is controlled by a thirst center in the hypothalamus. The center contains specific neurons known as **osmoreceptors**, which are located near the neurons that secrete antidiuretic hormone (ADH).
   b. The major **stimuli** to the thirst center are increased plasma osmolality and decreased blood volume
   (1) **An increase in the osmolality of the ECF**, as would be produced by ingestion of sodium chloride, causes the osmoreceptors to lose water, shrink, and depolarize. Impulses signal the cerebral cortex to give rise to the sensation of thirst, which is alleviated by the drinking of water.
   (2) A **decrease in blood volume** (and blood pressure), as would occur from hemorrhage, is sensed by the cardiovascular baroreceptors, and impulses are transmitted to the osmoreceptors in the hypothalamus to activate the thirst mechanism. Also, the release of renin by the kidneys results in the production of angiotensin, which acts directly on the brain to stimulate the sensation of thirst.
   (3) A **dry mouth and throat** causes the sensation of thirst.

3. **Hormonal regulation of water output**
   a. **ADH** is produced in response to the same osmotic and nonosmotic stimuli that cause the sensation of thirst. ADH results in water retention by the kidney and a decreased urinary output.
   (1) An **increase in plasma osmolality** stimulates the hypothalamic osmoreceptors and causes the reflex secretion of ADH. An increased plasma concentration of sodium ions (hypernatremia) and glucose (hyperglycemia) are potent stimuli to ADH release.
   (2) A **decrease in blood volume** by 10% to 15% is sensed by the hypothalamic osmoreceptors and results in an increased production of ADH.
   b. The **renin-angiotensin-aldosterone mechanism** controls the kidney reabsorption of sodium ions and excretion of potassium ions. Angiotensin stimulates aldosterone, which is secreted by the adrenal cortex to act on the distal convoluted renal tubules to increase sodium reabsorption. Because water follows the sodium osmotically, water retention occurs. The resultant increase in ECF volume inhibits renin production.

D. **Disorders of water balance**
1. **Dehydration** is a water deficit over a period of time that cannot be compensated by normal regulatory mechanisms. Thus, the body is in negative water balance. Water loss under abnormal or stress conditions occurs by hemorrhage, fever, burns, hyperventilation, vomiting, diarrhea, or excessive sweating.
   a. Excessive water loss from the ECF results in an increase in its osmolality. Intracellular water passes into the ECF by osmosis to keep the osmolalities equal. ADH is stimulated, which acts to conserve water, but the overall effect still is a reduction of total body water.
   b. The treatment of dehydration is by the oral ingestion of water or by the intravenous administration of a solution of the appropriate osmolality to repair the water deficit.
2. **Overhydration** (water intoxication) is a clinical state in which there is an overall excess of extracellular fluid or an excess in either the plasma or interstitial fluid compartment.

a. **Rapid, extra water intake** (for example, with a liter of tap water) results in the inhibition of ADH and a **water diuresis**, which is the excretion of a large volume of dilute urine. The increase in urinary excretion begins soon after water ingestion and the extra water is excreted within a few hours.
b. **Renal or cardiovascular disease** is associated with overhydration and is characterized by edema (accumulation of excessive interstitial fluid).

III. ELECTROLYTE BALANCE

A. **Overview**
1. The maintenance of water balance in the body is regulated by ECF volume and osmolality, which in turn is dependent on the ECF electrolyte balance because osmolality determines the "water-pulling power" of a solution.
2. Sodium ions account for most (90%) of the cations in the ECF. Thus, sodium and its attendant anions are responsible for the osmolality of the ECF.
3. Alterations in ECF cause changes in plasma volume and blood pressure. Thus, the mechanisms of blood volume and pressure regulation involve the control of the sodium content of the body.
4. Electrolyte concentrations within the body fluids are expressed in **milliequivalents per liter (mEq/L)**, which is a measure of the number of ions in solution times the number of electrical charges the ions carry in each liter.

B. **Sodium**
1. **Balance.** The major source of sodium is dietary ingestion. The intake varies from 4 g to 20 g of NaCl. Sodium is lost from the skin, kidneys, and gastrointestinal tract.
   a. **A positive sodium balance** occurs when input exceeds output. Because water is retained with the sodium, the ECF volume and plasma volume will increase. Significant sodium and water retention may result in weight gain and a generalized edema. Congestive heart failure or kidney disease are clinical conditions that lead to a positive sodium balance.
   b. **A negative sodium balance** occurs when output exceeds input. An increased sodium loss leads to a decreased ECF and plasma volume with accompanying low blood pressure and inadequate circulation.
2. **Sodium regulation** in the body is accomplished primarily by **kidney excretion** of sodium rather than by sodium intake. Factors influencing the renal mechanisms include alterations in blood volume, blood pressure, or plasma sodium.
   a. The **glomerular filtration rate (GFR)** regulates the amount of sodium filtered.
      (1) For example, decreased blood pressure results in reflex vasoconstriction of the afferent arterioles, which reduces blood flow into the glomeruli. Consequently the GFR decreases, the amount of sodium filtered decreases, the amount of sodium and water excreted is diminished, and the resultant salt and water conservation elevates the blood pressure.
      (2) Conversely, an increased blood sodium content and the accompanying increase in blood pressure will increase GFR and enhance sodium and water excretion.
   b. **Aldosterone stimulates sodium ion reabsorption** from the kidney distal and collecting tubules and from the sweat glands, salivary glands, and GI tract. It promotes potassium and hydrogen ion excretion at the distal convoluted tubule and collecting ducts. The control of aldosterone secretion has several components.
      (1) The **renin-angiotensin-aldosterone** system
         (a) **Renin** is released from the juxtaglomerular apparatus of the kidney in response to low blood volume, low concentration of sodium, decreased arterial pressure, and water loss (hypovolemia).

(b) Renin combines with the substrate **angiotensinogen** in plasma and converts it to the biologically inactive **angiotensin I**. Angiotensin I is converted in the lung to **angiotensin II**.

(c) Angiotensin II, which is enzymatically inactivated within minutes, has several immediate physiological actions.

(i) It stimulates the release of **aldosterone** from the zona glomerulosa of the adrenal cortex.

(ii) It stimulates **thirst**.

(iii) It stimulates the secretion of **ADH** and **ACTH**. ACTH maintains the growth of the zona glomerulosa.

(iv) It is a powerful **vasoconstrictor** and results in increased blood pressure.

(2) **Potassium.** A 10% increase in the concentration of plasma potassium ions (hyperkalemia) directly stimulates the release of aldosterone, which promotes K^+ secretion in the distal convoluted tubules when sodium is absorbed.

C. **Potassium**

1. **Balance.** Potassium is the major (95%) **intracellular cation**. It is normally ingested and excreted in balanced amounts, with about 10% of the daily intake excreted in the feces and 90% in the urine.

2. **Functions.** Potassium is important in the **electrical activity** of nerve and muscle tissue. It maintains **osmolality** inside the cells and is important in **cellular metabolism**. Potassium in the ECF influences its acid-base balance.

3. **Regulation** of the blood level of potassium is under the control of aldosterone. Other hormones that stimulate the cellular uptake of potassium are insulin and epinephrine.

4. **Potassium imbalances**

    a. Potassium deficit (**hypokalemia**) may arise through vomiting and diarrhea, high sodium intake, kidney disease, or by the use of diuretic drugs for hypertension and edema. **Hypokalemia** can cause cardiac arrhythmias.

    b. Potassium excess (hyperkalemia) results from inadequate renal excretion. Hyperkalemia can cause cardiac fibrillation and is life-threatening.

D. **Calcium and phosphate**

1. **Distribution**

    a. Calcium primarily is an extracellular electrolyte. Almost all of the calcium (99%) is present in the skeleton, where it is combined with phosphate in the hydroxyapatite crystal of the matrix. The remainder is in ECF and a number of tissues.

    b. Phosphate concentration is high in ICFs and low in ECFs. The product of the concentrations of calcium and phosphate in the plasma remains a constant despite an increase or decrease of either ion. Generally, alterations in the concentration of phosphate in the ECF have little or no physiological effect.

2. **Balance.** The factors affecting the amount of calcium in the plasma are the amount of calcium ingested, the amount absorbed from the digestive tract, and the amount excreted in the feces and the urine.

3. **Functions.** In addition to its structural role in bones and teeth, the calcium of ECF is closely regulated because it is important in cell motility, blood clotting, muscle contraction, nerve conduction, hormonal responses, and secretory processes. In contrast to phosphates, alterations of the calcium ion concentrations have significant effects.

4. **Regulation** of the calcium concentration in the ECF and blood plasma is primarily by hormonal mechanisms (see Chapter 10 IV A–D).

    a. **Parathyroid hormone** (from the parathyroid glands) stimulates the osteoclasts in bone to release bone calcium and phosphate into the ECF. It increases the absorption of calcium from the digestive tract and reabsorption

from the kidney tubules, and it decreases calcium excretion. A low blood calcium concentration stimulates the release of parathyroid hormone.
- b. **Calcitonin** (from the thyroid gland) is released in response to a high blood calcium concentration. Calcitonin inhibits osteoclasts and stimulates osteoblasts to result in bone deposition.
- c. **Vitamin D**, which is essential for new bone formation, is activated by parathyroid hormone. It increases calcium absorption from the digestive tract and reabsorption from the kidney tubules.
- d. Other modulators of blood calcium levels include the mechanical stress on bone, prolonged and intense muscular activity, changes in blood pH, and sex hormones.

E. **Other anions**, such as chloride and bicarbonate, are regulated along with the regulation of sodium ions and the acid-base balance of the body. Sulfates, nitrates, and lactates have a transport maximum (Tm). When their Tm is exceeded, excess ions are excreted.

IV. ACID-BASE BALANCE

A. The acid-base balance of the body fluids is the regulation of the hydrogen ion concentration, which is essential to normal cell function. The hydrogen ion concentration (expressed as pH) influences enzymatic activity, cell permeability, and cell structure (see Chapter 2 III B 3 a–c).
1. The acid-base status is evaluated in systemic arterial blood. The normal pH of arterial blood is 7.4. The normal pH of venous blood and interstitial fluid is slightly more acid because of the CO_2 content, which forms carbonic acid.
2. **Acidosis** is a condition characterized by a decrease in arterial blood pH to below 7.35. **Alkalosis** exists when the arterial pH is above 7.45. The range of pH that is compatible with life extends from approximately 7.0 to 7.70.

B. **Overview of acid-base regulation of blood pH**
1. **Definitions**
   a. An **acid** is any chemical compound that liberates hydrogen ions to a solution or to a base. Examples of acids in the body include hydrochloric acid, sulfuric acid, nitric acid, phosphoric acid, lactic acid, carbonic acid, acetic acid, or the ammonium ion (NH_4^+).
   b. A **base** is any chemical compound that accepts hydrogen ions. Examples of bases include sodium hydroxide; potassium hydroxide; and the ammonia, lactate, acetate, and bicarbonate ions.
   c. A **strong acid** (or base) is one that dissociates completely and rapidly when dissolved in water and produces the maximum number of hydrogen ions possible. Hydrochloric acid (HCl) is a strong acid and it completely dissociates, leaving little or no HCl.

   $$HCl \longleftrightarrow H^+ + Cl^-$$

   d. A **weak acid** (or base) is one that dissociates only slightly when dissolved in water and produces fewer hydrogen ions per unit of acid. Carbonic acid (H_2CO_3) is a weak acid and most of it remains undissociated in solution.

   $$H_2CO_3 \longleftrightarrow H^+ + HCO_3^-$$

   e. An **acid-base buffer** is a solution of two or more chemicals that prevents significant changes in hydrogen ion concentration (pH) when either an acid or a base is added to the solution.
      (1) A buffer system consists of a weak acid, such as carbonic acid, and a salt of that acid, such as sodium bicarbonate.
      (2) The purpose of a buffer is to substitute a weak acid for a strong acid or a strong base for a weak base.
2. **Sources of hydrogen ions in the body**

a. Most hydrogen ions are produced as by-products or end products during the complete catabolism of carbohydrates, fats, and proteins. Incomplete oxidation of carbohydrates and fats results in lactic acid, keto acids, and fatty acids. The oxidation of some amino acids yields phosphoric acid, and purine metabolism results in uric acid.
b. Another major source of hydrogen ions is through the production of carbon dioxide (CO_2) by tissue cells. The CO_2 combines with water (mainly in RBCs) to form carbonic acid (H_2CO_3), which dissociates to form hydrogen ions.

C. **Body defenses against changes in the hydrogen ion concentration**
1. **Acid-base buffer systems** in intracellular and extracellular fluids work very rapidly and take effect in seconds. There are four major buffer systems in the body fluids.
    a. The **carbonic acid–sodium bicarbonate system** is the main buffer in ECF.
        (1) Normally, the ratio of molecules of carbonic acid (H_2CO_3) to molecules of the bicarbonate base ($NaHCO_3$) in plasma is 1:20.

        $$\frac{H_2CO_3}{NaHCO_3} = \frac{1}{20} = ph\ 7.4$$

        (2) Any change in the hydrogen ion concentration will change the ratio (the proportion of carbonic acid to bicarbonate) and result in alkalosis or acidosis.
        (3) The buffer system functions to avoid alteration of the ratio as follows:
            (a) When HCl (a strong acid) is added to a solution containing $NaHCO_3$, the weak carbonic acid is formed.

            $$HCl + NaHCO_3 \longleftrightarrow H_2CO_3 + NaCl$$

            (b) When sodium hydroxide (a strong base) is added to the buffer solution, the weak sodium bicarbonate base is formed.

            $$NaOH + H_2CO_3 \longleftrightarrow NaHCO_3 + H_2O$$

            (c) The net result is the conversion of a strong acid to a weak acid or a strong base to a weak base.
    b. The **phosphate buffer system** functions in a similar manner to change a strong acid to a weak acid and a strong base to a weak base. Disodium hydrogen phosphate (Na_2HPO_4) is the weak base, and sodium dihydrogen phosphate (NaH_2PO_4) is the weak acid. These components operate intracellularly, primarily in RBCs and in the epithelium of kidney tubules.

    $$HCl + Na_2HPO_4 \longleftrightarrow NaH_2PO_4 + NaCl$$

    $$NaOH + NaH_2PO_4 \longleftrightarrow Na_2HPO_4 + H_2O$$

    c. The **protein buffer system** is the most powerful in the body. It includes both the intracellular proteins and the extracellular plasma proteins, which buffer carbonic and organic acids.
        (1) Proteins are excellent buffers because they have both the carboxyl group and can function as an acid and the amino group and can function as a base, depending on the medium surrounding the protein.
        (2) Most proteins in the body exist in a basic medium. They behave as acids and function as large anions.
    d. The **hemoglobin buffer system** in RBCs serves to buffer the H^+ generated during CO_2 transport between the tissues and the lungs. Hemoglobin is an example of an intracellular protein that acts as a weak acid to buffer the less weak carbonic acid. If it were not for the hemoglobin buffer system, venous blood would become too acidic.

2. **Respiratory regulation of pH** involves the alteration of pulmonary ventilation to eliminate CO_2 and limit the amount of carbonic acid formed. Respiratory regulation takes one to three minutes to be initiated and functions after acid-base buffers as a second line of defense against changes in pH.
   a. Carbon dioxide is continuously added to venous blood as a result of cell metabolism and is transported to the lungs (see Chapter 13 V B 1–5). When CO_2 dissolves in plasma it produces carbonic acid, which dissociates to form hydrogen ions and bicarbonate ions.

   $$CO_2 + H_2O \longleftrightarrow H_2CO_3 \longleftrightarrow H^+ + HCO_3^-$$

   The carbon dioxide is exhaled in the lungs, which causes the reaction to go to the left and the plasma does not become too acidic.
   b. Under normal conditions, the production of carbon dioxide is balanced by its elimination as the respiratory system functions in regulation of acid-base balance.
   c. When metabolic activity is increased by exercise, the arterial partial pressure of carbon dioxide (pCO_2) increases, the plasma carbonic acid level increases, the reaction goes to the right, and the plasma pH decreases (acidosis). Respiration is adjusted to eliminate more carbon dioxide.
      (1) The excess carbon dioxide molecules in the blood diffuse into the central nervous system to reach the central chemoreceptors. The CO_2 diffuses into the neurons and forms carbonic acid, which dissociates to release hydrogen ions.
      (2) The hydrogen ions stimulate the central chemoreceptors and result in an increased rate and depth of ventilation. The increased respiratory rate expels the CO_2, reduces carbonic acid, and **increases the pH**.
      (3) Conversely, if the plasma pH increases (alkalosis), the respiratory rate decreases in order to decrease the elimination of CO_2. Less CO_2 in the plasma drives the above reaction to the right and **decreases the pH**.
3. **Renal regulation of pH** occurs through the renal excretion of either an acid or an alkaline urine. The kidney regulates blood pH by eliminating more hydrogen ions and reabsorbing more bicarbonate ions when blood pH is more acid, and by eliminating fewer hydrogen ions and reabsorbing fewer bicarbonate ions when the blood pH is more alkaline. The renal function takes several hours to several days to compensate for changes in pH and operates through the following mechanisms:
   a. **Tubular secretion of hydrogen ions**
      (1) Carbon dioxide in the interstitial fluid diffuses into the tubular epithelial cells and combines with water to form carbonic acid, which ionizes into hydrogen ions and bicarbonate ions.
      (2) Hydrogen ions are actively transported out of the cells into the tubule lumen and are eliminated from the body in the urine.
   b. **Bicarbonate reabsorption and excretion**
      (1) For each hydrogen ion secreted from the epithelial cell into the lumen of the tubules, one sodium ion is actively reabsorbed into the epithelial cell from the lumen to maintain the electrochemical balance. Both the sodium ion and the bicarbonate ion are transported together from the epithelial cell into the interstitial fluid and pass into the blood.
      (2) Under normal physiological conditions, the rate of hydrogen ion secretion is equal to the rate of glomerular filtration of bicarbonate. The kidneys reabsorb all the filtered bicarbonate.
      (3) When the plasma pH is alkaline, decreased hydrogen ions are secreted from the tubular cells and fewer are excreted in the urine. The filtered bicarbonate is incompletely reabsorbed and more bicarbonate is excreted in the urine.
   c. **Buffer systems** that allow excessive hydrogen ions to be excreted into the urine
      (1) **The phosphate buffer pair**
         (a) The phosphate buffers are concentrated in the tubular fluid be-

cause they are not reabsorbed. They serve to remove the hydrogen ions from the tubular fluid and carry them into the urine
- (b) This mechanism permits the elimination of large amounts of secreted hydrogen ions without highly acidifying the urine, which would be damaging to the urinary tract.
- (2) **The ammonia and ammonium buffer pair**
  - (a) The tubular cells synthesize ammonia (NH_3) from glutamic acid. The ammonia diffuses into the tubule lumen and reacts with hydrogen ions to form ammonium ions (NH_4^-). The ammonium ions are excreted into the urine together with chloride and other anions in the filtrate.
  - (b) This process assists the phosphate buffering mechanism. In addition, the ammonium ion displaces a sodium or some other basic ion to form an ammonium salt, and frees the sodium ion to diffuse back into the tubule cell and combine with bicarbonate. The formation of ammonium ions thus permits the addition of more bicarbonate ions to the blood and increases blood pH.

D. **Disorders of acid-base balance**
  1. **Acidosis** depresses mental activity. When acidosis is severe (pH of the blood below 7.0), it leads first to disorientation and later to coma and death.
     a. **Respiratory acidosis** is the result of decreased pulmonary ventilation with less elimination of carbon dioxide by the lungs. The subsequent increase in arterial pCO_2 and carbonic acid increases the hydrogen ion content of the blood. Respiratory acidosis can be acute or chronic.
        (1) **Causes.** Clinical conditions that can cause carbon dioxide retention in the blood include pneumonia, emphysema, chronic obstruction of respiratory passages, stroke, or trauma. Certain drugs (barbiturates, narcotics, and sedatives) or drug abuse suppress the breathing rate and can result in respiratory acidosis.
        (2) **Compensatory factors**
           (a) As the carbon dioxide accumulates, an increased respiratory rate (hyperventilation) at rest occurs to rid the body of CO_2.
           (b) The kidney compensates for the increased acid by excreting more hydrogen ions to return blood pH to near normal levels.
        (3) If respiratory and renal adjustments of pH fail, symptoms of central nervous system depression occur.
     b. **Metabolic acidosis** occurs when normally produced metabolic acids are not eliminated at a normal rate or there is a loss of bicarbonate base from the body.
        (1) **Causes.** Metabolic acidosis is most commonly a consequence of ketoacidosis from diabetes mellitus or starvation, lactic acid accumulation from increased skeletal muscle activity, such as convulsions, or renal disease. Severe and prolonged diarrhea with loss of bicarbonate can lead to acidosis.
        (2) **Compensatory factors.** Hyperventilation in response to nervous stimulation is a clinical sign of metabolic acidosis. Together with renal compensation, an increased respiratory rate may restore blood pH to near normal levels. Uncompensated acidosis depresses the central nervous system and leads to disorientation and, eventually, to coma and death.
  2. **Alkalosis** increases overexcitability of the nervous system. When severe, alkalosis may lead to tetanic muscle contractions, convulsions, and death as a result of tetany of the respiratory muscles.
     a. **Respiratory alkalosis** occurs when CO_2 is eliminated too rapidly from the lungs and decreases in the blood.
        (1) **Causes.** Hyperventilation may be caused by anxiety, as a result of fever, from the effect of aspirin overdose on the respiratory center, by hypoxia due to lowered air pressure at high altitudes, or severe anemia.
        (2) **Compensatory factors.** When hyperventilation is due to anxiety,

the symptoms can be relieved by re-breathing expelled carbon dioxide. The kidney compensates for the tubular alkaline fluid by excreting bicarbonate ions and retaining hydrogen ions.
b. **Metabolic alkalosis** is a bicarbonate excess. It occurs when there is an excessive loss of hydrogen ions or an excessive increase in bicarbonate in body fluids.
   (1) **Causes.** Prolonged vomiting (loss of hydrochloric stomach acid), kidney dysfunction, treatment with diuretics that result in hypokalemia and ECF volume depletion, or the excessive ingestion of antacids can produce metabolic alkalosis.
   (2) **Compensatory factors**
      (a) The respiratory compensation is a decrease in pulmonary ventilation and a resulting increase in pCO_2 and carbonic acid.
      (b) Kidney compensation involves the excretion of fewer ammonium ions, more sodium and potassium ions, less conservation of bicarbonate ions, and more excretion of bicarbonate.

Study Questions

Directions: Each question below contains four suggested answers. Choose the **one best** response to each question.

1. A young male adult in good health weighs 150 pounds. How much does water contribute to his weight?
   (A) about 30 pounds
   (B) about 60 pounds
   (C) about 90 pounds
   (D) about 120 pounds

2. What separates the interstitial fluid compartment from the plasma compartment?
   (A) intracellular fluid
   (B) lymph
   (C) capillary cell membranes
   (D) epithelial cell membranes

3. A comparison of the composition of plasma and interstitial fluid would reveal that
   (A) The concentration of proteins in plasma is greater than it is in interstitial fluid.
   (B) Plasma contains a higher concentration of sodium and interstitial fluid contains a higher concentration of potassium.
   (C) The predominant anion in plasma is HCO_3^-, whereas the predominant anion in interstitial fluid is HPO_4^-.
   (D) Similar quantities of cells, proteins, inorganic ions, and organic ions are found both in plasma and interstitial fluid.

4. All of the following are stimulants to the thirst mechanism EXCEPT
   (A) an increase in the osmolality of the extracellular fluid
   (B) a decrease in plasma osmolality
   (C) a decrease in blood volume
   (D) the action of angiotensin on the brain center

5. An individual with an adrenocortical tumor that resulted in the hypersecretion of aldosterone would be likely to have all of the following conditions or symptoms EXCEPT
   (A) increased blood levels of potassium (hyperkalemia)
   (B) increased blood levels of sodium
   (C) decreased blood levels of renin
   (D) increased blood pressure

6. Which of the following is most likely to lead to a negative sodium balance (sodium depletion)?
   (A) a decreased ingestion of sodium
   (B) an excessive renal excretion of sodium
   (C) an abnormal shift of sodium ions from the extracellular fluid to the intracellular fluid
   (D) a decrease in the glomerular filtration rate for sodium

7. Carbonic acid is considered to be a weak acid because
   (A) It releases both cations and anions.
   (B) It dissociates into H^+ and HCO.
   (C) It does not ionize completely in solution.
   (D) It can liberate protons.

8. A solution contains a mixture of carbonic acid and sodium bicarbonate in the ratio of 1:20 parts. If a small quantity of hydrochloric acid were added to the solution, all of the following would result EXCEPT
   (A) The hydrogen ions released from the hydrochloric acid will be converted to carbonic acid.
   (B) The pH will change from 7.4 to 6.8.
   (C) Carbonic acid releases fewer hydrogen ions into solution than did hydrochloric acid.
   (D) A new salt and a weak acid are formed.

9. An individual who is trying to lose weight by subsisting on a prolonged starvation diet would be most likely to
   (A) excrete a more alkaline urine than normal
   (B) have a decreased rate of respiration (hypoventilation)
   (C) suffer from metabolic alkalosis
   (D) have an arterial blood pH of less than 7.35

10. All of the following are mechanisms by which the kidney regulates the pH of the blood EXCEPT
    (A) tubular secretion of hydrogen ions into the tubular lumen
    (B) tubular synthesis and secretion of ammonia into the tubular lumen
    (C) buffering of excessive hydrogen ions in the tubular fluid
    (D) excretion of excessive carbon dioxide molecules

Questions 11–16. Match each of the descriptive phrases with the most appropriate electrolyte. A letter may be used more than once.
   (A) calcium ion
   (B) potassium ion
   (C) sodium ion
   (D) bicarbonate ion

11. most abundant intracellular cation

12. most abundant extracellular cation

13. controlled by parathyroid hormones

14. an excess in the plasma may have life-threatening cardiac effects

15. important in buffer control of pH.

16. absorption from the digestive tract requires vitamin D

Answers and Explanations

1. **The answer is C.** (II A 1 b) In adults, the total body water in males is about 60% of body weight and in females is about 50% of body weight. The TBW decreases with age and obesity.

2. **The answer is C.** (II B 1, 2) Interstitial fluid and blood plasma are subcompartments of the extracellular fluid compartment and are separated by endothelial (capillary) cell membranes. Another subcompartment of the ECF is transcellular fluid, which is separated from the rest of the ECF by epithelial cell membranes.

3. **The answer is A.** (II B 3) Plasma and interstitial fluid are similar in their composition with regard to inorganic and organic ions but dissimilar with regard to protein, which is in greater concentration in plasma. Only plasma contains cells (RBCs, WBCs, and platelets).

4. **The answer is B.** (II C 2) Thirst is stimulated by increased ECF and plasma osmolality, a decrease in plasma volume, a dry mouth and throat, and circulating angiotensin, which acts directly on the hypothalamic center to cause the sensation of thirst.

5. **The answer is A.** (II C 3; III B 2 b) Aldosterone causes kidney reabsorption of sodium and excretion of potassium. The renal reabsorption of sodium is accompanied by the osmotic reabsorption of water. The water retention and dilution of the plasma leads to an increased ECF and plasma volume and an increased blood pressure. The production of renin from the kidney is inhibited by an increase in ECF volume and is stimulated by a decrease in volume.

6. **The answer is B.** (III B 2 a, b) Sodium balance in the body is controlled by kidney output of sodium rather than by sodium intake. Because the GFR regulates the amount of sodium filtered and excreted, a decrease in the GFR would reduce the amount of sodium excreted and enhance sodium and water conservation.

7. **The answer is C.** (IV B 1 a–e) An acid is an organic or inorganic compound in solution that dissociates and liberates hydrogen ions (protons). Carbonic acid is considered a weak acid because it does not dissociate (ionize) completely and thus produces fewer hydrogen ions per unit of acid than does a strong acid, such as hydrochloric acid.

8. **The answer is B.** (IV C 1 a) The carbonic acid-bicarbonate buffer system reduces the effects of strong acids or bases on the solution and the pH is unchanged or changed insignificantly.

9. **The answer is D.** (IV D 1, 2) A starvation diet would result in the increased production of ketoacids, which causes metabolic acidosis. In compensation, the kidneys would excrete urine that would be more acidic than usual and the respiratory rate would increase in order to restore the pH to near normal levels.

10. **The answer is D.** (IV C 3 a–c) The mechanisms involved in the renal regulation of pH include the tubular secretion of hydrogen ions, the exchange in the tubule of hydrogen ions for sodium ions, which are accompanied by bicarbonate ions as they are reabsorbed, the tubular secretion of ammonia into the tubular fluid to constitute the ammonia-ammonium buffer, and the phosphate buffer pair in the tubular fluid.

11-16. **The answers are 11–B, 12–C, 13–A, 14–B, 15–D, and 16–A.** (III B, C, D; IV A–C) The most abundant intracellular cation is potassium, while the most abundant extracellular ion is sodium. Calcium and phosphate ions are controlled by the parathyroid hormones. Hyperkalemia, or an excess of blood potassium, may cause cardiac fibrillation. The bicarbonate ion forms part of the carbonic acid-sodium bicarbonate buffer system most significant in acid-base balance. Calcium requires vitamin D to enhance absorption from the digestive tract.

18 The Reproductive Systems, Pregnancy, and Development

I. **INTRODUCTION.** The reproductive systems in males and females are concerned primarily with the perpetuation of the human species. In that respect, they are unlike all the other organ systems of the body, which are concerned with the homeostasis and survival of the individual. The processes of reproduction include **sexual maturation** (the physiological preparation for reproduction), the formation of **gametes** (spermatozoa and ova), **fertilization** (union of gametes), **pregnancy**, and **lactation**.

A. The **primary sex organs** are the **gonads**, which consist of the **testes** in the male and the **ovaries** in the female.

1. **Gametes** (sex cells) are produced by the gonads in a process called **gametogenesis**. The testes produce spermatozoa (spermatogenesis) and the ovaries produce ova (oogenesis).

2. **Male and female sex hormones** also are products of the gonads and function in all the processes of reproduction. These steroid secretions are responsible for the **prenatal development** of the reproductive organs, for the development and maintenance of the **secondary sex characteristics** (physical changes that occur at puberty), and for the **neuroendocrine activity of the hypothalamus.**
   a. **Androgens** (mainly **testosterone**) are the primary male sex hormones.
   b. **Estrogens** (mainly estradiol) and **progesterone** are the primary female sex hormones.

B. The **accessory sex organs** are the reproductive ducts and related glands that transport, nourish, and protect the gametes after they leave the gonads. The external genitalia are included with the accessory sex organs.

C. The male reproductive system consists of the testes, a duct system that includes the epididymis, ductus deferens, ejaculatory duct and the urethra; accessory glands; and the penis.

D. The female reproductive system consists of the ovaries, the oviducts, uterus, vagina, external genitalia, and the mammary glands (Color Plate 15).

II. **EMBRYONIC DEVELOPMENT**

A. **Sex determination** (see Chapter 4 II A–F)

1. The **genotype of normal females** is 46, XX (22 pairs of autosomes and two sex chromosomes, X and X). After meiotic division during oogenesis, each oocyte will contain the haploid number of chromosomes and only one X chromosome.

2. The **genotype of normal males** is 46, XY (22 pairs of autosomes and two sex chromosomes, X and Y). After meiotic division during spermatogenesis, each of the spermatozoa will contain the haploid number with either an X chromosome or a Y chromosome.
3. If an X-bearing sperm fertilizes an ovum, the resulting **zygote**, or fertilized egg, will have the full complement of autosomes, 22 + X from the sperm and 22 + X from the ovum, and will be a female (XX).
4. If a Y-bearing sperm fertilizes an ovum, the resulting zygote (XY) will have the full complement of autosomes, 22 + Y from the sperm and 22 + X from the ovum, and will be male (XY).
5. Thus, genetic sex of the offspring is determined at fertilization solely by the spermatozoan.

B. **Sex differentiation and development**
   1. **Internal reproductive organs.** The differentiation of the gonads into testes or ovaries and the development of the internal duct systems begins to develop in the fourth week of embryonic life and is essentially completed by the twelfth week after fertilization (Figure 18–1).
      a. **Primordial germ cells** (primitive sex cells), which become segregated from the other cells of the developing embryo at four weeks, migrate by ameboid movement during the sixth week to the undifferentiated **primitive gonads** in the posterior abdominal wall of the embryo.
         (1) The gonads develop within **gonadal (genital) ridges**. These are located near the **mesonephric (Wolffian) ducts**, which drain the embryonic kidneys.

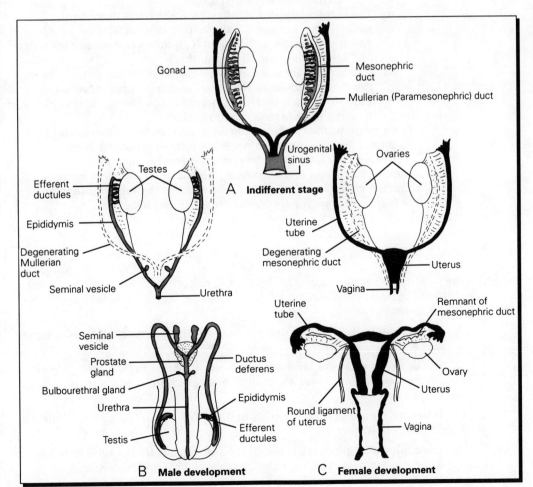

Figure 18–1. Schematic diagram showing the development of the male and female internal reproductive ducts and organs.

(2) **Mullerian (paramesonephric) ducts** develop independently alongside the mesonephric ducts.
   b. **Development of the testes.** If the genetic sex is **male**, the gonads develop into testes, which produce testosterone and a Mullerian duct inhibitor.
      (1) Testosterone results in the **incorporation of the mesonephric ducts** into the male internal reproductive duct system. On either side, portions of the mesonephric duct become the **epididymis, ductus deferens, ejaculatory duct**, and **seminal vesicle**.
      (2) Mullerian duct inhibitor causes the atrophy of the Mullerian ducts.
   c. Development of the ovaries. If the genetic sex is **female**, the gonads develop as ovaries. Without testosterone, the mesonephric ducts atrophy. The Mullerian ducts develop into the **uterine (Fallopian) tubes**, **uterus**, and upper part of the **vagina**. No hormonal influence is required for their development.
2. **External genitalia** (Figure 18-2). Until about the seventh week of embryonic life, the external genitalia are undifferentiated and consist of a **genital tubercle, urogenital folds**, and **labioscrotal swellings**.
   a. **In males**, under the influence of testosterone, the genital tubercle forms the **penis** and the labioscrotal swellings merge to become two **scrotal sacs**, which will contain the testes after they descend from the abdominal cavity.
      (1) **Descent of the testes**
         (a) The **gubernaculum** is a fibrous cord on the outside of each testis that connects it to the labioscrotal swellings. As the male embryo grows, the cord shortens.
         (b) During the seventh intrauterine month, paired **inguinal canals** in the abdominal wall develop to provide a passageway from the pelvic cavity into the scrotum.
         (c) The shortened gubernaculum pulls the testes downward through the inguinal canal into the scrotum. The ductus deferens, blood vessels, nerves, and lymphatics accompany the testis and are enclosed in the **spermatic cord**.
         (d) If the inguinal canal fails to close after testicular descent, or if the closure remains a weakened area, a potential site for **inguinal hernia** is created. A hernia is a protrusion of viscera through the weakened abdominal wall.
      (2) **Cryptorchidism** is failure of one or both testes to descend into the scrotal sac. The condition results in infertility because body core temperature is too high for normal sperm production.
   b. In **females**, without any hormonal influence, the genital tubercle become the **clitoris**, the urogenital folds become the **labia minora**, and the labioscrotal swellings become the **labia majora**.

III. THE MALE REPRODUCTIVE SYSTEM (Figures 18-3 and 18-4)

A. The **scrotum** is a loose sac of skin, fascia, and smooth muscle that encloses and supports the testes outside the body at an optimum temperature for the production of spermatozoa.
   1. Two **scrotal sacs**, each containing a single testis, are separated by an internal septum.
   2. The **dartos muscle** is a layer of smooth muscle fibers in the underlying fascia that contracts to cause a wrinkling of scrotal skin in response to cold or to sex excitement.
B. The **testes** are smooth, oval organs, about 4 cm to 5 cm (1.5 in. to 2 in.) long and 2.5 cm (1 in.) in diameter.
   1. The **tunica albuginea** is a connective tissue capsule that covers the testis and extends inward to partition it into about 250 lobules.

The Reproductive Systems, Pregnancy, and Development   351

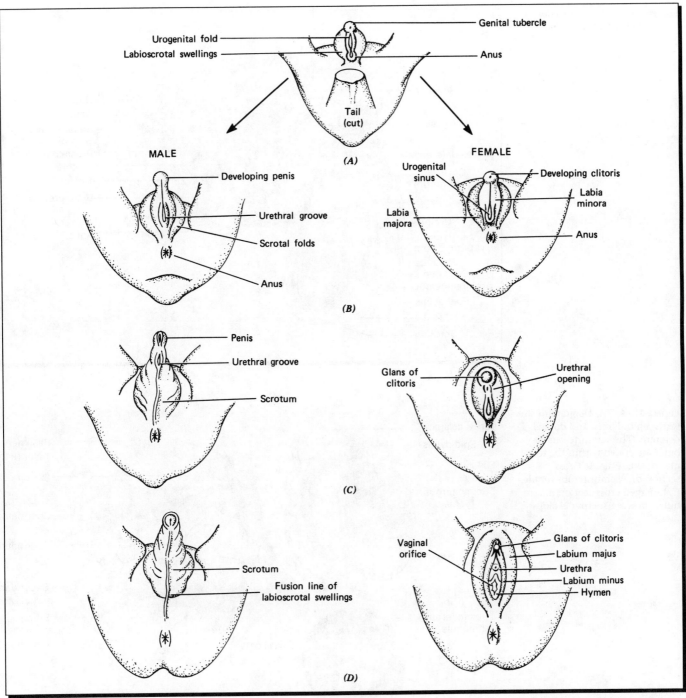

Figure 18-2. Diagrams illustrating development of male and female external genitalia. A, At seventh week of gestation. B, at tenth week of gestation. C, At about the twelfth week of fetal life. D, Shortly before birth. Reproduced by Permission. *Biology of Women*, 3E, by Ethel Sloane, Delmar Publishers Inc, Albany, NY. Copyright ©1993.

2. The **seminiferous tubules**, in which spermatogenesis takes place, are coiled within the lobules. The specialized germinal epithelium that lines the seminiferous tubules contains **stem cells** (spermatogonia), which give rise to sperm; **Sertoli cells**, which support and nourish developing sperm; and **interstitial cells** (of Leydig), which have an endocrine function.
   a. **Spermatogenesis** is the process of development from the spermatogonia to spermatozoa and takes approximately 64 (plus or minus four) days.
      (1) **Spermatogonia** are located adjacent to the basement membrane of the seminiferous tubules. They proliferate by mitosis and differentiate into **primary spermatocytes** (see Chapter 3 V C 1–2; D 1–4).

Figure 18–3. *Sagittal section of the male reproductive system.*

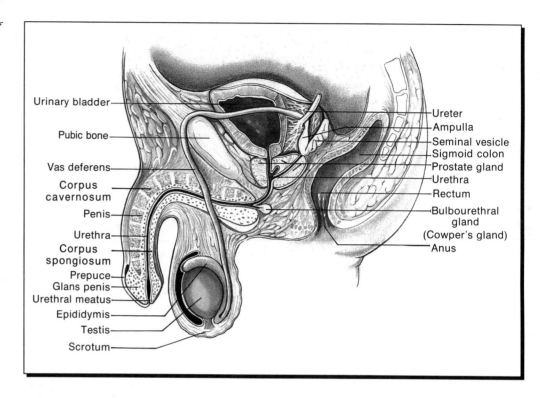

Figure 18–4. *A*, Diagram of the testis, epididymis, and ductus deferens. The seminiferous tubules are coiled within the lobules of the testis. *B*, Cross section of a seminiferous tubule. *C*, Enlarged cross section of a tubule, showing sperm development.

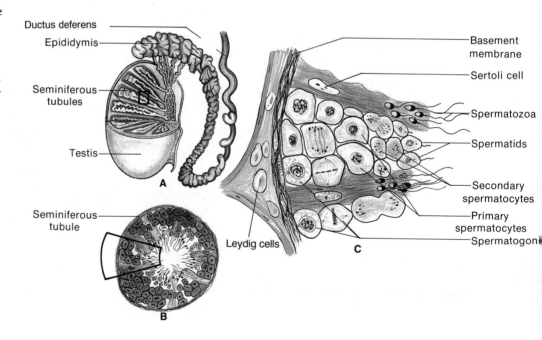

(2) Each primary spermatocyte undergoes maturational divisions through meiosis to form two **secondary spermatocytes**. The second meiotic division in secondary spermatocytes results in the production of four **spermatids**.
(3) The final stage of spermatogenesis is the maturation of a spermatid into a **spermatozoan** (sperm). Mature spermatozoa are 60 μm in length. They have a **head**, a **midpiece** and a **flagellum** (tail).
   (a) The head contains the nucleus. It is covered by an **acrosome** (headcap), which contains enzymes necessary for ovum penetration.
   (b) The midpiece contains mitochondria, which produce ATP necessary for movement.
   (c) The beating of the flagellum results in sperm motility (swimming).
b. The **Sertoli cells** extend from the epithelium to the lumen of the tubule. They have the following functions:
   (1) Sertoli cells mechanically **support and nurture** maturing spermatozoa.
   (2) Sertoli cells secrete **Mullerian duct inhibitor**, which is a glycoprotein produced during embryonic development of the male reproductive tract. It causes atrophy of the Mullerian ducts in genetic males (see section IV).
   (3) Sertoli cells secrete **androgen-binding protein** in response to follicle-stimulating hormone (FSH) released by the anterior pituitary. The protein binds testosterone and helps maintain its high concentration in the seminiferous tubules. Testosterone stimulates spermatogenesis.
   (4) Sertoli cells secrete **inhibin**, a protein that exerts a negative feedback effect on FSH secretion by the anterior pituitary gland.
   (5) Sertoli cells secrete **H-Y antigen**, which is a cell membrane surface protein important in inducing the differentiation of the testes in genetic males.
c. The **interstitial cells** (of Leydig) secrete androgens (testosterone and dihydrotestosterone). The interstitial cells disappear six months after birth and reappear at the onset of puberty under the influence of gonadotropins from the anterior pituitary.

C. **Ducts of the male reproductive tract** carry the mature sperm from the testes to the exterior of the body.
1. Within the testis, the sperm move to the lumen of the seminiferous tubules, which lead into **tubuli recti** (straight tubules). The tubuli recti lead into a network of canals call the **rete testis**, which lead into about 10 to 15 **ductuli efferentia** (efferent ducts), which emerge from the upper part of the testis.
2. The **epididymis** is a coiled tube about 20 ft (4 m to 6 m) long, which runs along the posterior side of the testis. It receives the sperm from the efferent ducts.
   a. The epididymis **stores sperm** and retains them for up to six weeks, during which time they become motile, fully mature, and are capable of fertilization.
   b. During sexual excitement, smooth muscle layers in the epididymal walls contract to propel the sperm into the ductus deferens.
3. The **ductus (vas) deferens** is the continuation of the epididymis. It is a straight tube located within the **spermatic cord**, which also contains blood and lymphatic vessels, nerves of the ANS, cremaster muscle, and connective tissue. Each ductus deferens leaves the scrotum to run upward into the abdominal wall by way of the inguinal canal. It passes behind the lower part of the urinary bladder to connect with the ejaculatory duct.
4. The **ejaculatory duct** on either side is formed by the junction of an enlargement (ampulla) at the end of the ductus deferens and the duct from a seminal vesicle. Each ejaculatory duct is about 2 cm long and penetrates the prostate gland to join the urethra as it exits from the urinary bladder.

5. The **urethra** extends from the bladder to the end of the penis and has three parts.
   a. The **prostatic urethra** starts at the base of the bladder. It passes through and receives the secretions of the prostate gland.
   b. The **membranous urethra** is 1 cm to 2 cm long. It is surrounded by the external urethral sphincter.
   c. The **penile (cavernous, spongy) urethra** is surrounded by spongy erectile tissue (corpus spongiosum). It enlarges into the **fossa navicularis** before it terminates as the **external urethral opening** in the glans penis.

D. Accessory glands
   1. The paired **seminal vesicles** are convoluted sacs that empty into the ejaculatory duct. The secrete a viscous, alkaline fluid rich in fructose, which serves to **nourish** and **protect** the sperm. Seminal vesicle secretion constitutes more than half the bulk of semen (the fluid in which sperm leave the body).
   2. The **prostate** gland surrounds the urethra as it exits the bladder. The prostate empties its secretion into the prostatic urethra by 15 to 30 prostatic ducts.
      a. The prostate secretes a milky, alkaline fluid that **neutralizes the acidity** of the vagina during intercourse and enhances sperm motility, which is optimum at pH 6.0 to 6.5.
      b. The prostate gland enlarges during adolescence and reaches its optimal size in men in their 20s. In many men, it continues to increase in size with age. By age 70, two thirds of all males have an enlarged prostate that interferes with urination.
   3. The paired **bulbourethral (Cowper's) glands** are small glands approximately the size and shape of a pea. They secrete an alkaline, mucous-containing fluid into the penile urethra that is **lubricating and protective** and adds to the bulk of semen.

E. The **penis** consists of three parts: a **root**, a **body**, and an expanded **glans penis**, which has abundant sensory nerve endings. It serves as both an outlet for urine and semen and as an organ for copulation.
   1. The skin of the penis is thin and has no hair except near the root of the organ. The **prepuce** (foreskin) is a loose-fitting, circular fold of skin that extends over the glans penis unless it is removed by circumcision. The **corona** is the proximal edge of the glans penis.
   2. The body of the penis is formed by three cylindrical masses of erectile tissue: two dorsal **corpora cavernosa** and one ventral **corpus spongiosum** around the urethra.
      a. Erectile tissue is a meshwork of irregular blood spaces (venous sinusoids), which are supplied by afferent arterioles and capillaries, drained by venules, and surrounded by fibrous connective tissue and smooth muscle.
      b. The corpora cavernosa are surrounded by dense connective tissue called the **tunica albuginea**.
   3. Mechanism of penile erection. Erection is a vascular function of the corpora cavernosa under the control of the ANS.
      a. When the penis is flaccid, sympathetic stimuli to the arterioles of the penis cause their partial constriction so that the flow of blood through the penis is even and only a small amount of blood enters the cavernous sinusoids.
      b. During sexual or mental stimulation, parasympathetic stimuli cause vasodilation of the arterioles entering the penis. More blood enters the penis than can be drained by the veins.
      c. The sinusoids of the corpora cavernosa become distended with blood and compress the veins, which are surrounded by the nondistensible tunica albuginea. The penis become turgid with blood and swells into the rigid state of erection.
      d. After ejaculation, sympathetic impulses produce arterial vasoconstriction and the blood is permitted to enter the veins to be carried away. The penis undergoes **detumescence**, or return to the flaccid state.

4. **Ejaculation** is accompanied by orgasm and is the culmination of the male sex act. The semen is ejected in a series of spurts.
   a. Sympathetic impulses from reflex centers in the spinal cord pass along lumbar spinal nerves (L1 and L2) to the genital organs and cause peristaltic contractions in the ducts of the testis, epididymis, and ductus deferens. These contractions move sperm along the tract.
   b. Parasympathetic impulses travel in the pudendal nerve and cause the bulbocavernosus muscles at the base of the penis to contract rhythmically.
   c. Simultaneous contractions in the seminal vesicles, prostate, and bulbourethral glands cause their secretion of seminal fluid that mixes with the sperm to form semen.

F. **Quantity and composition of semen**
   1. The volume of the ejaculate may range from 1 ml to 10 ml; the average is 3 ml. Semen is 90% water and contains 50 to 120 million sperm per ml; sperm constitute only 5% of the bulk of semen by volume.
   2. Semen is ejaculated as a viscous, yellowish-gray fluid with pH 6.8 to 8.8. It coagulates immediately after ejaculation and liquefies spontaneously within 15 to 20 minutes.
   3. The first portion of the ejaculate contains the spermatozoa, the epididymal fluids, and the secretions of the prostate and bulbourethral glands. The last part of the ejaculate contains the secretions of the seminal vesicles.
   4. Semen contains many of the substances found in blood plasma as well as additional substances such as prostaglandins, proteolytic enzymes, enzyme inhibitors, vitamins, and amounts of steroid hormones and gonadotropins in different concentration than in blood plasma.
   5. After ejaculation, spermatozoa live only 24 to 72 hours in the female tract. Sperm can be stored for several days at lowered temperatures or frozen to be stored for more than a year.

G. **Hormonal regulation of the male reproductive system**
   1. **Testicular hormones.** The major androgen produced by the testes is **testosterone**. The testes also secrete much smaller amounts of **androstenedione**, which is a precursor to estrogens in males, and **dihydro-testosterone (DHT)**, which is necessary for the prenatal growth and differentiation of the male genitalia.
      a. **In the male fetus**, the secretion of testosterone causes male differentiation of the internal ducts and the external genitalia, and stimulates the descent of the testes into the scrotum during the last two months of gestation. From birth until puberty, little or no testosterone is produced.
      b. **At puberty and thereafter**, testosterone is responsible for the development and maintenance of the following male secondary sex characteristics:
         (1) Testosterone increases the growth and development of the male genitalia.
         (2) Testosterone is responsible for the distribution of hair in the male pattern.
         (3) Testosterone causes enlargement of the larynx and an increase in the length and thickness of the vocal cords, which results in a low-pitched voice.
         (4) Testosterone increases the thickness and texture of the skin and results in a darker and ruddier skin hue. It also increases sweat gland and sebaceous gland activity and is involved in the development of acne (in both males and females).
         (5) Testosterone increases muscle and bone mass, increases the basal metabolic rate, increases the number of RBCs, and increases men's oxygen-carrying capacity.
   2. **Pituitary and hypothalamic hormones** control androgen production and testicular function.

a. **Pituitary gonadotropins.** Follicle stimulating hormone (FSH) has receptors on seminiferous tubule cells and is required for spermatogenesis. Luteinizing hormone (LH) has receptors on interstitial cells and stimulates the production and secretion of testosterone. LH may be called ICSH (interstitial-cell-stimulating hormone) in males.
b. **Hypothalamic gonadotropin releasing hormone (GnRH)** interacts with testosterone, FSH, LH, and inhibin in negative feedback mechanisms that regulate testosterone synthesis and secretion.
   (1) A lowered concentration of circulating testosterone stimulates hypothalamic production of GnRH, which stimulates the secretion of FSH and LH. FSH stimulates spermatogenesis in the seminiferous tubules and LH stimulates the interstitial cells to produce testosterone.
   (2) The increased blood level of testosterone exerts negative feedback controls on the secretion of GnRH and on the pituitary secretion of FSH and LH.
   (3) **Inhibin** is synthesized and secreted by Sertoli cells in response to FSH secretion. It acts by direct negative feedback on the pituitary gland to inhibit the secretion of FSH. Inhibin does not effect the release of LH (ICSH).
   (4) **Androgen-binding protein** is a polypeptide also produced by Sertoli cells in response to FSH secretion. The protein binds testosterone to maintain its concentration in seminiferous tubules 10 to 15 times greater than in the blood. It thus increases the cells' receptivity to testosterone's effects and serves to support spermatogenesis.
c. **Puberty** is initiated by the increase in GnRH secretion.
   (1) GnRH is inhibited through negative feedback by small quantities of circulating testosterone prior to puberty.
   (2) At puberty, maturation of the brain and a decreased sensitivity of the hypothalamus to testosterone inhibition leads to an increased secretion of GnRH, which increases pituitary FSH and LH secretion. This results in spermatogenesis, testosterone production, and the development of male secondary sex characteristics.
   (3) Increased levels of GnRH cause increased FSH and LH secretion by the anterior pituitary gland.

IV. THE FEMALE REPRODUCTIVE SYSTEM (Figure 18–5, Color Plate 15)

A. The **ovaries** are 3 cm to 5 cm long, 2 cm to 3 cm wide, and 1 cm thick. They are almond-shaped.
   1. **Location and attachments.** Each ovary lies on the side wall of the posterior pelvic cavity in a shallow depression, the ovarian fossa, and is held in position by **pelvic mesenteries** (folds of the peritoneum between the visceral peritoneum and the parietal peritoneum). The ovaries are the only organs in the pelvic cavity that are **retroperitoneal** (located behind the peritoneum).
   2. **Structure.** The ovaries are covered by germinal (surface) epithelium. The connective tissue of the ovary is called the **stroma** and contains an outer cortex and inner medulla.
      a. The **medulla** of the ovary is the innermost area. It contains blood and lymphatic vessels, nerve fibers, smooth muscle cells, and connective tissue cells.
      b. The **cortex** is the dense, outer layer of the stroma. It contains **ovarian follicles**, which are the functional units of the ovaries.
   3. Oogenesis—the development of the ovarian follicles
      a. **Prenatal oogenesis.** The oogonia proliferate during fetal life and give rise to 6 to 7 million **primary oocytes.**
         (1) Each primary oocyte is surrounded by a single layer of follicular cells and is known as a **primordial follicle.**
         (2) Primary oocytes remain in prophase I of meiosis during all of fetal life and after birth until puberty (see Chapter 3 V D 2 a).

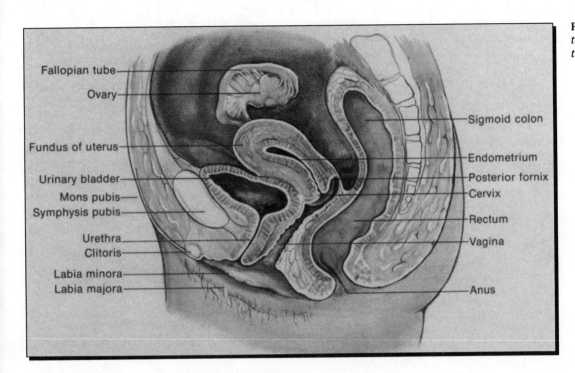

Figure 18-5. Sagittal section of the female reproductive system.

- (3) The number of primordial follicles declines throughout life by **atresia** (regression and degeneration of follicles).
- b. **Postnatal oogenesis**
  - (1) At birth, the number of primordial follicles in the ovaries has decreased to 2 million.
  - (2) By age seven, 300,000 primary oocytes remain; by puberty, 50,000 to 100,000 follicles have survived to provide the stock for future ovulations.
  - (3) In contrast to males, who continuously produce spermatogonia and primary spermatocytes, females are born with all the primary oocytes they will ever have. Of the declining pool of oocytes, only 350 to 400 (one each month) will mature and be ovulated during the reproductive years.
- c. **Postpubertal oogenesis.** At puberty, under the influence of pituitary gonadotropins and hypothalamic GnRH, cyclical development of primordial follicles is initiated. Each month, a number of primary follicles develop from some of the primordial follicles and one of them continues to maturity and ovulation (Figure 18–6).
  - (1) The **primary follicle**
    - (a) The primary oocyte is stimulated to enlarge. The surrounding follicular cells divide to form multiple layers of **granulosa** cells.
    - (b) A noncellular, clear **zona pellucida** layer is deposited between the oocyte and the granulosa cells.
    - (c) The stroma cells around the primary follicle form into two layers: The **theca interna**, which is composed of secretory cells that secrete estrogen, and the **theca externa**, which is an outer layer of connective tissue.
    - (d) Spaces form between the granulosa cells and become filled with **follicular fluid.** Later, the spaces coalesce to form an **antrum**, or cavity, in the follicle.
  - (2) The **secondary follicle**
    - (a) A growing follicle containing an antrum is known as a **secondary follicle.** About 20 to 50 primary follicles reach the antral stage, but only one will mature to ovulation.

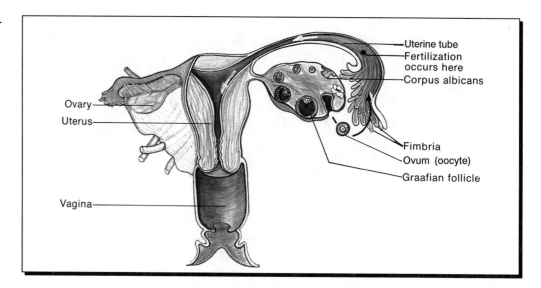

Figure 18–6. *Diagram illustrating oogenesis in the ovary and the path of the oocyte into the uterine tube after ovulation.*

(b) The **cumulus oophorus** is a mound of granulosa cells that surrounds and supports the oocyte in a secondary follicle. The **corona radiata** is formed by the granulosa cells that encircle the oocyte.
(c) The primary oocyte is pushed to one side of the antral cavity by accumulating antral fluid and projects into the cavity.
(3) The **mature (Graafian) follicle**
  (a) The dominant follicle that will ovulate takes 10 to 14 days to develop. It migrates to the surface of the ovary to form a bulge (stigma) before it ruptures (ovulates) through the ovarian tissue.
  (b) Before ovulation, the primary oocyte in the mature follicle completes the first meiotic division. The division of cytoplasm is unequal: a **secondary oocyte** receives half the chromosomes and almost all the cytoplasm, and a small **polar body**, which eventually disintegrates, receives the other half of the chromosome.
  (c) The secondary oocyte proceeds to metaphase of the second meiotic division and stops. If the oocyte is fertilized after ovulation, meiosis will continue.
(4) **Ovulation**
  (a) The oocyte breaks away from the surrounding cells and floats freely in the antrum surrounded by the **corona radiata.**
  (b) The oocyte is extruded from the surface of the ovary along with some follicular fluid and the attached corona radiata.
  (c) If the oocyte is not fertilized, it disintegrates within a few days.
(5) The **corpus luteum** (yellow body) is formed in the ovary at the site of the empty follicle.
  (a) The walls of the empty follicle collapse; the granulosa cells undergo structural and biochemical changes and become lutein cells.
  (b) The lutein cells of the corpus luteum produce estrogen and progesterone, reaching a peak of activity 5 to 7 days after ovulation. Unless fertilization occurs, the corpus luteum begins to regress and deteriorates by the 15th day after ovulation.
(6) The **corpus albicans** (white scar) is formed after connective tissue invades the disintegrating corpus luteum.

B. Two **uterine tubes** (fallopian tubes or oviducts) receive and transport the oocyte to the uterus after ovulation.
  1. Each uterine tube is 10 cm long and 0.7 cm in diameter and is supported by a part of the broad ligament of the uterus. It is attached at one end to the uterus and opens at the other end into the pelvic cavity.

a. The **infundibulum** is the funnel-shaped open end (ostium) of the uterine tube. It has motile finger-like processes (**fimbria**) that extend over the surface of the ovary to help sweep the ovulated oocyte into the tube.
b. The **ampulla** is the intermediate segment of the tube.
c. The **isthmus** is the segment nearest the uterus.
2. The wall of the uterine tubes is composed of circular and longitudinally arranged smooth muscle fibers, connective tissue, and a ciliated epithelial lining. The oocyte is moved along the tube to the uterus by the beat of the cilia and peristaltic contractions of smooth muscle. It takes 4 to 5 days for the oocyte to reach the uterus.
3. Fertilization generally occurs in the upper one third of the uterine tube.

C. The **uterus** is a single, hollow, muscular organ. The fertilized oocyte implants in the endometrial lining of the uterus and is nourished to develop and grow until birth.
  1. **Size and location.** The uterus is shaped like an inverted pear and in the nonpregnant state is about 7 cm long, 5 cm wide, and 2.5 cm in diameter (3 in. x 2 in. x 1 in.). It lies in the pelvic cavity between the rectum and the urinary bladder. Typically, the uterus is flexed forward (**anteflexed**) and **anteverted** so that it lies nearly horizontally over the urinary bladder. In some women, the uterus normally may be retroflexed and retroverted and rest against the rectum.
  2. **Support.** The uterus is primarily supported by a peritoneal fold, the **broad ligament**, which attaches the uterus to the pelvic wall. **Round ligaments** extend from the lateral angle of the uterus through the inguinal canal to the labia majora. The uterus also is anchored by the **cardinal** and **uterosacral** ligaments.
  3. **Structure**
     a. The walls of the uterus are composed of an outer **serosa** (perimetrium); a middle **myometrium** (smooth muscle layer); and an inner **endometrium**. The endometrium undergoes cyclic changes during menstruation and forms the site for implantation of a fertilized ovum. It is composed of two layers.
        (1) The **superficial layer (stratum functionale)** of the endometrium is thicker. It contains glands, responds to steroid hormones, and is almost completely lost during menstruation.
        (2) The **basal layer (stratum basalis)** remains the same throughout the cycle.
     b. The **fundus** of the uterus is the rounded portion superior to the entrance of the uterine tubes.
     c. The **body** of the uterus is the thick-walled expanded portion that encloses the uterine cavity.
     d. The **cervix** is the lower, constricted neck portion of the uterus. The **external os** is the opening of the cervix into the vagina; the **internal os** is the opening into the uterine cavity. The **endocervical canal** lines the passageway between the two openings.
     e. The **portio vaginalis** is the portion of the cervix that protrudes into the upper end of the vagina. The circular recesses that are formed at the junction are the anterior, posterior, and lateral **fornices** (singular, fornix).
  4. **Blood supply.** Arterial blood is supplied to the uterus through the uterine arteries (off the internal iliac arteries) and branches of the ovarian and vaginal arteries.
     a. In the wall of the uterus, the arteries become the **arcuate arteries**, which branch to penetrate the myometrium as the **radial arteries**. Continuations of the radial arteries into the endometrium are the spiral (coiled) arterioles. The blood supply to the endometrium is significant in the process of menstruation.
     b. Blood returns from the uterus through uterine veins that parallel the pathway of the arteries.

D. The **vagina** is a distensible, fibromuscular tube. It is the passageway for the delivery of an infant and for menstrual flow, and it functions as the female organ of copulation.
  1. **Size and location.** The vagina is 8 cm to 10 cm long. It passes upward to the uterus at an approximate 45° angle from the vestibule of the external genitalia. It is situated between the urinary bladder and urethra anteriorly and the rectum posteriorly.
  2. **Structure.** The **vaginal wall** consists of an outer adventitia, a smooth muscle coat, and a mucosa of nonkeratinized stratified squamous epithelium, which is known as the vaginal lining. Cells of the vaginal lining have membrane-bound receptors for estrogen.
      a. Before puberty and after menopause, when the blood estrogen concentration is low, the vaginal lining is thin and made up almost entirely of basal cells.
      b. During the reproductive years and under the influence of estrogen, the vaginal lining is thick and consists of as many as 40 layers of basal, intermediate, and superficial cells.
  3. **Vaginal fluid and discharge.** The vagina is moistened and lubricated by fluid from capillaries in the vaginal walls and secretions from cervical glands. The pH of the vaginal fluid is estrogen-dependent.
      a. During the reproductive years, the vaginal discharge is acid (pH 3.5 to 4.0). Under estrogen stimulation, the mucosal cells store glycogen, which is metabolized to lactic acid by normal vaginal bacteria.
      b. Before puberty and after menopause, less estrogen stimulation results in less glycogen accumulation in mucosal cells and an alkaline pH.
      c. The acid discharge and thick epithelium protect the vagina from infection from harmful bacteria. When estrogen levels are lower, as in prepubertal girls and menopausal women, the vagina is more vulnerable to infection. Infection also is more likely in women of reproductive age if the normal vaginal bacteria are disturbed or destroyed by chemical contraceptives or antibiotics.
E. The **external genitalia** collectively are called the **vulva**, or **pudendum** (Figure 18–7).

Figure 18–7. *External genitalia of the female.*

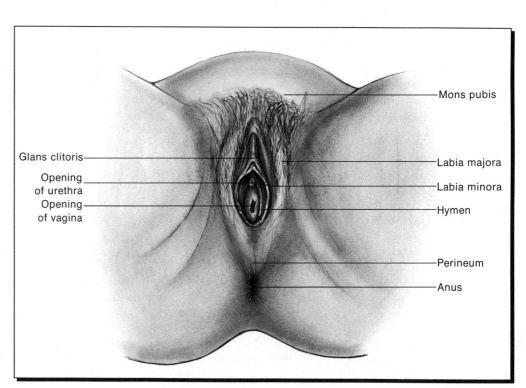

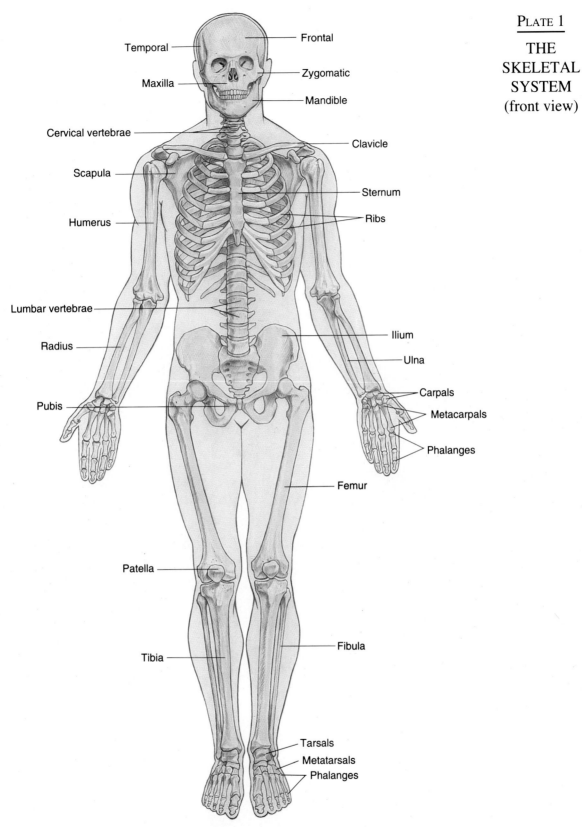

Plate 2

THE SKELETAL SYSTEM
(side and back views)

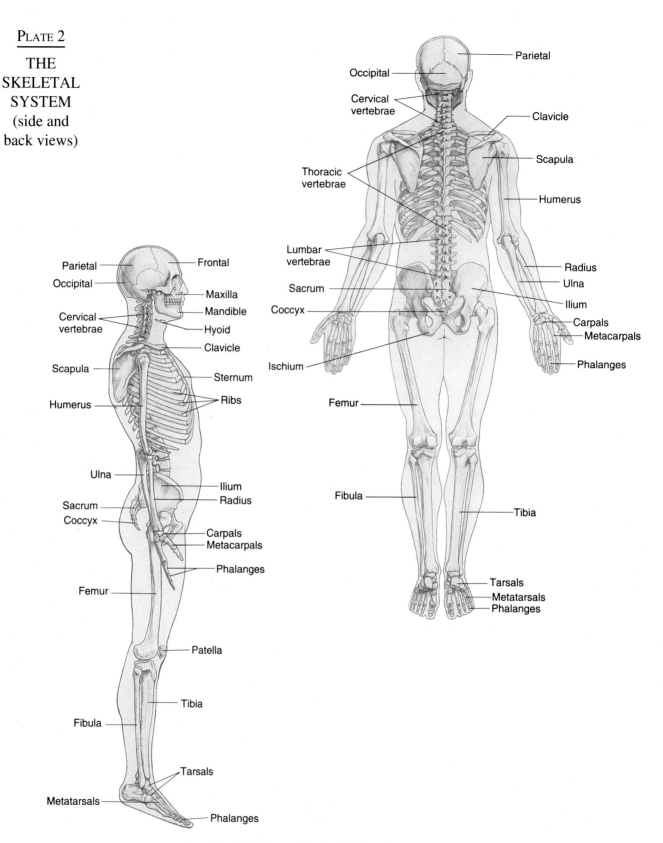

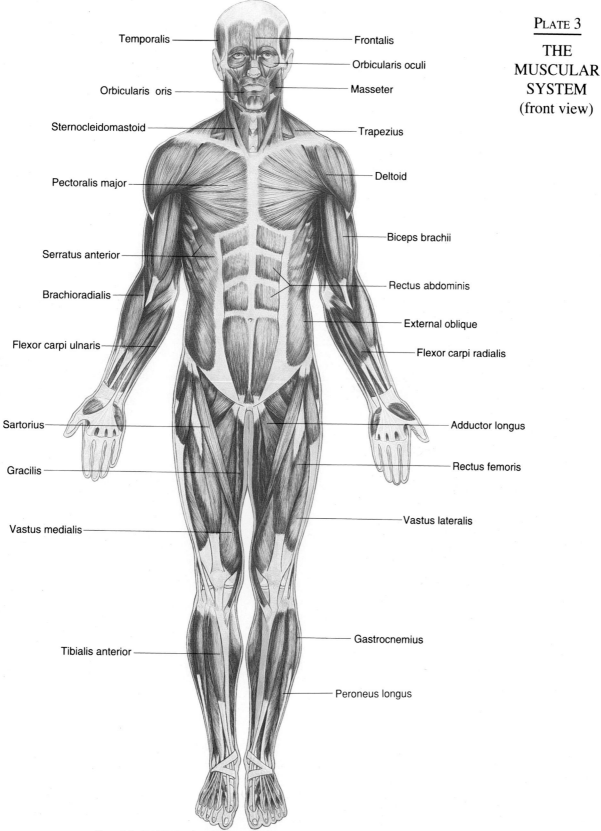

PLATE 3
THE MUSCULAR SYSTEM
(front view)

Copyright © 1991 Jones and Bartlett Publishers, Inc. Artist: Vincent Perez

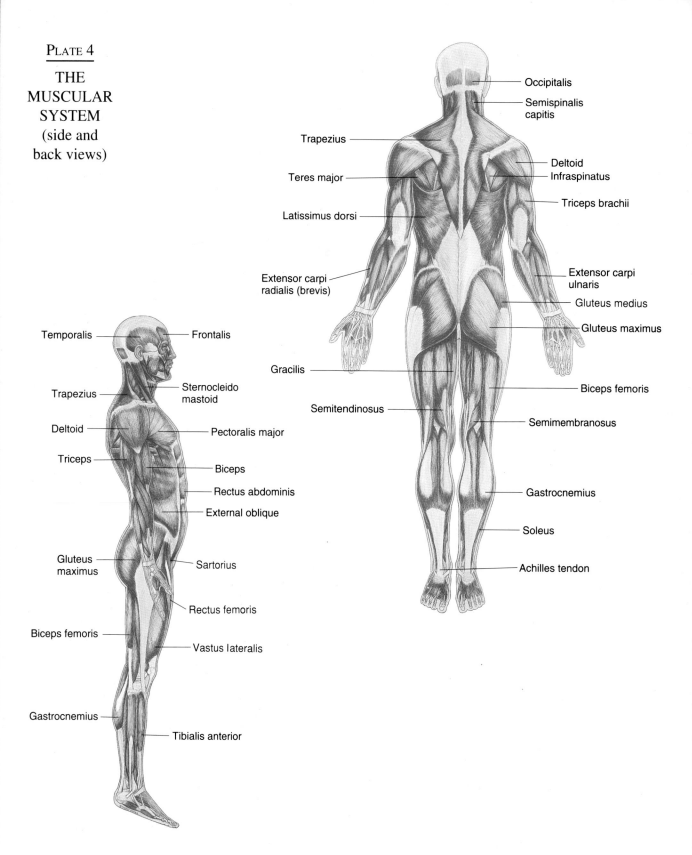

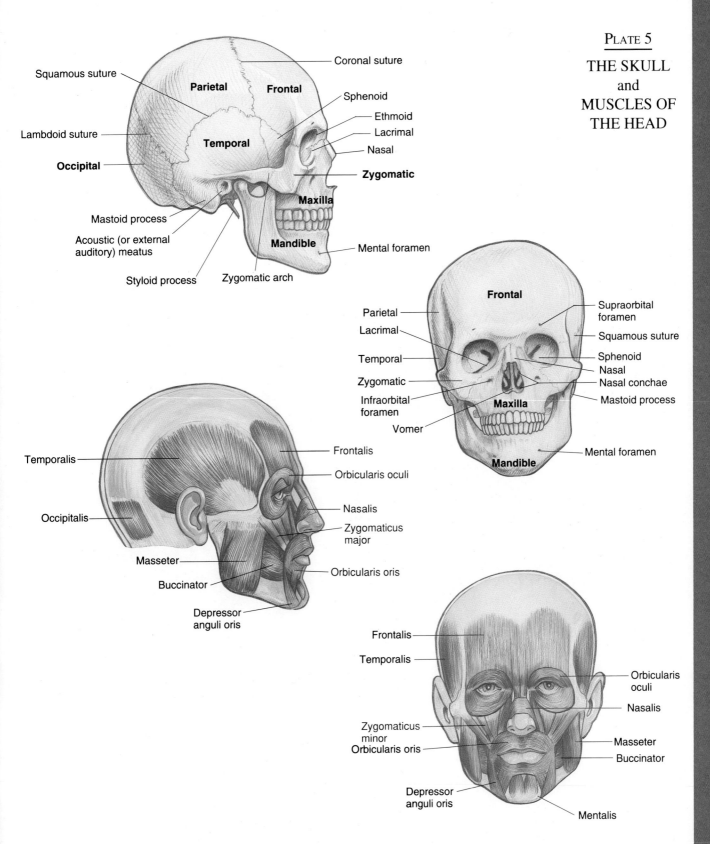

PLATE 5

THE SKULL and MUSCLES OF THE HEAD

Copyright © 1991 Jones and Bartlett Publishers, Inc. Artist: Vincent Perez

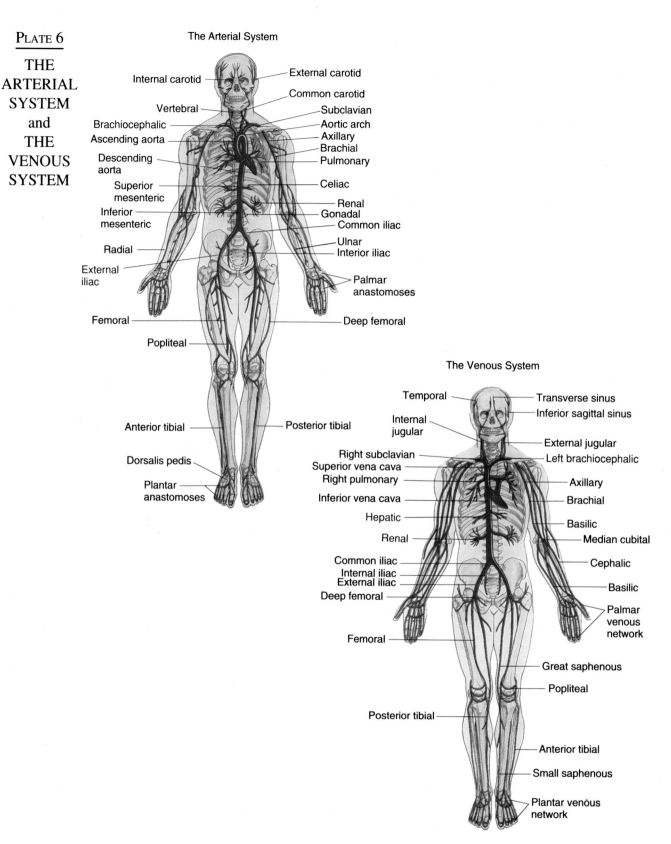

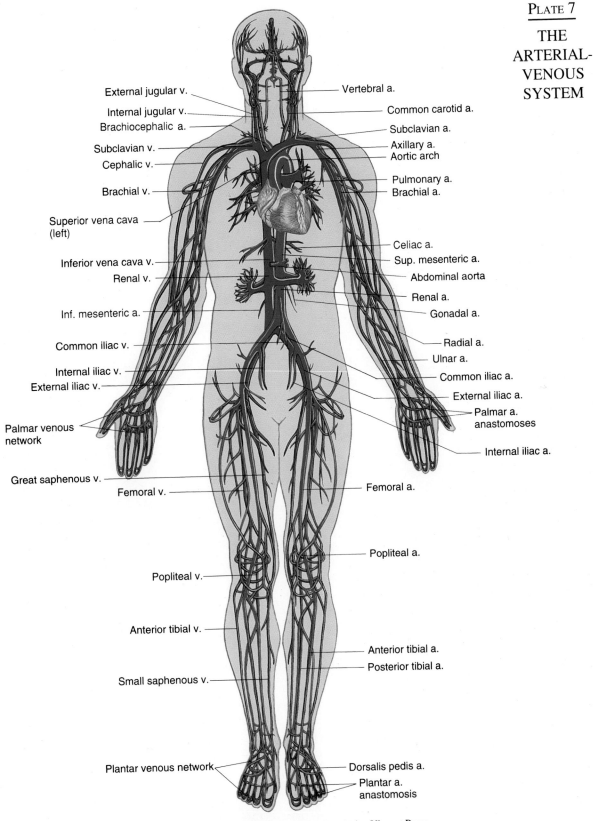

PLATE 7

THE ARTERIAL-VENOUS SYSTEM

Copyright © 1991 Jones and Bartlett Publishers, Inc. Artist: Vincent Perez

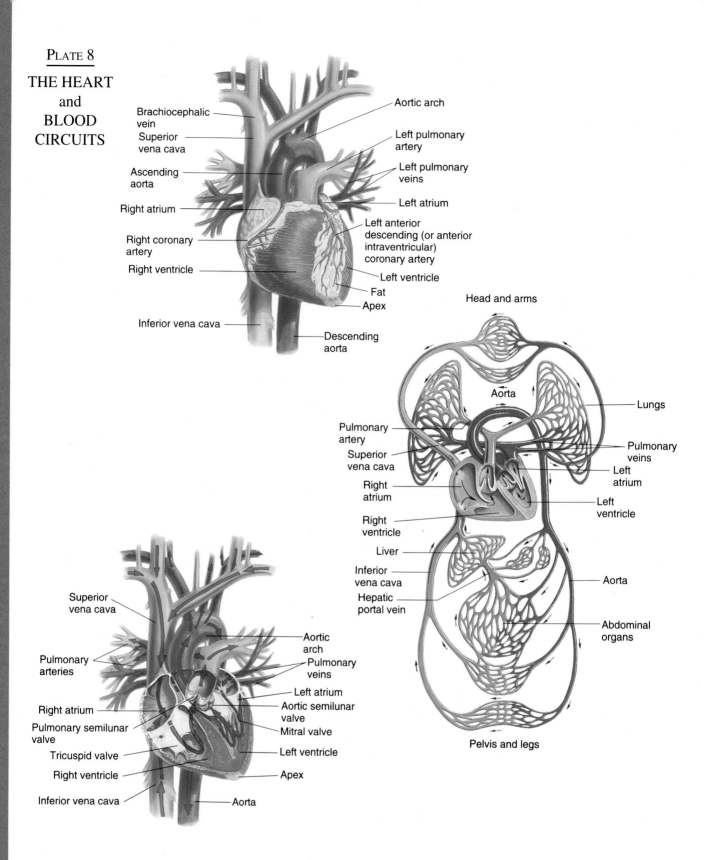

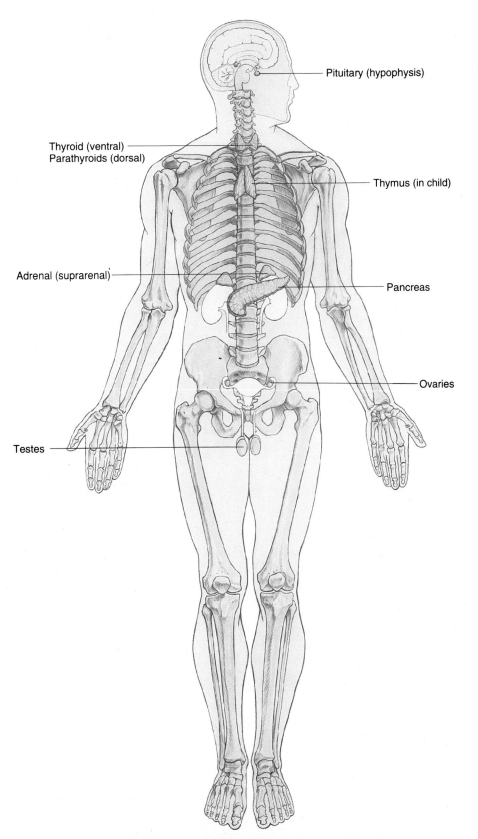

PLATE 9

THE ENDOCRINE SYSTEM

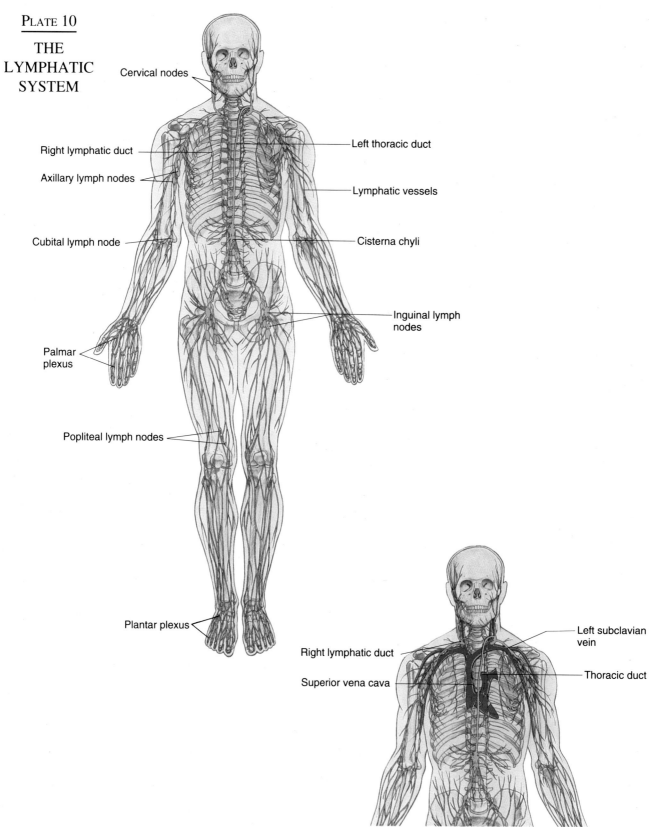

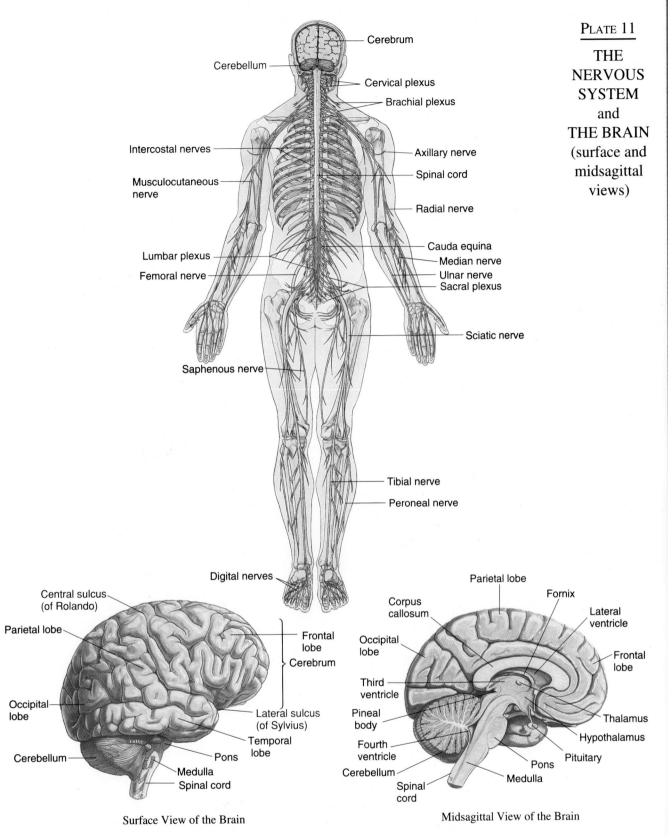

PLATE 12

THE VISCERA

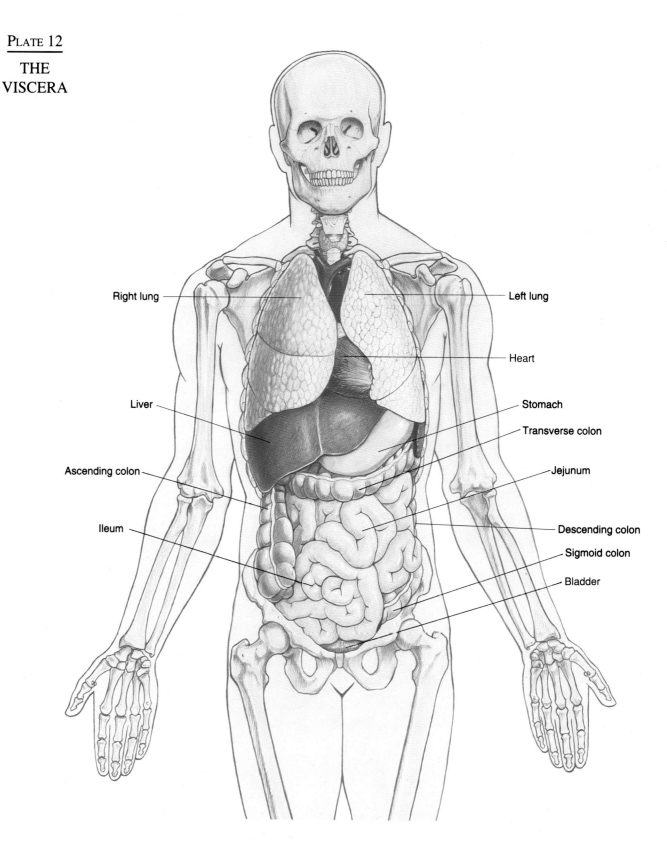

Copyright © 1991 Jones and Bartlett Publishers, Inc. Artist: Vincent Perez

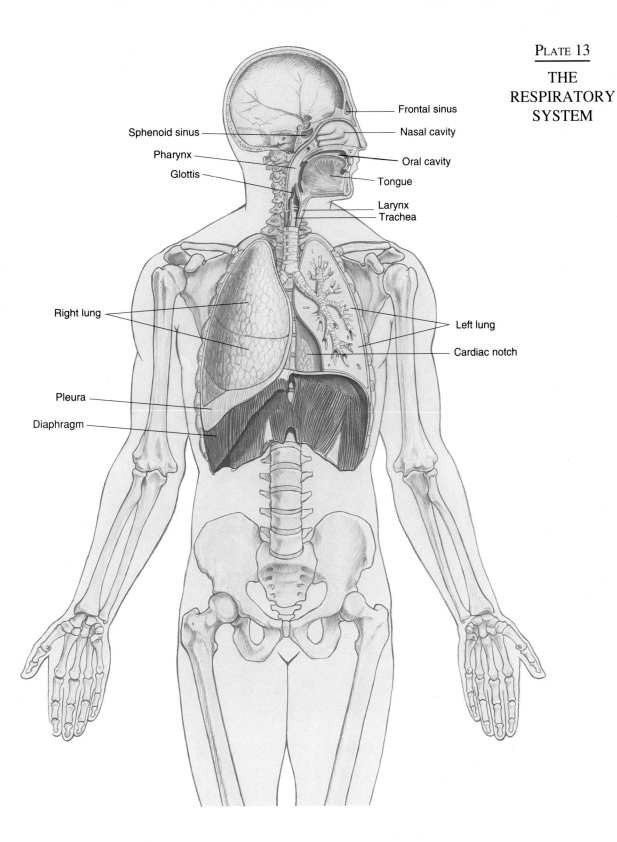

Plate 14
THE DIGESTIVE SYSTEM

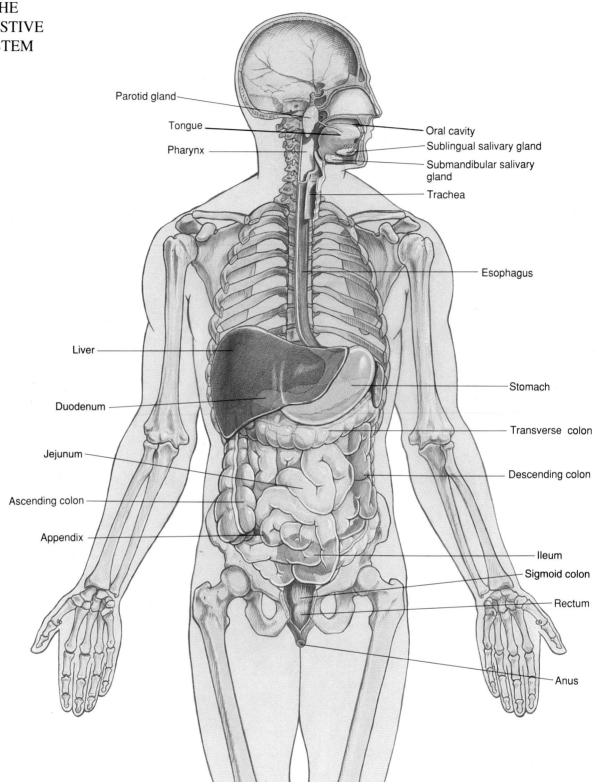

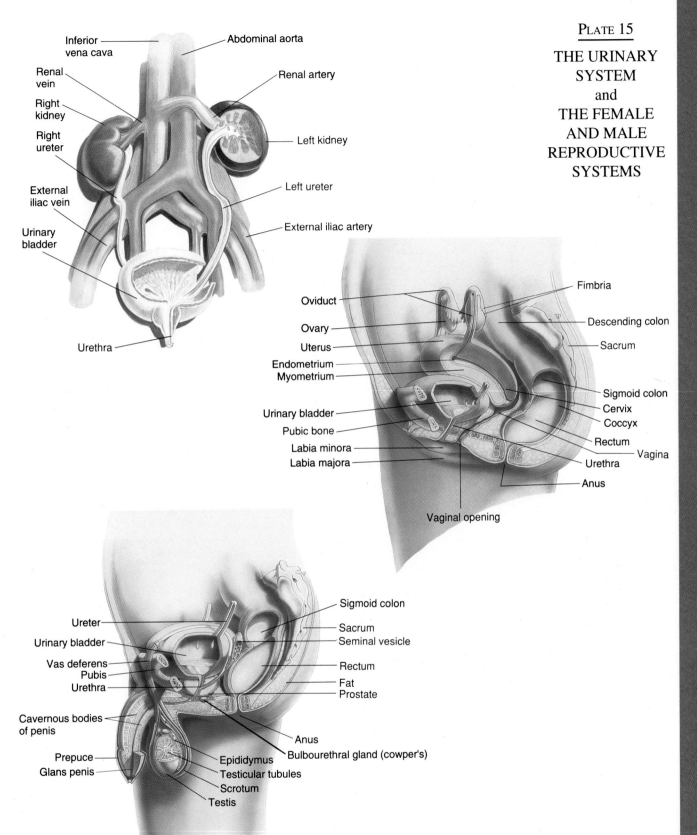

Plate 15

THE URINARY SYSTEM and THE FEMALE AND MALE REPRODUCTIVE SYSTEMS

Copyright © 1991 Jones and Bartlett Publishers, Inc. Artist: Vincent Perez

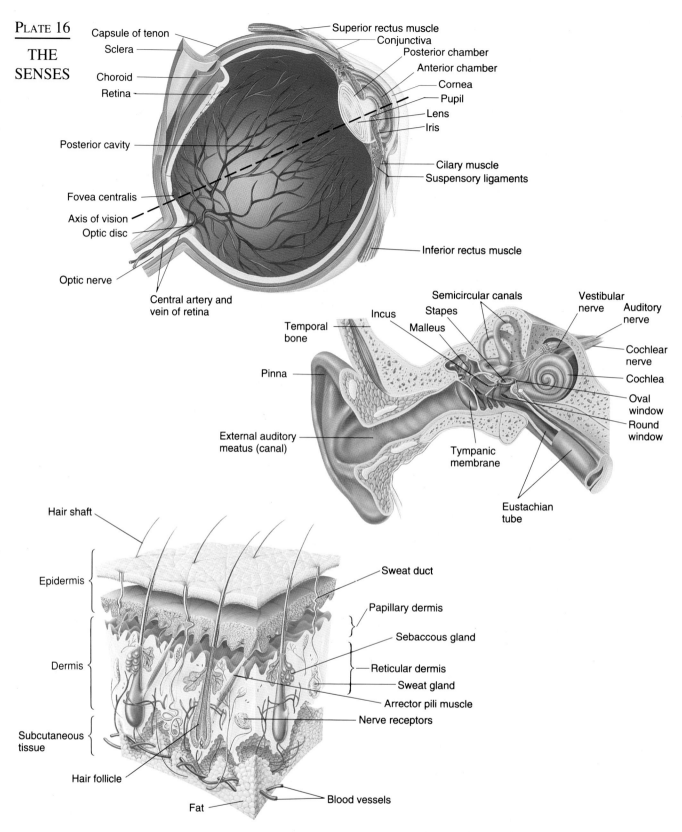

1. The **mons pubis** is the cushion of fatty tissue and skin that lies over the pubic symphysis. It is covered with pubic hair after puberty.
2. The **labia majora** (major lips) are two longitudinal folds of skin that extend down from the mons pubis and meet posteriorly in the **perineum**, which is the skin between their junction and the anus. The labia majora are homologous (similar in structure and origin) to the scrotum of the male.
3. The **labia minora** (minor lips) are two folds of skin between the labia majora. They are hairless, but they contain sebaceous glands and some sweat glands.
    a. The **prepuce** of the clitoris is the junction of the labia minora folds below the clitoris.
    b. The **frenulum** is the area of the folds below the clitoris.
4. The **clitoris** is homologous to the penis of the male, but it is smaller and has no urethral opening.
    a. The clitoris is composed of two **crura** (roots), a **shaft** (body) and a rounded **glans clitoris**, which has many nerve endings and is highly sensitive.
    b. The shaft contains two **corpora cavernosa** composed of erectile tissue. When engorged with blood during sexual excitement, they are responsible for clitoral erection.
5. The **vestibule** is the area surrounded by the labia minora. The vestibule encloses the urethral opening, the vaginal opening, and the ducts of Bartholin's (greater vestibular) glands.
    a. **Bartholin's glands** are homologous to the bulbourethral glands in the male. They produce a few drops of mucous secretion to help lubricate the vaginal orifice during sexual excitement.
    b. The **vestibular bulbs** are masses of erectile tissue deep within the substance of the labial tissue. They are equivalent to the corpora spongiosa of the penis.
6. The **urethral orifice** is the exit for urine from the urinary bladder. Its lateral margin contains ducts to two **paraurethral (Skene's) glands**, which are considered homologous to the prostate gland in the male.
7. The **vaginal opening** is below the urethral orifice. The **hymen**, a membrane that is quite variable in size and shape, encircles the vaginal opening.
8. The **perineum** (in both males and females) is the diamond-shaped area extending from the pubic symphysis anteriorly to the coccyx posteriorly and to the ischial tuberosities laterally.

F. **The mammary glands** (breasts) are present in both sexes. They become functional at puberty in response to estrogen in females and normally remain undeveloped in males. In pregnancy, mammary glands reach full development and function in milk production (lactation) after delivery of the infant.
   1. **Structure** (Figure 18–8). Each breast is an elevation of **glandular and adipose tissue** covered by skin on the anterior chest wall. The breasts lie over the pectoralis major muscles and are attached to them by a layer of connective tissue. Variation in the size of the breasts is due mainly to variation in the amounts of fatty and connective tissue and not to the amount of actual glandular tissue.
      a. The glandular tissue consists of up to 15 to 20 major **lobes**, each drained by its own **lactiferous duct**, which enlarges to a **lactiferous sinus** (ampulla) before emerging to perforate the nipple with 15 to 20 openings.
      b. The lobes are surrounded by adipose tissue and separated by the **suspensory ligaments of Cooper** (bundles of fibrous connective tissue). The suspensory ligaments extend from the deep fascia over the pectoralis muscles to the superficial fascia just under the skin.
      c. The major lobes are subdivided into 20 to 40 **lobules**, each of which further branches into small ducts that terminate in secretory **alveoli**. The **alveolar cells**, under hormonal influence during pregnancy and after delivery, are the glandular units that synthesize and secrete milk.

Figure 18-8. *Mammary gland: A*, lateral view; *B*, sagittal section.

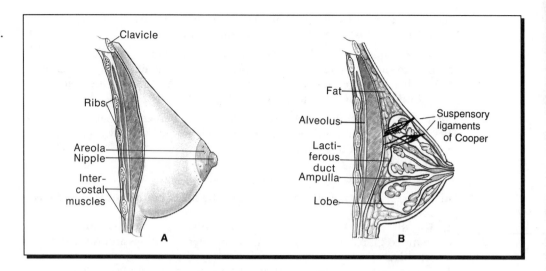

      d. The **nipple** has wrinkled, pigmented skin that extends outward for 1 cm to 2 cm to form the **areola**. The areolae contain large sweat and sebaceous glands, some associated with hair follicles and smooth muscle fibers that produce nipple erection when contracted. There is no muscle in the breast itself.

  2. **Blood supply and lymphatic drainage of the breasts**
      a. The arterial supply to the breasts is from the internal mammary artery, which is a branch of the subclavian artery. Additional contributions are from the thoracic branch of the axillary artery. Blood is drained from the breasts by superficial and deep veins, which lead into the superior vena cava.
      b. The lymphatic drainage from the central part of the mammary gland, the skin, nipple, and areola is lateral, toward the axilla. Thus, the lymph from the breasts passes through the axillary lymph nodes. This is clinically significant in the spread (metastasis) of breast cancer.

G. **Hormonal regulation of the female reproductive system.** The cyclic pattern of female reproductive function is precisely regulated by a balance of hypothalamic (GnRH), pituitary (FSH and LH), and ovarian (estrogen and progesterone) hormones. Both positive and negative feedback mechanisms are involved.

  1. **Menarche** (the first menstrual period) marks the onset of female sexual maturity.
      a. During childhood, the ovaries secrete small amounts of estrogen, which inhibits the hypothalamic release of GnRH.
      b. At puberty, the hypothalamus becomes less sensitive to estrogen and releases GnRH in pulsatile bursts. GnRH stimulates the anterior pituitary to release FSH and LH, which, in turn, stimulate the ovaries to produce estrogen and progesterone.
      c. **Estradiol** is the most biologically active and principal estrogen secreted by the ovaries. **Estrone** is a weaker estrogen formed by conversion from androstenedione. **Estriol** is the weakest estrogen.

  2. **Physiological effects of estrogen**
      a. It stimulates **growth of all reproductive organs**, particularly in the mucosa and muscle layers of the uterine tubes, uterus, and vagina. Estrogen stimulates growth of the ducts and alveoli of the mammary glands.
      b. It affects **total body configuration** by causing an increase in bone formation and an increased deposition of fat in all subcutaneous tissue, especially in the buttocks, thighs, and breasts.
      c. It has several **metabolic effects**, including a reduction in blood levels of cholesterol and low-density lipoproteins by comparison with males and postmenopausal women, and a facilitation of calcium metabolism.

d. It affects the hypothalamic **temperature-regulating** and **vasomotor centers**, which control the nerves that cause dilation and constriction of blood vessels.
   e. It causes the production of a clear, **watery cervical secretion**, which tends to facilitate sperm passage into the uterus.
3. Physiological effects of progesterone
   a. It stimulates further **growth of the uterine endometrium** to prepare it for implantation of a fertilized ovum. Progesterone inhibits uterine contractions so that an implanted ovum is retained.
   b. It stimulates the **growth and differentiation of the alveolar cells** of the mammary glands into milk-secreting cells. (The actual production of milk is a function of prolactin after the breasts have been prepared for lactation by estrogen and progesterone.)
   c. It increases the **viscosity of cervical mucus** and thus tends to retard the passage of sperm through the cervical os.
   d. It causes a slight **rise in basal body temperature** and an increase in sodium and water excretion from the kidneys.
4. The **menstrual cycle** designates the rhythmic fluctuation of the hypothalamic, pituitary, and ovarian hormones and the resulting morphological changes in the ovaries and the endometrium of the uterus.
   a. **Menstruation** (menses) is the monthly bleeding that occurs when part of the endometrium is sloughed off and expelled from the uterus through the vagina.
   b. The **length of the menstrual cycle** typically is described as being 28 days, although it is extremely variable. Cycles as short as 18 days or as long as 40 days are considered normal.
   c. The menstrual cycle incorporates both the **ovarian cycle** and the **endometrial (uterine) cycle**.
      (1) The ovarian cycle consists of the **follicular (preovulatory) phase**, which encompasses the period of follicular growth; the **ovulatory phase**, which culminates in ovulation; and the **luteal (postovulatory) phase**, which is the period of corpus luteum activity.
      (2) The endometrial cycle consists of the **menstrual phase**; the **proliferative phase**, which corresponds to the follicular phase in the ovary; and the **secretory (progestational) phase**, which corresponds to the luteal phase in the ovary.
5. **The ovarian cycle**
   a. At the beginning of the cycle (day 1 of the follicular phase), FSH and LH are secreted from the anterior pituitary upon signal from pulsatile secretion of hypothalamic GnRH.
   b. A group (20 to 25) of primary follicles (with FSH receptors on their granulosa cells and LH receptors on their theca cells) begin to secrete estrogen, grow, and develop an antrum. They become **secondary follicles**.
   c. Initially, the increase in plasma estrogen levels inhibits FSH and LH through negative feedback. The decline in FSH tends to inhibit further follicle development except in the dominant follicle selected for ovulation. The production of estrogen continues to increase.
   d. As the blood estrogen concentration continues to increase during the midfollicular phase, it causes a **positive feedback** stimulatory effect on the pituitary and **enhances** the production of LH.
   e. The **estrogen surge** is a high blood level of estrogen that is maintained for 50 hours. The surge results in a peak burst of LH, or **LH surge**.
   f. The LH surge has the following effects on the dominant follicle:
      (1) The oocyte completes the first meiotic division and moves on to become a **secondary oocyte** and the first polar body. Meiosis will continue only if the oocyte is fertilized.
      (2) **Enzymes and prostaglandins**, which are necessary for follicle rupture, are synthesized.

(3) **Ovulation**, or release of the oocyte and its associated cells into the body cavity to be picked up by the uterine tube, takes place within 24 to 38 hours after the LH surge. This occurs 13 to 15 days prior to the onset of menstruation (the start of the next cycle).

(4) The ruptured follicle cells undergo luteinization and transform into a **corpus luteum**, which produces progesterone and lesser quantities of estrogen.

g. The increasing progesterone and estrogen levels in the blood exert a strong **negative feedback effect on FSH and LH**. Without LH to sustain it, the corpus luteum deteriorates and the blood levels of estrogen and progesterone sharply decrease.

h. With the decline in blood levels of estrogen and progesterone, their negative feedback effect on the anterior pituitary diminishes. FSH and LH again increase to initiate a new cycle.

6. The **endometrial (uterine) cycle** occurs in order to prepare the uterine endometrium to nourish and maintain an ovum should it be fertilized. The events of the endometrial cycle are correlated precisely with the hormonal and morphological events of the ovarian cycle.
    a. The **menstrual phase** is the period of bleeding that lasts 4 to 5 days. Under the influence of estrogen from the developing follicles in the ovaries, the denuded endometrium is repaired by cell division in the basal layer while menstruation is still occurring.
    b. The **proliferative phase** lasts until ovulation occurs.
        (1) The endometrium proliferates from the basal layer and **again** becomes thick and well vascularized. Estrogen also causes development of progesterone receptors on endometrial cells.
        (2) Tubular glands grow in the superficial layer. The gland cells proliferate rapidly but do not accumulate much secretion.
        (3) The spiral arterioles project between the glands to supply the endometrial and glandular cells.
    c. During the **secretory (progestational) phase**, progesterone stimulates the continued growth of the superficial layer.
        (1) The glands enlarge and secrete nutrients (glycogen and fats) to sustain a developing embryo should fertilization occur.
        (2) The spiral arterioles become more convoluted. The endometrium is ready for implantation.
        (3) If fertilization does not occur, the endometrium regresses.
            (a) The corpus luteum degenerates; the progesterone and estrogen levels decline.
            (b) The spiral arterioles, now unsupported by hormones, constrict and dilate intermittently. The constriction reduces blood flow and causes ischemia and then death (necrosis) of surrounding tissue and glands.
            (c) When the arterioles dilate, blood escapes from the areas that have already disintegrated.
            (d) Endometrial tissue fragments, glandular secretions, mucus, and small amounts of blood are released into the uterine cavity.
            (e) The bleeding (menses) lasts for 4 to 5 days and the cycle starts again.

H. **Menopause** is the cessation of menstrual cycles. It is considered complete after amenorrhea (absence of menstruation) for one year. The **climacteric** is the period during which cycles are irregular before they stop.

1. The **average age at menopause** in the United States is 50 years. Irregular and anovulatory cycles may start at age 40.
2. The **cause of menopause** is the age-related loss of follicles due to atresia and monthly ovulation. The loss of follicles results in diminished estrogen and progesterone secretion.

- a. The decreased levels of estrogen and progesterone alter the hypothalamic-pituitary-ovarian hormone axis and the negative feedback mechanisms. The cycles stop, although small amounts of ovarian hormones continue to be secreted by the adrenal glands and the ovarian stroma.
- b. Because pituitary FSH and LH levels are not inhibited by the negative feedback of ovarian hormones, they remain high during menopause. The source of human gonadotropins used clinically is the urine of menopausal women.

3. The **symptoms of menopause** are related to the decreased levels of estrogen and progesterone, which affect a number of organ systems and body chemistry.
   - a. The estrogen-supported tissues (mammary glands, reproductive organs, and external genitalia) gradually decrease in size.
   - b. The vaginal lining thins, and vaginal secretions become more alkaline.
   - c. The vasodilation of blood vessels in the skin result in hot flashes and excessive sweating in approximately 75% to 80% of women during the climacteric. In some women, it persists for a number of years.
   - d. Irritability, insomnia, headache, joint pains, and heart palpitations are experienced by some women.
   - e. In about 25% of women, the accelerated loss of bone mass in postmenopause leads to osteoporosis. Osteoporosis is more likely in small women with less bone mass.

V. **FERTILIZATION** is the union of a spermatozoan and a secondary oocyte to form a diploid cell—a **zygote** containing both maternal and paternal chromosomes.

   A. **Spermatozoa.** During sexual intercourse, about 250 to 400 million sperm are deposited into the vagina when the male ejaculates semen.
   1. The spermatozoa, swimming under their own power and aided by muscular contractions of the uterus and uterine tubes, ascend through the cervix, uterine cavity, and the isthmus of the uterine tubes to the tubal ampulla. Only a small number of the ejaculated sperm reach the oocyte, which is in one ampulla.
   2. **Capacitation of sperm.** While in the fluid environment of the uterus and tube, the sperm undergo conditioning of the sperm cell membrane and acrosome that renders sperm capable of fertilization.
   3. **Acrosomal reaction.** Capacitated sperm release hydrolytic enzymes from the acrosome to digest the corona radiata cells and the zona pellucida of the oocyte and open the way for one sperm to penetrate the oocyte. When one sperm penetrates the oocyte cell membrane, the zona pellucida of the oocyte makes it impervious to other sperm.

   B. **Oocyte.** Usually a single oocyte is released from the ovary into one uterine tube at ovulation.
   1. When one sperm penetrates the oocyte cell membrane, the zona pellucida of the oocyte undergoes a chemical change and becomes impervious to the entrance of any other sperm.
   2. **The oocyte is activated** by penetration of the sperm to complete its second meiotic division and forms the **ovum** and the **second polar body**. The chromosomes acquire a new nuclear envelope, which is known as the **female pronucleus**.

   C. **Fusion of the pronuclei**
   1. The sperm that penetrates the oocyte loses its flagellum and its nuclear membrane disappears. With the formation of a new pronuclear envelope, it develops into the **male pronucleus**.
   2. The male pronucleus moves toward the female pronucleus to join it. Their nuclear membranes break down, their DNA replicates, and their chromosomes line up on the equatorial plate. The first mitotic cleavage division immediately follows.

D. The oocyte is capable of being fertilized for about 24 hours. Sperm are viable in the female reproductive tract for 48 to 72 hours. Therefore, for fertilization to occur, sexual intercourse must take place between three days before to one day following ovulation.

VI. **CONTRACEPTION** is the practice of preventing fertility, which is the ability to conceive and produce a child. All major contraceptive methods are techniques to avoid accidental pregnancy. Each varies widely in safety and effectiveness and has advantages and disadvantages.

A. **Hormonal contraceptives**
1. **Oral contraceptives** (OCs, birth control pills) are combinations of synthetic estrogen and synthetic progesterone taken by the female for 21 days of the menstrual cycle.
    a. Birth control pills inhibit ovulation, probably by suppressing LH.
    b. Secondary effects of the pills are the alteration of tubal transport and endometrial changes that inhibit implantation. Low-dose pills primarily function through these secondary effects without necessarily inhibiting ovulation.
2. **Norplant®** is a subdermal implant of a synthetic progesterone that provides contraception for five years.
3. **Depoprovera®** is an injectible contraceptive. A single injection of 150 mg of a synthetic progesterone provides contraception for three months. Both Norplant and Depoprovera may produce side effects such as irregular bleeding, weight gain, headaches, or nausea.

B. **Intrauterine devices** (IUDs) are inserted into the uterine cavity. Their exact mechanism of preventing pregnancy is unknown; they are believed to interfere with implantation of the fertilized ovum by altering the uterine environment. In the United States, most IUDs have been removed from the market because of adverse side affects.

C. **Surgical sterilization** in females is **tubal ligation**—the cutting, cauterizing, or tying off of the uterine tubes. In males, the procedure is **vasectomy**—the cutting, cauterizing, or tying off of the ductuli (vasa) deferentia.

D. **Barrier contraceptives** prevent sperm from contacting the oocyte.
1. **Physical barriers** mechanically obstruct the passage of the sperm through the cervix.
    a. A **vaginal diaphragm** or a **cervical cap** covers the cervix. Either of these devices must be used in conjunction with a spermicide.
    b. **Condoms** fit over an erect penis to trap the ejaculate and prevent semen from entering the vagina. They also partially protect against the spread of sexually transmitted diseases.
2. **Chemical barriers.** Foams, jellies, creams, suppositories, and the vaginal sponge contain spermicides that chemically destroy the sperm in the vagina.

E. The **rhythm** method of contraception is based on periodic abstinence from sexual intercourse during the fertile period of the menstrual cycle. When the rhythm method is combined with methods to recognize ovulation, such as changes in body temperature or cervical mucus, the technique is called **natural family planning** or the **sympto-thermal method**.

F. **Coitus interruptus** is withdrawal of the penis from the vagina before ejaculation. For most couples, it is less effective than other methods.

G. Other means of controlling fertility are being clinically tested. They include hormonal contraceptives for men, intravaginal rings that release hormones, and immunological methods such as antifertility vaccines.

VII. PREGNANCY AND EARLY DEVELOPMENT

A. **Pregnancy**
  1. **General aspects.** Pregnancy begins with fertilization of an egg cell by a sperm cell. It includes the period of **gestation** (embryonic and fetal development) and normally ends with **parturition**, or the delivery of a baby.
  2. **Duration.** The length of pregnancy is 266 days (38 weeks) from the time of fertilization until the time of delivery of the infant. Because the precise time of fertilization usually is unknown, the date of delivery is generally calculated from the onset of the last menstrual period. Assuming a 28-day cycle, parturition would occur in 280 days, or 40 weeks, or 10 lunar months, or 9 calendar months.
  3. **Hormones secreted during pregnancy**
      a. **Human chorionic gonadotropin (HCG)** is an LH-like hormone secreted by embryonic cells beginning about 10 days after fertilization.
          (1) HCG maintains the production of progesterone and estrogen by the corpus luteum in the ovary, which is known as the **corpus luteum of pregnancy**. In the absence of fertilization, the corpus luteum degenerates and ceases production of progesterone and estrogen.
          (2) HCG levels remain high for several months, after which the placenta takes over as the source of estrogen and progesterone.
          (3) HCG passes into the maternal circulation and is secreted into the pregnant woman's urine. It can be detected by immunoassay tests for the diagnosis of pregnancy.
      b. **Progesterone and estrogen** are secreted by the corpus luteum of pregnancy under HCG influence and, later, by the placenta. They have the following functions in pregnancy:
          (1) Progesterone and estrogen **maintain the uterine lining** for the implantation and retention of the conceptus (developing offspring).
          (2) The high levels of progesterone and estrogen in the maternal circulation **inhibit pituitary gonadotropins** and prevent the start of another menstrual cycle. Menstruation ceases during pregnancy and resumes after parturition. Breastfeeding may prolong the absence of menstruation.
          (3) Both estrogen and progesterone stimulate further **development of the mammary glands**.
          (4) Progesterone **inhibits uterine muscular contractions** during the course of pregnancy.
      c. **Human placental lactogen (HPL)**, also known as chorionic somatomammotropin, is secreted by the placenta. It stimulates **growth of the mammary glands** in preparation for lactation. HPL also has **metabolic effects** on proteins, fats, and glucose to provide energy for the pregnant woman and nutrients for the developing fetus.
      d. **Chorionic thyrotropin**, a hormone similar to TSH from the anterior pituitary, is secreted by the placenta and functions to **increase the rate of maternal metabolism**. Almost all hormones that affect metabolism, including growth hormone, ACTH, TSH, and insulin, are produced by the placenta.
      e. **Relaxin** is a polypeptide hormone secreted by the corpus luteum of pregnancy. It functions to relax (soften) the fibrocartilage in the **pubic symphysis** in preparation for passage of the fetus through the birth canal.
      f. **Prolactin** secretion by the anterior pituitary gland increases during pregnancy under stimulation by estrogen. In conjunction with HPL and estrogen, prolactin **stimulates growth of the ducts and alveoli in the mammary glands**. After delivery, prolactin results in **milk production**.
          (1) During pregnancy, high levels of estrogen stimulate the pituitary to secrete prolactin but inhibit its milk-producing action.
          (2) After delivery, the decline in estrogen and progesterone will cause prolactin levels to decrease unless breastfeeding occurs. Suckling reflexly maintains high levels of prolactin.

g. **Oxytocin**, which is released from the posterior pituitary gland (neurohypophysis), **stimulates uterine smooth muscle contractions** during the birth process. During lactation, oxytocin causes contraction of myoepithelial cells in the mammary gland so that milk becomes available at the nipple (**milk ejection**).
h. **Prostaglandins**, produced by the uterus, **stimulate uterine contractions** during birth.

4. **Parturition** is the birth of the baby. **Labor** is the collective term for the series of events that surround parturition.
   a. **The first stage of labor** is dilation of the cervix. It begins with regular uterine contractions and continues until the cervical os is completely dilated. The endocervical canal shortens (effaces) so that the external os is flush with the vagina.
   b. **The second stage of labor** is the actual birth. It starts with maximum dilation of the cervix and continues until the baby is expelled from the vagina.
   c. **The third stage of labor** is the expulsion of the placenta from the uterus.

5. **Lactation**, which is the production of milk by the mammary glands, normally occurs shortly after parturition. **Colostrum** is a fluid lacking milk fat, but it is rich in proteins, minerals, and immunoglobulins. It is produced for a few days before and after birth.

B. Prenatal development

1. The **preembryonic** (cleavage, germinal) period is a time of rapid mitotic cell division and includes the **first two weeks** after fertilization.
   a. **Initial cleavage divisions.** Within 18 to 39 hours after fertilization, the **zygote** undergoes mitotic cleavage to form two cells. After 60 hours, four cells are formed. By 72 hours postconception, eight cells have formed. During this time, the conceptus moves down the uterine tube to the uterus.
   b. **Morula.** After four cleavages, the 16 cells are called a morula ("little mulberry"). The conceptus reaches the uterine cavity in the morula stage.
   c. **Blastocyst.** In the morula, an off-center fluid-filled space, or **blastocoele**, appears, which transforms the morula into a **blastocyst**, or hollow ball of cells. The blastocyst remains free in the uterine cavity for several days before it implants in the uterine lining (Figure 18–9).
      (1) The **inner cell mass** is a mound of cells at one pole of the inner surface of the **blastocyst**. The embryo will develop from this area.
      (2) The **trophoblast** (trophectoderm) consists of the cells making up the outer wall of the blastocyst. Two populations of trophoblast cells will form the placenta and the extra-embryonic membranes that surround the embryo.
         (a) The **cytotrophoblast** is the inner layer of trophoblast nearer the embryo. It is made up of individual cells.
         (b) The **syncytiotrophoblast** is the multinucleated mass of the outer trophoblast. It arises from the cytotrophoblast and penetrates into the uterine wall to form the placenta.
   d. **Implantation** takes place from day 6 following ovulation and is completed about day 11. The blastocyst implants completely into the endometrium in order to gain access to the maternal blood supply.
      (1) The pole of the blastocyst that contains the inner cell mass attaches to the endometrium of the uterus. The trophoblast cells secrete proteolytic enzymes that erode the cells of the endometrium and blood vessels.
      (2) The **primitive villi** are finger-like extensions of the syncytiotrophoblast with a cytotrophoblast core, which protrude into the endometrium.
         (a) As the endothelium of the maternal arteries is broken down, blood-filled spaces (lacunae) in the villi enable the developing embryo to tap the maternal arteries for nourishment.

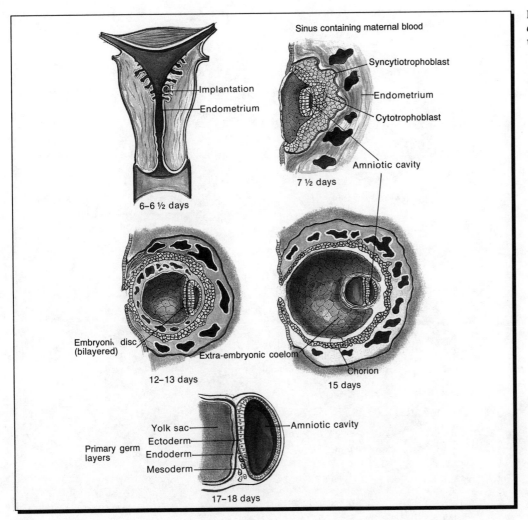

Figure 18–9. *Schematic drawing of implantation and embryonic development in the first 2 to 3 weeks.*

      (b) By the end of the third week, small blood vessels formed by ingrowing connective tissue arise in the primitive villi, which are then known as true **chorionic villi**. Chorionic villi constitute the fetal portion of the placenta.
    (3) The **decidua** (from "to shed") refers to the uterine endometrium after implantation.
      (a) The **decidua capsularis** is the portion of endometrium that roofs over the implanted blastocyst.
      (b) The **decidua parietalis** is continuous with the decidua capsularis. It is the endometrium that lines the uterine cavity except for the area of implantation.
      (c) The **decidua basalis** is the endometrium under the blastocyst. This will become the maternal portion of the placenta.
2. The **embryonic stage** extends from the second through the eighth week of development. It starts with the completion of implantation and includes the formation of the germ layers, embryonic membranes, and the placenta; the development of major internal organs; and the appearance of major external body structures (Figure 18–10).
   a. **Formation of germ layers.** The further differentiation of the inner cell mass results in the formation of a fluid-filled **amniotic cavity** and an **embryonic disc**. The embryonic disc is composed of two germinal layers.
    (1) The **ectoderm** (outer layer of the embryonic disc) will give rise to the entire nervous system, the special senses, skin, and some of the endocrine glands.

Figure 18–10. *Embryonic development to week eight of gestation. A, Schematic diagram of fetal membranes. B, Section through the uterus. C and D, Further development of the embryo.*

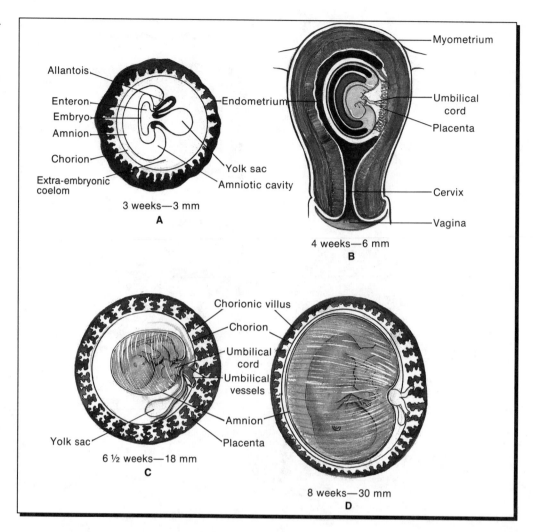

(2) The **endoderm** (inner layer of the embryonic disc) will give rise to the digestive and respiratory systems and parts of the reproductive system. The **yolk sac** is a cavity that forms in the endoderm below the embryonic disk.

(3) The **mesoderm** forms later between the ectoderm and the endoderm. It will give rise to the skeletal, urinary, circulatory, and reproductive systems.

    (a) The **extra-embryonic mesoderm**, which forms tissues that are not part of the body of the embryo, arises from cells between the trophoblast and the yolk sac.

    (b) The **intra-embryonic mesoderm**, which forms tissues inside the body of the embryo, arises from cells between the embryonic disk and the yolk sac.

b. **Formation of fetal (extra-embryonic) membranes.** The fetal membranes are discarded at birth and form from layers of cells that are not incorporated into the body of the embryo. They serve to protect and nourish the developing embryo and fetus.

(1) The **amnion** arises from extra-embryonic mesoderm and trophoblast. It forms the roof of the amniotic cavity, which becomes filled with amniotic fluid. Eventually, the amniotic cavity enlarges and the amnion grows down to completely surround the embryo and the umbilical cord.

(a) **Amniotic fluid** protects and cushions the embryo and fetus and allows free movement. After about 14 to 16 weeks of gestation, when fluid volume is sufficient, samples of amniotic fluid, which contain fetal cells and metabolic products, may be withdrawn to check for certain fetal abnormalities. The process is called **amniocentesis.**
(b) **Chorionic villi sampling** is another procedure for the prenatal detection of genetic defects. It may be performed between 7 to 10 weeks of gestation.
(2) The **yolk sac** forms in the endoderm and persists until the sixth week of gestation, when it becomes greatly reduced in size. It functions as an early respiratory and digestive organ. Part of the yolk sac is incorporated into the umbilical cord.
   (a) The **blood cells and blood vessels** are derived from mesoderm cells in the yolk sac.
   (b) The **primordial germ cells**, which migrate to the primitive gonads to eventually become spermatogonia and oogonia, originate in the yolk sac.
(3) The **chorion** is formed of trophoblast and extra-embryonic mesoderm and is the outermost membrane surrounding the developing embryo and fetus. It forms the chorionic villi, which form the fetal part of the placenta and are the source of HCG. The chorion fuses with the amnion to form the protective sac around the embryo and fetus.
(4) The **allantois** is a small, highly vascularized outgrowth from the yolk sac. It forms a structural base for the umbilical cord, which links the embryo to the placenta.
   (a) The proximal part of the allantois forms the **urinary bladder.** The distal part degenerates.
   (b) The allantoic blood vessels become the umbilical arteries and umbilical vein. Their branches throughout the villi link the fetus with the placenta.
c. The **placenta** arises from the union of chorionic villi and the uterine endometrium. It serves the digestive, respiratory, excretory, and metabolic functions of the fetus and is an endocrine organ. The fully formed placenta is a dark red disc, approximately 20 cm (7 in.) long and 2.5 cm (1 in.) thick. It weighs about 1 pound.
   (1) **Formation** of the placenta.
      (a) Originally, chorionic villi are present over the entire surface of the implanted embryo. As the developing embryo enlarges, the villi under the decidual capsularis portion of the endometrium disappear.
      (b) The chorionic villi under the embryo remain and become highly developed. They branch and enlarge and are known as the **chorion frondosum.** The chorion frondosum and decidua basalis portion of the endometrium together constitute the placenta. The embryo is attached by a connecting stalk (**umbilical cord**) to the placenta.
   (2) **Placental circulation**
      (a) The fetal capillaries in the terminal branches of the chorionic villi (chorion frondosum) are bathed by maternal blood in the blood sinuses of the decidua basalis of the uterine endometrium. The surfaces of the fetal and maternal tissue are separated by an intervillous space.
         (i) On the **maternal side**, blood enters the intervillous space from the eroded maternal arterioles. The maternal arterial blood is rich in oxygen and nutrients.
         (ii) On the **fetal side**, blood enters the villi from the umbilical arteries. Umbilical arterial blood is poor in oxygen and high in CO_2 and waste products.

(b) After the exchange of gases, nutrients, and wastes between the maternal and fetal blood in the villi capillaries, the oxygen and nutrient-rich blood returns to the fetus through the umbilical vein. Maternal blood is returned through uterine veins (See Chapter 11 VI D).
  (i) Fetal and maternal blood are in close contact, but they have no direct connection. Substances move between fetal and maternal blood by diffusion, active transport, and pinocytosis.
  (ii) Toward the end of pregnancy, the placenta allows maternal antibodies to pass into the fetal circulation. The antibodies provide temporary passive immunity to the fetus.
  (iii) Drugs, alcohol, environmental pollutants, viruses, and other disease-causing agents freely pass from the maternal circulation to the fetal circulation. Some of these substances are **teratogens**, or agents that may cause birth defects.

3. The **fetal stage** of development begins at the end of the eighth week during the first trimester (weeks 1 through 12) and continues to parturition. All body systems have developed by the eighth week; the remainder of the fetal period is concerned with further growth and differentiation of the organs.
   a. **Weeks 9 through 12 (third month).** The head slows in growth and the growth in body length accelerates. Ossification centers (see Chapter 7 I D) appear in most bones. The external genitalia have differentiated as male or female by the 12th week.
   b. **Weeks 13 through 16 (fourth month).** Facial features become well formed, and the hair, eyelashes, eyebrows, and nails develop. The appendages lengthen, the skeleton continues to ossify, and the fetus reaches 13 cm to 17 cm in length.
   c. **Weeks 17 through 20 (fifth month).** Growth decelerates somewhat. The legs attain their final relative proportions. Fetal movements (quickening) are felt by the mother. The skin is covered with fine hair (**lanugo**) and coated by a mixture of sebum and dead epidermal cells (**vernix caseosa**) for protection.
   d. **Weeks 21 through 25 (sixth month).** The fetus increases in body weight to about 900 gm. The skin is wrinkled and translucent. Because blood in the dermal vessels shows through the skin, it appears reddish.
   e. **Weeks 26 through 29 (seventh month).** The skin is less wrinkled as subcutaneous fat becomes deposited. The eyelids open. The fetus grows in length.
   f. **Weeks 30 through 33 (eighth month).** The testes in male fetuses descend into the scrotum. The skin is still somewhat wrinkled and reddish.
   g. **Weeks 34 through 38 (ninth month).** The fetus is "full-term" by the end of 38 weeks (or 40 weeks dated from the onset of the last menstrual period). Its average crown-rump length is 36 cm (14 in.) with a total length of 50 cm to 56 cm (20 in. to 22 in.), and it weighs about 7.5 pounds. The fetus is usually in **vertex presentation** with its head toward the cervix, ready for birth. If the buttocks are toward the cervix, the presentation is termed a **breech**.

Study Questions

Directions: Each question below contains four suggested answers. Choose the **one best** response to each question.

1. Which of the following embryonic structures gives rise to the female accessory ducts (uterine tubes, uterus, upper portion of vagina)?
   (A) Wolffian (mesonephric) ducts
   (B) Mullerian (paramesonephric) ducts
   (C) Gonadal ridge
   (D) Primordial germ cells

2. The descent of the testes from the abdominal cavity into the two scrotal sacs occurs during fetal life because
   (A) Additional room is required for the complete development of the testes.
   (B) The developing viscera in the abdomen move the fetal testes downward.
   (C) Normal spermatogenesis cannot occur at body temperature.
   (D) Testosterone can be produced by the testes only when they are located in the scrotum.

3. All of the following are functions of the Sertoli cells in the seminiferous tubules EXCEPT
   (A) They secrete an androgen-binding protein into the seminiferous tubule in response to FSH stimulation.
   (B) They secrete inhibin, which exerts a negative-feedback effect on the pituitary gland to inhibit FSH secretion.
   (C) They secrete Mullerian-duct stimulating hormone during embryonic sexual differentiation, which results in the development of the ductus deferens.
   (D) They provide support for developing spermatozoa.

4. A sperm cell moving from the lumen of the seminiferous tubule to the exterior of the body passes through all of the following structures EXCEPT the
   (A) seminal vesicle
   (B) epididymis
   (C) ductus deferens
   (D) urethra

5. Erection of the penis involves
   (A) engorgement of blood vessels under autonomic nervous system control
   (B) is dependent on a surge of LH released by the pituitary gland
   (C) the contraction of skeletal muscles in the penis
   (D) initiation only through tactile stimulus to the penis

6. Most of the primary follicles in the ovary normally
   (A) undergo atresia and disintegrate
   (B) mature and are ovulated
   (C) are lost in the menstrual flow each month
   (D) develop throughout the life span

7. The LH surge from the anterior pituitary gland
   (A) occurs just prior to ovulation
   (B) occurs just prior to menstruation
   (C) stimulates an estrogen surge from the ovaries
   (D) is responsible for follicle development in the uterus

8. A woman could still have menstrual periods after the surgical removal of
   (A) both ovaries
   (B) one ovary only
   (C) one ovary and the uterus
   (D) uterus only

9. All of the following occur during natural menopause EXCEPT
   (A) The number of ovarian follicles is reduced and eventually depleted.
   (B) Estrogen levels in the blood decrease.
   (C) Progesterone levels in the blood decrease.
   (D) FSH and LH levels in the blood decrease.

10. All of the following statements concerning pregnancy are accurate EXCEPT
    (A) The detection of human chorionic gonadotropin in the urine forms the basis for pregnancy tests.
    (B) The cyclic release of pituitary gonadotropins and ovarian steroids is continued.
    (C) The mammary gland tissue of the pregnant woman is stimulated to develop by placental hormones.
    (D) The corpus luteum of pregnancy maintains the uterus until the placenta is well established.

11. The stage of development at which implantation occurs is known as the
    (A) morula
    (B) zygote
    (C) blastocyst
    (D) fetus

12. Which of the following serves as an organ for nutrition, respiration, excretion, and hormone production for the developing embryo?
    (A) placenta
    (B) fetal liver
    (C) fetal lung
    (D) umbilical cord

Questions 13–16. For each descriptive statement below, choose the extra-embryonic structure with which it is most closely associated.
  (A) amnion
  (B) allantois
  (C) yolk sac
  (D) chorion

13. the outermost embryonic membrane that contributes to the formation of the placenta

14. the fetal membrane that enters into the formation of the urinary bladder and gives rise to fetal blood vessels

15. functions as a primitive digestive and respiratory organ before the placenta is fully developed

16. the innermost of the membranes enveloping the embryo; contains fluid in which the embryo is free to move and is protected from mechanical injury

Answers and Explanations

1. **The answer is B.** (II B c) In the undifferentiated stage, the primitive gonads are located near the Wolffian (mesonephric) ducts and the Mullerian (paramesonephric) ducts. Without testosterone and Mullerian duct inhibitor secreted by the fetal testes of a genetic male, the mesonephric ducts atrophy and the Mullerian ducts develop into the female internal reproductive tract.

2. **The answer is C.** (II B 2 a; III A) The scrotum maintains the temperature in the testes several degrees below the temperature of the body. Even the few degrees higher temperature in the abdomen prevents spermatogenesis by causing degeneration of the seminiferous tubules. Thus, failure of one or both testes to descend from the abdominal cavity to the scrotum results in an inability to produce sperm.

3. **The answer is C.** (III B 2 b) The Sertoli cell mechanically supports developing sperm and produces androgen-binding protein, inhibin, and mullerian duct inhibitor, which causes regression of the mullerian duct system.

4. **The answer is A.** (III C 1–5; D 1–3) The structures through which a sperm passes from the testis to the exterior include the epididymis, ductus deferens, ejaculatory duct, prostatic urethra, membranous urethra, and penile urethra. The secretions of the prostate gland empty into the prostatic urethra; the secretions of the seminal vesicles empty into the ejaculatory duct; and the secretions of the bulbourethral glands empty into the penile urethra.

5. **The answer is A.** (III E 3) The mechanism of penile erection includes autonomic nervous system innervation, arterial blood vessel constriction, and distention of the blood vessels in the corpora cavernosa. Neither hormones nor skeletal muscle contraction play a role in erection. An erection may be initiated by tactile sexual stimulation or by mental stimulation.

6. **The answer is A.** (IV A 3 c; G 5) Although the human female is born with millions of primordial follicles in the ovaries that can give rise to primary, secondary, and mature follicles, the vast majority become atretic. Although a number of follicles begin to develop each month, only one primary follicle matures to become the dominant follicle that is ovulated in each monthly ovarian cycle. Approximately 400 mature follicles are ovulated during the reproductive life of the human female.

7. **The answer is A.** (IV G 4,5) The LH surge is a peak burst of LH from the anterior pituitary gland, which causes ovulation. It occurs in the middle of the ovarian cycle and is stimulated by the positive feedback of a high blood level of estrogen that is sustained for 48 to 50 hours. Although FSH and LH stimulate follicle development at the beginning of a cycle, it is the LH surge that results in the rupture of the dominant follicle from the ovary.

8. **The answer is B.** (IV G 5,6) In the menstrual cycle, the estrogen and progesterone secreted by the ovaries stimulates the proliferation of the uterine endometrium so that implantation of a fertilized ovum can take place. Should fertilization not occur, the endometrial lining of the uterus is sloughed off during the menstrual period. Removal of both ovaries would completely eliminate the source of estrogen and progesterone, but with an intact uterus and one ovary to secrete the hormones, menstruation would still occur.

9. **The answer is D.** (IV H 1,2) Menopause is the cessation of menstrual cycles. It is caused by the depletion of ovarian follicles that started before birth and continues throughout the reproductive life of the female. In the absence of follicular development, the decreased estrogen and progesterone levels in the blood do not inhibit the pituitary FSH and LH levels and the cycles stop. The FSH and LH levels, however, remain high in the postmenopause.

10. **The answer is B.** (VII A 1–3) The high levels of estrogen and progesterone in the maternal circulation during pregnancy inhibit the cyclic release of pituitary gonadotropins and prevent the menstrual cycles. Pregnancy hormones include HCG, which maintains the corpus luteum of pregnancy and forms the basis for pregnancy diagnosis urine tests. Ovarian and placental estrogen and progesterone, human placental lactogen, and pituitary prolactin stimulate development of the ducts and alveoli in the mammary glands.

11. **The answer is C.** (VII B 1 c) The zygote is the fertilized ovum. The morula is the 16-cell stage. The conceptus is called a fetus after 56 days.

12. **The answer is A.** (VII B 2 c) The placenta is the digestive, respiratory, excretory, and metabolic organ for the developing embryo and fetus and also secretes both protein and steroid hormones. The amnion surrounds the embryo and covers the umbilical cord. The fluid in the amniotic cavity protects and cushions the developing embryo. The fetal lung is nonfunctional until birth; gas exchange between the maternal and the fetal circulation is accomplished by the placenta. The umbilical cord contains two umbilical arteries, which carry blood low in oxygen and high in carbon dioxide and waste products, and a single umbilical vein, which returns oxygenated blood from the placenta to the fetus.

13–16. **The answers are 13–D, 14–B, 15–C, and 16–A.** (VII B 2 b) The chorion is the outermost extra-embryonic membrane that consists of an external layer of trophoblast reinforced by mesoderm. It develops villi that become embedded in the uterine endometrium. The allantois is an outgrowth of the yolk sac that is associated with the development of blood cells and blood vessels and contributes to the formation of the umbilical cord. The yolk sac has an early digestive and respiratory function; it is nonfunctional after the sixth week of gestation and becomes incorporated into the umbilical cord. The amnion surrounds the embryo and covers the umbilical cord; it is filled with amniotic fluid.

Index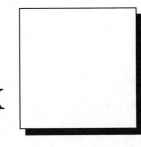

Abdominal
  aorta, 244
  muscles, 138
  pelvic region, 8
Abdominopelvic peritoneum, 284
Abducens nerves, 179
Abduction, 115
Abscess, 257
Absorption, 4
Absorptive (feasting) state, 311
Accommodation, 189
Acetabulum, 108
Acetylcholine (ACh), 123-124, 163, 183,
    214-216, 310, 340
Achondroplasia, 63
Acid-base balance, 269, 320, 341-345
Acidosis, 163, 309, 341, 344
Acids, 17, 19-20
Acinar, 292
Acne, 88
Acquired immunity, 255
Acquired immune deficiency syndrome
    (AIDS), 227, 266
Acromegaly, 209
Acromion process, 106
ACTH, 209
Actin, 121
Action potential, 160-161
Activation energy, 17, 29
Active
  immunity, 260-261
  transport, 41, 44-45, 159-160
Acute inflammation, 256
Adam's apple, 271
Addison's disease, 215, 266
Adduction, 115
Adenine, 27, 46
Adenohypophysis, 206-207
Adenoids, 270
Adenosine triphosphate, 27
ADH. *See* Antidiuretic hormone.
Adipose tissue, 74, 76-77
ADP, 45

Adrenal
  cortex, 308
  cortical hormones, 214
  diabetes, 215
  glands, 213-215
  medulla gland, 183, 308
  medulla hormones, 240, 243
Adrenocorticotropic hormone, 209
Adrenogenital syndrome, 215
Adventitia, 284
Aerobic reactions, 123
Afferent
  arteriole, 322-323
  lymph vessels, 249
  nerves, 155
  pathway, 165
Agammaglobulinemia, 218
Agglutination, 230, 260
Agonist, 130
Agranulocytes, 223, 226-227
AIDS, 227, 266
Alar nasal cartilages, 269
Albumins, 222
Aldosterone, 214-215, 327, 339-340
Alimentary tract, 283
Alkalosis, 163, 280, 341, 344-345
Allantois, 371
Allele, 60
Allergens, 266
Allergy, 265-266
All-or-none response, 161
Allosteric enzymes, 30
Alveolar
  bones, 100
  macrophages, 256
  process, 286
  ventilation rate, 275
  wall, 276
Alveoli, 273-274, 361
Ameboid movement, 226
Amenorrhea, 364
Amino acids, 21, 24, 164, 310
  derivatives, 204

  reabsorption, 325
Aminopeptidases, 293, 294
Ammonia and ammonium buffer pair, 344
Amniocentesis, 65, 371
Amnion, 370
Amniotic cavity, 369
Amphiarthroses, 113
Ampullae, 195, 359
Amygdaloid body, 172
Amylase, 285, 295
Amylopectin, 21
Amylose, 21
Anabolic reactions, 16
Anabolism, 4, 16, 45, 301
Anaerobic reactions, 122-123
Anal canal, 297
Anaphase, 49-50, 52, 64
Anaphylaxis, 266
Anatomical
  dead space, 275
  position, 5-7
Anatomy, defined, 1
Androgen, 96, 210, 215, 348
  binding protein, 353
Android pelvis, 110
Anemia, 225, 266, 280, 344
Angiotensin, 243, 340
Angiotensinogen, 340
Anions, 15
Ankle
  bones of, 112-113
  muscles, 150-151
Anoxia, 163, 280
Antagonists, 130
Anterior lobe hormones, 208-210
Antibody, 259
  antigen interaction, 260
Anticodon, 53
Antidiuresis, 210
Antidiuretic hormone, 210, 243, 327, 338
Antigens, 230, 259, 261, 263
Antithrombin, 228
Antrum, 357

377

Anuria, 332
Aorta, 232, 234, 239, 243-245
Aortic
    semilunar valves, 233
    valve, 238
Apical foramen, 287
Aplastic anemia, 225
Apneustic center, 279
Apocrine
    secretion, 73
    sweat glands, 88
Appendicular skeleton, 92, 105-113
Appositional growth, 78
Aqueduct of Sylvius, 174
Aqueous
    fluid, 188
    humor, 187
Arachnoid
    layer, 167
    villi (granulations), 168-169
Arbor vitae, 174
Arcuate
    arteries, 245, 323
    veins, 324
Areolar (loose) connective tissue, 75
Arm, muscles of, 141-142
Arrector pili muscle, 88
Arrhythmias, 235
Arteries, 221, 240-241, 243-245
Arterioles, 241
Arteriovenous anastomoses, 241
Arthritis, 116
Articular
    cartilage, 94-95
    discs, 114
Articulating processes, 101
Articulations (joints), 113-116
Arytenoid cartilages, 271
Asphyxia, 280
Association areas and tracts, 171
Astigmatism, 190
Astrocytes, 158
Atalectasis, 274
Atherosclerosis, 228, 240
Atlas, 103
Atmospheric pressure, 273
Atomic
    mass, 12
    nucleus, 11, 28
    number, 12
    structure, 11
    weight, 12, 19
Atoms, 3, 11-12, 19
ATP, 45, 302, 305
    ATP-ADP system, 302
Atresia, 357
Atria, 232
Atrioventricular node, 235
Atrophy, 127
Auditory
    association area 171
    canal, 191
    ossicles, 92, 100, 192
Auerbach's plexus, 284
Auricle, 191
Autoimmune diseases, 266
Autoimmunity, 258

Autonomic
    nervous system, 182-185
    stimulation, effects, 184-185
Autosomal
    abnormalities, 64–65
    dominant alleles, 60-61
    genes, 63
Autosomes, 58
Avogadro's number, 19
Axial skeleton, 92, 97-105
Axillary
    artery, 244
    nodes, 250
    vein, 246
Axis, 103
Axon, 81, 123, 156-157, 162, 210

Baldness, 88
Ball-and-socket joint, 114
Baroreceptors, 239
Barr body, 58
Barrier contraceptives, 366
Bartholin's glands, 361
Basal
    bodies, 40
    ganglia, 171-172
    lamina, 69
    metabolic rate (BMR), 312
Basement membrane, 69, 322
    of alveolar wall, 276
Bases, 17, 19-20, 341
Basilar membrane, 193
Basilic vein, 246
Basophils, 75, 223, 226-227
B cells, 261-263
Beta
    oxidation, 309
    receptors, 242
Bicarbonate, 278
Bicuspid (mitral) valve, 233
Bile, 292, 294-295
    canaliculi, 294
    duct, 294
    salts, 24, 294
Bilirubin, 225, 293
Biliverdin, 225, 294
Binominal nomenclature, 2
Biochemistry, 17-27
Bipedality, 3
Bipolar neurons, 188
Birth control pills, 366
Black lung, 280
Bladder, 320, 330-331, 371
Blastocoele, 368
Blastocyst, 368
Blind spot (optic disk), 188
Blood, 79, 221-231
    amino acid concentration, 320
    brain barrier, 158
    cell formation of, in bone marrow, 92
    cells, 223
    clot, 228, 340
    clotting factors, 294
    colloid osmotic (oncotic) pressure, 248
    flow, 240-241,
    glucose, 216, 320
    groups, 230-231

    hematopoiesis, 222-223
    hydrostatic pressure, 247
    plasma, 222, 340
    pressure, 210, 237, 239-243, 269, 320, 327, 340
    proteins, 25
    storage, 294
    sugar, 293, 308, 311
    supply to kidney, 323-324
    transfusions, 230-231
    typing, 230-231
    vessels, 221, 240-241
    viscosity, 242
    volume, 222, 327, 338
Body
    cavities, 7
    fluid compartments, 336-337
    fluids, 336-339
    temperature, 221, 312, 316-317. See also Temperature
Bohr effect, 278
Bolus, 287
Bomb calorimetry, 312
Bond positions, 23
Bone
    classification, 96-97
    collar, 95
    growth, 94-96
    markings, 97
    marrow, 223
    matrix, 93
    muscles and, 129
    remodeling, 96
    repair, 96
    salts, 93-94
    tissue, 78-79
Bony labyrinth, 192
Bowman's capsule, 322
Boyle's law, 276
Brachial
    artery, 244
    plexus, 181
    vein, 246
Brachiocephalic
    artery, 244
    vein, 246
Bradycardia, 238
Bradykinins, 243
Brain, 3, 167-175
    lateralization, 171
    stem, 183
Breastbone, 96
Breastfeeding, 210, 367
Breasts, 361-362
Breathing, 273-275
Breech presentation, 372
Broca's area, 170
Brodmann classification, 171
Bronchial
    arteries, 244
    tree, 272
Bronchitis, 280
Bronchomediastinal trunk, 249
Brunner's glands, 292
Buffering, 221
Bulbourethral (Cowper's) glands, 354
Bulk transport, 45

Bursae, 114
Bursitis, 116

Calcaneus, 112
Calcitonin, 211, 213, 341
Calcium, 212-213. 296, 340-341
    imbalance, 128
    pump, 44
Callus, 96
Calmodulin, 127, 205
Calorie, 312
Calyces, 321
cAMP, 204-205
Canaliculi, 78, 93-94
Canal of Schlemm, 187
Cancellous (spongy) bone, 78, 93
Capillary, 221, 241
    basement membrane, 276
    endothelium, 276, 322
    exchange, 247-250
    membrane permeability, 250
    network, 208
    permeability, 257
Capitate bone, 108
Capitulum, 107
Capsule wrappings, 76
Carbaminohemoglobin, 224, 278
Carbohydrates, 21, 36, 302-308
    absorption, 296
    digestion, 290
    sources, 313
Carbon, 20
Carbonic
    acid-sodium bicarbonate system, 342
    anhydrase, 224, 278
Carbon monoxide, 278, 280
Carbonyl group, 21
Carboxyl group, 21, 24
Carboxypeptidase, 293
Carcinogens, 47
Cardiac
    cycle, 237-238
    muscle, 80, 120, 128
    output, 239-240, 242
    physiology, 235-240
    portion of the stomach, 288
    reserve, 239
    veins, 234
Cardioaccelerator reflex center, 238, 242
Cardioinhibitory reflex center, 239, 242
Cardiovascular
    disease, 240
    system, 4, 221
Carotene, 86
Carotid artery and sinus, 239, 244
Carpals, 96
Carpus, 107–108
Carrier-mediated
    active transport, 44-45
    diffusion, 43-44
Carriers, genetic, 62
Cartilage, 76-78, 95, 113, 116, 269, 271
Cartilaginous joints, 113
Caruncle, 186
Catabolism, 4, 16, 45-46, 301
Catalysts, 17, 29
Cat-cry syndrome, 65

Catecholamines, 164
Cations, 14
Caudad, 7
Cauda equina, 176
Caudate nucleus, 171
Cavernous urethra, 331
CCK, 243, 292-294
Cecum, 297
Celiac artery, 245
Cells, 3, 34-54
    components, 35-41
    connective tissue, 75-76
    contractivity, 40
    cycle and mitosis, 47-48
    division, 40, 46-52
    functions, 34-35
    growth and repair, 39
    immunity, 255
    membrane, 35-36, 41-45
    metabolism, 45-46
    motility, 340
    nucleus, 35, 39-40
    numbers, 34
    respiration, 4, 46, 269, 301
    shapes, 34
    size, 34
    specialization, 34
Cellulose, 21
Cementum, 287
Central nervous system (CNS), 81, 155
    and peripheral nervous system, 167-185
Centrioles, 40
Centromere, 48
Centrosome, 40
Cephalic
    position, 5
    stage, 289
    vein, 246
Cerebellum, 174, 180
Cerebral
    aqueduct (of Sylvius), 168
    cortex, 169-172, 279
    dominance, 171
    peduncles, 174
    tracts, 171
Cerebrospinal fluid, 168-169
Cerebrum, 169-170
Ceruminous glands, 88
Cervical
    cap, 366
    enlargement, 176
    nerves, 180
    nodes, 249
    plexus, 181
    vertebrae, 101
Cervix, 359
Cheeks, 285
Chemical
    bonds, 13
    control, 279-280
    equilibrium, 16
    reactions, 16-17, 28-29
    structure, 3
    symbol, 11
    synapses, 162
Chemiosmotic model for electron transport
    and ATP synthesis, 307

Chemistry, basic, 11-16
Chemoreceptors, 185
Chemotaxic ability, 226
Chemotaxis, 256-257
Chloride shift, 278
Chlorine reabsorption, 325
Cholecystokinin (CCK), 243, 292-294
Cholesterol, 24, 36, 228, 308-309
Cholinergic fibers, 183
Cholinesterase, 124
Chondroblasts, 78
Chondrocytes, 77-78, 95
Chordae tendineae, 233
Chorion, 371
Chorionic
    thyrotropin, 367
    villi, 369
    villi sampling, 66, 371
Choroid,
    layer of the eye, 187
    plexuses, 168
Chromatids, 48
Chromatin, 39-40
Chromosomal abnormalities, 64-66
Chromosomes, 39, 47, 58
Chronic
    inflammation, 256
    obstructive pulmonary diseases, 280
Chylomicrons, 308
Chyme, 289-290, 293, 297
Chymotrypsin, 293
Cilia, 40-41, 71, 256, 270, 359
Ciliary
    body, 187
    glands of Moll, 88
    muscles, 189
Cingulate gyrus, 173
Circuits in central nervous system, 164
Circulation, 4
    placental, 371-372
Circulatory
    pathways, 243-247
    system, 221-250
Circumduction, 115
Cisterna chyli, 249
Cisternae, 37-38
Citric acid cycle, 46, 305-307
Claustrum, 172
Clavicle, 105
Clearance of blood plasma, 326
Climacteric, 364
Clitoris, 361
Clonal selection theory, 263
Clot, 257
Clotting. See also Blood
    abnormalities, 228-230
    factors, 228-229
Coccygeal
    nerves, 180
    plexus, 182
Coccyx, 103
Cochlea, 180, 192-193
Codominance, 62
Codon, 52
Coenzymes, 27, 30, 302
Cofactors, 30
Coitus interruptus, 366

Collagen, 25, 93, 95
Collagenous fibers, 75
Collateral ganglia, 183
Colliculi, 174
Colloid, 222
    osmotic pressure, 324
    solute particles, 42
    thyroid, 212
Colon, 297
Color
    blindness, 62–63
    vision, 191
Colostrum, 368
Columnar cells, 70
Commissure, 158, 171
Communication, and integument, 84
Compact (dense) bone, 78, 93
Competitive inhibitors, 30
Complement
    fixation, 260
    system, 258
Complementary base pairing, 46
Compliance, 274
Concentration gradients, 42
Condensation reactions, 21, 24
Condensing vacuoles, 38
Condoms, 366
Conduction, 316
Conductivity, 4, 155
Condyles, medial and lateral, 110-111
Condyloid joint, 115
Cones of retina, 188, 190-191
Conjunctiva, 186
Connective tissue, 69, 74-79
Contraception, 366
Contractile cells, 119
Contraction period, 125
Contralateral, 7
Control center of feedback system, 5
Conus medullaris, 176
Convection, 316
Converging circuit, 164
Cornea, 187-188
Corniculate cartilages, 271
Cornonary sinus, 234
Corocoid process, 106
Corona, 354
Coronal
    plane, 5
    suture, 98
Coronary
    arteries, 234, 244
    circulation, 234
    sinus, 232, 245
Coronoid process, 107-108
Corpora
    cavernosa, 354, 361
    quadrigemmina, 173
Corpus
    albicans, 358
    callosum, 169
    luteum, 358, 364, 367
    spongiosum, 331, 354
Cortex
    of adrenal, 213
    of brain, 169
    of hair, 87
    of ovary, 356

Corticosterone, 214
Corticotropin, 209
    releasing hormone, 214
Cortisol, 214
Cortisone, 214, 225
Costal notches, 103-105
Costocervical trunk, 244
Co-transport, 44-45, 325
Coumarin compounds, 228
Countercurrent
    exchanger mechanism, 329
    multiplier system, 327-328
Coupled
    reaction, 29
    transport, 44-45
Covalent
    bonds, 13-14, 28
    molecule, 15
Craniad, 5
Cranial
    bones, 92
    cavity, 7
    nerves, 178-180
Craniosacral division, 183
Cranium, 97-100
Creatinine, 320, 332
Cretinism, 212
Cribriform plate, 99, 270
Cricoid cartilage, 271
Cri-du-chat syndrome, 65
Crista, 37, 99, 195
Crossed extensor reflex, 166
Crown of tooth, 286
Cryptorchidism, 350
Crypts of Lieberkuhn, 292
Crystalloid solute particles, 42
Crystals, 41
Cubital vein, 246
Cuboid, 113
Cuboidal cells, 70
Cuneiform cartilages of larynx, 271
Cuneiform bones of foot, 112
Cupula, 195
Cushing's disease, 215
Cuticle, 87
Cyclic
    nucleotides, 27
    guanosine monophosphate, 205
Cystic
    artery, 245
    fibrosis, 63
Cystitis, 331
Cytogenetics, 58
Cytokinesis, 48, 50
Cytology, 1
Cytolysis, 260
Cytoplasm, 35, 46
    components, 37-41
    inclusions, 41
Cytosine, 27, 46
Cytoskeleton, 40
Cytotoxic reactions, 266
Cytotrophoblast, 368
Cytoxic T cells, 264

Dalton's law, 275
Dark and light adaptation, 190
Dartos muscle, 350

Dead space air, 275
Decidua, 369
Decomposition reaction, 16
Deep veins, 245, 247
Defecation, 283
Dehydration, 336, 338
    reactions, 24
    synthesis, 21, 27
Delayed hypersensitivity
    reactions, 266
    T cell, 265
Deltoid tuberosity, 107
Dendrites, 81, 156, 162, 176
Dense connective tissue, 74, 76
Dental arch, 286
Dentin, 287
Deoxyhomoglobin, 224
Deoxyribonuclease, 293
Deoxyribonucleic acid (DNA), 26-27, 39-40, 52
    differences from RNA, 52
    polymerase, 47
    repair, 47
    replication, 46-47
Deoxyribose, 46
Depolarization, 160
Depoprovera, 366
Dermal papillae, 86-87, 89
Dermis, 86
Dermatitis, 266
Desmosomes, 71
Detoxification, 294
Detrusor muscle, 330
Development, embryonic, 348-350
Diabetes, 266, 344
    adrenal, 215
    insipidus, 210
    mellitus, 216-217, 309, 330
    non-insulin-dependent, 216
Diabetic nephropathy, 332
Dialysis, 42
Diapedesis, 226, 257
Diaphragm, 3, 273, 288
Diaphragma sellae, 168
Diaphysis, 93, 96
Diarthroses, 113-116
Diastole, 237-238, 241
Diencephalon, 172-173
Dietary components and guidelines, 313
Diffusion, 41-42, 159-160, 247, 296
Digestive
    system, 4, 283-297
    tract, 283-284
Digital arteries, 244
Dihybrid cross, 63
Dihydrotestosterone, 355
Dipeptidases, 293, 294
Dipeptides, 24
Diploe, 96
Diplosome, 40
Disaccharides, 21, 293, 295
Dislocation, 116
Distal convoluted tubule (DCT), 323
Distal position, 7
Diuretics, 345
Diverging circuit, 164
DNA. See Deoxyribonucleic acid
Dominant gene, 60

Dopamine, 164, 210
Dorsal
    cavity, 7
    root, 177
    venous arch, 247
Dorsalis pedis artery, 245
Dorsoflexion, 115
Double
    bonds, 14
    helix, 27
    Y syndrome, 66
Down's syndrome, 65
DPG, 278
Ductuli efferentia, 353
Ductus
    arteriosus, 234
    deferens, 353
    venosus, 247
Duodenal
    gastrin, 290
    hormones, 292
Duodenum, 290-291
Dura mater, 168, 206
Dwarfism, 209
Dynamic equilibrium, 193, 195
Dysuria, 332

Ear, 191-196
    canal, 99
    ossicles, 3, 97
Eardrum, 191
Eccrine sweat glands, 88
Ectoderm, 369
Ectopic foci, 235-236
Eczema, 266
Edema, 250, 274, 339
Edward's syndrome, 65
Effective filtation pressure (EFP), 248, 324
Effectors, 5, 155, 165
Effector T cells, 264-265
Efferent
    arteriole, 322-323
    lymph vessel, 249
    nerves, 155
    neurons, 176
    pathway, 165
    tracts, 172
Egestion, 283
Ejaculation, 354-355
Ejaculatory duct, 353
Elastic
    arteries, 240
    cartilage, 78
    connective tissue, 76
    fibers, 75
Elastin, 75
Electrical synapses, 163
Electrocardiography, 236-237
Electrolyte, 329
    absorption, 296
    balance, 339-341
Electron, 11-12, 28
    acceptor, 16
    donor, 16
    field, 11
    system, 307
Electronegativity, 14
Elements, 11-12

Embryonic
    brain development, 167
    connective tissue, 74
    development, 348-350
    disc, 369
    stage, 369-372
Emmetropia, 189
Emphysema, 280
Enamel, 287
Endergonic reactions, 29
Endocardium, 232
Endocervical canal, 359
Endochondral (intracartilaginous) ossification, 94-96
Endocrine
    activities regulated, 203
    characteristics of, 202
    glands, 37, 69, 72, 202-203
    hormones, 204
    system, 4, 202-218
Endocytosis, 45
Endoderm, 370
Endogenous pyrogens, 257
Endolymph, 192-193
Endometrial (uterine) cycle, 363-364
Endometrium, 359, 363
Endomysium, 121
Endoneurium, 158
Endoplasmic reticulum (ER), 37, 156
Endorphins, 209
Endosteum, 79, 93
Endothelium, 69
Energy, 11, 184-185, 216, 301
    chemical reactions and, 28-29
    enzymes and, 28-30
    balance, 312
    sources for muscle contraction, 122-123
    storage, 45
Enteroendocrine glands, 292
Enterogastric feedback reflex, 290
Enterokinase, 293, 295
Enzymes, 17, 29–30, 292, 295, 302, 363
    activation, 204
Eosinophils, 223, 226
Ependyma, 167
Ependymal cells, 158
Epicardium, 232
Epicondyles, 107, 111
Epidermis, 72, 84-85
Epididymis, 353
Epidural space, 168
Epigastric region, 8
Epiglottis, 271, 287
Epimysium, 121
Epinephrine, 164, 183-184, 213-214, 240, 308, 310
Epineurium, 158
Epiphyseal plate, 95
Epiphyses, 94, 96
Epiphysis cerebri, 217
Epithalamus, 173
Epithelial
    cells, specializations, 71
    tissues, 69-73
    types, characteristics and distribution, 71-72
Epithelium, 73, 283
Epitope, 259

Eponychium, 87
Epstein-Barr virus, 227
Equal diffusion, 42
Equilibrium, 174, 180, 191-196
Erectile tissue, 354
Erythroblastosis fetalis, 231, 266
Erythrocytes, 223-224, 276
    production, 224
    survival and destruction, 225
Erythropoietin, 224
Esophageal
    arteries, 244
    sphincter, 287
Esophagus, 287
Essential
    amino acids, 311
    fatty acids, 309
Estradiol, 348, 362
Estriol, 362
Estrogen, 24, 96, 209-210, 348, 360, 362-365, 367
Estrone, 362
Ethmoid, 269
    bone, 99, 270
    sinuses, 99
Eustachian (auditory) tube, 191, 270
Evaporation, 316
Eversion, 115
Exchange reaction, 16
Excitability, 119
Excitation-contraction coupling, 124
Excitatory post-synaptic potentials, (EPSPs), 162
Excretion, 4, 84
Exergonic reactions, 28-29
Exocrine
    cells, 292
    glands, 69, 72-73
Exocytosis, 45, 73
Exogenous pyrogens, 257
Exophthalmic goiter (Graves' disease), 212
Expiration, 273
Expiratory
    neurons, 279
    reserve volume, 274
Extensibility, 119
Extension, 115
External
    genitalia, 350
    occipital protuberance, 98
    os, 359
    respiration, 269
    urethral orifice, 320, 331
Exteroceptors, 185
Extrapyramidal tracts, 178
Extrasystole, 235
Eye, 186-191
    cavities, 187-188
Eyeball, muscles of, 133, 179
Eyebrows, 186
Eyelids, 186

Fabry's disease, 39
Facial
    bones, 92, 100-101
    expression muscles, 131
    nerves, 180
Facilitated diffusion, 43-44

Falciform ligament, 284, 293
Fallopian tubes, 358-359
False ribs, 105
Farsightedness, 189-190
Fascicles, 120, 129, 158
Fasciculi, 177-178
Fast-twitch white fibers, 127
Fat
   absorption, 296
   anabolism, 309
   catabolism, 309
   digestion, 290
   emulsification, 294
   metabolism, 308-310
   soluble vitamins, 294
   sources, 313
   synthesis, 293
Fatty acids, 23, 295, 309, 311
Feedback
   control mechanisms, 205
   system, 4-5
Female reproductive system, hormonal regulation, 362-364
Femoral
   artery, 245
   nerve, 181
   vein, 247
Femur, 109-111
Fermentation, 305
Ferritin, 296
Fertilization, 59-60, 348, 359, 365-366
Fetal
   bypasses through the heart, 234
   circulation, 247
   development, 372
   hemoglobin, 278
   membranes, 370
Fever, 257-258, 317
Fiber, 21
Fibrillation, 235-236
Fibrils, 40
Fibrin, 25
Fibrinogen, 222, 228, 257
Fibroblasts, 74-75
Fibrocartilage, 78
Fibrous
   connective tissue sheaths, 120
   joints, 113
   proteins, 25
   skeleton of the heart, 233
   tunic of eye, 186
Fibula, 111-112
Filtration, 44, 247
   slits, 322
Filum terminale, 176
Final terminations, 157
Fingernails, 84, 86-87
Fingerprints, 86
First filial (F1) generation, 61
Fissures, 169-170, 176
Fixators, 130
Flagella, 40-41
Flat bones, 96
Flavin adenine dinucleotide (FAD), 302
Flexion, 115
Flexor reflex, 166
Floating ribs, 105

Fluid mosaic model, 35
Flutter, atrial, 235-236
Follicles, 87, 211
Follicle stimulating hormone (FSH), 356
Follicular fluid, 357
Foot
   bones, 112-113
   muscles, 150-151
Foramen
   magnum, 98
   ovale, 234
Forced expiratory volume (FEV), 275
Forebrain, 167
Foreign body giant cells, 256
Fornix
   brain, 173
   vagina, 359
Fossa
   navicularis, 331, 354
   ovalis, 234
Fourth ventricle, 168
Fovea, 188
Fovea capitis, 110
Frank-Starling law of the heart, 239
Free fatty acids, (FFA), 308
Frenulum, 361
Frontal
   association area, 171
   bone, 98
   plane, 5
   tuberosities, 98
Fructose, 302
   reabsorption, 325
Fulcrum, joints as, 129-130
Functional
   groups of atoms, 20-21
   residual capacity, 275
Fundic glands, 289
Fundus, 288, 359
Funiculi, 177
Furuncles, 88

Galactose, 302
Gallbladder, 294-295
Gamete, 59, 348
Ganglia, 177
Ganglion, 158
   cells, 188
   chain, 183
Gap junctions, 71, 163
Gas
   exchange, 275-276
   properties, 275-276
   solubility, 276
   volume, 276
   transport by blood, 276-278
Gastric
   artery, 245
   emptying, 290
   glands, 288-289
   inhibitory peptide (GIP), 290, 292
   pits, 288
   secretion, 290
   stage, 289
Gastrin, 289-290
Gastroduodenal artery, 245
Gastrointestinal

   hormones, 216
   tract, 283
Gaucher's disease, 39
Gene, 46, 60
   activation, 205
   regulatory proteins, 52
Genetic
   code, 46, 52
   disorders, 63-64
   material, 40
Genetics, 58-66
   behavioral, 58
   developmental, 58
   terminology, 58-61
Genome, 46
Genotype, 60-62, 348-349
Germ layers, 369
Germinal matrix, 87
Germinativum, 85
Gestation, 367
Gigantism, 209
Gingiva, 287
GIP, 292
Gladiolus, 103
Glandular epithelium, 69, 72-73
Glans clitoris, 361
Glaucoma, 187
Glenoid cavity, 105
Glia, 157-158
Gliding joint, 115
Globin, 224-225
Globular proteins, 25
Globulins, 222
Globus pallidus, 172
Glomerular filtration, 324
   barrier, 322
   rate (GFR), 324-325
Glomerulonephritis, 332
Glomerulus, 322-323
Glossopharyngeal nerves, 180
Glottis, 271-272, 287
Glucagon, 215-216, 243, 292, 310
Glucocorticoids, 214-215, 33
Gluconeogenesis, 308, 310-311
Glucose, 46, 216, 302
   reabsorption, 325
Glycerol, 23, 295, 309
Glycogen, 21, 41, 293
Glycogenolysis, 311
Glycolipids, 36
Glycolysis, 46, 302-305
Glycolytic pathway, 122-123
Glycoproteins, 36
Glycosuria, 217
Goblet cells, 72, 292
Goiter, 212
Golgi
   apparatus, 38-39
   complex, 156
   tendon organs, 185
Gomphoses, 113
Gonadal
   dysgenesis, 66
   ridges, 349
Gonadocorticoids, 215
Gonadotropins, 209-210, 217, 356
   releasing hormone, 210

Gonads, 348-349
Graded muscle responses, 125-126
Granulocytes, 223, 226
Granuloma, 257
Granulosa, 357
Graves' disease, 212
Gray
    commissure, 176
    matter, 167, 169, 176
Greater sciatic notch, 109
Ground substance, 75, 93
Growth, 4, 212
    hormone, 96, 208-209, 216, 225, 310-311
    hormone-releasing-hormone, 208
Guanine, 27, 46
Gubernaculum, 350
Gustation, 196-197
Gynecoid pelvis, 109
Gyrus, 170, 178

Hair, 3, 84, 87-88
    cells, 193, 195
    follicles, 256
Hamate bone, 108
Hand, 3, 108
    muscles, 143-145
Haptens, 259
Haustra, 297
Haversian
    canals, 93
    system, 78-79
Head muscles, 135
Hearing, 191-196
Heart, 3, 221, 231-234
    block, 235
    chambers, 232
    conducting system, 235-236
    murmurs, 238
    rate, 238-239
    sounds, 238
    surface markings, 233
    valves, 232-233
Heat
    loss, 89, 313
    production, 126, 294, 313
    retention, 89
Heavy (H) chains of antibody, 259
Helicotrema, 193
Helper T cells, 264-265
Hematocrit, 224
Hematoma, 96
Hematopoiesis, 92
Hematuria, 332
Heme, 225
Hemiazygous vein, 246
Hemocytoblasts, 223
Hemodialysis, 332
Hemodynamics, 240-243
Hemoglobin, 25, 223-224, 276, 278
    buffer system, 342
    saturation with oxygen, 277
Hemolysis, 231
Hemophilia, 63, 230
Hemorrhage, 224
Hemorrhagic anemia, 225
Hemostasis, 227-228
Henry's law, 276

Heparin, 227-228
Hepatic
    artery, 245
    ducts, 294
    portal system, 246-247
    portal vein, 293, 302
Hepatitis, 294
Hepatopancreatic ampulla, 292
Hereditary material, 35
Herniated (slipped) disc, 102–103
Heterozygous genes, 60
Hexoses, 21
Hiatus, 287
Hilus, 321
Hindbrain, 167
Hinge joint, 114
Hip
    bone, 108-109
    muscles, 146-148
Hippocampal gyrus, 173
Hippuric acid, 329
Histamine, 227, 243, 257, 260
Hives, 266
Holocrine
    glands, 88
    secretion, 73
Homeostasis, 4-5, 185, 320, 336
Homozygous genes, 60
Hormonal
    action, 204-205
    contraceptives, 366
    mechanisms, 317
Hormones, 25, 96, 173, 202, 292, 310, 329, 355
    biochemistry, 204
    receptor complex, 204
    secretion, regulation of, 205-206
Horns of gray matter, 176
Housemaid's knee, 116
Human
    chorionic gonadotropin, 367
    placental lactogen, 367
Humerus, 105–106
Humoral immunity, 255, 262
Hunchback, 103
Hyaline
    cartilage, 77, 94, 96
    membrane disease, 274
H-Y antigen, 353
Hydrogen, 20, 342
    bonds, 13, 15
    electron transport system, 46
Hydrogenation, 23
Hydrolytic enzymes, 39
Hydrophobic
    interactions, 16
    molecules, 18
Hydrostatic pressure, 250, 324, 337
Hydroxyapatite, 93
Hydroxyl group, 20
Hyoid bone, 92, 101
Hypercapnia, 280
Hypercholesteremia, 64
Hyperextension, 115
Hyperglycemia, 217
Hyperinsulinism, 217
Hyperkalemia, 340

Hyperopia, 189-190
Hyperosmotic solutions, 43
Hyperparathyroidism, 213
Hyperpolarization, 162
Hypersensitivity, 265-266
Hyperthyroidism, 212
Hypertonic
    to plasma, 326
    solutions, 43
Hypertrophy, 127
Hyperventilation, 280, 344
Hypochondriac regions, 8
Hypodermis, 86
Hypogastric region, 8
Hypoglossal nerves, 180
Hypoglycemia, 217
Hypokalemia, 340
Hypoosmotic solutions, 43
Hypoparathyroidism, 213
Hypophysis pituitary gland, 206-211
Hypophysis, 206-211
Hypothalamic
    gonadotropin releasing hormone, 356-357
    hormones, 355, 362
    thermoregulatory centers, 316
    thyrotropin-releasing hormone, 209, 212
    hypophyseal portal system, 208-209, 214
Hypothalamus, 172-173, 210-211, 338, 348
    pituitary relationships, 207-208
Hypothryroidism, 212
Hypotonic
    to plasma, 326
    solutions, 43
Hypoxia, 225, 280, 344

Ileocecal valve, 297
Ileum, 291
Iliac
    arteries, 245
    regions, 8
    spine, 109
    vein, 242-247
Ilium, 108-109
Immune
    complex reactions, 266
    response, 249, 261, 263
    response, damaging effects, 265-266
    system, 258-266
Immunioglobulins, 259-260
Immunodeficiencies, 266
Immunodeficiency diseases, 218
Immunoelectrophoresis, 260
Implantation, 368
Inactivation, 163
Incomplete dominance, 62
Incus (anvil), 100, 192
Independent Assortment, Mendel's Law of, 61, 63
Inferior position, 7
Inflammation, 256-257, 260
Inflation reflex, 279
Infraorbital foramen and margins, 100
Infundibulum, 173, 206-207, 359
Ingestion, 283
Inguinal canals, 350

Inguinal
   hernia, 350
   nodes, 250
   regions, 8
Inheritance, 61-64
Inhibin, 353
Inhibitory post synaptic potentials (IPSPs), 162
Inner
   ear, 192-193
   meningeal layer, 168
Innominate bones, 108
Inorganic compounds, 17-20
Inositol triphosphate, 205
Insoluble nonpolar fatty acid tails, 36
Inspiration, 273
Inspiratory
   capacity, 275
   neurons, 279
   reserve volume, 274
Insulin, 215-216, 292, 309-311
   dependent diabetes mellitus, 216
Integral proteins, 36
Integrative activities, 155
Integumentary system, 4, 84-89
Interatrial septal defect, 234
Intercalated discs, 80, 120
Intercellular
   "ground" substance, 74
   junctions, 71
Intercostal
   arteries, 244
   muscles, 273
Interferons, 258
Interleukin-2, 265
Interlobar
   arteries, 323
   veins, 324
Intermediate fibers, 127
Internal
   os, 359
   respiration, 269
Interoceptors (visceroceptors), 185
Interosseus membrane, 107
Interphase, 48
Interstitial
   cells of Leydig, 351
   fluid, 337
   growth, 78
Interventricular
   foramen (of Munro), 168
   septum, 232
Intestinal
   enzymes, 292, 295
   glands, 292
   lipase, 295
   stage, 290
   wall, 292
Intracellular
   digestion, 39
   fluid (ICF) compartment, 336-337
   receptor systems, 205
Intramembranous ossification, 94-96
Intraocular pressure, 187
Intrapleural pressure, 273
Intrauterine devices (IUDs), 366
Intrinsic factor, 289

Inversion, 115
Invertebral disc, 104
Involuntary
   control of respiration, 279
   muscle, 79
Iodine, 211-212
Iodopsin, 190-191
Ion
   channels, 162
   gates, 159-160
Ionic
   bonds, 13-15
   compounds, 15
Ionization of water, 19
Ions, 13-14, 240
Ipsilateral, 7
Iris, 187
Iron,
   as nutrient, 316
   deficiency anemia, 225
Irregular bones, 96-97
Irritability, 4, 155
Ischemic heart disease, 240
Ischium, 109
Islets of Langerhans, 215, 292
Isomaltase, 295
Isometric contraction, 126
Isosmotic solutions, 43
Isotonic
   contraction, 126
   solutions, 43
Isotopes, 12
Isovolumetric contraction and relaxation, 238

Jaundice, 294
Jejunum, 291
Joints, 92, 113-116
   classification, 113
   disorders, 115–116
   function, 113
Jugular
   (suprasternal) notch, 103
   veins, 246
Juxtaglomerular apparatus, 323

Karyotype, 59
Keratin, 25, 85
Ketone bodies, 309, 329
Ketonemia, 217
Ketonuria, 217
Ketosis, 309
Kidneys, 210, 213, 308, 310, 320-324, 338, 344-345
   stones, 332
Kilocalorie, 312
Kinase, 204-205
Kinetic energy, 28
Kinins, 257
Klinefelter's syndrome, 66
Knee
   joint, 112
   muscles, 148-149
Korotkoff's sounds, 243
Krause's end bulbs, 185
Krebs cycle, 46
Kupffer cells, 256
Kyphosis, 103

Labia
   majora, 361
   minora, 361
Labial frenulum, 285
Labor and delivery, 211, 368
Lacrimal
   apparatus, 186
   bones, 101
Lactase, 295
Lactation, 210, 348, 368
Lacteals, 248-249, 291
Lactic acid, 123, 308
Lactiferous duct and sinus, 361
Lacunae, 79, 94
Lamina,
   propia, 73, 283
   of vertebra, 101
Langerhans cells, 256
Lanugo, 372
Large intestine, 296–297
Larynx, 180, 271-272, 287
Latent period, 125
Lateral
   canthus, 186
   position, 7
   ventricles, 168
Leg muscles, 148-151
Lens, 187-188
Lenticular (lentiform) nucleus, 172
Leukemia, 227
Leukocytes, 76, 226-227
Leukocytosis, 258
Leukotrienes, 257
Levers, bones as, 129-130
Ligaments, 76, 114, 187, 359
Ligamentum
   arteriosum, 234
   teres, 247
   venosum, 247
Light (L) chains of antibody, 259
Light and dark adaptation, 190
Limbic system, 173
Linear acceleration, 194
Lingual
   frenulum, 285
   tonsils, 285
Linked genes, 61
Lipids, 21-24.
Lipogenesis, 309-310
Lipoproteins, 308-309
Lips, 285
Liver, 211, 216, 228, 293-294, 302, 308-311
Lobules, 361
Lock-and-key model of enzyme action, 29
Locus, 60
Long bones, 96
Loop of Henle, 323, 327, 329
Loose (areolar) connective tissue, 74
Lordosis, 103
Lower limb, 109-113
Lumbar
   enlargements, 176
   nerves, 180
   plexus, 181
   puncture, 169
   regions, 8
   and sacral arteries, 245

vertebrae, 103
Lunate bone, 108
Lung, 272-275
    cancer, 280
    capacity, 275
    volumes, 274-275
Lunula, 87
Lupus erythematosus, 266
Luteinizing hormone, 209-210, 356
Lymph, 248, 362
Lymphatic
    capillaries, 248-249
    obstruction, 250
    system, 4, 221, 248-250
    tissue, 292
Lymphoblasts, 223
Lymphocytes, 223, 227
Lymphoid tissue, 285, 297
Lymphokines, 261, 265
Lysosomes, 39

Macromolecules, 3
Macrophages (histiocytes), 75, 96, 225, 256, 261, 263
Maculae, 195
Macula lutea, 188
Male reproductive system hormonal regulation, 355-356
Malleolus, 111-112
Malleus (hammer), 100, 192
Maltase, 295
Mammary glands, 3, 84, 88, 210, 361-362, 367
Mandible, 100-101, 286
Mandibular branch, 179
Manubrium, 103
Margination, 257
Marrow cavity, 93
Masculinization, 215
Mass number, 12
Mast cells, 75
Mastication, 287
    muscles, 132
Mastoid portion of temporal bone, 99
Matrix, 74
Mature (Graafian) follicle, 358
Maxilla, 270, 286
Maxillary
    bones, 100
    branch, 179
    sinuses, 100
Meatus, 191, 270
Mechanoreceptors, 185
Medial
    canthus, 186
    position, 7
Mediastinum, 7
Medula, of hair, 87
Medulla oblongata, 174-175, 180, 238-239, 286, 356
Medullary, 213-214
    recycling of urea, 328
    respiratory centers, 279
Megakaryoblasts, 223
Meiosis, 47, 50-52, 59, 64
Meissner's
    corpuscles, 185

plexuses, 284
Melanin, 188, 217
Melanocytes, 86
    stimulating hormone, 209
Melatonin, 217
Membrane potential, 44, 158-160
Membranes, 73
Membranous
    labyrinth, 192-193
    urethra, 331, 354
Memory B cells, 263
Memory cells, 264
Menarche, 362
Mendelian genetics, 61-63
Meninges, 167-168, 176, 181
Menisci, 111, 114
Menopause, 364-365
Menstrual cycle, 363, 367
Merkle's disks, 185
Merocrine secretion, 73
Mesangial cells, 256
Mesencephalon, 167
Mesenchyme, 74, 94
Mesenteric
    artery, 245
    vein, 246
Mesenteries, 284
Mesocolon, 284
Mesoderm, 370
Mesonephric (wolffian) ducts, 349-350
Mesothelium, 69
Metabolic
    acidosis, 344
    activities, 35
    alkalosis, 345
    control, 30
    pathway, 301-302
    rate, 212, 258, 312-313
    water, 338
Metabolism, 4, 45, 84, 213, 293, 301-308, 367
    cell, 45-46
    hormonal regulation, 302
Metacarpus, 108
Metaphase, 49-51
Metatarsal bones, 112-113
Methyl group, 20
Microbodies, 39
Microfilaments, 40
Microglia, 158, 256
Microtubules, 40
Microvilli, 34, 71, 291, 323
Micturition, 331
Midbrain, 167, 173-174
Middle ear, 191
    cavities, 7
Midsagittal plane, 5
Milk
    ejection, 211
    production, 367
    secretion, 210
Milliosmole, 43
Mineralocorticoids, 214
Minerals, 93, 294, 313, 315
Minute respiratory volume, 275
Mitochondria, 37, 45-46, 123, 156
Mitosis, 47-50, 64
Mitotic cleavage, 368

Mixed nerves, 158
Moderator band, 232
Modiolus, 193
Molarity, 19
Mole, 19
Molecular
    genetics, 58
    weight, 42
Molecules, 3, 13, 19
Monoamines, 164
Monoblasts, 223
Monocytes, 223, 227
Monohybrids, 61
Mononuclear phagocytic system (MPS), 256
Mononucleosis, 227
Monosaccharides, 21, 295
Monosomy, 64
Monosynaptic ipsilateral reflex arc, 165
Monounsaturated fats, 313
Mons pubis, 361
Morula, 368
Motor
    nerves, 123, 165, 178-180
    tracts, 177-178
Movement, 4
Mucoid tissue, 74
Mucosa, 283-284, 330
Mucous
    membranes, 73, 256, 283-284
    neck cells, 289
    secretion, 285
Mucus, 270, 289
    producing glands, 292
    production, 289
    secreting glands, 73
Mullerian (paramesonephric) ducts, 350
Multiple
    alleles, 62-63
    motor unit summation, 126
    sclerosis, 266
Multisynaptic reflex, 166
Multiunit smooth muscle, 128
Murmurs, 238
Muscle
    actions, 129-151
    atrophy, 127
    contraction, characteristics of, 124-127
    contraction, chemistry, 122
    contraction, molecular basis, 121-122
    contraction, nervous control, 123-124
    fatigue, 126
    fiber types, 127
    fibers, 119
    hypertrophy, 127
    length-tension relationships, 126-127
    properties of, 119
    relaxation, 122
    tendon arrangements, 129
    tissue, 69, 79-80
    tissue, classification of, 119
    tone, 126
    twitch, 125
    types of, 120
Muscles, 76
    abdominal wall, 138
    attachment of, 129

eyeball, 133
facial expression, 131
hip, 146-148
leg, 150-151
mastication, 132, 179, 285
names, 130
neck and vertebral column, 135
neck, thorax, and shoulder, 140-141
pelvis, 138-139
perineum, 139
respiratory system, 137
thigh, 148-149
throat and neck, 134
tongue, 132, 180
vertebral column, 136
wrist and hand, 143-145
Muscular
activity, 317
arteries, 241
system, 4, 119-151
system functions, 119
system structure and physiology, 119-128
Muscularis
externa, 284
mucosae, 73, 283
Mutation, 47
Myasthenia gravis, 266
Myelin, 157
Myelinated
axons, 183
fibers, 156, 161-162
Myeloblasts, 223
Myocardial infarction, 240
Myocardium, 232
Myofibrils, 121
Myofilaments, 121
Myoglobin, 127
Myometrium, 359
Myopia, 189
Myosin, 25, 121
Myxedema, 212

NADH, 305
Nails, 3, 86-87
Nares, 269-270
Nasal
bones, 100
cavity, 7, 269-270
conchae, 99, 101, 270
septum, 269
Nasolacrimal duct, 186
Natural
family planning, 366
immunity, 255
killer (NK) cells, 265
Navicular bone, 107, 112
Nearsightedness, 189
Neck muscles, 134-135, 140-141
Negative feedback, 206
mechanisms, 5
Nephron, 322-323
Nerve, 158
endings, 185
fiber, 123
impulse, 158-164
impulse velocity, 161-162
signal, 161

Nervous
control of respiration, 279
system, 4, 155-197
system cells, 156-158
system organization, 155-156
tissue, 69, 80-81
Net diffusion, 42
Neural
arch, 101
crest, 167
plate, 167
tube, 167
Neurilemma, 156-157
Neurofibrils, 156
Neuroglial cells, 81, 157-158
Neurohormones, 204
Neurohypophysis, 206-207
Neuromodulators, 204
Neuromuscular spindles, 185
Neuronal pools and circuits, 164
Neurons, 80-81, 123, 156-157, 162
classification of, 157
groups, 158
Neuropeptides, 164
Neurotransmitters, 163-164, 183-184, 204
Neutral fat, 21-23
Neutralization, 260
Neutron, 11
Neutrophils, 223, 226, 256
Nicotinamide adenine dinucleotide (NAD), 302
Night blindness, 190
Nissl bodies, 156
Nitrogen balance, 311
Nitrogenous bases, 27, 46, 329
Non-insulin-dependent diabetes mellitus, 216
Noncompetitive inhibitors, 30
Nondisjunction, 64
Nonessential amino acids, 311
Nonpolar
covalent bonds, 14
molecules, 18
Nonspecific defenses, 255-258
Norepinephrine (noradrenaline), 164, 183-184, 213-214, 239-240, 243
Norplant, 366
Nose, 269-270
Notochord, 3
Nuclear envelope and pores, 39
Nuclei of cranial nerves, 174
Nucleic acids, 26-27
Nucleolus, 40, 156
Nucleoplasm, 40
Nucleotides, 27, 46
Nucleic acids, 26–27
Nucleus, 156, 158
atomic, 11
Nutrition, 312-313

Obdurator foramen, 109
Occipital bone and condyles, 98
Oculomotor nerves, 178-179
Odor reception, 270
Olecranon process, 107
Oleic acids, 23
Olfaction, 197
Olfactory

epithelium, 197
nerves, 178
Oligodendrocytes, 158
Oligodendroglia, 158
Oliguria, 332
Omenta, 284
Oocyte, 365
Oogenesis, 348, 356-358
Ophthalmic branch, 179
Opsin, 190
Opsonization, 260
Optic
chiasma, 172, 188
disk, 188
foramen, 186
nerves, 178
tracts, 188
Oral
cavity, 7, 285-287
contraceptives, 366
Orbit, 186
Orbitals, 7, 12
Organ of Corti, 193
Organelles, 35, 37–40, 156
Organic compounds, 17, 20-27
Orgasm, 355
Oropharynx, 271, 287
Os, 359
Osmolality, 43, 210
Osmoreceptors, 338
Osmosis, 42-43, 247
Osmotic pressure, 43, 248, 337
Ossa coxae, 108
Osseous tissue, 78-79
Ossification, 94-96
Osteoblasts, 78, 94
Osteoclasts, 78, 93, 96, 213
Osteocytes, 78–79, 93-94
Osteogenesis, 94-96
Osteoid, 94
Osteon, 78
Osteoporosis, 365
Otoliths, 195
Outer periosteal layer, 168
Oval window, 192
Ovarian
arteries, 245
cycle, 363-364
follicles, 209, 356
hormones, 362
Ovaries, 348, 350, 356
Overhydration, 338
Oviducts, 358-359
Ovulation, 210, 358, 364
Oxidation, 301
reduction reaction, 16, 301-302
Oxidative
phosphorylation, 46, 307
respiration, 305-307
Oxygen
capacity, 276
debt, 123
deficiency, 280
hemoglobin dissociation curve, 277-278
saturation, 277
transport, 276-278
Oxyhemoglobin, 224, 276
Oxyphil cells, 212

Oxytocin, 210-211, 243, 368

Pacemaker of the heart, 235
Pacinian corpuscles, 185
Pain receptors (nociceptors), 185
Palate, 270, 287
Palatine
    bones, 100, 270
    tonsils, 271
Palmar arches, 244
Palmitic acid, 23
Palpebrae, 186
Palpebral fissure, 186
Pancreas, 215-217, 292-293
Pancreatic
    enzymes, 292–293
    exocrine secretion, 293
    lipase, 293
    polypeptide, 216
    proteolytic enzymes, 293
Papillae, 285
Papillary
    muscles, 232-233
    layer, 86
Parafollicular cells, 211
Parallel circuit, 164
Paranasal sinuses, 101, 270
Parasagittal plane, 5
Parasympathetic
    fibers, 239
    impulses, 284, 294, 355
    nerves, 156
    stimulation, 286
    system, 182-183
Parathyroid
    glands, 212-213
    hormone, 96, 212, 340–341
Paraventricular nucleus, 210
Parental (P) generation, 61
Parietal
    bones, 98
    cells, 289
    membrane, 8, 232
    pericardium, 8
    peritoneum, 8, 284
    pleura, 8
Parotid glands, 285
Partial pressures, 275-276
Parturition, 367-368
Parurethral (Skene's) glands, 361
Passive
    immunity, 261
    transport, 41
Patau's syndrome, 65
Patella, 97, 111
Patellar reflex, 165
Patent (open) ductus arteriosus, 234
Pathogens, 255
Pathology, 2
Pectoral girdle, 105–106
    muscles, 140
Pectoralis
    major, 273
    muscles, 361
Pedicels, 322
Pedicles, 101
Pelvic
    diaphragm, 138-139
    girdle, 108-109
    mesenteries, 356
    splanchnic nerves, 331
Pelvis, 108–109
Penile
    erection, 354
    urethra, 354
Penis, 354
Pennate muscles, 129
Pentoses, 21, 27
Peptide bonds, 24
Pericardial
    arteries, 244
    cavity, 7, 232
Pericardium, 8, 231-232
Perichondrium, 78
Peridontal membrane, 287
Perikaryon, 80-81, 156
Perilymph, 192
Perimetrium, 359
Perimysium, 121
Perineum, 361
    muscles, 139
Perineurium, 158
Perinuclear space, 39
Periosteum, 79, 93, 95
Peripheral chomoreceptors, 280
Peripheral nervous system (PNS), 81, 155-156, 178-185
    and central nervous system, 167-185
Perirenal fat, 321
Peristalsis, 283, 287, 291, 297
Peritoneal cavity, 7, 284
Peritoneum, 8
Peritubular capillaries, 323
Pernicious anemia, 225
Peroneal
    artery, 245
    vein, 247
Peroxisomes, 39
Perpendicular plate, 99, 269
Perspiration, 89, 316, 338
Petrous portion of temporal bone, 99
Peyer's patches, 292
pH, 19-20, 30, 278, 320, 341-345
Phagocytosis, 40, 45, 225-226, 256-257, 260
Phalanges, 108, 112-113
Pharyngeal tonsils, 270
Pharynx, 180, 270-271, 287
Phenotypes, 60, 62
Phenylketonuria (PKU), 63
Phospate, 46, 212-213, 340-341
    buffer pair, 343-344
    buffer system, 342
    group, 21, 27
Phosphodiesterase, 205
Phospholipids, 21, 23-24, 36, 308
Photoreceptors, 185
Phrenic
    arteries, 244
    nerve, 181
Physiological dead space, 275
Pia mater, 167-168
Pigment granules, 41
Pili, 87-88
Pilosebaceous unit, 88
Pineal body, 173, 217
Pinealocytes, 217

Pinna, 191
Pinocytosis, 40, 45
Pisiform bone, 108
Pitch, 193
Pituitary, 356
    gland, 206-211, 243
    gonadotropins, 357, 367
    hormones, 355, 362
    hypothalamus feedback system, 212
    hypothalamus relationships, 207-208
Pivot joint, 114
Placenta, 3, 260, 371-372
Placental circulation, 371-372
Plane, 5
Plantar
    arch, 245
    arteries, 245
    flexion, 115
Plasma, 278
    cells, 76, 261
    clearance, 326
    colloid osmotic pressure, 250
    membrane, 35-36, 42
    osmolality, 338
    proteins, 222, 248, 258, 294
Plasmalemma, 35-36
Platelet, 223, 227
    factors, 229
    plug, 227-228
Pleura, 8, 272
Pleural cavities, 7
Plexuses, 181-182
Plicae circulares, 291
Pneumonia, 280
Pneumontaxic center, 279
Pneumothorax, 274
Podocytes, 322
Polar
    bodies, 59
    covalent bonds, 14
    phosphate ends, 36
Polarization of the axon membrane, 159
Polycythemia, 225-226
Polygenic inheritance, 64
Polypeptide, 24, 53
Polyphagia, 217
Polyribosomes, 37
Polysaccharides, 21
Polysynaptic reflex, 166
Polyunsaturated fats, 313
Polyunsaturation, 23
Polyuria, 217
Pons, 174
    respiratory centers, 279
Popliteal
    artery, 245
    vein, 247
Population genetics, 58
Porta hepatis, 293
Portal canals, 293
Portal lobule, 293
Portio vaginalis, 359
Positive feedback, 206
    mechanisms, 5
Postabsorptive (fasting) state, 311
Postcapillary venules, 241
Posterior, 5
    lobe hormones, 210-211

muscles, 136
position, 5
Postganglionic neurons, 182-183
Postsynaptic neuron, 162
Potassium, 340
Potential energy, 12
Precapillary sphincter, 241
Precipitation, 260
Precocious puberty, 215
Preembryonic period, 368
Preganglionic
fibers, 183
neuron, 182
Pregnancy, 210, 348, 367-372
hormones of, 367-368
Premature ventricular contraction, 235
Premotor area, 170
Prepatellar bursitis, 116
Prepuce, 354, 361
Presbyopia, 189
Pressoreceptors, 239
Pressure
gradient, 241
wave, 193, 243
Presynaptic neuron, 162
Primary
auditory area, 170
cortical nodules, 249
follicle, 357
motor area, 170
olfactory area, 170
sensory area, 170
taste (gustatory) area, 171
visual area, 170
Prime mover, 130
Primordial germ cells, 349
P-R interval, 237
Probability and inheritance, 62
Proerythroblasts, 223
Progesterone, 24, 210, 348, 363-365, 367
Projection fibers, 171
Prolactin, 210, 367
inhibiting hormone, 210
releasing factor, 210
Pronation, 115
Pronuclei, 365
Proopiomelanocortin, 209
Propagation of the nerve impulse, 161
Prophase I, 48, 50
Proprioceptors, 185
input, 279
Prosencephalon, 167
Prostacyclin, 228
Prostaglandin, 227
Prostaglandins, 204, 227, 243, 257, 363, 368
Prostate gland, 354
Prostatic urethra, 331, 354
Proteases, 293
Protein, 24-26, 36, 204, 308,
absorption, 296
anabolism 310-311
assembly, 53
buffer system, 342
catabolism, 310
denaturation, 25-26
digestion, 289
digestion, 290

metabolism, 310-311
sources, 313
structure, 24-26
synthesis, 37, 52-54, 156, 208, 310
Prothrombin, 228
Proton, 11
Protraction, 115
Proximal
convoluted tubule, 323, 325
position, 7
Pterygoid processes, 100
Puberty, 355-357, 362
Pubic symphysis, 367
Pubis, 109-110
Pudendal artery, 245
Pudendum, 360-361
Pulmonary
arteries, 243
carcinoma, 280
circuit, 233-234
gas exchange, 276
semilunar valve, 238
trunk, 243
veins, 232, 243
ventilation, 269, 273-275
Pulmonic semilunar valves, 233
Pulp chamber, 287
Pulse, 243
Punctum, 186
Punnett Square, 61, 63
Pupils, 187, 189
Purines, 27
Purkinje fibers, 235
Pus, 257
Putamen, 172
P wave, 236
Pyelonephritis, 332
Pyloric
glands, 289
portion, 288
Pyramids, 174
Pyriform lobe, 173
Pyrimidines, 27
Pyrogens, 257

QRS complex, 236
Quadrants, 8
Quantum, 13

Radial
artery, 244
notch, 107
tuberosity, 107
vein, 246
Radiation, 316
Radiographic anatomy (radiology), 2
Radioimmunoassay, 260
Radioisotopes, 12
Radiology, 2
Radius, 106–107
Ramus, 109, 181, 183
Receptors, 5, 155, 165
Recessive genes, 60, 63
Rectum, 297
Red nucleus, 174
Redox reaction, 16
Reduced hemoglobin, 276

Reduction, 16, 302
Reflexes, 165-167
Refraction, 188
Refractory periods, 125, 161
Regeneration of injured nerves, 157
Reissner's membrane, 193
Relaxation period, 125
Relaxin, 367
Renal
arteries, 245, 323
autoregulation, 324-325
capsule, 321
corpuscle, 322
failure, 332
fascia, 321
parenchyma, 321
pelvis, 321
plasma threshold, 325
regulation of pH, 343
shutdown, 231
sinus, 321
Renin, 320
angiotensin-aldosterone mechanism, 338-340
angiotensin mechanism, 214
Repolarization, 161
Reproductive system, 4, 350-365
female, 356-365
male, 350-356
Residual volume, 275
Respiration, 240
control, 279-280
Respiratory
acidosis, 344
alkalosis, 344-345
center, 174
distress syndrome, 274
epithelium, 270
membrane, 276
problems, 280
regulation of pH, 343
reflexes, 279
system, 4, 269-280
Responsiveness, 4
Resting potential, 158-160
Rete testis, 353
Reticular
cells, 256
connective tissue, 77
fibers, 75
formation, 175
layer, 86
Reticuloendothelial system (RES), 256
Reticulum, 75
Retina, 188, 190
Retinaldehyde, 190
Retinene, 190
Retraction, 115
Retroperitoneal organs, 284
Reverberating (oscillating) circuit, 164
R groups, 24
Rh system, 231
Rheumatoid arthritis, 116, 266
Rhodopsin, 190
Rhombencephalon, 167
Rhythm method of contraception, 366
Rhythmic segmentation, 291
Ribonuclease, 293

Ribonucleic acid (RNA), 26-27, 39, 40, 52
  differences from DNA, 52
  polymerase, 52
  mRNA, 53
  tRNA, 53
Ribosomes, 37, 53, 156
Ribs, 96, 103-105
Rigor mortis, 122
RNA. See Ribonucleic acid
Rods, 188, 190
Root canals, 287
Rostral position, 5
Rotation, 115
Roughage, 21
Round window, 192
Ruffini's corpuscles, 185

Saccule, 192-193
Sacral
  nerves, 180
  plexus, 181
Sacrum, 103
Saddle joint, 114
Sagittal
  plane, 5
  suture, 98
Saltatory conduction, 162
Saliva, 256, 285-286
Salivary glands, 180, 285
Salivatory nuclei, 286
Saphenous vein, 247
Sarcolemma, 79, 119, 123
Sarcoplasm, 79, 119
Sarcoplasmic reticulum, 37, 79, 119, 124, 128
Scala
  media, 193
  tympani, 193
  vestibuli, 193
Scalene muscles, 273
Scapula, 105
Scar formation, 257
Sciatic
  nerve, 181
  notch, 109
Sclera, 186
Scoliosis, 103
Scotopsin, 190
Scrotal sacs, 350
Scrotum, 350
Sebaceous glands, 84, 88, 256
Sebum, 88
Secondary follicle, 357-358
Secretin, 243, 292-294
Secretory
  granules, 41
  products, 38
Sedimentation rate, 224
Segregation, Mendel's Law of, 61
Sella turcica, 100
Semen, 355
Semicircular
  canals, 192
  ducts, 193
Semiconservative replication, 47
Semilunar
  cartilages, 111
  valves, 233

Seminal vesicles, 354
Seminiferous tubules, 351
Sensor, 5
Sensory
  input, 155
  nerves, 165, 178-180
  receptors, 185-197
  tracts, 177-178
Septal area, 173
Septomarginal trabecula, 232
Seromucous glands, 73
Serosa, 284, 330, 359
Serotonin, 164, 227, 257
Serous
  fluid, 73
  membranes, 8, 73
  secreting glands, 73
  secretion, 285
Sertoli cells, 351, 353
Serum, 228
  sickness, 266
Sesamoid bones, 97, 113
Setpoint, 4
Sex
  characteristics, 356
  chromosome abnormalities, 65
  chromosomes, 58
  determination, 59-60, 348-349
  differentiation and development, 349-350
  hormones, 96, 348
  influenced traits, 63
  linked genes, 63
  organs, 348
  steroids, 215
Sexual maturation, 348
Sharpey's fibers, 93
Shells, 12-13
Short bones, 96
Shoulder, muscles of, 140-141
Sickle cell anemia, 225
Silicosis, 280
Simple
  diffusion, 42
  epithelia, 70
  series circuit, 164
  squamous epithelium of the alveolar wall, 276
Sinoatrial node, 235
Sinuses, 100-101, 270
Skeletal
  anatomy, 97-113
  muscle, 80, 120-127
  system, 4, 92-116
  system functions, 92-93
Skin , 84-86, 255-256
  color, 86
  derivatives, 86—88
  glands, 84, 88
  layers, 85-86
Skull, 3, 92, 96, 98-101
Sliding filament
  hypothesis, 121
  mechanism, 128
Small arteries, 241
Small intestine, 290-296
  absorption, 295-296

Smell, 197
Smooth muscle, 79-80, 120, 127-128
Sodium, 339-340
  chloride, 327-329
  ions, 325-326
  potassium ion pump, 44, 159-160, 325
Somatostatin, 208, 216, 292
Somatotropic hormone, 208
Sound waves, 193
Specialized connective tissue, 74
Special senses, 185
Specific dynamic action (SDA), 312
Speech, 269-270
  muscles, 180
Spermatic cord, 350, 353
Spermatogenesis, 351-353, 356
Spermatogonia, 351
Spermatozoa, 59, 209, 349, 353, 365
Sphenoid bone, 100, 270
Sphincter of Oddi, 292, 294-295
Sphygmomanometer, 243
Spicules, 95
Spina bifida, 103
Spinal
  cavity, 7
  cord, 3, 175-178, 180
  nerves, 176-177, 180-182
  reflexes, 279
  tap, 169
  tracts, 178-179
Spine, 106
Spinous process, 101
Splanchnic nerves, 183
Splenic
  artery, 245
  vein, 246
SPONCH, 11
Sprains, 115
Squamous
  cells, 70
  epithelium of the alveolar wall, 276
  portion, 98
Stapedius muscle, 192
Stapes (stirrup), 100, 192
Staphylococcal infections, 116
Starling's law, 248
Starvation, 309, 325
Static equilibrium, 193-195
Stearic acid, 23
Stem cells, 351
Stensen's duct, 285
Stereocilia, 71
Sterilization, 366
Sternoclavicular joint, 105
Sternocleidomastoid muscles, 180, 273
Sternum, 103-105
Steroids, 21, 24, 204, 215
Stomach, 288-290
Storage diseases, 39
Stratified epithelia, 70
Stratum
  basale, 85, 359
  corneum, 85-86
  functionale, 359
  granulosum, 85
  lucidum, 85
  spinosum, 85

Stroma, 356
Styloid process, 99, 107
Subarachnoid space, 168-169
Subclavian
    artery, 244
    vein, 246
Subcutaneous layer, 86
Subdural space, 168
Sublingual glands, 285
Submaxillary
    glands, 285
    nodes, 249
Submucosa, 284, 330
Substance P, 292
Substantia nigra, 174
Substrates, 17, 29
Sucrase, 295
Sudoriferous glands, 88
Sulci, 169-170, 233
Sulfhydryl group, 21
Summation, 162
Superciliary arches, 98
Superficial
    position, 7
    veins, 245
Superior
    orbital fissure, 186
    position, 5
Supination, 115
Supporting connective tissue, 77-79
Supraoptic nucleus, 210
Supraorbital
    foramen, 98
    margins, 98
Suprarenal (adrenolumbar) arteries, 245
Supratrochlear nodes, 249
Surfactant, 273-274
Surgical neck, 107
Surveillance system, 265
Suspensory ligaments of Cooper, 361
Sutures, 113
Swallowing, 287
    center, 287
    muscles, 180
Sweat glands, 84, 88, 317
Sweating, 316, 338
Sympathetic
    impulses, 284, 355
    nerves, 155
    stimulation, 286, 325
    system, 182-183
Symphyses, 113
Sympto-thermal method, 366
Synapses, 50, 162-163
Synaptic
    cleft, 123
    fatigue, 163
    vesicles, 123
Synarthroses, 113
Synchondroses, 113
Syncytiotrophoblast, 368
Syndesmoses, 113
Synergists, 130
Synovial
    joints, 113–114
    membrane, 114
Synthesis
    cell products, 37
    reactions, 16
Systemic circuit, 233-234
Systole, 237-238, 241

Tachycardia, 238
Taeniae coli, 297
Talus bone, 111-112
Target tissues, 202
Tarsal plate, 186
Tarsals, 96
Taste buds, 196-197, 285
Taste, 180, 196-197
Taxonomic classification of humans, 2-3
Tay-Sachs disease, 39, 63
T cells, 227, 255, 261–266
    lymphocytes, 218
    receptors, 261
    response, 263-265
    suppressor, 265
Tear glands, 180
Tears, 186, 256
Teeth, 3, 286-287
Telophase, 50-52
Temperature, 42, 221, 278, 312, 316-317
    enzymes and, 30
    regulation, 84, 313, 316-317, 363
    skin and, 89
Temporal bones, 98
Tendons, 76
Tensor tympani muscle, 192
Tentorium cerebelli, 168
Teratogens, 372
Terminal
    boutons, 162
    cisternae, 120
    hairs, 87
Termination, 54
Testes, 350-353
Testicular
    arteries, 245
    hormones, 355
Testosterone, 24, 311, 348, 350, 355-356
Tetanic contraction, 126, 213
Tetraiodothyronine, 211
Tetrapeptidases, 294
Thalamus, 172
Theca interna, 357
Thermoreceptors, 185, 316
Thigh, muscles of, 146-149
Third ventricle, 168
Thirst, 338, 340
Thoracic
    aorta, 244
    artery, 244
    cage, 104
    cavity, 7
    duct, 249
    muscles, 137
    nerves, 180
    vertebrae, 103
Thoracolumbar division, 182
Thorax, 273
    muscles of, 140-141
Threshold stimulus, 124
Throat, muscles of, 134
Thrombocytes, 227
Thrombocytopenia, 230
Thromboplastin, 228
Thromboxane, 227, 257
Thrombus, 228
Thumb, 3
Thymine, 27, 46
Thymosins, 218
Thymus
    gland, 217-218
    derived lymphocytes (T cells), 255
Thyrocalcitonin, 213
Thyrocervical trunk, 244
Thyroglobulin, 211-212
Thyroid
    cartilage, 271
    gland, 211-213
    hormone, 96, 225, 311-312
    stimulating hormone, 209, 212
Thyroiditis, 266
Thyrotropin, 209
Thyroxine, 211, 310
Tibia, 111
Tibial
    arteries, 245
    vein, 247
Tidal volume, 274
Tissue, 3, 69-81
    fluid colloid osmotic pressure, 248
    regeneration, 257
    transplant rejection, 266
Toenails, 84, 86-87
Toilet training, 331
Tongue, 285
    muscles of, 132
Tonicity, 43-44
Tonus, 126
Total body water (TBW), 336
Total solute concentration (osmolality), 337
Toxic chemicals, 30, 329
Trabeculae carneae, 232
Trabecular network, 95
Trace elements, 316
Trachea, 272
Tracts, 158, 177-178
Transamination, 310
Transcellular fluid, 337
Transcription, 52–53
Transferrin, 296
Transfusion reactions, 230-231
Translation, 53
Translocation, 53, 64
Transport
    maximum, 325, 341
    vesicles, 38
Transverse
    arch, 112
    plane, 5
    processes, 101
Trapezium bone, 108, 180
Treppe, 126
Tricuspid valve, 232-233
Trigeminal nerves, 179
Triglycerides, 23, 308
Trigone, 331
Triiodothyronine, 211
Trioses, 21
Tripeptidases, 294
Tripeptide, 24

Triple bonds, 14
Triplets, centriole, 40
Triquetral bone, 108
Trisomy, 64
   13, 65
   18, 65
   21, 64–65
   X, 66
Trochanter, 110
Trochlea, 107
Trochlear nerves, 179
Trophoblast, 368
True ribs, 105
Trunk, muscles of, 135
Trypsin, 293, 295
Trypsinogen, 293
T-tubules, 119, 124, 128
Tubal ligation, 366
Tubercle, 105
Tuberculosis, 280
Tubular
   nerve cord, 3
   reabsorption, 325-326
   secretion, 326
Tubuli recti, 353
Tunica albuginea, 350
Turbinates, 101, 270
Turner's syndrome, 66
T wave, 237
Tympanic
   membrane, 191
   portion, 99

Ulna, 106-107
Ulnar
   artery, 244
   nerve, 107
   notch, 107
   vein, 246
Ultrasound scanning, 66
Umbilical
   arteries, 247
   cord, 371
   region, 8
   vein, 247
Unmyelinated
   axons, 176
   fibers, 157, 161
Unsaturation, 23
Unstriated muscle, 79
Upper limb, 106-108
Uracil, 27
Urea, 293, 310, 320, 328, 332
   reabsorption, 326
Ureters, 320
Urethra, 320, 331, 354
Urethral
   opening, 354
   orifice, 361
   sphincter, 331
Uric acid, 320
Urinary
   bladder, 320, 330-331, 371
   calculi, 332
   system, 4, 320-332
   system disorders, 332
Urination, 331

Urine, 256, 310, 320, 324-326, 329-330, 343-344
   concentration and dilution, 326-329
   volume, 326-327
Uterine tubes, 358-359
Uterus, 359
Utricle, 192-193
Uvula, 271, 287

Vagina, 360
Vaginal
   diaphragm, 366
   opening, 361
Vagus nerves, 180
Valence, 13
Valves, 240
Van der Waals interaction, 15
Vasa recta, 327, 329
Vascular tunic (uvea), 187
Vas deferens, 353
Vasectomy, 366
Vasoactive
   chemical factors, 257
   intestinal peptide, 292
Vasoconstriction, 227, 242, 258, 317, 340
Vasodilation, 242, 257, 317
Vasomotion, 241
Vasomotor center in the medulla, 242
Vasopressin, 210, 243
Veins, 221, 241, 245-247
Vellus hairs, 87
Velocity of the nerve impulse, 161-162
Vena cava, 232, 239, 245-248
Venous sinuses, 168, 245
Ventral
   cavity, 7
   corticospinal tracts, 178
   root, 177
Ventricles, 169, 232
Ventricular folds, 271
Venules, 241
Vermiform appendix, 297
Vermis, 174
Vernix caseosa, 372
Vertebrae, 97, 101-103
Vertebral
   artery, 244
   cavity, 7
   column, 92, 101
   column curvatures, 103
   column, muscles of, 135-136
   disorders, 103
   structure, 101–102
Vertebra prominens, 103
Vertex presentation, 372
Vestibular
   apparatus, 194-195
   branch, 180
   bulbs, 361
   membrane, 193
Vestibule, 192, 285, 361
Vestibulocochlear nerves, 180
Vibrations, 193
Vibrissae, 270
Villi, 291, 368
Visceral
   afferent (sensory) fibers, 182

   efferent (motor) system, 182
   membrane, 8, 232
   pericardium, 8
   peritoneum, 8, 284
   pleura, 8
   portion, 73
Vision, 186-191
Visual
   association area, 171
   pathways to the brain, 188
   purple, 190
Vital capacity, 275
Vitamin, 296, 313-315
   absorption, 296
   A, deficiency, 190
   D, 341
   K, 228
Vitreous humor, 188
Vocal cords, 271-272
Volkmann's canals, 78-79
Voluntary
   muscle, 79
   system, 279
Vomer bone, 101, 269
Vomiting, 345
Vulva, 360-361

Water
   absorption, 296
   balance, 337-338
   diuresis, 339
   properties, 17-18
   reabsorption, 326
   retention, 210
Wave summation, 125
Waxes, 21, 23
Wernicke's speech area, 171
Wharton's
   ducts, 285
   jelly, 74
White matter, 167, 169
Wings, greater and lesser of pterygoid, 100
Withdrawal reflex, 166
Wormian bones, 100
Wrist, 108
   muscles of, 143-145

X chromosomes, 58, 65, 348
Xiphoid process, 103
X-rays, 47

Y chromosomes, 58-59, 65, 349
Yolk sac, 370-371

Zona
   fasciculata, 213-215
   glomerulosa, 213-214, 340
   pellucida, 357
   reticularis, 213, 215
Zonula
   adherens, 71
   occludens, 71
Zygomatic
   bones, 100
   process, 98, 100
Zygote, 60, 349, 368
Zymogen granules, 38